Cultural Heritage Science

The preservation and interpretation of our cultural heritage is one of the major challenges of today's society. Cultural Heritage Science is a highly interdisciplinary book series covering all aspects of conservation, analysis and interpretation of artworks, objects and materials from our collective cultural heritage. The series focuses on science and conservation in three main fields

- Art technology
- Active conservation and restoration
- Preventive conservation and risk management

The series addresses conservators and conservation scientists at museums, institutes, universities and heritage organizations. It also provides valuable information for curators and decision makers at museums and heritage organizations. All titles in the book series will be peer reviewed. Titles will be published as printed books and as eBooks, opening up the opportunity to include electronic supplementary material (videos, high-resolution figures, special data formats, and access to databases).

Howell G. M. Edwards • Philippe Colomban

Blue by Fire: A Marker of the Technical History of Glass and Ceramics

Springer

Howell G. M. Edwards
School of Chemistry and Biosciences
University of Bradford
Bradford, West Yorkshire, UK

Philippe Colomban
MONARIS UMR8233
Sorbonne Université, CNRS
Paris, France

ISSN 2366-6226 ISSN 2366-6234 (electronic)
Cultural Heritage Science
ISBN 978-3-031-77121-7 ISBN 978-3-031-77122-4 (eBook)
https://doi.org/10.1007/978-3-031-77122-4

This Springer imprint is published by the registered company Springer Nature Switzerland AG
The registered company address is: Gewerbestrasse 11, 6330 Cham, Switzerland

Blue Book Dedication

The authors, Howell GM Edwards and Philippe Colomban, acknowledge the contribution made by their colleague William Jay to this text in the form of background draft material and concepts prior to his untimely death in January 2024, which have been incorporated into several case studies here.

Foreword

Blue by Fire: A Marker of the Technical History of Glass and Ceramics

There have been many books on ceramics enquiring into individual factories into makers, types of decoration, and the variety of ceramics produced. This contribution, *Blue by Fire: A Marker of the Technical History of Glass and Ceramics*, by the respected ceramic researchers, Professor Howell Edwards of the University of Bradford, England, and Professor Philippe Colomban of the Sorbonne University, Paris, provides an exciting additional approach to this world of ceramics, metals, enamels, glass, and their decoration using cobalt blue.

These writers with their feet firmly based in science, address a vast array of ceramics, glass, beads, and enamelled metalware from porcelains, to stone ware, Delft wares, Iznik tiles, and Chinese export wares, where the beguiling colour of cobalt blue appears as either under-glaze, within glaze, or on-glaze linking each of these groups. The colour association of blue and white is arguably the most cherished and regarded combination on ceramics stretching from South East Asia, Japan, China, the Middle East, and the china cabinets of devotees in Paris, London, and the United States. This colour combination is to be found in many homes where interior decorators build their colour scheme (ours too) accordingly around blue and white.

Cobalt blue is an oxide pigment that holds its colour during firing whether in oxidising or reducing conditions and over a range of kiln temperatures. Consequently, it has been both welcomed and adopted by potters, metal, bead, and glass workers from many corners of the world. This book provides a dazzling array of subject matter related to this pigment. Never has there been such an exhaustive enquiry into its origins and sources, and application. Topics addressed include the Chinese use of this pigment, blue pigments in antiquity, Mediaeval pigments and their transition into the Renaissance, zaffre and smalt, and modern analytical techniques used in its characterisation. American readers will also applaud the section dealing with

America's earliest blue and white porcelains from Cain Hoy, South Carolina. In addition, there are sections dealing with enamelled metalwares, beads, and glass coupled with marine archaeology, including the Tek Sing treasure, and the San José treasure, the supposed Holy Grail of shipwrecks. An important component of this book is a discussion on the related Egyptian green pigments, and the historical evidence for its usage in pottery and glazes.

Of critical importance to the museum curator, dealer, and collector is a section on trace elements associated with the various cobalt ore sources thus opening up yet another avenue for the correct sourcing, attribution, and the identification of spurious artefacts.

An underlying aspect of this welcome contribution is that it reinforces a somewhat cryptic change in the decorative arts world that has been occurring over the last 20 or 30 years relating to how we view, study, collect, and attribute ceramics. Traditionally the criteria for both the understanding and attribution of ceramics were to be found in notions of the artistic pursuit and that there was little or no role for repugnant science to intervene in this venture. However, it is now being recognised that a sound attribution and understanding of porcelain and other types of ceramic artwork requires a holistic approach. This approach, or Trinity, involves a combination of documentary and oral evidence, connoisseurship, and science. The authors do not disappoint in that they discuss the role of analysis in this holistic attribution of ceramics, metals, beads, and glass and the techniques now available that are best suited for this work to assist connoisseurs in their correct and definitive attribution.

Without doubt this contribution is the most scholarly and exhaustive account dealing with cobalt blue on ceramics, metalware, beads, and glass yet to appear in the literature, taking us around the world on a magic cobalt blue carpet. A highly recommended addition to the literature and an important source for all serious scholars in the decorative arts.

Kerikeri, New Zealand Dr. W. Ross H. Ramsay

Acknowledgements

The authors wish to acknowledge with gratitude the kindness of the following institutions and staff therein for allowing access to their specimens and artefacts for analysis or for according them permission to reproduce images used in this book.

Anonymous Collectors

Museum Geelvinck Hinlopen Huis, Amsterdam, The Netherland
Gardiner Museum of lCeramic Art, Toronto, Canada
Art Institute, Chicago, USA
Palace Museum, Beijing, China
National Museum of Ireland, Dublin
Cité de la Céramique, Sèvres, France
DCAC, Prefecture, Mayotte, France
Hotel Cabu Collection, Orléans, France
LRMH, Champs-sur-Marne, France
Marie-Christine David Gallery, Paris, France
Musée des arts décoratifs, Paris, France
Musée Cernuschi, Paris, France
Musée Chinois, Château de Fontainebleau, Fontainebleau, France
Musée d'Histoire Naturelle, Nimes, France
Musée du Louvre, Département des objets d'art, Paris, France
Musée du Louvre, Département Islam, Paris, France
Musee Ernest Cognacq, St Martin-de-Re, France
Musée national des arts asiatiques-Guimet, Paris, France
Nicolas Fournery Gallery, Paris, France
Philippe Magloire Gallery, Paris, France
Sainte-Chapelle, Monum', Paris, France
Royal Heritage, Shatin, Hong-Kong
E & H Manners Ltd., Kensington, London
Kingschina Ltd., Mayfair, London

National Maritime Museum, Greenwich, London
Victoria & Albert Museum, London
Los Angeles County Museum, Los Angeles, USA
Metropolitan Museum of Art, New York, USA
Californian Palace of the Legion of Honor, San Francisco, USA
National Museum, Belgrade, Serbia
Asian Civilisations Museum, Singapore
AMAFA & Ezemvelo KZN Wildlife, uKhahlama-Drakensberg Park, South Africa
Sewefontein Studio and Gallery, Komani, South Africa
University of Pretoria Museum, Pretoria, South Africa
Ariana Museum, Genève, Swiss
Fondation Baur, Musée des arts d'Extrême-Orient, Genève, Swiss
Institut National du Patrimoine, Tunis, Tunisie
Edirne City Museum, Edirne Foundation, Selimiye Fondation Museum, Edirne, Turkey
Iznik Tile Kiln Excavation Directorate, Bursa, Türkiye
City of Birmingham Museum of Art, Birmingham, UK
Brampton Museum & Art Gallery, Newcastle-under-Lyme, UK
Burghley House Trust, Stamford, UK
Hai Duong Provincial Museum, Hai Duong, Vietnam
Quang Nam Provincial Museum, Quang Nam, Vietnam
Quang Ngai Provincial Museum, Quang Ngai, Vietnam

Contents

About the Authors

Howell G. M. Edwards, M.A., B.Sc., D.Phil., C.Chem., FRSC is Professor Emeritus of Molecular Spectroscopy at the University of Bradford. He read Chemistry at Jesus College in the University of Oxford and after completing his B.A. and B.Sc. degrees he studied for his doctorate in Raman spectroscopy at Oxford and then became a Research Fellow at Jesus College, University of Cambridge. He joined the University of Bradford as a Lecturer in Structural and Inorganic Chemistry, becoming Head of the Department of Chemical and Forensic Sciences, and was awarded a Personal Chair in Molecular Spectroscopy in 1996. He has received several international awards (Sir Harold Thompson Award; Charles Mann Award; Emanuel Boricky Medal; Norman Sheppard Award) in a spectroscopic career which has resulted in the publication of 1400 research papers in Raman spectroscopy and the characterisation of materials, along with six books on the application of this analytical technique to art, archaeology, and forensic science. He has had a lifelong interest in the porcelains of William Billingsley, especially those from the Derby, Nantgarw, and Swansea factories and has authored eight books on porcelain: *Swansea and Nantgarw Porcelains: A Scientific Reappraisal*; *Nantgarw and Swansea Porcelains: An Analytical Perspective*; *Porcelain to Silica Bricks: The Extreme Ceramics of William Weston Young, 1776–1847*; *eighteenth and nineteenth Century Porcelain Analysis: A Forensic Provenancing Assessment*; *Porcelain Analysis and Its Role in the Forensic Attribution of Ceramic Specimens*; *Raman Spectroscopy in Cultural Heritage Preservation* (with Philippe Colomban and Peter Vandenabeele), *Armorial Porcelain: The Genesis* (with Rachel and Morgan Denyer) and *Welsh Armorial Porcelains: Nantgarw and Swansea Crested China*, published by Springer Publishing, Dordrecht, The Netherlands, and two specialist texts, namely: *The Pendock-Barry Porcelain Service: A Forensic Evaluation* (with R. Denyer and M. Denyer) and *The Farnley Hall Service: A Unique Survivor in Nantgarw Porcelain*. He has also produced several monographs: *William Billingsley: The Enigmatic Porcelain Artist, Decorator and Manufacturer*; *Nantgarw*

Porcelain: The Pursuit of Perfection (with M.D. Denyer); *Swansea Porcelain: The Duck-Egg Translucent Vision of Lewis Dillwyn* and *Derby Porcelain: The Golden Years, 1780–1830.* Howell Edwards is Honorary Scientific Adviser to the de Brecy Trust on the scientific evaluation of their artworks and paintings.

Philippe Colomban, MSc, Ph.D. is CNRS Research Professor Emeritus (Directeur de Recherche CNRS émérite) since 2018 at the Sorbonne Université (Laboratory UMR SU-CNRS 8233: from MOlecule to NAno-objects: Reactivity, Interactions and Spectroscopy – MONARIS). After an MSc in Ceramics, Glass, and Cement Engineering (ENSCI-Sèvres) in 1975 and an MSc in Solid State Physics (DEA) in 1976, Philippe Colomban obtained in 1979 his Ph.D. Hab. degree (*Docteur es-Sciences Physiques*) from the Université Pierre-et-Marie-Curie (now part of Sorbonne Université). He was one of the first in the world to develop Sol-Gel routes for advanced optoelectronic PLZT ($Pb_{1-y} La_y[Zr_xTi_{1-x}]O_3$) ceramics. He studied then the structure of superionic conductors and ion mobility at different scales for battery design. From 1989 to 1993 he was responsible for new projects at the Materials Department of *ONERA* (Office National des Etudes et Recherches Aérospatiales), the French Establishment for Aerospace Research and Development (materials for rockets and missiles, aircraft engines, sol-gel routes, ceramic- and polymer-matrix composites, fibres, nanoparticle-based microwave absorbing materials, and functionally graded composites), and worked as a Consultant at ONERA for 10 additional years. From 1994 to present, as CNRS Research Professor, and as director of the UMR7075 Laboratory from 2002 to 2010, his research interests include materials science and spectroscopy XRF (X-ray Fluorescence), Raman, IR, and neutron spectroscopy: *in situ – operando* analysis, fuel cells, electrolytes, fibres/composites). Attention was given to the correlation between Raman parameters and mechanical and electrical (ionic, electronic) properties as well as the identification of the technology used in ancient ceramics, glasses, textiles, paintings, and buildings. Philippe Colomban has (co-)authored 3 books, has published more than 500 peer-reviewed papers, and has 10 patents, and he has presented more than 120 invited talks. He is an Associate Editor of the *Journal of Raman Spectroscopy*, the *npj Heritage Science* and member of many Editorial Boards of *Materials Science*, *Spectroscopy*, and *Cultural Heritage* journals for decades.

Chapter 1
Colour Blue in Different Cultures

Abstract The use of colour, its symbolism and availability in ancient societies. The employment of colouring techniques using the Arts of Fire to obtain the colour blue for ceramics, glasses and enamelled artefacts in the Occident, Orient, Africa and the Americas. Natural dyestuffs and plant extracts.

Keywords Indigo · Woad · Technologies of extraction · Requirements for achieving a blue colour

1.1 Colours

The use of colours depends on their symbolism and their availability, whatever the type of coloured object. We are going to establish the state of the art of knowledge concerning colouring techniques using the Arts of Fire to obtain the colour blue, that is to say the preparation of ceramics, glasses and enamelled pieces, often in imitation of the colour of natural objects, gems but also covering plants and their extracts and animals or their secretions/extracts.

Colour has been used since the dawn of humanity to decorate the body, adornments, clothing and habitat. This colouring uses the two chemical synthesis pathways, the liquid pathway with the production of essences, dyes and the solid pathway using fire [3, 4]. The first way will be used to dye fabrics in particular; the second to prepare pigments and stable objects, which thanks to this stability are preserved over millennia and are the best testimonies of past civilizations.

Nature offers us a wide range of colours, the blue of the sky, the deep sea or the ice, the green of the leaves or the water of lakes and mountain rivers, the yellow of the fields of rice or rapeseed, the red of blood or fire, etc., all the primary colours being found in the rainbow. The appreciation and the terms designating the different colours have long been subjective—and still remain so—as reflected by the greater or lesser wealth of qualifiers depending on the language but also depending on the community. Thus in some Asian languages green and blue are variants of the same

H. G. M. Edwards, P. Colomban, *Blue by Fire: A Marker of the Technical History of Glass and Ceramics*, Cultural Heritage Science,
https://doi.org/10.1007/978-3-031-77122-4_1

colour and in the West only turquoise is attached to blue. Isaac Newton (1642–1727) [17] then Johann Wolfgang von Goethe (1749–1832) [16] among others less known established colour ranges. Closer to home, Johannes Itten (1888–1967) completed this approach [11].

We distinguish the basic colours, blue (or cyan), red (or magenta) and yellow called primaries and the secondary ones obtained by combination of the first (orange, green and purple), visible in the rainbow, i.e. the spectral range seen by our eyes, from violet (~380 nm) to deep red (~720 or even 780 nm) but with a maximum sensitivity between 495 and 555 nm (yellow-green) [7]. Blue, or rather blues are located between ~400 and 490 nm, blue *stricto sensu* between 478 and 483 nm. We owe to James Clark Maxwell (1831–1879) the demonstration that light is an electromagnetic wave [15] and to Louis de Broglie (1892–1987) [5] its corpuscular character (photons).

The CIE 1976 L*a*b* colour space, generally called CIELAB, is a colour space particularly used for the characterization of surface colours [14]. Three quantities characterize colours: lightness L* derives from the luminance of the surface; the two parameters a* and b* express the difference in colour with respect to that of a gray surface of the same lightness:

- the lightness L* which takes values between 0 (black) and 100 (reference white)
- the parameter a* represents the value on an axis going from green to red
- the parameter b* represents the value on an axis going from blue to yellow

Many colour charts have been established, one of the most widely used is that of the British Standard Institution [18].

1.2 Occident

In antiquity in the West, the colour blue occupies a modest part, with the exception of its use in Egyptian frescoes and the glazed brick walls of Babylon. Some archaeologists report that blue pigments resulting from atmospheric dust were used during prehistory. This situation will last until the beginning of the Middle-Ages. This has been ascribed to the technical difficulties of producing the colour blue, including in nature as shown by the limited number of naturally occurring blue gems and minerals. Michel Pastoureau [13] notes that while the Greeks and Romans had two natural blue dyes, woad (*la guède* in French) and indigo, they preferred the colours white and red for clothing (purple was the symbol of political authority). The woad was also widely used by the Celts and the Germans, barbarian peoples ('savages') who were rejected for their lack of culture. Indigo, a dye from India, was expensive and was little used. This virtual absence of the colour blue among the Greeks and Romans is linked by Pastoureau to the scientific conceptions of the time. For Epicurus (342–270 B.C.) it is matter that emits rays towards our eyes; for Pythagoras (580–475 B.C.) it is the eye that emits the rays towards the coloured areas. For Plato (428–348 B.C.) the perception of colour results from the meeting between the rays coming from the eye and those emitted by the coloured zone. This implies that when the eye does not look at a coloured object, the colour does not exist! Also at this time the blue colour was not identified in the rainbow. For Pliny (23–79) blue is an unappreciated, even a bad colour: for a woman, a blue colour of the eyes reflects a lack of virtue, for a man, a barbaric character [13].

In the Christian world three colours make sense: white was a sign of purity and joy, black a sign of penance and red a sign of sacrifice and divine love (reminiscent of the blood of Christ). Blue entered the Christian World between the ninth and twelfth centuries with the stained glass windows of Romanesque churches and cathedrals (Fig. 1.1).

Michel Pastoureau attributes the development of this colour to that of the representation of the Virgin, dressed in blue (Fig. 1.2 left). Illuminated manuscripts use lapis lazuli to paint the clothes of the Virgin or of angels (Fig. 1.2 right). The queens of France then dressed in blue and Louis IX (1214–1270) was the first king to dress in blue. The Arms of France use blue (Fig. 1.3) and it is the same for those of many noble families. In the romances of courtly chivalry, blue represents courage, loyalty and fidelity. At the end of the Middle Ages blue replaced red as the favourite colour. The culture of woad (the colouring molecule is indigotine as it is for indigo, although the botanical sources are different) then developed strongly during the Middle-Ages. It should be realised that at that time a mixture of colours was considered to be taboo: it was tantamount to cheating. The classification of colours before the

Fig. 1.1 Example of the use of blue in the stained glass windows of the Sainte-Chapelle (Palatine church of the King of France, Paris, thirteenth century) with the arms of Blanche of Castile, mother of King Saint Louis

Fig. 1.2 Virgin and Child (enamel on metal, Coll. Museum of Decorative Arts, Paris) and illuminated manuscript. (Coll. Bibliothèque Nationale, Paris)

Fig. 1.3 Examples of sophisticated enamelled objects: left, porcelain dish with the arms of Louis XV, King of France (ca. 1720–1740; body made at Jingdezhen, enamelled decoration made at Canton, Private Coll.); right: enamelled glass bottle dedicated to Tuquztimur. (Syria or Egypt, middle of the fourteenth century, Musée du Louvre, Department of Islamic Arts, Paris Coll.)

eighteenth century is rather different from ours: green, blue and black were considered to be part of the same group. Saturation is more important than hue!

Blue is also associated with the colours yellow and green. In the fifteenth century with the development of Protestantism (Luther 1483–1546), the colour blue was highly valued and for seventeenth century Protestants the colours were associated with moral values. A dress code was established: widows, clerics and magistrates must wear black; the infirm and non-Christians must wear a discriminating colour (yellow or green). Blue, having no symbolism, can be democratized. Art, for Calvinists who are among the most rigorous of Protestant groups, should only serve to honour God, and therefore colour is secondary. On the contrary, the Catholic Counter-Reformation favours polychrome and in particular gold. We must also put ourselves in the context of times before the twentieth century and the absence of powerful lighting. Most of the sets were only lit by candlelight.

In the eighteenth century the classification of colours is made into red, yellow and blue, white and black not being considered to be colours. The development of the importation of indigo, cheaper than woad and pastel produced in Europe, then the development of Prussian blue (>1709), an even less expensive pigment, allowed a more generalized use of blue: the colour blue became then a symbol of progress, dreams and freedom. The novel "*The Sorrows of Young Werther*" by Goethe contributes to the success of this colour during the Romantic era. A number of expressions in several languages associate blue with melancholy, romanticism (examples include "*feeling blue*" in English, "*note bleue*" in French, the song "*Les mots bleus*" by Christophe, a French singer in the 1980s, etc.). Blue also became a colour with political symbolism. Before 1789 it is the colour of the King of France along with white. American colonial revolutionaries used blue, white and red for their flag, even though these colours were already in use by the British. Even earlier the Scottish Jacobean Pretenders to the English throne (the Jacobites, Bonnie Prince Charlie), descendants of Mary Queen of Scots and in opposition to the Stuarts and Hanoverians, adopted the colour of blue on their standard and headgear in battle: the flag of Scotland today is still the white Cross of St Andrew on a blue background [9]. The French Revolution also made this choice, red being with blue the colour of Paris and many national flags still use these three colours: France, UK, USA, Netherland, Luxemburg, Russia, Cuba, Croatia, Serbia, Iceland, Dominican Republic, Panama, Paraguay, Taiwan, etc. In Europe, the colour blue was first the colour of progressive republicans, then of centrists and now that of conservatives. It is the colour of the European flag but also of the UN and UNESCO. In the USA, blue remains the colour of the Democratic Party. It is also the colour of "blue" jeans or denim (jeans or denim coming from the respective rendering by the Americans of Genoa or Nimes because Levi Strauss used fabrics from these cities).

1.3 East, Africa and Americas

To our knowledge, the colour blue in its historical and symbolic context has not been studied in most other cultural areas. The importance of the colour blue in Maya frescoes and paintings is well established, with the particular nature of Maya blue, a form of indigo encapsulated in palygorskite clay [6, 8, 10] which gives it remarkable stability. The symbolism of blue, in particular the blue of the sky for the Mongols, has been associated with the development of blue ceramics under the Yuan Dynasty and Lajvardina pottery in Iran in the thirteenth and fourteenth century in imitation of lapis lazuli [1, 2]. A high level of quality is achieved at the same period in the enamelling of glass object (Fig. 1.3). The wide use in Africa for millennia of glass beads, mainly imported from Asia or the Near and Middle East, then after the implantation of the Portuguese also from Europe as jewellery and adornment or for hoarding gives a photograph of the colours appreciated at different times [12, 19]: turquoise is the dominant colour for most times but blue *stricto sensu* is only significant around the tenth century and after the sixteenth century depending on the imports, respectively from the Near and Middle East and then from Europe.

This rough overview shows that the possibility of creating colour is the *sine qua non* of its use.

References

1. Colomban, P.: Lapis lazuli as unexpected blue pigment in Iranian Lâjvardina ceramics. J. Raman Spectrosc. **34**(6), 420–423 (2003). https://doi.org/10.1002/jrs.1014
2. Colomban, P.: Routes du lapis lazuli, lâjvardina et échanges entre arts du verre, de la céramique et du livre. Taoci. **4**, 145–152 (2005). https://www.researchgate.net/publication/259998622_Routes_du_Lapis_lazuli_Lajvardina_et_echanges_entre_arts_du_verre_de_la_ceramique_et_du_livre. Accessed 4 Mar 2024
3. Colomban, P.: The destructive/non-destructive identification of enamelled pottery and glass artifacts and associated pigments – a brief overview. Arts. **2**(3), 77–110 (2013). https://doi.org/10.3390/arts2030077. http://www.mdpi.com/2076-0752/2/3/77/
4. Colomban, P.: Natural nanosized raw materials and sol-gel technology: the base of pottery since millennia. In: Dillmann, P., Bellot-Gurlet, L., Nenner, I. (eds.) Nanoscience and Cultural Heritage, pp. 59–73. Springer-Atlantis, Amsterdam (2016). https://doi.org/10.2991/978-94-6239-198-7
5. de Broglie, L.: A tentative theory of light quanta. Lond. Edinb. Dublin Philos. Mag. J. Sci. **47**(278), 446–458 (1924). https://doi.org/10.1080/14786442408634378
6. del Río, M.S., Doménech, A., Doménech-Carbó, M.T., de Agredos Pascual, M.L.V., Suárez, M., García-Romero, E.: The Maya blue pigment. Develop. Clay Sci. **3**, 453–481 (2011). https://doi.org/10.1016/B978-0-444-53607-5.00018-9
7. Dillon, J.: New trends in photobiology: the photophysics and photobiology of the eye. J. Photochem. Photobiol. B Biol. **10**(1-2), 23–40 (1991). https://doi.org/10.1016/1011-1344(91)80209-Z
8. Doménech, A., Doménech-Carbó, M.T., Edwards, H.G.M.: On the interpretation of the Raman spectra of Maya Blue: a review on the literature data. J. Raman Spectrosc. **42**(1), 86–96 (2011). https://doi.org/10.1002/jrs.2642
9. Edwards, H.G.M.: Welsh Armorial Porcelain: Nantgarw and Swansea Crested China. Springer-Nature, Dordrecht (2022). https://doi.org/10.1007/978-3-030-97439-8_8
10. Giustetto, R., Llabrés i Xamena, F.X., Ricchiardi, G., Bordiga, S., Damin, A., Gobetto, R., Chierotti, M.R.: Maya Blue: a computational and spectroscopic study. J. Phys. Chem. B. **109**(41), 19360–19368 (2005). https://doi.org/10.1021/jp048587h
11. Itten, J.: The Elements of Colour, vol. 4. Wiley (1970). 9780471289296 (ISBN10: 0471289299)
12. Koleini, F., Colomban, P., Pikirayi, I., Prinsloo, L.C.: Glass beads, markers of ancient trade in Sub-Saharan Africa: methodology, state of the art and perspectives. Heritage. **2**(3), 2343–2369 (2019). https://doi.org/10.3390/heritage2030144
13. Pastoureau, M.: Bleu, histoire d'une couleur. Editions du Seuil, Paris (2006). EAN 9782020204750
14. Robertson, A.R.: The CIE 1976 colour-difference formulae. Colour Res. Appl. **2**(1), 7–11 (1977). https://doi.org/10.1002/j.1520-6378.1977.tb00104.x
15. Sengupta, D.L., Sarkar, T.K.: Maxwell, hertz, the Maxwellians, and the early history of electromagnetic waves. IEEE Antennas Propagat. Mag. **45**(2), 13–19 (2003). https://ieeexplore.ieee.org/abstract/document/1203114

16. Von Goethe, J.W.: Goethe's Theory of Colours. Translated from the German y Eastlake C.L., J. Murray, London. https://books.google.fr/books?hl=fr&lr=&id=OIoIAAAAMAAJ&oi=fnd&pg=PR35&dq=Goethe+colour+scale&ots=xZzDDXOj8c&sig=ufUH9C-GgWzIejLQY1MSr1W_L4A&redir_esc=y#v=onepage&q&f=false (1840). Accessed 2 Feb 2023
17. Westfall, R.S.: The development of Newton's theory of colour. Isis. **53**(3), 339–358 (1962). https://www.journals.uchicago.edu/doi/10.1086/349596
18. Whitfield, T.W.A., O'Connor, M., Wiltshire, T.J.: The British building-colour standards: a model for international application. Colour Res. Appl. **11**(3), 215–222 (1986). https://doi.org/10.1002/col.5080110309
19. Wood, M.: Interconnections: Glass beads and trade in southern and eastern Africa and the Indian Ocean-7th to 16th centuries AD. Doctoral dissertation, Department of Archaeology and Ancient History. https://www.google.com/url?sa=t&rct=j&q=&esrc=s&source=web&cd=&ved=2ahUKEwik0vaSkdqEAxUbdqQEHcC_CMgQFnoECA8QAQ&url=http%3A%2F%2Fwww.diva-portal.org%2Fsmash%2Fget%2Fdiva2%3A463276%2FFULLTEXT01.pdf&usg=AOvVaw3zS8ZaFNtCCUwfM0CRvVP9&opi=89978449 (2012). Accessed 4 Mar 2024

Chapter 2
Analytical Methods

Abstract Analytical methods for the analysis of colours and coloured pigments: elemental, structural and mineralogical and molecular characterisation of the composition of pigments. The origins of colour in minerals: electronic transitions, light scattering. Opals and gemstones.

Keywords Colour generation · Opals · Light scattering · Analytical techniques · Raman · XRD · SEM · Mineralogy

2.1 Classification of Analytical Methods

The purpose of an analytical method is to provide answers to a formulated question. We can classify the methods according to this criterion: we therefore have the methods (Table 2.1) providing information on: (i) the composition, using the various elementary analysis techniques, keeping in mind that most methods do not allow (or require very specific procedures) to identify all the elements, in particular the lightest ones such as Na, Li, B, C, O and H, and that the volume analysed is very variable; the comparison of the volume analysed within heterogeneous material is therefore crucial; (ii) the nature of the phases, that is to say the identification of their crystalline (and in this case their symmetry) or vitreous structure, (iii) the colour and more precisely the absorption profile versus energy and the type of element responsible for the chromophore, and (iv) the microstructure (stratigraphy, heterogeneity).

Some of these methods make it possible to produce surface maps, which is useful for complex coloured decorations on substrates. The adverse effects of the variation in depth upon which the measurement is made must be taken into account for these

H. G. M. Edwards, P. Colomban, *Blue by Fire: A Marker of the Technical History of Glass and Ceramics*, Cultural Heritage Science,
https://doi.org/10.1007/978-3-031-77122-4_2

Table 2.1 Analytical methods and their main characteristics

Questions	Analyzed area	Techniques	Preservation	Remarks
Composition	>10 grain size	LA-ICP-MS NAA XRF	Micro-destructive Non-invasive Non-invasive	All element, isotopes Radioactive Semi-quantitative
Composition localised	< grain size	SEM/TEM-EDS μXRF	Destructive Non-invasive	– Semi-quantitative
Identification of phases	Crystal	Raman (ATR-)FTIR OM (polarized)	Non-invasive – Destructive	Variable efficiency – –
	Amorphous (glassy)	Raman (ATR-)FTIR OM (polarized)	Non-invasive – Destructive	Variable efficiency Surface Large grains
Colours	Variable (macro/micro)	UV-visible-nIR Transmission/reflexion	Destructive	–
Porosity	Macro	Hg porometry Water absorption Gas absorption (BET)	Destructive Sampling Sampling	–
	Micro	SEM/OM	Destructive	Local analysis
Stratigraphy	Close to surface	OCT	Non-invasive	
		OM/SEM on Fracture or polished section	Destructive	–

measurements. The observation of microstructures (porosity, grain size, texturing, etc.) is therefore essential to derive a complete picture of the substrate or the applied coloured pigments.

Another classification criterion is linked to the preservation or not of the object or the sample taken from the object studied: this is critical for the analysis of a sample of a microscopic or macroscopic volume, with or without the destruction of this sample, or alternative analysis without sampling. This criterion is decisive for the analysis of Cultural Heritage objects, some of which may be unique or of great value. For example, the first elementary analysis techniques practiced by Ferchault de Réaumur in the middle of the eighteenth century [52] required the dissolution of large samples and series of dissolution-precipitations then drying, or even dehydration, to finally produce gravimetric analyses of the different characteristic salts of each chemical element. As late as 1922 it was reported that H.A. Eccles and B. Rackham requested several pieces from a porcelain service to analyze the composition of the paste, which required the destruction of major parts of the samples such as plates, dishes and saucers: it was sometimes possible to use already damaged items to facilitate this! [47] The mass of ‘consumed’ material was reduced to a few grams in the 1960s with the replacement of wet chemical gravimetric analysis by spectroscopic techniques and this quantity has continually decreased until it has been limited to a few tens of mg or even less today with the latest generation of analytical instrumentation.

The analysis of coloured material presents other specific problems for the analyst. We are able to distinguish between the chromophore (which provides the observed colour) and the matrix housing the chromophore. An opacifying phase is often added to magnify the colour if the chromophore absorbs too much light. In fact, too high a concentration of chromophore can frequently darken or blacken the colour. This knowledge dates back to Antiquity, where it was possible to differentiate between the matrix (the body, *corpo* in Latin) and the dye (the soul, *anima* in Latin) [105]. The chromophore can itself be in small, or even very small quantities if the inherent colouring power is strong: this can be a point defect (an electronic defect associated with a chemical structural defect, for example, a vacancy), an isolated ion presenting electronic transitions in the visible region of the electromagnetic spectrum (mainly ions with an incomplete 3d (transition metals) or 4f (rare earths) electronic structure) or which occurs through the interaction with other ions (such as complexes involving charge transfer, as Fe-S group for amber colouring), The chromophores can be of very small dimensions (less than one nm for defects), metal or semiconductor nanoparticles of dimensions varying between a few nm to a hundred nm, and grains of a specific phase coloured by the phenomena mentioned above (producing a pigment) whose dimension is between one micron and a few tens of microns [50].

An opacifier is a phase having an optical index that is much higher than the matrix in which it is dispersed. Typically, the necessary quantities of a chromophore range from less than 0.1% wt for an ion with very high colouring power, cobalt ions for example, to more than 5–10% wt for an ineffective colourant such as Fe^{2+} ions and most opacifiers (SnO_2, CaF_2, TiO_2, etc.). The quantity of colouring agent must be maximum in a ‘painted’ polychrome enamel (having a thickness from a few tens of microns to two or three hundred microns), much greater than that which is necessary for a monochrome glaze whose thickness can reach more than one mm and *a fortiori* a stained glass window whose thickness exceeds several millimetres. In the latter case, a high transparency must be maintained. The procedures for detecting colouring agents will therefore be very different depending on the type of object being analysed and the homogeneity of the specimen which can vary from homogeneous (solid glass, metal, thick monochrome glaze) through to heterogeneous (polychrome painted enamel or coloured by the dispersion of another phase) materials. Determination of the volume (surface and depth) of the analyte specimen in which the analysis is carried out is hence essential for the interpretation of the results.

2.2 Colourimetric Analysis and Valence State Analysis (Speciation)

*2.2.1 UV-Visible Spectrum and CIEL L*a*b* Colorimetry*

The first technique for analysing a coloured object involves visual observation (i.e. the eye), implemented or not by enhancement optics, such as binoculars or a microscope using optical (magnification up to ~2500 times) or electronic (magnification up to more than 10^6 times) assistance. The work is revolutionized by the recent availability of portable devices, handheld digital microscopes, which allow the examination—and taking photos—at variable and high magnifications (1000×) at a moderate cost [123, 133]. This makes it possible to visualize the heterogeneity of the distribution of chromophores (on the surface of specimens) on site. Specialized devices adapting to smartphones are also now widely used [85, 88, 118].

The traditional technique used to understand coloured matter is the examination of the optical spectrum in the near UV-visible range (~200–700 nm), generally extending to the near infrared (~700–1100 nm). For liquids the examination is simple by transmission with a spectrometer covering the standard range from 190–800 or 1100 nm, or even 175–3300 nm for the most sophisticated instruments, using transparent cells for the different energy ranges. For solids, three approaches are possible, either measurement by transmission after dispersion in the state of fine powder in a liquid or an inert, translucent and colourless solid matrix [117], or after the formation of a thin layer of pigment on a transparent support [54, 72], or measurement by reflection, generally diffuse reflection using a spherical mirror (integrating sphere) which collects the maximum amount of scattered light [1, 117, 134] with appropriate optics (FORS) [90, 109]. Portable spectrometers are now commonly used [2, 130, 104] for Cultural Heritage specimen analyses.

The different components of light absorption are (with the knowledge or assumption of the chemical elements present) attributed to chromophores by comparison with the literature. These chromophores can be transition metal ions (incompletely filled 3d shells allowing inter-layer electronic transitions), rare earth or lanthanide ions (incompletely filled 4f shells), metal complexes, and ions with other electronic and/or polaronic defects, etc. [56, 70, 78, 102].

Colour is defined by several models [75], the most used being the "CIE L*a*b* colour model" [48] [http://www.colourbasics.com/ColourSpace/]: the L*a*b* colour model determines the colour depending on its position in a 3D colour space, the L* component is the lightness of the colour (when L* = 0 means black and when L* = 100 refers to white), and the chroma a*(which for positive values indicates red and for negative values indicates green) and the hue b* (where positive values refer to yellow while negative values refer to blue). Whereas in the study of coloured objects the qualification L*a*b* is almost essential, this quantity being specific to the object and its surface state, studies on materials and their chromophores mainly use UV-visible-NIR curves which are directly linked to electronic transitions and their physico-chemical origin.

Technological developments in solid CCD (Charge Coupled Device) sensors, converting photons into an electrical signal, which are sensitive to almost the entire electromagnetic spectrum, have led since the 2010s to the replacement of the collection of a spectrum that is characteristic of a coloured zone by the recording of a set of spectra for all the points (pixels) comprising an area and to the construction from the set of spectra attached to each point (optical image) for different wavelengths a 'hyperspectral' file which can be analyzed on demand, providing false colour images to highlight certain particularities [39].

2.2.2 *Hyperspectral Imaging*

Reflectance hyperspectral imaging, introduced in the 1980s by Goetz and co-workers, led to a revolution in the field of remote sensing of the earth and near planets [65]. In the subsequent decades, thanks to rapid advances in CCD solid-state sensor technology, reflectance hyperspectral imaging, which one time was only available to large laboratories, was extended to new fields of application [57], such as biology [97], agri-foods [40], pharmaceutical products [129], the environment [128], and to cultural heritage [63, 89, 110, 127] and covering a very large spectral range both on very high and very low energy sides [8, 89]. In the 2000s, the potential of this non-invasive technology for the study of artworks became evident and, consequently, the methodology is becoming more widely used in the art conservation science field.

The objective is not only 'technical' to identify the distributions and the type of colouring agent but to identify the stages used in creating the decoration, any alterations, the stratigraphic superposition, and later restoration, etc. in order to better characterize the construction process used in the painting of a decoration. For this, false colour images are produced from hyperspectral data [55].

2.2.3 *EXAFS, XANES, XPS: Identification of the Electronic Configuration of the Chromophore Element*

The availability of synchrotron sources from the 1980s led to the development of electronic absorption spectroscopies called XAFS (X-ray absorption fine structure spectroscopy), EXAFS (Extended XAFS) and XANES (X-ray absorption near-edge structure) whose spectra make it possible to identify the electronic configuration (also called speciation) of the major elements constituting the matrix (for example Si and Al [32] or minor ones such as the chromophores Cu°, Cu^{+} or Cu^{2+} [7, 74]). They can be applied in air and have virtually no restrictions on the type and size of the sample, they are applicable to most of the elements of interest, even at very low concentrations, and, finally, to a wide variety of materials, ranging from hard matter such as glass, ceramics, and metals, to softer materials such as paintings, bone, wood, and paper with a submicron operating resolution [23, 51, 77, 113]. The spectral features observed close to the absorption edge, referred to as X-ray absorption near-edge structure (XANES), depend on the local molecular environment (oxidation state, coordination numbers, site symmetry) of a given absorbing atom, and also provide the basic mechanism for imaging with chemical specificity. EXAFS oscillations provide information on the identity of the surrounding atoms, specific bond distances, and coordination numbers.

The X-ray photoelectron spectroscopy (XPS) technique provides similar information but for the immediate surface of objects [86]. The technique is especially useful for the study of surface corrosion [26, 76].

2.2.4 *Mossbauer Spectroscopy*

The differentiation between Fe^{2+} and Fe^{3+} ions and the determination of the relative proportions of ions and metallic matter (Fe°) has been carried out for many decades by Mossbauer spectroscopy [11]. However, the use of the technique in the study of cultural heritage remains very limited [12, 96].

2.3 Elemental Analysis

Knowledge of the composition of the material of objects is essential in any physical or chemical study and this is necessary at different scales to assess the state of conservation or to document the manufacturing process of the object. Knowledge of the volume actually analysed must be evaluated taking into account the physics of the radiation-matter interaction and heterogeneity of the material. We will see that for many analytical techniques this evaluation is difficult and often has not been considered by experimenters.

The most widely used techniques currently are mass and X-ray fluorescence spectroscopy excited either by an X-ray beam (XRF: X-ray Fluorescence), by an electron beam (SEM-EDS: Scanning Electron microscopy-Energy Dispersion Spectroscopy), or by a synchrotron source (μXRF), or a particle beam accelerator (PIXE (Particle Induced X-ray Emission) and PIGE (Particle Induced Gamma ray Emission)). This was not always the case. In the 1960s and 1970s neutron activation, X-ray fluorescence on sodium borate glass pellets and flame emission spectroscopies were dominant. We see the importance of the development of tools on the evolution of experimental practice. It is likely that strictly non-invasive techniques (such as portable XRF) will be dominant in the future, with the preservation of objects and refusal to sacrifice even small samples becoming an imperative criterion.

The strong points of mass spectroscopy are its high sensitivity and its ability to take into account almost all chemical elements [62]. The high resolution of the latest generations of instruments makes it possible to separate isotopes of an element, important information to establish the geological origin of some minerals [24, 80, 107, 126]. Its weak point is its destructive nature: the material is put into solution or volatilized after sampling [3, 62, 101, 135].

Techniques based on excitation of the emission of an X-ray fluorescence spectrum use a beam of particles, photons, electrons or ions, which makes it possible to maintain a non-destructive or quasi-non-destructive character (the creation of defects in the analysis area is difficult to avoid and may be definitive) to the determination of the composition. Certain techniques make it possible to vary the surface analysed over a large scale. This has long been an advantage of scanning electron microscopy but it is also possible with synchrotron sources [37, 81] and also with pXRF [34, 35, 82]. However, depending on the technique and its implementation conditions (whether conducted in air, under vacuum, etc.) the number of elements escaping the analysis is variable: light elements with a mass lower than that of sodium (Z = 23), including the characteristic XRF peaks are low energy and are difficult to ascertain or are not measurable due to the absorption of low energy radiation by molecules in the air. In PIXE, the limit is fluorine element (Z = 9). In SEM-EDX (under standard high vacuum), the limit is carbon (Z = 6) but differentiation between intrinsic carbon and that coming from contamination a challenge.

The resolution of the instruments does not always make it possible to separate the characteristic peak(s) and the conversion of the signal into composition requires assumptions about the homogeneity and depth of the volume analysed, which is not always supported experimentally [35, 44]. When it is possible to take microsamples, EDS analysis using an electron, scanning or transmission microscope allows for the in-depth characterisation, particularly through the final machining of the sample in the microscope (FIB: Focused Ion Beam machining) to produce thinning, thereby allowing the precise study of crystalline or amorphous phases of submicron size [7, 38, 79, 84, 106, 119, 122].

Techniques using particular particle beams such as protons (PIXE: Particle-induced X-ray Emission), which can also emit in a particular energy range (PIGE: Particle-induced gamma emission) offer particular advantages (where the extracted beam allows the non-invasive analysis of objects) but the instruments are very expensive and are therefore rare [19, 38, 125, 139]. Protons can also interact with the nucleus of the atoms in the sample through elastic collisions, Rutherford backscattering, often repelling the proton at angles close to 180 degrees. The backscatter gives information on the sample thickness and composition [33, 69, 94]. This technique is therefore useful for the study of thin layers. If protons (or deuterons) are most often used in ion beam source, helium-3 and helium-4 ions are also used [132].

Determination of the elemental composition by neutron activation uses gamma rays emitted by radionuclides formed by irradiation in an atomic pile under the effect of neutron radiation. For the quantification of the elemental content, standardization is required. The former method is based on the irradiation of calibrators of elements to be determined simultaneously with the samples. In the latter "standardless" method, the samples are irradiated with neutron flux monitors for the evaluation of neutron flux parameters, and for the quantification of the elemental contents, k0-factors are employed [41, 53, 94, 96]. This technique has become uncommon with the disappearance of nuclear reactors dedicated to experimental studies and their replacement by Synchrotrons or Spallation Sources. It had the advantage of being able to be applied non-destructively to large objects, of measuring practically all chemical elements whatever their content, but the radioactivity induced for some of the elements required long periods (several months) of storage thereafter to facilitate the reduction of the induced radioactivity below the regulatory thresholds allowing the transport and handling of the artifacts. New miniaturized, and therefore mobile, irradiation devices can offer renewed interest in the application of the technique in certain cases [98].

Atomic Absorption Spectroscopy (AAS) also called Flame Atomic Absorption Spectroscopy (Flame AAS or FAAS) is one of the earliest elemental analytical techniques to be developed (in the 1950s) and was first commercially released as an analytical technique in the 1960s. It uses the principle that atoms (and ions) can absorb light at a specific, unique wavelength. When this specific wavelength of light is provided, the energy (light) is absorbed by the atom. The material is dissolved and introduced into a flame. The amount of light absorbed is measured and the concentration of the element in the sample can then be calculated [12, 21, 93].

In the LIBS technique, a laser pulse volatilizes a fraction of material (as in LA-ICP-MS) [13, 45, 59, 61]. Analysis of the plasma radiation emission spectrum provides access to the elementary atomic composition of the sample. The technique is mobile but micro-destructive as it involves the erosion of the surface but analysing a sample from the surface can be used to study corroded objects and successive layers.

2.4 Structural Analysis

The determination of an elementary composition, especially since it is very often incomplete due to the difficulty or even the impossibility of measuring elements such as F, Li, and B but also H, C, N and O, requires the use of other techniques, firstly thermodynamic analyses (Differential Thermal (Scanning) Analysis (DTA/DSC) and Thermogravimetric Analysis (TGA)) [10, 29] characterising for example the melting or decomposition temperatures as well as those of phase transitions and others that identify the arrangement of atoms and chemical bonds, namely X-ray diffraction techniques [6, 46, 112], electrons [121, 122] or neutron diffraction [53, 116]. Added to this are vibrational and infrared spectroscopies [5, 22, 42, 83, 92, 111], Raman [28, 49, 58], neutron [30, 73] and electronic (HREELS) [87].

To be representative, diffraction techniques are generally carried out on a powder, which requires sampling. It follows that the minority phases (less than 2–5%) are difficult to detect for the usual counting times. The development of microbeams (synchrotron, rotating anode sources) combined with the recording of maps (mapping) allows the identification of minority phases if they are present in the analysis zone. The crystallographic databases are very old and generally complete

(JCPDS [100]) and allow the identification of specific peaks by comparison of their spectra, visually or using digital database. The development of the Rietveld method leads to an easy determination with different software of the crystal lattice parameters (symmetry group, atomic positions) [108].

Spectroscopic methods also allow a determination of the structure but this requires the analysis of low energy vibrational modes (in the spectral window between 10 and 300 cm^{-1}). Indeed, the success of vibrational methods, in particular infrared spectroscopy, comes from the sensitivity of the technique to the main 'chemical functionalities' (X-O, X-N, X-C, X-H), to the good relationship between content and intensity of the signal, allowing appropriate dosages and the low cost of instrumentation, associated with a spectral window excluding the lowest energies. In addition, the technique is not disturbed by low crystallinity and is valid for all materials, including amorphous materials and glasses. This has led to neglecting all the information potentially contained in the spectrum below 300 cm^{-1} [16].

Recent developments in infrared instrumentation (the availability of separators with large spectral windows replacing the use of several components/instruments/samples for the different spectral ranges, improvements in sensors and signal processing, etc.) at the cost of an increase in costs make possible access to a large part of the field making it possible to discriminate allotropic phases, even using procedures that do not require the production of a sample for transmission measurement (involving thin layers on transparent substrates, emulsions or dispersion of powders in a transparent matrix). Reflection procedures without contact if the surface is sufficiently flat and smooth or in contact with a crystal of high optical index (diamond or Ge), known by the acronyms of ATR-FTIR (Attenuated Total Reflexion –Fourier Transform InfraRed) and DRIFT (Diffuse Reflectance Infrared Fourier Transform Spectroscopy) are very successful due to their simplicity of implementation. The µATR-FTIR procedure uses a microcrystal of a few tens of μm^2 which allows mapping to be carried out. However, one should be aware that the signal obtained has a complex origin and is a mixture of 'pure' reflection and transmission processes which depend on many factors (such as refractive indices, grain size, contact quality) and that even after data processing (Kubelka-Munk or Kramers-Krönig transformations) the spectrum obtained remains a combination and this leads in particular to the deformation of the shape of the bands, the displacement of the position of the vertex and the difficulty in identifying the broad components [4, 17, 27].

Raman spectroscopy, due to the very local interaction with the laser excitation beam via the polarizability of the bond, an interaction that is much more local than the induced dipole at the origin of infrared spectroscopy, presents spectra made up of bands having much finer widths than infrared spectra. This allows recording of the low energy spectrum, down to 100–200 cm^{-1} for low or medium cost instruments and down to 3–10 cm^{-1} for sophisticated instruments. This allows the precise study to be made of network modes and therefore structures. The use of microscopes for Raman analysis has led to the replacement of the 90 degree illumination/scattering geometry, which became generally adopted from the 1960s, by backscattering in the 2000s which favours the examination of the surface of the sample, a surface which however is not always representative of the whole object. An intermediate procedure called SORS (Spatially off-set Raman Spectroscopy) [18, 36, 103] allows one to assess the bulk contribution to the analysis. The question of determining the volume of material contributing to the signal must be considered in any analysis; this problem is unfortunately not considered by many experimenters. It is essential in the production of maps by Raman mapping and for the definition of the spatial resolution. The definition of optical resolution (Airy's law) [49, 66, 137] assumes a homogeneous medium for its behaviour with respect to visible light and depends mainly on the numerical aperture of the optics. However, the absorption and the Raman signal of a particular phase can vary by several orders of magnitude and the focusing of the light is poorer along the z axis. It follows that the signal from a phase underlying with an intense Raman spectrum will mask that of those on the surface and/or neighbouring ones [18]. The actual resolution in a Raman mapping image cannot therefore be deduced from the data indicated by the supplier of the optics used. When possible, optics optimized for the spectral domain should be preferred.

It is important recalling that Raman spectrometer record also luminescence spectrum. It consists in the 'Raman spectrum background' and content many information on the material impurities (and hence the provenance of raw materials), defects, etc. [31, 78].

2.5 Other Analyses

Other techniques may be useful for characterizing and differentiating coloured material. Firstly, ^{14}C radiocarbon dating can now be carried out on small quantities [71, 95], such as for example inclusions in a material [9, 43].

Magnetic resonance techniques (Nuclear Magnetic Resonance (NMR), Electron Paramagnetic Resonance (EPR)) [15, 20, 99] are rarely used for the study of the coloured matter of objects and materials of cultural heritage although the EPR technique is one of the most effective for study of certain defects in colouring materials of a very variable nature: paper, illuminated manuscripts, stained glass, frescoes, paintings, [25, 60, 64, 67, 68, 98, 114, 120, 124, 138]. Indeed, although attempts have been made to miniaturize NMR instruments [15, 20, 115, 136], these techniques generally work on larger samples.

Micromechanical devices measuring Vickers microhardness (regular pyramid indenter) or Knoop (the section of the pyramid is a lozenge to measure the anisotropy of the material) were formerly available as accessories for all optical microscopes [131], the measurement of microhardness being made from the examination of the imprint formed in the (polished) surface of the sample. These inexpensive devices are replaced by instrumental systems measuring directly and reversibly the applied force and the displacement of the indenter, giving access to the mechanical parameters (Young's modulus, plasticity, etc.) [14, 91].

References

1. Abbott, L.C., Batchelor, S.N., Smith, J.R.L., Moore, J.N.: Resonance Raman and UV–visible spectroscopy of black dyes on textiles. Forensic Sci. Intern. **202**(1–3), 54–63 (2010). https://doi.org/10.1016/j.forsciint.2010.04.026
2. Aceto, M., Agostino, A., Fenoglio, G., Idone, A., Gulmini, M., Picollo, M., Ricciardi, P., Delaney, J.K.: Characterisation of colourants on illuminated manuscripts by portable fibre optic UV-visible-NIR reflectance spectrophotometry. Anal. Methods. **6**(5), 1488–1500 (2014). https://doi.org/10.1039/C3AY41904E
3. Alberici, R.M., Simas, R.C., Sanvido, G.B., Romão, W., Lalli, P.M., Benassi, M., Cunha, I.B., Eberlin, M.N.: Ambient mass spectrometry: bringing MS into the "real world". Anal. Bioanal. Chem. **398**, 265–294 (2010). https://doi.org/10.1007/s00216-010-3808-3
4. Anderson, T.H., Weaver, F.W., Owen, N.L.: Anomalies in diffuse reflectance infrared spectra of wood and wood polymers. J. Mol. Struct. **249**(2–4), 257–275 (1991). https://doi.org/10.1016/0022-2860(91)85072-B
5. Arrizabalaga, I., Gómez-Laserna, O., Aramendia, J., Arana, G., Madariaga, J.M.: Applicability of a diffuse reflectance infrared Fourier transform handheld spectrometer to perform in situ analyses on cultural heritage materials. Spectrochim. Acta A Mol. Biomol. Spectrosc. **129**, 259–267 (2014). https://doi.org/10.1016/j.saa.2014.03.096
6. Artioli, G.: Science for the cultural heritage: the contribution of X-ray diffraction. Rendiconti Lincei. **24**, 55–62 (2013). https://doi.org/10.1007/s12210-012-0207-z
7. Aucouturier, M., Darque-Ceretti, E.: The surface of cultural heritage artefacts: physicochemical investigations for their knowledge and their conservation. Chem. Soc. Rev. **36**(10), 1605–1621 (2007). https://doi.org/10.1039/B605304C
8. Bandyopadhyay, A., Sengupta, A.: A review of the concept, applications and implementation issues of terahertz spectral imaging technique. IETE Tech. Rev. **39**(2), 471–489 (2022). https://doi.org/10.1080/02564602.2020.1865844
9. Berger, D., Soles, J.S., Giumlia-Mair, A.R., Brügmann, G., Galili, E., Lockhoff, N., Pernicka, E.: Isotope systematics and chemical composition of tin ingots from Mochlos (Crete) and other late bronze age sites in the eastern Mediterranean Sea: an ultimate key to tin provenance? PLoS One. **14**(6), e0218326 (2019). https://doi.org/10.1371/journal.pone.0218326
10. Bevis, J.A., Bottom, R., Duncan, J., Farhat, I., Forrest, M., Furniss, D., MacNaughton, B., Nazhat, S., Saunders, M., Seddon, A.: Principles and Applications of Thermal Analysis. Blackwell, Oxford (2008) ISBN: 978-1-405-13171-1
11. Bianchi, C.L., Djellabi, R., Ponti, A., Patience, G.S., Falletta, E.: Experimental methods in chemical engineering: Mössbauer spectroscopy. Can. J. Chem. Eng. **99**(10), 2105–2114 (2021). https://doi.org/10.1002/cjce.24216
12. Bitossi, G., Giorgi, R., Mauro, M., Salvadori, B., Dei, L.: Spectroscopic techniques in cultural heritage conservation: a survey. Appl. Spectrosc. Rev. **40**(3), 187–228 (2005). https://doi.org/10.1081/ASR-200054370
13. Botto, A., Campanella, B., Legnaioli, S., Lezzerini, M., Lorenzetti, G., Pagnotta, S., Poggialini, F., Palleschi, V.: Applications of laser-induced breakdown spectroscopy in cultural heritage and archaeology: a critical review. J. Anal. At. Spectrom. **34**(1), 81–103 (2019). https://doi.org/10.1039/C8JA00319J
14. Blau, P.J.: Microindentation Techniques in Materials Science and Engineering: A Symposium Sponsored by ASTM Committee E-4 on Metallography and by the International Metallographic Society, Philadelphia, PA, 15–18 July 1984 (No. 889). Astm International (1986). https://doi.org/10.1520/STP889-EB
15. Blümich, B.: Introduction to compact NMR: a review of methods. TrAC Trends Anal. Chem. **83**, 2–11 (2016). https://doi.org/10.1016/j.trac.2015.12.012
16. Bourson, P., Simon, G., Chapron, D., Kokanyan, N., Colomban, P.: Spectroscopie et spectrométrie Raman, ch. In: Simon, G. (ed.) Spectroscopies Vibrationnelles, pp. 109–154. E.A.C, Paris (2020). https://doi.org/10.17184/eac.4199

17. Buffeteau, T., De Sousa Meneses, D., Dussauge, M., Tassaing, T.: Spectroscopie infrarouge. In: Simon, G. (ed.) Spectroscopies vibrationnelles. Théorie, aspects pratiques et applications, pp. 47–84. Editions des archives contemporaines, France, ISBN: 9782813002556 (2020). https://doi.org/10.17184/eac.4197
18. Caggiani, M.C., Colomban, P.: Raman microspectroscopy for cultural heritage studies. Phys. Sci. Rev. **3**(11), 20180007 (2018). https://doi.org/10.1515/psr-2018-0007
19. Calligaro, T., Dran, J.C., Salomon, J., Walter, P.: Review of accelerator gadgets for art and archaeology. Nucl. Instrum. Methods Phys. Res. Sect. B. **226**(1–2), 29–37 (2004). https://doi.org/10.1016/j.nimb.2004.04.176
20. Capitani, D., Di Tullio, V., Proietti, N.: Nuclear magnetic resonance to characterize and monitor cultural heritage. Prog. Nucl. Magn. Reson. Spectrosc. **64**, 29–69 (2012). https://doi.org/10.1016/j.pnmrs.2011.11.001
21. Carter, S., Clough, R., Fisher, A., Gibson, B., Russell, B.: Atomic spectrometry update: review of advances in the analysis of metals, chemicals and materials. J. Anal. At. Spectrom. **37**(11), 2207–2281 (2022). https://doi.org/10.1039/D2JA90050E
22. Casadio, F., Toniolo, L.: The analysis of polychrome works of art: 40 years of infrared spectroscopic investigations. J. Cult. Herit. **2**(1), 71–78 (2001). https://doi.org/10.1016/S1296-2074(01)01107-4
23. Chadwick, A.V., Berko, A., Schofield, E.J., Jones, A.M., Mosselmans, J.F.W., Smith, A.D.: Application of microfocus x-ray beams from synchrotrons in heritage conservation. Int. J. Archit. Herit. **6**(2), 228–258 (2012). https://doi.org/10.1080/15583058.2010.528825
24. Chivers, T., Elder, P.J.: Ubiquitous trisulfur radical anion: fundamentals and applications in materials science, electrochemistry, analytical chemistry and geochemistry. Chem. Soc. Rev. **42**(14), 5996–6005 (2013). https://doi.org/10.1039/C3CS60119F
25. Christiansen, M.B., Sørensen, M.A., Sanyova, J., Bendix, J., Simonsen, K.P.: Characterisation of the rare cadmium chromate pigment in a 19th century tube colour by Raman, FTIR, X-ray and EPR. Spectrochim. Acta A Mol. Biomol. Spectrosc. **175**, 208–214 (2017). https://doi.org/10.1016/j.saa.2016.12.028
26. Coccato, A., Moens, L., Vandenabeele, P.: On the stability of mediaeval inorganic pigments: a literature review of the effect of climate, material selection, biological activity, analysis and conservation treatments. Herit. Sci. **5**, 12 (2017). https://doi.org/10.1186/s40494-017-0125-6
27. Colomban, P.: Hydrogen bonding in hydrogen-substituted lithium aluminosilicates. J. Mol. Struct. **270**(1992), 407–416 (1992). https://doi.org/10.1016/0022-2860(92)85043-G
28. Colomban, P.: The on-site/remote Raman analysis with mobile instruments: a review of drawbacks and success in cultural heritage studies and other associated fields. J. Raman Spectrosc. **43**(11), 1529–1535 (2012). https://doi.org/10.1002/jrs.4042
29. Colomban, P.: Proton and protonic species: the hidden face of solid state chemistry. How to measure H-content in materials? Fuel Cells. **13**(1), 6–18 (2013). https://doi.org/10.1002/fuce.201200088
30. Colomban, P.: Proton conductors and their applications: a tentative historical overview of the early researches. Solid State Ionics. **334**, 125–144 (2019). https://doi.org/10.1016/j.ssi.2019.01.032
31. Colomban, P.: Full spectral range Raman signatures related to changes in Enameling technologies from the 18th to the 20th century: guidelines, effectiveness and limitations of the Raman analysis. Materials. **15**, 3158 (2022). https://doi.org/10.3390/ma15093158
32. Colomban, P., Jones, D.J., Grandjean, D., Flank, A.M.: EXAFS and XANES study of (Si, Ge) mullite gels and glasses prepared by slow hydrolysis of alkoxides. J. Non Crystall. Solids. **147–148**, 135–140 (1992). https://doi.org/10.1016/S0022-3093(05)80606-4
33. Colomban, P., Calligaro, T., Vibert-Guigue, C., Edwards, H.G.: Gilded ceramics and tiles: technologies and adhesion. ArcheoSciences. **29**(1), 7–20 (2005). https://doi.org/10.4000/archeosciences.424

34. Colomban, P., Gironda, M., Vangu, D., Kırmızı, B., Zhao, B., Cochet, V.: The technology transfer from Europe to China in the 17th–18th centuries: non-invasive on-site XRF and Raman analyses of Chinese Qing dynasty Enameled masterpieces made using European ingredients/recipes. Materials. **14**, 7434 (2021). https://doi.org/10.3390/ma14237434
35. Colomban, P., Gironda, M., Simsek Franci, G., d'Abrigeon, P.: Distinguishing genuine imperial Qing dynasty porcelain from ancient replicas by on-site non-invasive XRF and Raman spectroscopy. Materials. **15**(16), 5747 (2022). https://doi.org/10.3390/ma15165747
36. Conti, C., Botteon, A., Colombo, C., Pinna, D., Realini, M., Matousek, P.: Advances in Raman spectroscopy for the non-destructive subsurface analysis of artworks: micro-SORS. J. Cult. Herit. **43**, 319–328 (2020). https://doi.org/10.1016/j.culher.2019.12.003
37. Cotte, M., Genty-Vincent, A., Janssens, K., Susini, J.: Applications of synchrotron X-ray nano-probes in the field of cultural heritage. Comptes Rendus Physique. **19**(7), 575–588 (2018). https://doi.org/10.1016/j.crhy.2018.07.002
38. Crina Anca Sandu, I., de Sá, M.H., Pereira, M.C.: Ancient 'gilded' art objects from European cultural heritage: a review on different scales of characterization. Surf. Interface Anal. **43**(8), 1134–1151 (2011). https://doi.org/10.1002/sia.3740
39. Cucci, C., Delaney, J.K., Picollo, M.: Reflectance hyperspectral imaging for investigation of works of art: old master paintings and illuminated manuscripts. Acc. Chem. Res. **49**(10), 2070–2079 (2016). https://doi.org/10.1021/acs.accounts.6b00048
40. Dale, L.M., Thewis, A., Boudry, C., Rotar, I., Dardenne, P., Baeten, V., Pierna, J.A.F.: Hyperspectral imaging applications in agriculture and agro-food product quality and safety control: a review. Appl. Spectrosc. Rev. **48**(2), 142–159 (2013). https://doi.org/10.1080/05704928.2012.705800
41. Das, D.D., Sharma, N., Chawla, P.A.: Neutron activation analysis: an excellent nondestructive analytical technique for trace metal analysis. Crit. Rev. Anal. Chem. 1–17 (2023). https://doi.org/10.1080/10408347.2023.2178841
42. Dazzi, A., Prater, C.B.: AFM-IR: technology and applications in nanoscale infrared spectroscopy and chemical imaging. Chem. Rev. **117**(7), 5146–5173 (2017). https://doi.org/10.1021/acs.chemrev.6b00448
43. Defrasne, C., Chalmin, E., Bellot-Gurlet, L., Thirault, E., André, G.: From archeological layers to schematic rock art? Integrated study of the Neolithic pigments and pigmented rocks at the Rocher du Château (Western Alps, Savoie, France). Archaeol. Anthropol. Sci. **11**, 6065–6091 (2019). https://doi.org/10.1007/s12520-019-00882-9
44. Demirsar Arli, B., Simsek Franci, G., Kaya, S., Arli, H., Colomban, P.: Portable X-ray fluorescence (p-XRF) uncertainty estimation for glazed ceramic analysis: case of Iznik tiles. Heritage. **3**(4), 1302–1329 (2020). https://doi.org/10.3390/heritage3040072
45. Detalle, V., Bai, X.: The assets of laser-induced breakdown spectroscopy (LIBS) for the future of heritage science. Spectrochim. Acta B At. Spectrosc. **191**, 106407 (2022). https://doi.org/10.1016/j.sab.2022.106407
46. Duran, A., Herrera, L.K., de Haro, M.J., Justo, A., Perez-Rodriguez, J.L.: Non-destructive analysis of cultural heritage artefacts from Andalusia, Spain, by X-ray diffraction with Göbel mirrors. Talanta. **76**(1), 183–188 (2008). https://doi.org/10.1016/j.talanta.2008.02.025
47. Eccles, H.A., Rackam, B.: Introduction and scope. In: Chalmers, J.M., Edwards, H.G.M., Hargreaves, M.H. (eds.) Infrared and Raman spectroscopy in forensic science British Museum Report, ch. 1, p. 5. Wiley, Chichester (1922)., ISBN: 978-0-470-74906-7 (2012)
48. Edge CIEL https://www.science.org/doi/abs/10.1126/science.284.5415.760
49. Edwards, H.G.M., Vandenabeele, P., Colomban, P.: Raman spectroscopy in cultural heritage preservation. Springer Nature (2022). https://doi.org/10.1007/978-3-031-14379-3
50. Enríquez, E., Reinosa, J.J., Fuertes, V., Fernández, J.F.: Advances and challenges of ceramic pigments for inkjet printing. Ceram. Int. **48**(21), 31080–31101 (2022). https://doi.org/10.1016/j.ceramint.2022.07.155

51. Farges, F., Cotte, M.: X-ray absorption spectroscopy and cultural heritage: highlights and perspectives. In: X-Ray Absorption and X-Ray Emission Spectroscopy: Theory and Applications, pp. 609–636. Springer, Singapore (2016). https://doi.org/10.1002/9781118844243.ch21
52. Ferchault de Réaumur, A.R.: Idée générale des différentes manières dont on peut faire la porcelaine et quelles sont les véritables matières de celle de la Chine, Mémoires Académie des Sciences, Paris. (1727). https://www.google.com/url?sa=t&source=web&rct=j&opi=89978449&url=https://www.academiesciences.fr/pdf/dossiers/Reaumur/Reaumur_pdf/p185_203_vol3589m.pdf&ved=2ahUKEwiT1qrxkoCGAxVERqQEHQ67CnoQFnoECBMQAQ&usg=AOvVaw2_33YgOseTCyfmDzSB5APN. Accessed 25 Dec 2024
53. Festa, G., Romanelli, G., Senesi, R., Arcidiacono, L., Scatigno, C., Parker, S.F., Marques, M.P.M., Andreani, C.: Neutrons for cultural heritage—techniques, sensors, and detection. Sensors. **20**(2), 502 (2020). https://doi.org/10.3390/s20020502
54. Fiorito, P.A., Gonçales, V.R., Ponzio, E.A., de Torresi, S.I.C.: Synthesis, characterization and immobilization of Prussian blue nanoparticles. A potential tool for biosensing devices. Chem. Commun. **3**, 366–368 (2005). https://doi.org/10.1039/B412583E
55. Fischer, C., Kakoulli, I.: Multispectral and hyperspectral imaging technologies in conservation: current research and potential applications. Stud. Conserv. **51**(sup1), 3–16 (2006). https://doi.org/10.1179/sic.2006.51.Supplement-1.3
56. Fritsch, E., Rossman, G.R.: An update on color in gems. Part 3: Colors caused by band gaps and physical phenomena. Gems Gemology. **24**(2), 81–102 (1988)
57. Gao, L., Smith, R.T.: Optical hyperspectral imaging in microscopy and spectroscopy–a review of data acquisition. J. Biophotonics. **8**(6), 441–456 (2015). https://doi.org/10.1002/jbio.201400051
58. Garcia Moreno, R., Strivay, D., Gilbert, B.: Maya blue–green pigments found in Calakmul, Mexico: a study by Raman and UV-visible spectroscopy. J. Raman Spectrosc. **39**(8), 1050–1056 (2008). https://doi.org/10.1002/jrs.1972
59. Gaudiuso, R., Dell'Aglio, M., De Pascale, O., Senesi, G.S., De Giacomo, A.: Laser induced breakdown spectroscopy for elemental analysis in environmental, cultural heritage and space applications: a review of methods and results. Sensors. **10**(8), 7434-7468 (2010). https://doi.org/10.3390/s100807434
60. Germinario, L., Giannossa, L.C., Lezzerini, M., Mangone, A., Mazzoli, C., Pagnotta, S., Spampinato, M., Zoleo, A., Eramo, G.: Petrographic and chemical characterization of the frescoes by Saturnino Gatti (Central Italy, 15th century). Appl. Sci. **13**(12), 7223 (2023). https://doi.org/10.3390/app13127223
61. Giakoumaki, A., Melessanaki, K., Anglos, D.: Laser-induced breakdown spectroscopy (LIBS) in archaeological science—applications and prospects. Anal. Bioanal. Chem. **387**, 749–760 (2007). https://doi.org/10.1007/s00216-006-0908-1
62. Giussani, B., Monticelli, D., Rampazzi, L.: Role of laser ablation–inductively coupled plasma–mass spectrometry in cultural heritage research: a review. Anal. Chim. Acta. **635**(1), 6–21 (2009). https://doi.org/10.1016/j.aca.2008.12.040
63. France, F.G.: Advanced spectral imaging for noninvasive microanalysis of cultural heritage materials: review of application to documents in the US Library of Congress. Appl. Spectrosc. **65**(6), 565–574 (2011). https://opg.optica.org/as/abstract.cfm?uri=as-65-6-565
64. Godet, M., Binet, L., Schöder, S., Brunel-Duverger, L., Thoury, M., Bertrand, L.: X-ray irradiation effects on Egyptian blue and green pigments. J. Anal. At. Spectrom. **37**(6), 1265–1272 (2022). https://doi.org/10.1039/D2JA00020B
65. Goetz, F.H., Vane, G., Solomon, B.N., Rock, N.: Imaging spectrometry for earth remote sensing. Science. **228**, 1147–1152 (1985)
66. Gouadec, G., Colomban, P.: Raman spectroscopy of nanomaterials: how spectra relate to disorder, particle size and mechanical properties. Prog. Cryst. Growth Charact. Mater. **53**(1), 1–56 (2007). https://doi.org/10.1016/j.pcrysgrow.2007.01.001

67. Gourier, D., Binet, L., Gonzalez, V., Vezin, H., Touati, N., Calligaro, T.: Effect of analytical proton beam irradiation on lead-white pigments, characterized by EPR spectroscopy. Nucl. Instrum. Methods Phys. Res., Sect. B. **415**, 64–71 (2018). https://doi.org/10.1016/j.nimb.2017.10.032
68. Griscom, D.L.: Trapped-electron centers in pure and doped glassy silica: a review and synthesis. J. Non-Cryst. Solids. **357**(8–9), 1945–1962 (2011). https://doi.org/10.1016/j.jnoncrysol.2010.11.011
69. Guerra, M.F., Calligaro, T.: Gold cultural heritage objects: a review of studies of provenance and manufacturing technologies. Meas. Sci. Technol. **14**(9), 1527 (2003). https://doi.org/10.1088/0957-0233/14/9/305
70. Halliburton, L.E.: ESR and optical characterization of point defects in quartz. *International journal of radiation applications and instrumentation. Part a*. Appl. Radiat. Isot. **40**(10–12), 859–863 (1989). https://doi.org/10.1016/0883-2889(89)90007-5
71. Hendriks, L., Hajdas, I., Ferreira, E.S., Scherrer, N.C., Zumbühl, S., Küffner, M., Carlyle, L., Synal, H.A., Günther, D.: Selective dating of paint components: radiocarbon dating of lead white pigment. Radiocarbon. **61**(2), 473–493 (2019). https://doi.org/10.1017/RDC.2018.101
72. Hintz, H., Egelhaaf, H.J., Peisert, H., Chassé, T.: Photo-oxidation and ozonization of poly (3-hexylthiophene) thin films as studied by UV/VIS and photoelectron spectroscopy. Polym. Degrad. Stab. **95**(5), 818–825 (2010). https://doi.org/10.1016/j.polymdegradstab.2010.02.004
73. Hudson, B.S.: Vibrational spectroscopy using inelastic neutron scattering: overview and outlook. Vib. Spectrosc. **42**(1), 25–32 (2006). https://doi.org/10.1016/j.vibspec.2006.04.014
74. Hunault, M., Bauchau, F., Loisel, C., Hérold, M., Galoisy, L., Newville, M., Calas, G.: Spectroscopic investigation of the colouration and fabrication conditions of medieval blue glasses. J. Am. Ceram. Soc. **99**(1), 89–97 (2016). https://doi.org/10.1111/jace.13783
75. Ibraheem, N.A., Hasan, M.M., Khan, R.Z., Mishra, P.K.: Understanding colour models: a review. ARPN J. Sci. Technol. **2**(3), 265–275 (2012). http://www.researchgate.net/publication/266462481_Understanding_Color_Models_A_Review. Accessed 4 Mar 2024
76. Ingo, G.M., Riccucci, C., Pascucci, M., Messina, E., Giuliani, C., Biocca, P., Tortora, L., Fierro, G., Di Carlo, G.: Combined use of FE-SEM+ EDS, ToF-SIMS, XPS, XRD and OM for the study of ancient gilded artefacts. Appl. Surf. Sci. **446**, 168–176 (2018). https://doi.org/10.1016/j.apsusc.2018.01.278
77. Janssens, K., Cotte, M.: The use of XAS and related methods in cultural heritage investigations. International tables for crystallography. Vol. I. In: Chantler, C.T., Boscherini, F., Bunker, B. (eds.) X-Ray Absorption Spectroscopy and Related Techniques. Wiley (2023). https://doi.org/10.1107/S1574870720004802
78. Jasinevicius, R.: Characterization of vibrational and electronic features in the Raman spectra of gem minerals. Master's degree thesis, University of Arizona. https://scholar.google.com/scholar?hl=en&as_sdt=0%2C5&q=electronic+defect+optical+characterization+colour+gems&btnG= (2009). Accessed 3 Mar 2024
79. Joosten, I.: Applications of microanalysis in the cultural heritage field. Microchim. Acta. **161**, 295–299 (2008). https://doi.org/10.1007/s00604-007-0902-0
80. Kamiya, Y., Honda, T., Ohbuchi, A., Miyakoshi, T.: Simultaneous organic and inorganic analysis of coloured oriental lacquerware by pyrolysis-gas chromatography/mass spectrometry. Int. J. Polym. Sci. **2015**, 725467 (2015). https://doi.org/10.1155/2015/725467
81. Kanngießer, B., Malzer, W., Mantouvalou, I., Sokaras, D., Karydas, A.G.: A deep view in cultural heritage—confocal micro X-ray spectroscopy for depth resolved elemental analysis. Appl. Phys. A. **106**, 325–338 (2012). https://doi.org/10.1007/s00339-011-6698-0
82. Kantarelou, V., Karydas, A.G.: A simple calibration procedure of polycapillary based portable micro-XRF spectrometers for reliable quantitative analysis of cultural heritage materials. X-Ray Spectrom. **45**(2), 85–91 (2016). https://doi.org/10.1002/xrs.2661
83. Kendix, E.L., Prati, S., Mazzeo, R., Joseph, E., Sciutto, G., Fagnano, C.: Far infrared spectroscopy in the field of cultural heritage. *Preservation*. Science. **7**, 8–13 (2010)

84. Khramchenkova, R., Safina, I., Drobyshev, S., Batasheva, S., Nuzhdin, E., Fakhrullin, R.: Advanced microscopy techniques for nanoscale diagnostic of cultural heritage: scanning electron microscopy for investigation of medieval coins and frescos from the republic of Tatarstan. In: Nanotechnologies and Nanomaterials for Diagnostic, Conservation and Restoration of Cultural Heritage, pp. 1–23. Elsevier, Amsterdam (2019). https://doi.org/10.1016/B978-0-12-813910-3.00001-X
85. Kim, J.H., Joo, H.G., Kim, T.H., Ju, Y.G.: A smartphone-based fluorescence microscope utilizing an external phone camera lens module. Biochip J. **9**, 285–292 (2015). https://doi.org/10.1007/s13206-015-9403-0
86. Krishna, D.N.G., Philip, J.: Review on surface-characterization applications of X-ray photoelectron spectroscopy (XPS): recent developments and challenges. Appl. Surf. Sci. Adv. **12**, 100332 (2022). https://doi.org/10.1016/j.apsadv.2022.100332
87. Lazar, S., Botton, G.A., Wu, M.Y., Tichelaar, F.D., Zandbergen, H.W.: Materials science applications of HREELS in near edge structure analysis and low-energy loss spectroscopy. Ultramicroscopy. **96**(3–4), 535–546 (2003). https://doi.org/10.1016/S0304-3991(03)00114-1
88. Lee, K.C., Lee, K., Jung, J., Lee, S.H., Kim, D., Lee, S.A.: A smartphone-based Fourier ptychographic microscope using the display screen for illumination. ACS Photonics. **8**(5), 1307–1315 (2021). https://doi.org/10.1021/acsphotonics.1c00350
89. Legrand, S., Vanmeert, F., Van der Snickt, G., Alfeld, M., De Nolf, W., Dik, J., Janssens, K.: Examination of historical paintings by state-of-the-art hyperspectral imaging methods: from scanning infra-red spectroscopy to computed X-ray laminography. Herit. Sci. **2**, 1–11 (2014). https://doi.org/10.1186/2050-7445-2-13
90. Leona, M., Casadio, F., Bacci, M., Picollo, M.: Identification of the pre-columbian pigment mayablue on works of art by noninvasive UV-vis and Raman spectroscopic techniques. J. Am. Inst. Conserv. **43**(1), 39–54 (2004). https://doi.org/10.2307/3179850
91. Loubet, J.L., Georges, J.M., Meille, G.: Vickers Indentation Curves of Elastoplastic Materials, pp. 72–89. ASTM International (1985). https://doi.org/10.1520/STP32952S
92. Liu, G.L., Kazarian, S.G.: Recent advances and applications to cultural heritage using ATR-FTIR spectroscopy and ATR-FTIR spectroscopic imaging. Analyst. **147**(9), 1777–1797 (2022). https://doi.org/10.1039/D2AN00005A
93. Machado, R.C., Andrade, D.F., Babos, D.V., Castro, J.P., Costa, V.C., Sperança, M.A., Garcia, J.A., Gamela, R.R., Pereira-Filho, E.R.: Solid sampling: advantages and challenges for chemical element determination—a critical review. J. Anal. At. Spectrom. **35**(1), 54–77 (2020). https://doi.org/10.1039/C9JA00306A
94. Macková, A., Kučera, J., Kamenik, J., Havránek, V., Šmit, Ž., Giuntini, L., Kasztovszky, Z.: Nuclear physics for cultural heritage. Il Nuovo Cimento C. **42**(2–3), 1–8 (2019). https://doi.org/10.1393/ncc/i2019-19053-6
95. Madariaga, J.M.: Analytical chemistry in the field of cultural heritage. Anal. Methods. **7**(12), 4848–4876 (2015). https://doi.org/10.1039/C5AY00072F
96. Magdy, M.: Analytical techniques for the preservation of cultural heritage: Frontiers in knowledge and application. Crit. Rev. Anal. Chem. **52**(6), 1171–1196 (2022). https://doi.org/10.1080/10408347.2020.1864717
97. Manley, M.: Near-infrared spectroscopy and hyperspectral imaging: non-destructive analysis of biological materials. Chem. Soc. Rev. **43**(24), 8200–8214 (2014). https://doi.org/10.1039/C4CS00062E
98. Marchese, N., Cannuli, A., Caccamo, M.T., Pace, C.: New generation non-stationary portable neutron generators for biophysical applications of neutron activation analysis. Biochimica et Biophysica Acta (BBA)-General Subjects. **1861**(1), 3661–3670 (2017). https://doi.org/10.1016/j.bbagen.2016.05.023
99. Azzoni, C.B., Di Martino, D., Marchesi, V., Messiga, B., Riccardi, M.P.: Colour attributes of medieval panes: electron paramagnetic resonance and probe microanalyses on stained glass window from Pavia Carthusian monastery. Archaeometry. **47**(2), 381–388 (2005). https://doi.org/10.1111/j.1475-4754.2005.00208.x

100. McMurdie, H.F., Morris, M.C., Evans, E.H., Paretzkin, B., Wong-Ng, W., Zhang, Y., Hubbard, C.R.: Standard X-ray diffraction powder patterns from the JCPDS research associateship. Powder Diffract. **2**(1), 41–52 (1987). https://doi.org/10.1017/S0885715600012239
101. Monge, M.E., Harris, G.A., Dwivedi, P., Fernandez, F.M.: Mass spectrometry: recent advances in direct open air surface sampling/ionization. Chem. Rev. **113**(4), 2269–2308 (2013). https://doi.org/10.1021/cr300309q
102. Moore, E.A., Smart, L.E.: Optical properties of solids. In: Solid State Chemistry, pp. 283–314. CRC Press (2020). https://doi.org/10.1201/9780429027284
103. Mosca, S., Dey, P., Salimi, M., Gardner, B., Palombo, F., Stone, N., Matousek, P.: Spatially offset Raman spectroscopy—how deep? Anal. Chem. **93**(17), 6755–6762 (2021). https://doi.org/10.1021/acs.analchem.1c00490
104. Mounier, A., Le Bourdon, G., Aupetit, C., Lazare, S., Biron, C., Pérez-Arantegui, J., Almazán, D., Aramendia, J., Prieto-Taboada, N., De Vallejuelo, S.F.O., Daniel, F.: Red and blue colours on 18th–19th century Japanese woodblock prints: in situ analyses by spectrofluorimetry and complementary non-invasive spectroscopic methods. Microchem. J. **140**, 129–141 (2018). https://doi.org/10.1016/j.microc.2018.04.023
105. Neri, E., Morvan, C., Colomban, P., Guerra, M.F., Prigent, V.: Late Roman and byzantine mosaic opaque "glass-ceramics" tesserae (5th–9th century). Ceram. Int. **42**(16), 18859–18869 (2016). https://doi.org/10.1016/j.ceramint.2016.09.033
106. Nicolopoulos, S., Das, P.P., Bereciartua, P.J., Karavasili, F., Zacharias, N., Gómez Pérez, A., Galanis, A.S., Rauch, E.F., Arenal, R., Portillo, J., Roqué-Rosell, J.: Novel characterization techniques for cultural heritage using a TEM orientation imaging in combination with 3D precession diffraction tomography: a case study of green and white ancient Roman glass tesserae. Herit. Sci. **6**(1), 64 (2018). https://doi.org/10.1186/s40494-018-0229-7
107. Nord, A.G., Billström, K.: Isotopes in cultural heritage: present and future possibilities. Herit. Sci. **6**, 25 (2018). https://doi.org/10.1186/s40494-018-0192-3
108. Oishi, R., Yonemura, M., Nishimaki, Y., Torii, S., Hoshikawa, A., Ishigaki, T., Morishima, T., Mori, K., Kamiyama, T.: Rietveld analysis software for J-PARC. Nucl. Instrum. Methods Phys. Res., Sect. A. **600**(1), 94–96 (2009). https://doi.org/10.1016/j.nima.2008.11.056
109. Picollo, M., Aceto, M., Vitorino, T.: UV-vis spectroscopy. Phys. Sci. Rev. **4**(4), 20180008 (2018). https://doi.org/10.1515/psr-2018-0008
110. Picollo, M., Cucci, C., Casini, A., Stefani, L.: Hyper-spectral imaging technique in the cultural heritage field: new possible scenarios. Sensors. **20**(10), 2843 (2020). https://doi.org/10.3390/s20102843
111. Poliszuk, A., Ybarra, G.: Analysis of cultural heritage materials by infrared spectroscopy. In: Infrared Spectroscopy: Theory, Developments and Applications, pp. 519–536. Nova Publishers, New York (2014). https://www.researchgate.net/publication/261697816_Analysis_of_Cultural_Heritage_Materials_by_Infrared_Spectroscopy. Accessed 4 Mar 2024
112. Possenti, E., Conti, C., Gatta, G.D., Merlini, M., Realini, M., Colombo, C.: Synchrotron radiation μ X-ray diffraction in transmission geometry for investigating the penetration depth of conservation treatments on cultural heritage stone materials. Anal. Methods. **12**(12), 1587–1594 (2020). https://doi.org/10.1039/D0AY00010H
113. Quartieri, S.: X-Ray Absorption Spectroscopy (XAS, EXAFS, XANES). The Encyclopedia of Archaeological Sciences, 1–6 (2018)
114. Raulin, K., Gobeltz, N., Vezin, H., Touati, N., Ledé, B., Moissette, A.: Identification of the EPR signal of S 2− in green ultramarine pigments. Phys. Chem. Chem. Phys. **13**(20), 9253–9259 (2011). https://doi.org/10.1039/C0CP02970J
115. Rehorn, C., Blümich, B.: Cultural heritage studies with mobile NMR. Angew. Chem. Int. Ed. **57**(25), 7304–7312 (2018). https://doi.org/10.1002/anie.201713009
116. Rinaldi, R.: Neutron scattering in mineral sciences: preface. Eur. J. Mineral. **14**(2), 195–202 (2002). https://doi.org/10.1127/0935-1221/2002/0014-0195

117. Rocha, F.S., Gomes, A.J., Lunardi, C.N., Kaliaguine, S., Patience, G.S.: Experimental methods in chemical engineering: ultraviolet visible spectroscopy—UV-vis. Can. J. Chem. Eng. **96**(12), 2512–2517 (2018). https://doi.org/10.1002/cjce.23344
118. Rosi, T., Malgieri, M., Onorato, P., De Ambrosis, A., Oss, S.: Studying colours with a smartphone. Il Nuovo Cimento C. **40**(2), 1–9 (2017). https://doi.org/10.1393/ncc/i2017-17104-8
119. Schreiner, M., Melcher, M., Uhlir, K.: Scanning electron microscopy and energy dispersive analysis: applications in the field of cultural heritage. Anal. Bioanal. Chem. **387**(3), 737–747 (2007). https://doi.org/10.1007/s00216-006-0718-5
120. Schreurs, J.W., Brill, R.H.: Iron and sulfur related colours in ancient glasses. Archaeometry. **26**(2), 199–209 (1984). https://doi.org/10.1111/j.1475-4754.1984.tb00334.x
121. Sciau, P., Salles, P., Roucau, C., Mehta, A., Benassayag, G.: Applications of focused ion beam for preparation of specimens of ancient ceramic for electron microscopy and synchrotron X-ray studies. Micron. **40**(5–6), 597–604 (2009). https://doi.org/10.1016/j.micron.2009.02.012
122. Sciau, P.: Transmission electron microscopy: emerging investigations for cultural heritage materials. In: Advances in Imaging and Electron Physics ch. 2, vol. 198, pp. 43–67. Elsevier, San Diego (2016). https://doi.org/10.1016/bs.aiep.2016.09.002
123. Secondulfo, A.P., Somma, P.I., Aprile, A.A., Rossi, P., Melchiorre, C., Cennamo, P., Trojsi, G., Spinelli, M., Amoresano, A., Carpentieri, A.: Molecular characterisation of the tempera painting on canvas' The Martyrdom of Gorcum. In Diagnosis for the Conservation and Valorization of Cultural Heritage XIII Convegno Internazionale 2022. https://www.google.com/url?sa=t&rct=j&q=&esrc=s&source=web&cd=&ved=2ahUKEwjCio2QhdqEAxUtcKQEHbrYBI0QFnoECBcQAQ&url=https%3A%2F%2Fwww.unisob.na.it%2Fateneo%2Frestauro%2Fattivita8_202212b.pdf&usg=AOvVaw34aQIYvC1BOcJlBdS8bgdj&opi=89978449 (2022). Accessed 4 Mar 2024
124. Skovmøller, A., Therkildsen, H.: The polychromy of roman polished marble portraits. In: The Polychromy of Roman Polished Marble Portraits, pp. 891–900. L'Erma di Bretschneider, Rome (2015). https://research.ku.dk/search/result/?pure=da%2Fpublications%2Fthe-polychromy-of-roman-polished-marble-portraits(732799fa-0eb3-419a-88e2-0903b18efda2)%2Fexport.html. Accessed 4 Mar 2024
125. Sottili, L., Giuntini, L., Mazzinghi, A., Massi, M., Carraresi, L., Castelli, L., Czelusniak, C., Giambi, F., Mandò, P.A., Manetti, M., Ruberto, C.: The role of PIXE and XRF in heritage science: the INFN-CHNet LABEC experience. Appl. Sci. **12**(13), 6585 (2022). https://doi.org/10.3390/app12136585
126. Spangenberg, J.E., Lavrič, J.V., Meisser, N., Serneels, V.: Sulfur isotope analysis of cinnabar from Roman wall paintings by elemental analysis/isotope ratio mass spectrometry–tracking the origin of archaeological red pigments and their authenticity. Rapid Commun. Mass Spectrom. **24**(19), 2812–2816 (2010). https://doi.org/10.1002/rcm.4705
127. Striova, J., Dal Fovo, A., Fontana, R.: Reflectance imaging spectroscopy in heritage science. La Rivista del Nuovo Cimento. **43**, 515–566 (2020). https://doi.org/10.1007/s40766-020-00011-6
128. Stuart, M.B., McGonigle, A.J., Willmott, J.R.: Hyperspectral imaging in environmental monitoring: a review of recent developments and technological advances in compact field deployable systems. Sensors. **19**(14), 3071 (2019). https://doi.org/10.3390/s19143071
129. Su, X., Wang, Y., Mao, J., Chen, Y., Yin, A., Zhao, B., Zhang, H., Liu, M.: A review of pharmaceutical robot based on hyperspectral technology. J. Intell. Robot. Syst. **105**(4), 75 (2022). https://doi.org/10.1007/s10846-022-01602-7
130. Tibúrcio, C., Valadas, S., Cardoso, A., Candeias, A., Barreira, C., Miguel, C.: On the use of EDXRF and UV–vis FORS to unveil the production of two illuminated manuscripts from the fifteenth century portuguese royal court. Microchem. J. **153**, 104455 (2020). https://doi.org/10.1016/j.microc.2019.104455
131. Tournié, A., Ricciardi, P., Colomban, P.: Glass corrosion mechanisms: a multiscale analysis. Solid State Ionics. **179**(38), 2142–2154 (2008). https://doi.org/10.1016/j.ssi.2008.07.019

132. Trocellier, P., Berger, P., Wilde, M.: Nuclear reaction analysis. In: Encyclopedia of Analytical Chemistry: Applications, Theory and Instrumentation, pp. 1–17. Wiley, Chichester (2006). https://doi.org/10.1002/9780470027318.a6208.pub3
133. Van de Voorde, L., Van Pevenage, J., De Langhe, K., De Wolf, R., Vekemans, B., Vincze, L., Vandenabeele, P., Martens, M.P.J.: Non-destructive in situ study of “mad meg” by Pieter Bruegel the Elder using mobile X-ray fluorescence, X-ray diffraction and Raman spectrometers. Spectrochim. Acta B At. Spectrosc. **97**, 1–6 (2014). https://doi.org/10.1016/j.sab.2014.04.006
134. Verma, G., Mishra, M.: Development and optimization of UV-vis spectroscopy-a review. *World*. J. Pharm. Res. **7**(11), 1170–1180 (2018)
135. Wagner, B., Syta, O., Sawicki, M.: A moderate microsampling in laser ablation inductively coupled plasma mass spectrometry analysis of cultural heritage objects: a review. Lasers in the conservation of artworks. In: Proceedings of LACONA XI, pp. 155–178. NCU Press, Toruń (2017). https://www.google.com/url?sa=t&rct=j&q=&esrc=s&source=web&cd=&ved=2ahUKEwj92aHtjNqEAxUkVqQEHQhGAg0QFnoECA0QAQ&url=https%3A%2F%2Fwydawnictwo.umk.pl%2Fupload%2Ffiles%2FOPEN%2520ACCESS%2FLasers%2520in%2520the%2520Conservation%2FLaCona_11-14.pdf&usg=AOvVaw1VTq-t3fvTtqLjHqX9Sy5u&opi=89978449. Accessed 4 Mar 2024
136. Webber, J.B.W.: A review of the use of simple time-domain NMR/MRI for material-science: time-domain benchtop and mobile NMR/MRI for liquid and solid material science, and nano-to micro-pore-size measurement. SN Appl. Sci. **3**(10), 809 (2021). https://doi.org/10.1007/s42452-021-04791-2
137. Werner, C., Sauer, M., Geis, C.: Super-resolving microscopy in neurosciences. Chem. Rev. **121**(19), 11971–12015 (2021). https://doi.org/10.1021/acs.chemrev.0c01174
138. Zoleo, A., Vecchia, F., Brustolon, M.: Characterization of ancient and modern papers by CW-EPR spectroscopy. Appl. Magn. Reson. **35**(2), 213–220 (2009). https://doi.org/10.1007/s00723-008-0155-z
139. Zucchiatti, A.: Ion beam analysis for the study of our cultural heritage. A short history and its milestones. Nucl. Instrum. Methods Phys. Res., Sect. B. **452**, 48–54 (2019). https://doi.org/10.1016/j.nimb.2019.05.059

Chapter 3
Blue-Forming Minerals

Abstract Blue minerals and cobalt blues, their extraction from ores, geological occurrence, colour effects in crystal lattices and scattering, opals, lapis lazuli, Maya blue and palygorskite. Impurities that can be used for the characterisation of regional variations. Use of cobalt blues in ceramic decoration.

Keywords Lapis lazuli · Cobalt ores · Extraction of pigments from ores · Maya blue · Colour centres in gems · Geological formation of blue coloured minerals

3.1 Blue Minerals

Blue-coloured minerals are rarer than those presenting the colours for which the sensitivity of our sight is maximal (that is yellow to green). Is this partly due to the colour strength required due to our low visual sensitivity in this wavelength range leading to the sensation of blue colour? In fact, most of the colouring of minerals is based on the absorption of certain wavelengths by layers or electronic defects (called chromophores, they are linked to a local composition, an impurity or defects in the crystallographic structure) [89]. Only a few inorganic substances such as opals or labradorite develop colours—including blue—by a physical phenomenon, the diffraction of light by microstructural heterogeneities, in this case for opal these are due to the variation in the diameter of nanosized (hydroxylated) silica beads, arranged in almost compact layers [16, 65]. Many living beings use this mode of colouring by diffraction of light by quasi-periodic organic structures, including the famous Morpho butterflies with magnificent blue wings, and the plumage of certain birds, beetles and cuttlefish [13, 82, 92].

Table 3.1 lists the main blue-coloured minerals. The formulae are simplified because in natural compounds the chemical elements presenting similar chemical properties (i.e. belonging to the same column of Mendeleev's Periodic Table) partially substitute the main elements of the mineral in variable quantities, and do

H. G. M. Edwards, P. Colomban, *Blue by Fire: A Marker of the Technical History of Glass and Ceramics*, Cultural Heritage Science,
https://doi.org/10.1007/978-3-031-77122-4_3

Table 3.1 Main blue minerals

Name (group)	Formula	Origin of colour
Diamond	C	Default (B)
Sapphire	Al_2O_3	Fe-Ti defect
Spinel	$(Mg,Fe)Al_2O_4$	Fe^{2+}
Kyanite	Al_2SiO_5	Fe-Ti defect
Sillimanite	Al_2SiO_5	Fe-Ti defect
Zircon	$ZrSiO_4$	V defect
Olivine	$(Mg,Fe,Co)SiO_4$	Fe^{2+}, Co^{2+}
Ringwoodite	$(Mg_{1.7}Fe_{0.22})H_{0.16}SiO_4$	Fe^{2+}
Chalcedony (nanocrystal/amorphous) (Pietersite)	$SiO_{2-x}OH_x$ nH_2O	Diffraction
Opal (nanocrystal/amorphous)	$SiO_{2-x}OH_x$ nH_2O	Diffraction
Jadeite (pyroxene)	$NaAlSi_2O_6$	Fe^{2+}?
Labradorite (feldspar)	$(Na,Ca)(Al,Si)_4O_8$	Diffraction
Sapphirine	$(Mg,Al)_8(Al,Si)_6O_{20}$	Fe-Ti defect?
Tanzanite (zoïsite)	$Ca_2Al_3(SiO_4)_3OH$	V defect
Hibonite (magnetoplombite)	$CaAl_{12}O_{19}$:[Ni,Ti,La]	Dopants
Aquamarine (beryl)	$Be_3Al_2Si_6O_{18}$	NO_3 defect
Sodalite (haüyne, nosean, lazurite)	$(Na,Ca)_8(Al,Si)_{12}O_{24}(S,Cl)_2$	S_2^-, S_3^-
Topaze (nesosilicate)	$Al_2SiO_4(F,OH)_2$	Defect
Azurite	$Cu_3(CO_3)_2(OH)_2$	Cu^{2+}
Vivianite	$Fe_3(PO_4)_2$ $8(H_2O)$	Fe^{2+}
Smithsonite	$ZnCO_3$	Doping
Apatite	$Ca_5(PO_4)_3(OH,Cl,F)$	Mn doping
Halite	NaCl	K doping
Fluorite	CaF_2	Defect

this over different scales: in a few hundred microns the composition varies. Neighbouring structures are mentioned together.

Most of these minerals are precious or semi-precious gems, of great hardness, high hardness being a necessary characteristic for cutting and obtaining a beautiful lustre [9]. First is diamond, the rarest being blue. For instance, the King of France Louis XIV bought from Jean-Baptiste Tavernier, an exceptional blue diamond brought from India in 1668. The diamond was then stolen during the French Revolution (in 1792), found in 1795 then all trace of it disappeared and a slightly smaller diamond (we assume it to have been recut!), the 'Hope', named after the banker owner, appeared in 1812 in London. Resold in 1901, this diamond has since been in the Smithsonian Institute Museum [33]. The blue colour originates from defects caused by doping with a few boron atoms. Indeed, as noted above, impurities and structural defects are the cause of the blue colour of many minerals. Thus, dual doping with iron and titanium ions colours the alumina blue, which gives a sapphire. The same phenomenon has been proposed for the allotropic forms of mullite (kyanite and sillimanite) but here it seems to be charge transfer between defects [79]. The Fe^{2+} ion colours the glaze of certain porcelains [38] and vivianite blue [34]. It is the doping with vanadium ions which colours certain zircon blue (for example from Cambodia) and this effect is widely used to colour enamels with synthetic zircon [24, 32]. Iron ion but also cobalt or nickel ion coupled to titanium and/or manganese ions also colour in blue aluminates with magnetoplumbite ($CaAl_{12}O_{19}$) and beta-alumina ($NaAl_{11}O_{17}$) structures [21, 60]. It is also vanadium which gives the blue colour of tanzanite, an aluminosilicate [73]. Dopants such as manganese or potassium promote, respectively, the colour for apatite [78] and halite [96] via the formation of coloured centres.

The polysulfide ions S_2^- and S_3^- are at the origin of the colour of the minerals haüyne and lazurite, the main constituents of the rock called lapis lazuli. The heterogeneous nanostructure of certain phases gives the blue colour by diffraction (opals, labradorite).

Very few of these minerals exist in sufficient quantity to be used as a pigment; these are mainly apatite and the minerals colouring lapis lazuli (a rock made of different minerals). They are used either for producing pictorial layers at room temperature (or after a treatment at medium temperature), for brightening (bleaching) paper, for textiles [5, 46], or for the colouring of glasses and enamels, that is to say encapsulated by heat treatment in a glass, an amorphous silicate shaped by fusion [22]. These are cobalt ores, lapis lazuli or similar forms. Particular clay, palygorskite (also called attapulgite), can be intercalated with a plant extract dye, indigo: the pigment obtained is Maya blue. The compound obtained then has a good stability due to the encapsulation of the dye in a mineral matrix. We will be particularly interested in lapis lazuli, its constituents and homologue, Maya blue and cobalt ores, these three types of compounds playing a significant role in the production of blue coloured materials in association with more or less complex heat treatment.

3.2 Blue-Forming Minerals: Cobalt Ores

3.2.1 Geology and Mining Places

The element cobalt is rare in nature (27th rank) and it is even rarer to find minerals in which this element is the majority metal. The composition of the 'sources of blue' will therefore depend both on the geological context of the mining site but also on the selection-purification-refining process whose effectiveness was evolved during the eighteenth and nineteenth centuries [22, 69] and from its use in metallurgy, battery electrodes, pigments, etc. Cobalt is a transition metal like copper, nickel, iron, manganese and therefore the chemical properties of these elements are rather similar and many mining sites are common sources. Nowadays, cobalt is a by-product of copper or nickel mining, or even a by-product of iron or gold mining [58]. A single mine, in Morocco, as well as artisanal operations in rare countries mainly produce cobalt. It was the same in the past where cobalt was a by-product of the exploitation mainly of silver, the main numeracy agent, and/or incidentally of bismuth, the presence of the latter element being essential to lower the melting point of the characters of the Gutenberg' lead printing press and hardened by the addition of antimony and tin [6].

Transition metals precipitate together in the form of oxy-hydroxides of Cu, Zn, Mn, Fe, Ni and Co, to form fields of metal nodules on the ocean floor [51], fields capable of being exploited in the future. This type of geological formation, more or less transformed during tectonic movements and orogenesis, is at the origin of a first type of mining site based mainly on asbolane (Table 3.2) [22]. Hydrothermal circulations give rise to a second type of cobalt-based minerals, namely the veins of the 5 elements: Ag-Co-Ni-Bi-As [56, 66]. One, or several, of these typical elements may be missing, and elements such as uranium (U), copper (Cu), lead (Pb), zinc (Zn), antimony (Sb), mercury (Hg) and others may also be present. Carbonates such as calcite and siderite are the most common gangue (non-economic) minerals, although barite, quartz and fluorite may also be present. Ore grades are variable but can be extremely high (up to 10% Co and 1000's g/t Ag) [93]. Elements such as Ag, Bi, and to a certain extent As, occur as native elements associated with cobalt-containing minerals [74], while nickel and cobalt occur as arsenides and/or sulfides [1]. Five-element veins have been mined since at least medieval times [56], when their high grades of easily recognized and refined native Ag made them a relatively accessible source of the precious metal. The last type of cobalt-rich mineral site corresponds to final erosion with, for example, the alums of the salt lakes in the Egyptian world [22, 57].

Most known five-element vein exploited deposits occur in Europe and North America. In Europe well-known districts include Erzgebirge (literally 'Ore Mountains') at the border between Saxony (now Germany) and Bohemia (now Czech Republic) and Schwarzwald at the border between France and Bade-Wurtemberg (now Germany) [56]. The most famous sites are for the Erzgebirge Johanngeorgenstadt, Freiberg, Schneeberg, Annaberg, Marienberg, Johachimsthal/Jachymov and for the Schwarzwald, Wittichen, Wieden, Saint-Blasien, Sainte-Marie-aux-Mines. Other similar sites exist in Thuringia and the Harz Mountains and also in Hungary (Dobschau, Carpathians Mountains), Norvège (Oslo), Spain (Giftain Valley also called now Gistain in Spanish and Chistau in Aragon) of

Table 3.2 Co-based minerals precursors of blue colour in glass

Name	Formula	Colour
Linnaeite	Co_3S_4	Black
Cattierite	CoS_2	Black
Cobaltite	CoAsS	White to reddish
Skutterudite	$(Co,Ni)As_3$	White to gray
Erythrite (vivianite)	$Co_3(AsO_4)_2\ 8H_2O$	Red
Heterogenite	$Co_{1-x}OH_y$	Black
Wupatkite	$(Co,Mg,Ni)Al_2(SO_4)_4\ 22\ H_2O$	Pink
Asbolane	$(Ni,Co)_{2-x}Mn^{4+}(O,OH)_4 \cdot nH_2O$	Black
Palygorskiste	$(Mg,Al)_2Si\ _4O\ _{10}(OH)\cdot 4(H_2O)$	White

Pyrenean Mountains, Aragon), Great-Britain (Cornwall), Switzerland, Austria and France Alpine, Sardinia and Russia [54, 56, 94].

On the contrary, in Asia, around the Himalayas where the geological context is much more recent, it is mainly primary type sites which are exploited giving ores rich in manganese and iron together with cobalt [22].

3.2.2 Ore Processing

As mentioned above, in ancient times, sites rich in the precious native elements—gold, silver and copper—were the first to be exploited (Chalcholithic period, 4000–5000 BCE) [72]. In the mines of the Erzgebirge the toxicity of arsenic was recognized very early on and it is believed that the name cobalt comes from the name given in German to an evil genius (mountain spirits or 'goblins' called '*Kobolds*') inhabiting the silver mines; '*Kobold*' or '*Kobelt*', undoubtedly linked to the toxicity of cobalt salts rich in arsenic [22, 67, 70]. An alternative explanation is that it comes from the name '*Cabul*', a city in Palestine where King Solomon had a cobalt mine. In Arabic, Persian and Chinese cobalt ore is called '*Solomon's stone*' (*sang-e soleymâni* or *âbgineye soleymâni*, as *su-ni-po* (蘇尼坡) or *su-ma-li* (蘇麻里) [22]. Sites where a metal is native and almost pure are very rare and alternatively sites where the element is associated with another element which can be easily eliminated by a simple chemical or thermal treatment have been sought. The first step is the visual selection of the 'good' stones, according to their colour and appearance and the elimination by crushing of unwanted parts.

The extraction of metals requires physicochemical processes that are all the more sophisticated if the content of the desired element is low. In silver mines, part of the slag, rich in cobalt, was recovered in the fifteenth century to give *saffre*, which in the sixteenth century was mixed with potassium glass to give *smalt* [28, 70]. The simplest treatment is the grill which makes it possible to eliminate most volatile elements such as sulphur and arsenic quite well, which can also be recovered by condensation as a by-product and to recover/separate materials with low temperature fusion, e.g. bismuth [68, 69]. The other elements are concentrated in the slag. Dissolutions under the action of acids (or bases) are used to eliminate unwanted elements such as iron [22, 90] before progress in chemistry in the nineteenth century led to a more effective purification through the production of carbonates, sulphates, nitrates or even 'pure' oxides [27, 76, 90]. It follows that the identification and measurement of the contents of elements associated with cobalt is done to link the final compound coloured by cobalt ions to a geological context making it possible to attest to the use of one of the potential sources through its context, historical or geographical. However, we will see that it is obvious that the selection-purification process is as important as the geological context to eliminate elements associated with cobalt in ores [22]. In fact, this process eliminates only partially certain elements but can also incorporate others, mainly during grinding operations, the wear of the grinding agents significantly causing pollution [20, 24].

3.3 Elements Associated with Cobalt

Analytical strategies, in particular the analysis technique(s) (the volume analysed varies depending on the techniques and its definition is often difficult), the constraints relating to work on samples or not also limits the detection efficiency (see. Chap. 5) and the conclusions must take into account numerous parameters, which is an approach that is often neglected. The main elements likely to identify the sources of cobalt, or rather to identify the uses of different sources—it is often difficult to do more—are therefore on the one hand manganese which is characteristic of primary sites (for example Asian) and on the other the elements characteristic of the secondary lode sites: firstly, arsenic but also silver, bismuth, copper, uranium, indium, etc. Certain elements such as nickel present in both types of site are not always definitive although these have been used by certain authors [44, 45]. Other rarer transition metals such as chromium may be significant [22].

Iron which is present in most raw materials cannot be used. Cobalt does not have a specific isotope that can indicate a geological context as has been done for a decade for lead mines thanks to the sensitivity of mass spectroscopy [31, 59, 61]. However, some of the associated elements such as silver, nickel, bismuth, chromium and manganese have stable (or very long-lived) isotopes which should make it possible in the future to characterize mining sites more precisely as is the case for lead sites [2, 35, 39].

Different studies have tried to identify associated elements allowing the identification of common sources. Kaczmarczyk was the first to try to identify Egyptian objects made with the same source of cobalt and proposed places for these sources, namely Mesopotamia and Egypt [57]. The two articles by Gratuze et al. [44, 45], mainly concerning glasses from the Mediterranean world, demonstrated from mass spectrometry measurements (now a microdestructive technique) the association of certain elements, namely Sb-Cu-Mn, Zn-Pb-In, Ni and As-Ni-Bi for different periods and origins, in relation to mining expected from the Eastern Mediterranean and the Erzgebirge sites. This work has been regularly updated [47]. More recently, based on non-destructive pXRF measurements made with mobile instruments, Colomban et al. showed that the simultaneous consideration of several elements in ternary diagrams made it possible to classify better the sources of cobalt used and their intrinsic variability both for the productions of the European and Mediterranean and Asian worlds and that the elements used in previous reviews were not sufficient to achieve this [22, 23, 25, 26].

3.4 Lapis Lazuli

Lapis lazuli (the name comes from the Persian *lâjvard, azur*) is a rock used since ancient times (objects have been found at Ur, more than 2500 BCE) with the site of *Sar-e-sang* in the Badakhshan Mountains (Panchir valley, Hindu Kouch,

Afghanistan), was almost its sole source until the seventeenth century. This site is mentioned by Marco Polo [18, 19]. Notes that Buffon in his 'Natural History of Minerals' of 1787 [7] indicates other sites near Lake Baikal (Irkutsk and Slyudyanka river), in Pakistan (Pamir province), in Burma (Mogok) and in Chile (Las Flores de los Andes, Ovalle). More recently new sites in Pakistan, Iran, Sinai, Canada, California, Azerbaijan and Italy have been identified. Chile is, along with Afghanistan, now the main producer.

The blue minerals, lazurite and haüyne are considered to have been synthesized at 500–550 °C under 1.5 kbar with a final sulphur-rich hydrothermal intrusion [85, 95]. Sites of neighbouring rocks, that is to say those also containing similar alumino-silicate phases containing sulphate ions and not sulphur ions, are also known and undoubtedly more numerous.

Indeed Buffon already mentioned that heat treatment improved the colour of lapis lazuli [7] and recent work shows that these natural phases—non-blue—can be used as a glaze dye [8, 19, 64]. However, it is only recently with the work of Clark, Curli and Laganarra [15] concerning luxury medieval Frederician glass and pottery and then Colomban concerning common *Lajvardina* pottery (Iran) [17] that the importance of lapis lazuli as a colouring agent for glasses and enamels has been recognized. Other authors, e.g. [36], noted that they were unable to detect cobalt in blue Islamic glasses by elemental analysis but they attributed this to too low a cobalt content, below the detection threshold. Lazurite and hauyne, the blue minerals colouring the rock called lapis lazuli, have a complex composition $(Na,Ca)_8(Al,Si)_{12}O_{24}(S,Cl)_2$ and a sodalite type structure, related to feldspathoids [19]. The composition of the noseans is close. These silicate networks define large sites ('cages') where polysulfide ions, S_2^- or S_3^-, remain trapped. Certain SiO_4 tetrahedra, the main structural unit, are replaced by ClO_4 or SO_4 homologues and depending on the heat treatment, the latter decompose as follows.

$$2SO_4 \Rightarrow S_2 + 4O_2$$

which leads to other species, in particular polysulfide ions. Thermal history—geological, but also the treatments carried out by artisans—is therefore crucial [7, 8]. Thus, it was shown that the colour of lapis lazuli grains was stable, or even improved by rapid firing at 900 °C [19], in agreement with eighteenth century writings (Table 3.3).

Table 3.3 Blue and associated phases related to lapis lazuli

Name	Formula
Lazurite	$(Na,Ca)_8(AlSiO_4)_6(SO_4,S,Cl,OH)_2$
Hauyne	$Na_3Ca(Si_3Al_3)O_{12}(SO_4)$
Nosean (or noselite)	$Na_8Al_6Si_6O_{24}(SO_4)$
Zeolite	$[(SiO_2)(AlO_2)_x]M_{x/n}^{n+} \cdot w\,H_2O.M$

3.5 Maya Blue

Indigo dye (*Xiuhquilitlin Náhuatl*, in the language of the Aztecs) is extracted in Mesoamerica from the plant *Indigofera sufructicosa* [11, 12, 77] according to a recipe described in two texts from the beginnings of the colonial era, the *Florentine Codex, a Historia General de las Cosas de Nueva España* by Fray Bernardino de Sahagún [62], and the *Historia Natural de Nueva España* by Hernandez [52]. Indigo (formula $C_{16}H_9N_2O_2$) has been known and used on several continents, at least since 2000 BCE in India [3, 4, 14]. Some scholars date the use of indigo back to 4000 BCE in Peru [86].

The preparation can be summarized as follow. The leaves of the indigo plant are dried, crushed, treated with hot water (oxidation and fermentation then precipitation by the action of a base, soda (lye)) but other European plants (pastel, woad) or from Asia contain the same dye (e.g. *Indigofera tinctoria, Indigofera arrecta*, etc.). Synthetic indigo was obtained in 1880 by Adolf von Baeyer and commercialized in 1900 [37, 43]. Numerous studies concern this dye [83].

There are no documents giving the recipe for the preparation of Mayan blue but recent work concludes that the *Náhuatl* term *texotli* corresponds to it [30]. In the 1960s the understanding of its nature and production was resolved. Gettens identified the host network as a palygorskite (from a site in Russia) also called attapulgite (from a site in the USA!), or sepiolite (from its resemblance to a cuttlefish bone, a white and fibrous clay [40]. This clay has a very large specific surface area which allows it to adsorb liquids. Shepard was the first to identify indigo as the chromophore of Maya blue [84]. Then Van Olphen succeeded to synthesise Maya blue in a controlled way, by mixing indigo and palygorskite [91]. Palygorskite is a magnesium aluminum phyllosilicate with the ideal composition $Si_8O_{20}(Al_2Mg_2)(OH)_2(OH_2)_4{\bullet}4H_2O$. Empty channels are filled with water (zeolitic and therefore labile, leaving around 120 °C), while the magnesium and aluminium cations are linked to water molecules (structural water) which only leave at a high temperature (~700–800 °C) [42, 53, 87, 88]. Some varieties ($Si_{12}O_{30}Mg_8(OH)_4(OH_2)_4(H_2O)_8$), have larger channels [41].

Quantitatively, approximately 1% wt of indigo is mixed with palygorskite powder and heated for ~24 h between 130 and 200 °C [11, 12, 29, 48]. Different characterization techniques are used to identify Maya blue: UV-visible-NIR absorption, Raman spectroscopy, XRD, NMR [10, 55, 80, 81]. Analyses have concerned codices [63, 75], maps [49, 50, 71] and murals [63].

References

1. Anonymous. https://www.geologyforinvestors.com/cobalt-commodity-overview/ (2017). Accessed 19 Aug 2023
2. Arribas, A., Mathur, R., Megaw, P., Arribas, I.: The isotopic composition of silver in ore minerals. Geochem. Geophys. Geosyst. **21**(8), e2020GC009097 (2020). https://doi.org/10.1029/2020GC009097

3. Balfour-Paul, J.: Indigo in South and South-East Asia. Text. Hist. **30**(1), 98–112 (1999). https://doi.org/10.1080/00404969.2019.1660062
4. Balfour-Paul, J.: Indigo in the Arab World. Routledge (2012) ISBN 9781138972599
5. Barandon, J.-N., Irigoin, J.: Papiers de Hollande et papiers d'Angoumois de 1650 à 1810. Leur differentiation au moyen de l'analyse par activation neutronique. Archaeometry. **21**, 101–106 (1979). https://doi.org/10.1111/j.1475-4754.1979.tb00245.x
6. Berger, D.: Post-medieval printing type from Mainz and Oberursel, Germany, and the composition of early German type metal. Historical Metallurgy. **49**(2), 110–125 (2017). https://hmsjournal.org/index.php/home/article/view/67/65. Accessed 4 Mar 2024
7. Buffon, G.L.L.: Histoire Naturelle des Minéraux, tome septième, pp. 251–259. Imprimerie Royale, Paris (1787). https://gallica.bnf.fr/ark:/12148/bpt6k1067244v.image
8. Caggiani, M.C., Colomban, P., Mangone, A., Valloteau, C., Cambon, P.: Mobile Raman spectroscopy analysis of ancient enamelled glass masterpieces. Anal. Methods. **5**, 4345–4354 (2013). https://doi.org/10.1039/C3AY40648B
9. Cannela, A.-F.: Gemmes, verre coloré, Fausses pierres précieuses au moyen âge, Le quatrième livre du «Trésorier de Philosophie naturelle des pierres précieuses» de Jean d'Outremeuse. Bibliothèque de la Faculté de Philosophie et Lettres de l'Université de Liège-Fascicule. CCLXXXVIII. Librairie Droz S.A, Genève (2006). https://hdl.handle.net/2268.1/4133
10. Chiari, G., Giustetto, R., Ricchiardi, G.: Crystal structure refinements of palygorskite and Maya Blue from molecular modelling and powder synchrotron diffraction. Eur. J. Mineral. **15**(1), 21–33 (2003). https://doi.org/10.1127/0935-1221/2003/0001-0021
11. Chiari, G., Giustetto, R., Carson, Y.D.: Azul maya: una maravillosa nanotecnología precolombina. Boletín de Monumentos Históricos. **12**, 39–48 (2008a). https://www.researchgate.net/publication/269112452_AZUL_MAYA_una_maravillosa_nanotecnologia_precolombina/link/5481f2630cf25dbd59e9436b/download. Accessed 4 Mar 2024
12. Chiari, G., Giustetto, R., Druzik, J., Doehne, E., Ricchiardi, G.: Pre-columbian nanotechnology: reconciling the mysteries of the Maya blue pigment. Appl. Phys. A Mater. Sci. Process. **90**(1), 3–7 (2008b). https://doi.org/10.1007/s00339-007-4287-z
13. Chiou, T.-H., Mäthger, L.M., Hanlon, R.T., Cronin, T.W.: Spectral and spatial properties of polarized light reflections from the arms of squid (*Loligo pealeii*) and cuttlefish (*Sepia officinalis* L.). J. Exp. Biol. **210**(20), 3624–3635 (2007). https://doi.org/10.1242/jeb.006932
14. Clark, R.J., Cooksey, C.J., Daniels, M.A., Withnall, R.: Indigo, woad, and Tyrian purple: important vat dyes from antiquity to the present. Endeavour. **17**(4), 191–199 (1993). https://doi.org/10.1016/0160-9327(93)90062-8
15. Clark, R.J., Curri, M.L., Laganara, C.: Raman microscopy: the identification of lapis lazuli on medieval pottery fragments from the south of Italy. Spectrochimica Acta Part A Mol. Biomol. Spectrosc. **53**(4), 597–603 (1997). https://doi.org/10.1016/S1386-1425(96)01768-4
16. Colomban, P.: Gel technology in ceramics, glass-ceramics and ceramic-ceramic composites. Ceram. Int. **15**(1), 23–50 (1989). https://doi.org/10.1016/0272-8842(89)90005-9
17. Colomban, P.: Lapis lazuli as unexpected blue pigment in Iranian Lâjvardina ceramics. J. Raman Spectrosc. **34**(6), 420–423 (2003). https://doi.org/10.1002/jrs.1014
18. Colomban, P.: Routes du lapis lazuli, lâjvardina et échanges entre arts du verre, de la céramique et du livre. In Proc Conf "Chine–Méditerranee: routes et échanges de la céramique avant le XVIe siècle", Paris, France, Taoci, 4, 26–27 (2005). https://www.researchgate.net/publication/259998622_Routes_du_Lapis_lazuli_Lajvardina_et_echanges_entre_arts_du_verre_de_la_ceramique_et_du_livre. Accessed 4 Mar 2024
19. Colomban, P.: Rocks as blue, green and black pigments/dyes of glazed pottery and enamelled glass artefacts–a review. Eur. J. Mineral. **25**(5), 863–879 (2013). https://doi.org/10.1127/0935-1221/2013/0025-2305
20. Colomban, P.: Natural nanosized raw materials and sol-gel technology: the base of pottery since millenniums. In: Dillmann, P., et al. (eds.) Nanoscience and Cultural Heritage. Atlantis Press, Paris (2016). https://doi.org/10.2991/978-94-6239-198-7_3

21. Colomban, P., Vivien, D.: EPR study of ordering in stoichiometric β-Aluminate. Phys Stat. Sol. (a). **76**, 565–574 (1983). https://doi.org/10.1002/pssa.2210760220
22. Colomban, P., Kırmızı, B., Simsek Franci, G.: Cobalt and associated impurities in blue (and green) glass, glaze and enamel: relationships between raw materials, processing, composition, phases and international trade. Minerals. **11**, 633 (2021). https://doi.org/10.3390/min11060633
23. Colomban, P., Ngo, A.-T., Fournery, N.: Non-invasive Raman analysis of 18th century Chinese export/armorial Overglazed porcelain: identification of the different enameling technology. Heritage. **5**, 233–259 (2022a). https://doi.org/10.3390/heritage5010013
24. Colomban, P.: Full spectral range Raman signatures related to changes in enameling technologies from the 18th to the 20th century: guidelines, effectiveness and limitations of the Raman analysis. Materials. **15**, 3158 (2022). https://doi.org/10.3390/ma15093158
25. Colomban, P., Simsek Franci, G., Gironda, M., d'Abrigeon, P., Schuhmacher, A.-C.: pXRF data evaluation methodology for on-site analysis of precious artifacts: cobalt used in the blue decoration of Qing dynasty Overglazed porcelain enameled at Customs District (Guangzhou), Jingdezhen and Zaobanchu (Beijing) workshops. Heritage. **5**, 1752–1778 (2022b). https://doi.org/10.3390/heritage5030091
26. Colomban, P., Simsek Franci, G., Burlot, J., Gallet, X., Zhao, B., Clais, J.B.: Non-invasive on-site pXRF analysis of coloring agents, Marks and enamels of Qing Imperial and non-Imperial porcelain. Ceramics. **6**, 447–474 (2023). https://doi.org/10.3390/ceramics6010026
27. Davis, J.R. (ed.): Nickel, Cobalt, and their Alloys. ASM international, Materials Park (2000) ISBN: 978-0-87170-685-0
28. Delamare, F.: Aux origines des bleus de cobalt: les débuts de la fabrication du saffre et du smalt en Europe occidentale. Compt. Rend. Séances Académie Inscript. et Belles-Lettres. **153**(1), 297–315 (2009). https://doi.org/10.3406/crai.2009.92472
29. Doménech-Carbó, A., Holmwood, S., Di Turo, F., Montoya, N., Valle-Algarra, F.M., Edwards, H.G.M.: Composition and color of Maya Blue: reexamination of literature data based on the dehydroindigo model. J. Phys. Chem. C. **123**(1), 770–782 (2019). https://doi.org/10.1021/acs.jpcc.1c07932
30. Dupey García, É.: The materiality of color in pre-Columbian codices: insights from cultural history. Anc. Mesoam. **28**(1), 21–40 (2017). https://doi.org/10.1017/S0956536116000493
31. Durali-Mueller, S., Brey, G.P., Wigg-Wolf, D., Lahaye, Y.: Roman lead mining in Germany: its origin and development through time deduced from lead isotope provenance studies. J. Archaeol. Sci. **34**(10), 1555–1567 (2007). https://doi.org/10.1016/j.jas.2006.11.009
32. Eppler, R.A., Eppler, D.R.: Glazes and Glass Coating. The American Ceramic Society, Westerville (2000)
33. Farges, F., Sucher, S., Horowitz, H., Fourcault, J.M.: The French blue and the Hope: new data from the discovery of a historical lead case. Gems Gemology. **spring**, 4–19 (2009). https://www.gia.edu/doc/SP09A1.pdf
34. Figueiredo, M.O., Silva, T.P., Veiga, J.P.: The blue of iron in mineral pigments: a Fe *K*-edge XANES study of vivianite. Appl. Phys. A Mater. Sci. Process. **99**, 357–361 (2010). https://doi.org/10.1007/s00339-010-5637-9
35. Foltyn, K., Erlandsson, V.B., Zygo, W., Melcher, F., Pieczonka, J.: New perspective on trace element (Re, Ge, Ag) hosts in the Cu-Ag Kupferschiefer deposit, Poland: insight from a LA-ICP-MS trace element study. Ore Geol. Rev. **143**, 104768 (2022). https://doi.org/10.1016/j.oregeorev.2022.104768
36. Freestone, I.C., Stapleton, C.P.: Composition and Technology of Islamic enamelled glass of the 13–14th centuries (ch 24). In: Ward, R. (ed.) Gilded and Enamelled glass from the Middle East, pp. 122–128. British Museum Press, London (1998). https://www.academia.edu/4569075/Composition_and_Technology_of_Islamic_Enamelled_Glass_of_13_14th_Centuries_with_C_P_Stapleton_. Accessed 4 Mar 2024
37. Gaboriaud-Kolar, N., Nam, S., Skaltsounis, A.L.: A colorful history: the evolution of indigoids. Progr. Chem. Org. Nat. Prod. **99**, 69–145 (2014). https://doi.org/10.1007/978-3-319-04900-7_2

38. Gao, Z., Chen, S., Shen, Z.: Applied investigation of Moessbauer effect for the famous ancient chinese porcelains (CNIC--01107). China (1996). https://inis.iaea.org/search/search.aspx?orig_q=RN:28032858
39. Gentelli, L.: Provenance determination of silver artefacts from the 1629 VOC wreck Batavia using LA-ICP-MS. J. Archaeol. Sci. Rep. **9**, 536–542 (2016). https://doi.org/10.1016/j.jasrep.2016.08.044
40. Gettens, R.J.: Maya Blue: an unsolved problem in ancient pigments. Am. Antiq. **27**(4), 557 (1962). https://doi.org/10.2307/277679
41. Giustetto, R., Seenivasan, K., Bordiga, S.: Spectroscopic characterization of a sepiolite-based Maya blue pigment. Periodico di Mineralogia. **79**, 21–37 (2010). https://doi.org/10.1016/j.clay.2011.01.027
42. Giustetto, R., Wahyudi, O., Corazzari, I., Turci, F.: Chemical stability and dehydration behavior of a sepiolite/indigo Maya Blue pigment. Appl. Clay Sci. **52**(1), 41–50 (2011). https://doi.org/10.1016/j.clay.2011.01.027
43. Głowacki, E.D., Voss, G., Leonat, L., Irimia-Vladu, M., Bauer, S., Sariciftci, N.S.: Indigo and Tyrian purple–from ancient natural dyes to modern organic semiconductors. Israel J. Chem. **52**(6), 540–551 (2012). https://doi.org/10.1002/ijch.201100130
44. Gratuze, B., Soulier, I., Barrandon, J.N., Foy, D.: De l'origine du cobalt dans les verres. Revue d'Archéométrie. **16**, 97–108 (1992). https://doi.org/10.3406/arsci.1992.895
45. Gratuze, B., Soulier, I., Blet, M., Vallauri, L.: De l'origine du cobalt: Du verre à la céramique'. Revue d'Archéométrie. **20**, 77–94 (1996). https://doi.org/10.3406/arsci.1996.939
46. Gratuze, B.: Glass characterisation using laser ablation inductively coupled plasma mass spectrometry methods, Ch. 3.1. In: Janssens, K. (ed.) Modern Methods for Analysing Archaeological and Historical Glass, vol. 2, pp. 201–234. Wiley, Hoboken (2013). https://doi.org/10.1002/9781118314234.ch9
47. Gratuze, B., Pactat, I., Schibille, N.: Changes in the signature of cobalt colorants in late antique and early Islamic glass production. Fortschr. Mineral. **8**(6), 225 (2018). https://doi.org/10.3390/min8060225
48. Grazia, C., Buti, D., Amat, A., Rosi, F., Romani, A., Domenici, D., Sgamellotti, A., Miliani, C.: Shades of blue: non-invasive spectroscopic investigations of Maya Blue pigments. From laboratory mock-ups to Mesoamerican codices. Herit. Sci. **8**, 1 (2020). https://doi.org/10.1186/s40494-019-0345-z
49. Haude, M.E.: Identifcation of colorants on maps from the early colonial period of New Spain (Mexico). J. Am. Inst. Conserv. **37**(3), 240–270 (1998). https://doi.org/10.1179/019713698806082822
50. Haude, M.E.: Identifcation of colorants on maps from Magaloni Kerpel D. Painters of the New World: the process of making the Florentine codex. In: Wolf, G., Connors, J., Waldman, L.A. (eds.) Colors Between Two Worlds. The Florentine Codex of Bernardino de Sahagún, pp. 46–47. The Harvard University Center for Italian Renaissance Studies, Villa I Tatti-Florence (2012)
51. Havlik, T., Laubertova, M., Miskufova, A., Kondas, J., Vranka, F.: Extraction of copper, zinc, nickel and cobalt in acid oxidative leaching of chalcopyrite at the presence of deep-sea manganese nodules as oxidant. Hydrometallurgy. **77**(1–2), 51–59 (2005). https://doi.org/10.1016/j.hydromet.2004.10.009
52. Hernández, F.: Obras completas: Historia natural de nueva Espana. Universidad Nacional de México, Mexico (1959). https://books.google.fr/books/about/pt_1_2_Historia_natural_de_nueva_Espana.html?id=ID89AQAAIAAJ&redir_esc=y. Accessed 4 Mar 2024
53. Hojati, S., Khademi, H.: Thermal behavior of a natural sepiolite from Northeastern Iran. J. Sci. Islam. Rep. Iran. **24**(2), 129–134 (2013). https://jsciences.ut.ac.ir/article_32079.html. Accessed 4 Mar 2024
54. Horn, S., Gunn, A.G., Petavratzi, E., Shaw, R.A., Eilu, P., Törmänen, T., Bjerkgård, T., Sandstad, J.S., Jonsson, E., Kountourelis, S., Wall, F.: Cobalt resources in Europe and the potential for new discoveries. Ore Geol. Rev. **130**, 103915 (2021). https://doi.org/10.1016/j.oregeorev.2020.103915

55. Hubbard, B., Kuang, W., Moser, A., Facey, G.A., Detellier, C.: Structural study of Maya blue: textural, thermal and solid-state multinuclear magnetic resonance characterization of the palygorskite-indigo and sepiolite-indigo adducts. Clay Clay Miner. **51**(3), 318–326 (2003). https://doi.org/10.1346/CCMN.2003.0510308
56. Kissin, S.A.: 5 element (Ni-Co-As-Ag-Bi) veins. Geosci. Can. **19**, 113–124 (1992). https://journals.lib.unb.ca/index.php/GC/article/view/3768
57. Kaczmarczyk, A. 1986. The source of cobalt in ancient Egyptian pigments. In Proceedings of the 24th International Archaeometry Symposium Olin, J.S., Blackman, M.J., Eds.; Smithsonian Institution Press: Washington, DC, USA, pp. 369–376
58. L'élémentarium Database. https://lelementarium.fr/element-fiche/cobalt/. Accessed 1 Nov 2023
59. Le Roux, G., Weiss, D., Grattan, J., Givelet, N., Krachler, M., Cheburkin, A., Rausch, N.: Kober, B.: Shotyk, W.: Identifying the sources and timing of ancient and medieval atmospheric lead pollution in England using a peat profile from Lindow bog, Manchester. J. Environ. Monit. **6**(5), 502–510 (2004). https://doi.org/10.1039/B401500B
60. Li, J., Medina, E.A., Stalick, J.K., Sleight, A.W., Subramanian, M.A.: Colored oxides with hibonite structure: a potential route to non-cobalt blue pigments. Progr. Solid State Chem. **44**(4), 107–122 (2016). https://doi.org/10.1016/j.progsolidstchem.2016.11.001
61. Lin, J., Liu, Y., Yang, Y., Hu, Z.: Calibration and correction of LA-ICP-MS and LA-MC-ICP-MS analyses for element contents and isotopic ratios. Solid Earth Sci. **1**(1), 5–27 (2016). https://doi.org/10.1016/j.sesci.2016.04.002
62. López Austin, A., García Quintana, J.: Historia general de las cosas de Nueva España. Consejo Nacional para la Cultura y las Artes: Mexico 2000. https://historicas.unam.mx/publicaciones/publicadigital/libros/484/484_04_03_HistoriaGeneral.pdf. Accessed 4 Mar 2024
63. Magaloni Kerpel, D.: Materiales y técnicas de la pintura mural maya. In: (coord.) Staines Cicero L, editor. La pintura mural prehispánica en México. II, Tomo III Estudios, pp. 135–206. UNAM, Mexico (2001). https://repositorio.unam.mx/contenidos/materiales-y-tecnicas-de-la-pintura-mural-maya-69234?c=pzqMMX&d=false&q=*:*&i=1&v=1&t=search_0&as=0. Accessed 4 Mar 2024
64. Mangone, A., Caggiani, M.C., Forleo, T., Giannossa, L.C., Acquafredda, P.: A possible natural and inexpensive substitute for lapis lazuli in the Frederick II era: the finding of Haüyne in Blue Lead-Tin glazed pottery from Melfi Castle (Italy). Molecules. **28**(4), 1546 (2023). https://doi.org/10.3390/molecules28041546
65. Marlow, F., Muldarisnur, M., Sharifi, P., Brinkmann, R., Mendive, C.: Opals: status and prospects. Angewandte Chem. Inter. Edit. **48**, 6212–6233 (2009). https://doi.org/10.1002/anie.200900210
66. Markl, G., Burisch, M., Neumann, U.: Natural fracking and the genesis of 5-element veins. Mineral. Deposita. (2016). https://doi.org/10.1007/s00126-016-0662-z
67. Mimoso, J.M.: Origin, early history and technology of the blue pigment in azulejos. In: Proceedings of the GlazeArch2015, Lisbon, Portugal, 2–3 July 2015; pp. 357–376. Available online: http://azulejos.lnec.pt/AzuRe/links/07%20Origin_of_blue_pigment.pdf (2015). Accessed 20 Jan 2021
68. Matin, M., Pollard, A.M.: From ore to pigment: a description of the minerals and an experimental study of cobalt ore processing from the Kāshān Mine, Iran. Archaeometry. **59**(4), 731–746 (2017). https://doi.org/10.1111/arcm.12272
69. Molera, J., Climent-Font, A., Garcia, G., Pradell, T., Vallcorba, O., Zucchiatti, A.: Experimental study of historical processing of cobalt arsenide ore for colouring glazes (15–16th century Europe). J. Archaeol. Sci. Rep. **36**, 102797 (2021). https://doi.org/10.1016/j.jasrep.2021.102797
70. Morral, F.R.: From Cabul to cobalt: a historical view of the "mischievous metal". J. Chem. Educ. **34**(4), 185 (1957). https://doi.org/10.1021/ed034p185

71. Newman, R., Derrick, M., Miller, M.E., Mundy, B.E.: Analytical report of the pigments and binding materials used on the Beinecke map. In: Miller, M.E., Mundy, B.E. (eds.) Painting a Map of Sixteenth-Century Land, Writing, and Native Rule Beinecke Rare Book and Manuscript Library, pp. 91–100. Yale University Press, New Haven (2012) ISBN 978-0-300-18071-8
72. Patterson, C.C.: Native copper, silver, and gold accessible to early metallurgists. Am. Antiq. **36**(3), 286–321 (1971). https://doi.org/10.2307/277716
73. Pluthametwisute, T., Wanthanachaisaeng, B., Saiyasombat, C., Sutthirat, C.: Cause of color modification in tanzanite after heat treatment. Molecules. **25**, 3743 (2020). https://doi.org/10.3390/molecules25163743
74. Pottier, F., Micheli Patterson, C.C.: Native copper, silver, and gold accessible to early metallurgists. Am. Antiq. **36**(3), 286–321 (1971). https://doi.org/10.2307/277716
75. Pottier, F., Michelin, A., Tournié, A., Goubard, F., Histace, A., Lavédrine, B.: Preliminary investigation on the Codex Borbonicus: macroscopic examination and coloring materials characterization. In: Dupey García, É., Vázquez de Agredos-Pascual, M.L. (eds.) Painting the Skin: Pigments on Bodies and Codices in Pre-Columbian Mesoamerica, pp. 157–174. University of Arizona Press, Tucson (2019) ISBN: 9780816539093
76. Proust, M.: XLIX. Materials for a history of the prussiates. Philos. Mag. **32**(127), 336–357 (1808)
77. Reyes-Valerio, C.: De Bonampak al Templo Mayor: el azul maya en Mesoamérica. Coleccion America Nuestra. America antigua. **40** (1993) ISBN 968-23-1893-9
78. Ribeiro, H.B., Guedes, K.J., Pinheiro, M.V.B., Greulich-Weber, S., Krambrock, K.: About the blue and green colours in natural fluoroapatite. Phys. Stat. Solidi (c). **2**(1), 720–723 (2005). https://doi.org/10.1002/pssc.200460274
79. Rosman, G.R., Grew, E.S., Dollase, W.A.: The color of sillimanite. Am. Mineral. **67**(7–8), 749–761 (1982). https://pubs.geoscienceworld.org/msa/ammin/article-abstract/67/7-8/749/41414/The-colors-of-sillimanite
80. Sánchez del Rio, M., Martinetto, P., Reyes-Valerio, C., Dooryhee, E., Suárez, M.: Synthesis and acid resistance of Maya Blue pigment. Archaeometry. **48**(1), 115–130 (2006). https://doi.org/10.1111/j.1475-4754.2006.00246.x
81. Sánchez del Río, M., Doménech, A., Doménech-Carbó, M.T., Vázquez de Agredos Pascual, M.L., Suárez, M., García-Romero, E.: The Maya blue pigment. In: Emilio, G., Arieh, S. (eds.) Developments in clay science, vol. 3, pp. 453–481. Elsevier, Amsterdam (2011). https://www.researchgate.net/publication/248393332_The_Maya_Blue_Pigment. Accessed 4 Mar 2024
82. Seago, A.E., Brady, P., Vigneron, J.P., Schultz, T.D.: Gold bugs and beyond: a review of iridescence and structural colour mechanisms in beetles (Coleoptera). J. R. Soc. Interface. **6**, S165–S184 (2008). https://doi.org/10.1098/rsif.2008.0354.focus
83. Serrano-Andrés, L., Roos, B.O.: A theoretical study of the indigoid dyes and their chromophore. Chem. Eur. J. **3**(5), 717–725 (1997). https://doi.org/10.1002/chem.19970030511
84. Shepard, A.O.: Maya Blue: alternative hypotheses. Am. Antiq. **27**(4), 565–566 (1962). https://doi.org/10.2307/277680
85. Smith, G.D., Klinshaw II, R.J.: The presence of trapped carbon dioxide in lapis lazuli and its potential use in geo-sourcing natural ultramarine pigment. J. Cult. Herit. **10**(3), 415–421 (2009). https://doi.org/10.1016/j.culher.2008.12.001
86. Splitstoser, J.C., Dillehay, T.D., Wouters, J., Claro, A.: Early pre-Hispanic use of indigo blue in Peru. Sci. Adv. **2**(9), 1501623 (2016). https://doi.org/10.1126/sciadv.1501623
87. Suárez, M., García-Romero, E.: Advances in the crystal chemistry of sepiolite and palygorskite. Dev. Clay Sci. **3**, 33–65 (2011). https://doi.org/10.1016/B978-0-444-53607-5.00002-5
88. Tian, L., Wang, L., Wang, K., Zhang, Y., Liang, J.: The preparation and properties of porous sepiolite ceramics. Sci. Rep. **9**(1), 7337 (2019). https://doi.org/10.1038/s41598-019-43918-9
89. Tilley, R.: Colour and the optical properties of materials, 3rd edn. Wiley, Chichester (2019). SBN: 978-1-119-55468-4

90. Valmont de Bomare, M.: Minéralogie, ou Nouvelle Exposition du Règne Minéral, 2nd edn, pp. 76–96. Vincent Imprimeur-Libraire, Paris (1774). https://gallica.bnf.fr/ark:/12148/bpt6k486073x. Accessed 4 Mar 2024
91. Van Olphen, H.: Maya Blue: a clay-organic pigment? Science. **154**(3749), 645–646 (1966). https://doi.org/10.1126/science.154.3749.645
92. Vukusic, P., Sambles, J.R., Lawrence, C.R., Wootton, R.J.: Quantified interference and diffraction in single Morpho butterfly scales. Proc. R. Soc. Lond. **B.266**, 1403–1411 (1999). https://doi.org/10.1098/rspb.1999.0794
93. Watson, A..: https://www.geologyforinvestors.com/what-is-cut-off-grade-an-introduction-to-overlooked-yet-critical-concept-in-mining-and-resource-investing/ (2018). Accessed 19 Aug 2023
94. Wesch, H., Wiethege, T., Spiethoff, A., Wegener, K., Müller, K.M., Mehlhorn, J.: German uranium miner study: historical background and available histopathological material. Radiat. Res. **152**(6s), S48–S51 (1999). https://doi.org/10.2307/3580113
95. Wyart, J., Bariand, P., Filippi, J.: Lapis-lazuli from Sar-e-Sang, Badakhshan, Afghanistan. Gems Gemol. **17**, 184–190 (1981). https://www.gia.edu/gems-gemology/winter-1981-lapis-afghanistan-wyart. Accessed 4 Mar 2024
96. Zelek, S.M., Stadnicka, K.M., Tobola, T., Natkaniec-Nowak, L.: Lattice deformation of blue halite from Zechstein evaporate basin: Klodawa Salt Mine, Central Poland. Mineral. & Petrol. **108**, 619–631 (2014). https://doi.org/10.1007/s00710-014-0323-9

Chapter 4
Blue Pigments in Antiquity

Abstract A discussion is made of ancient pigments used in prehistoric times, the Roman period, Mediaeval and Renaissance through to the wide range offered by modern chemically synthesised alternatives. In particular, the confusion caused by the multiple Roman descriptors which were differentiated by original pigment quality and geological sourcing and taken on into mediaeval scriptoria and thence into the Renaissance is highlighted. The use of dyestuffs, plant extracts and minerals, sometimes in combination, and their incorporation into artists' palettes. Fugitive pigments and their degradation.

Keywords Prehistoric · Roman · Mediaeval · Renaissance pigments · Synthetic pigments of the eighteenth and nineteenth centuries · Fugitive pigments · Nomenclature problems · Dyestuffs · Plant extracts

From prehistoric times, natural pigments were adopted for the earliest cave art, reaching superb heights of achievement with only a limited palette in Lascaux and Altamira; the latter has been called the Sistine Chapel of Cave Art. Altamira Cave paintings have been dated to the Palaeolithic, some 36 kya, and comprise mainly haematite, yellow ochre and charcoal pigments with some purple hues. Few examples of cave art actually contain blue pigment, which does not occur widely in nature. The first historical documentation of pigments [29], used in wall paintings and other artworks is that of the Greek philosopher, Theophrastus (371–287 BCE), whose work was cited by both Roman chroniclers Vitruvius (75–15 BCE) and Pliny (23–79 CE) [22, 31]. Theophrastus cites *kyanos* as the only blue pigment used in the ancient world [29], but uses this same label for the mineral pigment azurite (basic copper carbonate, $2CuCO_3.Cu(OH)_2$) that was obtained from the copper mines of Cyprus and also lapis lazuli from Badakshan in Afghanistan, both of which were naturally occurring, and Egyptian blue, a synthetic blue glass frit that was highly prized for its deep blue colour, the other two naturally occurring pigments being considered much inferior substitutes in this respect. Although Theophrastus claims

H. G. M. Edwards, P. Colomban, *Blue by Fire: A Marker of the Technical History of Glass and Ceramics*, Cultural Heritage Science,
https://doi.org/10.1007/978-3-031-77122-4_4

that Egyptian blue was only obtained from sources in Egypt at that time, Vitruvius provides details of its manufacture at Pozzuoli near Vesuvius by Vestorius –and he does refer to it specifically there as "Alexandrian frit", although the manufacture of Egyptian blue in Egypt actually predated the foundation of Alexandria! Cobalt ores are rare, yet in the 18th Egyptian Dynasty (1550–1290 BCE) for a short time glazes coloured with cobalt in place of Egyptian blue and lapis lazuli are to be found [30]. Ptolemaic glassmakers and potters (ca. First Century CE) discovered that lapis lazuli grains can be used as a pigment to colour enamel [6–8]: the technique continued to be used in the Roman Empire and then by Islamic glassmakers and was rediscovered by Böttger at Meissen in the early eighteenth century [9]. Only a small amount of cobalt (*ca.* less than 0.1%) is required to colour a glaze for ceramic purposes and the saturation of a glass with cobalt ions leads to the precipitation of Co_2SiO_4 olivine and $CoAl_2O_4$ spinel in silica-rich and alumina-rich substrates as found in soft-paste and hard-paste porcelain glazes. Cobalt-containing glass has been recycled since antiquity, as demonstrated by the discovery of Roman shipwrecks containing broken pieces of blue glass. The source of cobalt in ancient Egypt is believed to be the alums based upon $KAl_2(SO_4)_2(OH)_6$ from the Kharga Oasis in the Western Desert [19, 23] which thermally decomposed around 800 °C to form a cobalt spinel, $CoAl_2O_4$: between the twelfth and seventh centuries BCE, cobalt glass was not used in ancient Egypt, then reappearing in the Ptolemaic 26th Dynasty but obtained from a different source [17].

Much discussion has centred around the reason for the production of Egyptian blue around 2550 BCE but evidence is now emerging that the chemical knowledge possessed by the ancient Egyptians was greater than has been appreciated hitherto and was directed at the synthesis of pigments, cosmetics and the use of resins in the mummification rituals associated with the preservation of human skeletal remains [5]. The synthesis of Egyptian blue, the first synthetic inorganic pigment, is a classic example of this that is ascribed to the desire of the Egyptians to imitate the precious gemstones and their colours that were associated with their gods, and especially the blue colour exemplified by lapis lazuli which was reserved for the colouration of the hair of their deities. Other natural associations were magnetite (associated with the god Horus) and natron (associated with the god Osiris). The calcium copper silicate that comprises Egyptian blue required the stoicheiometric mixing of silica, lime and copper shards heated with a flux such as borax—a tweaking of the furnace temperature resulted in the production of an Egyptian green pigment—and it is now believed that this was not an accidental procedure but rather the result of some experimentation that was carried out in specially designed "laboratories" or workshops, such as the example found in the Temple at Dendera., which reinforces the association of the pigment colours and their religious significance [2, 4]. In concord with this hypothesis is the realisation that these Egyptian procedures were known to the Greeks and predated Greek science, which has hitherto been credited with their recorded discoveries, mainly arising from the survival of the appropriate historical documentation.

4.1 Roman Pigments

The first comprehensive analytical study of Roman pigments was made by Sir Humphry Davy in 1815 [10] who determined the chemical nature of pigments used on wall painting shards obtained from Pompeii, then being excavated by his friend, Canova. A listing of Roman pigments is given in Table 4.1, which details the Latin names, alternatives and chemical composition of the important mineral and plant derived pigments from the works of Pliny and Vitruvius [22, 31] and from later chemical analyses. Strong colours were accorded the classification of *floridi colores* (for example, minium, vermilion, armenium, and chrysocolla) by Pliny and Vitruvius and their weaker counterparts were termed *floridi et austere* (for example, ochres, green earth, Melian white chalk).

Table 4.1 Pigments used by the Romans (and Greeks), alternative nomenclature and their chemical composition[a]

Pigment	Alternative name	Colour	Composition	Synthetic/ Natural
Ceruse	Basic lead white	White	$2PbCO_3.Pb(OH)_2$	Synthetic
Melian white	Chalk[b]	White	$CaCO_3$; $CaMg(CO_3)_2$	Natural
	Melinium			
Kaolin	China clay	White	$Al_2Si_2O_5(OH)_4$	Natural
	Kaolinite			
Paretonium	Calcite	White	$CaCO_3$	Natural
Elephantinum	Ivory black	Black	C	Synthetic
Tryginum	Carbon black	Black	C	Synthetic
	Ferrum carbonatum			
	Atramentum			
	Indicum			
Sepia officialinus	Cuttlefish ink	Black	C	Natural
Pyrolusite		Black	MnO_2	Natural
Cinnibaris	Cinnabar	Red	HgS	Natural
	Vermilion			
Cinnibaris indica	Dragon's blood	Red	*Dracaena* aromatic extract[c]	Natural
Minium	Red lead	Red	Pb_3O_4	Synthetic
	Sandárach			
	Cerussa usta			
	Verrulia			
	Minium secondarium			

(continued)

Table 4.1 (continued)

Pigment	Alternative name	Colour	Composition	Synthetic/ Natural
Sinopia	Sinope	Red	Fe_2O_3	Natural
	Sinopedium			
Litharge	Lead oxide	Red	PbO	Natural
	Syricum			
Rubina	Cicerculum	Red	Fe_2O_3	Natural
	Rubrica			
Madder	Rubiae redux	Red	Plant extract	Natural
Haematite	Sanguine	Red	Fe_2O_3	Natural
	Haematites			
Carmine	Kermes	Red	Insect extract	Natural
	Coccos			
	Granum scolecium			
	Vermiculum rubrum			
Realgar	Sandaraca	Red-orange	As_4S_4	Natural
Auripigmentum	Orpiment	Yellow	As_2S_3	Natural
	Operment			
	Arsenikon			
Saffron		Yellow	Plant extract	Natural
Massicot	Lead oxide	Yellow	PbO	Natural
Jarosite		Yellow	$Fe_3(SO_4)_2(OH)_5.2H_2O$	Natural
Ochre	Goethite	Yellow	FeO(OH)	Natural
Creta viridis	Green earth	Green	Mixture	Natural
	Verona green			
	Theodotion			
	Celadonite, glauconite			
	Prasitus			
	Appianum			
Chrysocolla	Malachite	Green	$CuCO_3.Cu(OH)_2$	Natural
Verdigris	Basic copper acetate	Green	$Cu(CH_3CO_2)_2.Cu(OH)_2$	Synthetic
Armenium	Azurite	Blue	$2CuCO_3.Cu(OH)_2$	Natural
Lapis lazuli	Ultramarine	Blue	$(Na,Ca)_8(AlSiO_4)_6(SO_4,S,Cl)_2$	Natural
	Azzurum ultramarinum			
	Outremer			
Indigo	Indigofera	Blue	Plant extract	Natural
	Hysginium			
	indikon			
	Pigmentum indicium			

(continued)

Table 4.1 (continued)

Pigment	Alternative name	Colour	Composition	Synthetic/ Natural
Woad	Isatis tinctoria	Blue	plant extract	Natural
	Vitrium			
	Glastum			
	Guadium			
Egyptian blue	Alexandrian frit	Blue	$CaCuSi_4O_{10}$	Synthetic
	cuprorivaite			
	Smalto			
	Caeruleum			
Verditer	Coelon	Blue	$2CuCO_3.Cu(OH)_2$	Natural
	Lomentum			
	Tritum			
Tyrian purple	Murexide	Purple	Dibromoindigo	Natural
Ostrum		Purple	Red ochre plus blue	Synthetic
Caput mortuum		Purple	Fe_2O_3	Natural
Purpurinum	Optimi hysginum	Purple	?	Synthetic ?

[a]References: Theophrastus, Pliny, Vitruvius, Harley

[b]Chalk was sometimes mixed with powdered glass to provide a glossy white pigment

[c]The Romans cite seven organic, non-mineral, dyes that were extracted from plant and animal sources in the ancient world: madder $C_{14}H_8O_4$, dragon's blood $C_{16}H_{14}O_6$, saffron $C_{44}H_{64}O_{24}$, indigo and woad $C_{16}H_{10}N_2O_2$, carmine (kermes) $C_{22}H_{20}O_{13}$ and Tyrian purple $C_{16}H_8Br_2N_2O_2$. The first five are plant—based extracts and the last two were obtained from insects (beetles) and shellfish, respectively. All have aromatic ring structures and several have fused rings such as indoles and anthracenes

References

Harley, R.D. **2001**. Artists' Pigments: A Study in English Documentary Sources, 2nd Edition. Archetype Publications: London.

Feller, R.L. **1986**. Editor, Artists' Pigments: A Handbook of Their History and Characterisation. Volume 1. Oxford University Press: New York and Oxford Fitzhugh, E.W. Green Earth, 141–168.

Roy, A. **1993**. Editor, Artists' Pigments: A Handbook of Their History and Characterisation. Volume 2 Oxford University Press: New York and Oxford, Gettens, R.J., Fitzhugh, E.W. Azurite and Blue Verditer, pp. 23–36. Plesters, J. Ultramarine Blue: Natural and Synthetic, pp. 37–65. Gettens, R.J., Kuhn H., Chase, W.T. *Lead White*, pp. 67–81. Kuhn, H. Verdigris and Copper Resinate, pp. 131–158. Gettens, R.J., Feller R.L., Chase, W.T. Vermilion and Cinnabar, pp. 159–182. Gettens, R.J., Fitzhugh, E.W. Malachite and Green Verditer, pp. 183–202.

Fitzhugh, E.W. **1997**. Editor, Artists' Pigments: A Handbook of Their History and Characterisation. Volume 3. Oxford University Press: New York and Oxford 1997. Riederer, J. Egyptian Blue, pp. 23–46. Fitzhugh, E.W. Orpiment and Realgar, pp. 47–79. Schweppe, H. Indigo and Woad, pp. 81–108. Schweppe, H., Winter, J. Madder and Alizarin, pp. 109–142.

Berrie, B.W. **2007**. Editor, Artists' Pigments: A Handbook of Their History and Characterisation. Volume 4. National Gallery of Art: Washington and Archetype Publications: London. Winter, J., Fitzhugh, E.W. Pigments Based on Carbon, pp. 1–37. Helwig, K. *Iron Oxide Pigments: Natural and Synthetic*, pp. 38–109.

A rather significant difference between the texts of Theophrastus and those of Pliny and Vitruvius is that the Romans had different Latin names for the same pigment that was nevertheless perceived to be of different qualities: for example, blue *verditer* was described variously as *coelon, lomentum* and *tritum,* which reflected the decreasing pigment quality of specimens acquired from different geographical sources in terms of their depth of colour when the material was adopted as a pigment. It can therefore be rather difficult sometimes to discern a new pigment from a quality graded alternative when the historical nomenclature is not precise. Another problem is that synthetic mixtures of natural pigments were sometimes used and these too then had different descriptors applied to them, such as *ostrum,* which has been tentatively ascribed to a purple colour achieved by mixing red ochre and a blue pigment (possibly, Egyptian blue). Likewise, two different Latin names were provided for carbon black obtained from different sources, namely *elephantinum* for charcoal from calcined ivory and *trygintum* for charcoal derived from vine twigs and branches (i.e. effectively differentiating its mammalian source from one of a plant botanical origin): a further variety was carbon extracted from cuttlefish ink sacs which was known as *sepia officialinus.* Even so, much confusion was created by the frequent use of loose terminology in the early literature and the adoption of even more terms such as *fuligo pictoria, atramentum, indicum, ferrum carbonatum, terra negra* and *plumbago,* which it is believed now all describe different sources of graphite, lamp black, charcoal, carbon black, soot, ivory/bone black and black earths [12, 32].

A similar problem is associated with the haphazard and loose historical terminology that was applied to several mineral pigments; for example, mercury (II) sulfide, HgS, was known as *cinnabar,* but also was termed *minium,* which was later used interchangeably for lead (II) lead (IV) tetroxide, Pb_3O_4. Also, *cinnabar* should not be confused with the similar term *cinnibaris,* which was used specifically for dragon's blood pigment extracted from dragon trees (Calamus) of the botanical genus *Dracaenae cinnabari* or *Daemonorops draco* that were sourced by the Romans from Socotra and Tenerife, respectively. Both Pliny and Vitruvius adopt distinctive Latin names for minerals that are derived from different geographical locations or from a range of processes because of the quality of the pigment that could be obtained therefrom: a typical example of this is provided by the generic green earths, which were described variously as *creta viridis, theodotion, apianum, prasitus* and *prason,* yet all of these are chemically mixtures in different proportions of celadonite, chlorite and glauconite—the most prized of these for pigment purposes being mined at Verona [14]. The latter was later recognised and became known as *terre verte di Verona* in the Renaissance. Pliny's *apianum* was judged to be the best substitute for *malachite (chrysocolla).* The Romans also adopted a limited number of organic dyes which were extracted from plants insects and shellfish, numbering seven in all, but all these were realised to be fugitive colours and certainly were not suitable for lasting artworks and decoration: indigo (from the East) and woad (from the West) were both indigo based blue dyes extracted from the plants *Indigofera tinctoria* and *Isatis tinctoria*, respectively, but the most highly prized organic dye was undoubtedly *Tyrian purple* (the purple of the Caesars), which was extracted from the glands of the Murex shellfish and is dibromoindigo in chemical constitution.

For the theme of this current text, we need to focus our discussion upon blue pigments that were suitable for ceramic decoration: these immediately can be classified into two types of decorative pigment, namely, those which are suitable for use in an overglaze cold application and those which required a superficial glaze and subsequent firing at elevated temperatures in a glost kiln, in the ceramics literature these are referred to later as paints and enamels, and the decorating artist was differentiated as a painter and an enameller. The essential difference between these two categories of pigment is that the latter needed to be stable at the high temperatures of around 800–900 °C pertaining to the glost kiln firing process: hence, temperature sensitive mineral pigments such as carbonates and some sulfates and sulfides would not be applicable for this procedure, such as the blue pigment azurite, basic copper carbonate, $2CuCO_3.Cu(OH)_2$, which at 300 °C degrades thermally to the black and brown copper oxides, CuO and Cu_2O, or the blue pigment brochantite, $CuSO_4.5H_2O$, which loses water at around 150 °C to give white $CuSO_4$ and loses sulfur dioxide at higher temperatures to give black CuO. The additional problem for overglaze ceramic decoration, of course, is its sensitivity to deterioration during use, especially for household artefacts that required exposure to hot water and which necessitated the use of cutting tools or scrubbing. The use of the ancient fugitive and water soluble organic pigments described in Table 4.1, therefore, would certainly not be appropriate for overglaze or underglaze ceramic decoration. We therefore need to consider carefully those ancient temperature stable blue pigments cited in this list, which can be simply enumerated as follows: lapis lazuli and Egyptian blue. Having said that, there is no evidence that lapis lazuli was used by the Romans in ceramic decoration [21], primarily because of its cost, although there are several reports of its adoption in wall paintings. Of course, with the evolution of later ceramics, especially high-fired porcelains, the advent of cobalt blue in its several forms and smalt, both being termed a blue glass in composition, was a necessary step for their effective decoration as became manifest in the successful production of the Yuan and Ming Dynasty porcelains from the fifteenth century onwards [20, 24].

4.2 Chinese Pigments

The earliest shipwreck known with a cargo of ceramics including early porcelain is that of an unnamed Arab dhow which sank off Belitung Island in the Singapore Straits in about 830 CE carrying a mixed cargo of Tang Dynasty (618–907 CE) ceramics, which included 300 pieces salvaged of elegant, high value white Xing porcelain (from the Heibei kilns), some green Yue and Gongxian celadons and an assortment of some 60,000 Tang Dynasty stonewares from the Changsha (Hunan) kilns. The wreck confirmed the trade route that existed at that time between China

Fig. 4.1 Oldest example of a Chinese blue and white decorated ceramic in existence: a Tang Dynasty glazed stoneware dish from the Chaysha (Hunan) kilns, *ca.* 830 CE, with double lozenge and foliate motifs. Recovered from the wreck of an Arab dhow which sank off Belitung Island in the Singapore Straits carrying a cargo of assorted ceramic artefacts to Arabia. (Reproduced with permission of the Asian Civilisations Museum, Singapore, from their Tang Shipwreck Collection, accession number 2005.1-00473)

and the Arabs of the Abbasid Dynasty in Africa across the Indian Ocean, long before these ceramics were known in Europe. One of the items from the salvaged cargo, which is now known collectively as "*The Tang Treasures from the Sea*", is a fine stoneware bowl decorated in cobalt blue over a white glaze slip—this is apparently the oldest example of a blue and white decorated Chinese ceramic in existence and this is now preserved in the Asian Civilisations Museum in Singapore (Fig. 4.1). Shipwrecks which have yielded porcelains and other ceramic wares from their cargoes are listed in Table 9.1 in Chap. 9 of this book, where a selection of 22 shipwrecks is given: in many cases the name of the ship is unknown and the wreck and salvaged cargo is then accorded an identification based upon its recovery site. In some cases, as for example in the case of the *San Jose,* which was sunk off Cartagena in the Caribbean in 1708, the Kangxi porcelain has been photographed lying apparently intact on the seabed but has not yet been raised archaeologically because of an ongoing international dispute as to the ownership of the wreck and especially the gold bullion associated with it estimated at a current value of $20 Billion.

4.2.1 The Use of Cobalt Blues in China for Ceramic Decoration

Whereas most texts on artists' pigments refer to the discovery of cobalt blue by Louis Jacques Thenard (1777–1857) in 1803–1804 [28], for example Roy [25] and Harley [16], and its subsequent rapid adoption by artists within a year or two, they generally do not mention its use in antiquity, particularly by the Chinese ceramics

Table 4.2 Comparison of Blue Glass Frits used in porcelain body and Glaze recipes and their decoration

Glass Frit	Alternative Name(s)	Year	Chemical Name/Formula[a]	Composition	Reference
Cobalt blue	Bristol blue	1803	Cobalt aluminate/	81% Al_2O_3	[28]
	Thenard's blue		$CoAl_2O_4$	15% CoO	[25]
	Vienna blue			1% (Na,K)O	[12]
				3% H_2O	
Smalt	Kaiserblau	1540	Cobalt silicate	70%SiO_2	[20]
	Bleu d'email			15%K_2O	
	Azzuro di smalto			5% Al_2O_3	
				10% CoO	
Egyptian blue	Blue frit	3000BCE	Calcium copper silicate/	70%SiO_2	[24]
	Caeruleum		$CaCuSi_4O_{10}$	8% CaO	[12]
	Cuprorivaite			16% CuO	
				3% Na_2O	
				2% Al_2O_3	
				2% Fe_2O_3	
Han blue	Chinese blue	1045BCE	Barium copper silicate/	43% SiO_2	
	Effenbergerite		$BaCuSi_4O_{10}$	37%BaO	
				19% CuO	[12]
Bristol blue	Cobalt blue	1780	Cobalt lead silicate	15% CoO	
				24% PbO	
				61%SiO_2	

[a]It is not possible to give a chemical formula for several blue pigments as they are glasses of an indefinite composition: for example, smalt and Bristol Blue. It should be noted that Bristol blue differs significantly in composition from smalt and cobalt blue in that it contains lead oxide but has no alumina. Yet, ceramic writers in the eighteenth and nineteenth centuries used these terms indiscriminately and interchangeably!

enamellers which pre-dated its formal nineteenth century synthesis by almost two thousand years. Actually, "cobalt blue" is a rather generic term which covers several distinct pigments of different chemical composition and even in the nineteenth century it was used to describe synthetic mineral species such as smalt (a potassium cobalt aluminosilicate glass), cobalt aluminate, and Bristol blue (a cobalt lead silicate). Reference to the list of shipwrecks and their porcelain cargoes in Table 4.2 will indicate that Chinese decorated cobalt blue and white porcelains from the Yuan and Ming Dynasties were extant long before the formal chemical synthesis of cobalt blue was reported, using cobalt ores, essentially Co_2O_3, which had been sintered with aluminium oxide clays at temperatures approaching 1300 °C to produce their cobalt blue pigment. The Chinese, it is believed, first imported cobalt ores from Persia via the Silk Road, probably during the ninth century CE, and an example of this early decoration is shown in Fig. 4.1: the sources of cobalt ores for the preparation of their cobalt blues for porcelain decoration varied with the imperial Dynasties and the earliest blue and white wares were limited in production by the cost of the expensive cobalt blue pigment, especially during the Song Dynasty (960–1279 CE). The Chinese in the following Yuan Dynasty (1279–1368 CE) differentiated between different qualities of the pigment: a Samarra blue (*sumali qing*) and a Sumatra blue (*suborin qing*) were rich in iron oxide and were much prized for their deep blue colour. The emperors of the later early Ming Dynasty (1368–1644 CE), for example Hongwu (1368–1398 CE), were initially faced with a disruption in the importation of these cobalt ores because of imposed restrictions in foreign trade and later Ming emperors such as Yongle (1403–1425 CE) and Yuande (1426–1435 CE) had to resort to the use of an imported smalt until locally discovered sources of cobalt were found. Hence, Chenghua (1465–1487 CE) and Zhengde (1506–1521 CE) used extensively local cobalt ores, which imparted a softer blue colour to the porcelain decoration. Mineral blue (*shi qing, shizi qing*) was sourced from Jiangxi and was used for the poorer quality mingao wares, whereas muslim blue (*huiqing*) was the best quality sourced from Xinjiang and Yunnan, which afforded a deeper bluish-purple hue; the latter was preferred by the emperor Xuande, whereas the emperor Chengzhou preferred a paler blue decoration. The last Ming emperor Chongzhu (1611–1644 CE) used local cobalt ore from Shaoxing Jinhua and Quzhou but the first emperor of the Qing Dynasty, Kangxi, (1662–1722 CE) preferred the bright and intense blue pigment derived from the cobalt ores sourced from Zhegiang (Zheliao). It has been claimed that the Chinese used no fewer than seven different types of cobalt blue of varying quality during this period, five of which were locally mined and two of which were imported from the Middle East. It has also been discovered that during the later Kangxi reign and onwards into the early Qing Dynasty, the French Jesuits were instrumental in supplying cobalt ore from Erzgebirge in Saxony to the Chinese court and this was especially valued for its depth of colour and adhesion—its use on Chinese porcelains can be recognised analytically by the presence of arsenic in the cobalt blue pigment derived from the mineral skutterudite, a cobalt arsenide $CoAs_3$, in the raw cobalt ore. Native Chinese cobalt ores that were sourced locally are rich in manganese and iron but have a relatively weaker colour intensity and a greyish-blue tone. The presence of a high iron content also results in a surface oxidation deposit which can counteract the perception of the strong blue colouration beneath.

There is historical evidence that trade between the Arabs in the Middle East, particularly the Persian potters who used cobalt blue to decorate their tiles and frit-wares, and China was established during the Tang Dynasty and that the green jade-like coloured glazed Chinese celadons were exchanged for the cobalt blue pigment that had been evolved in the Persian potteries in Kashan: as later transpired for an alleged alchemical property of porcelain itself, the celadons were believed to change colour magically in the presence of poisoned food! The Mongol emperors of the Yuan Dynasty traditionally believed that they descended ancestrally from a "blue wolf" and "white fallow doe" and they were therefore much intrigued by the concept of representing this appropriately as decoration on their finest blue-and-white porcelains [33]. The imported cobalt blue pigments, known as Sumali blue from Samarra and Mohammedan blue (Sunibo blue) from Sumatra, were used along with local pigments from Jiangxi, Zhejiang and Yunnan, up to the emperor Wanli towards the end of the Ming Dynasty.

It is necessary that we differentiate between the different "cobalt blues" and related pigments that have been used to decorate ceramics which have often been a source of confusion in the earlier literature: Table 4.2 details the cobalt blue pigments in the form of glass frits used historically that have been mentioned above, some alternative names and cites their chemical composition and sourcing, where appropriate. Included in the list in Table 4.2 is the Chinese pigment Han blue, which although not strictly a cobalt blue pigment as being considered here is the barium equivalent of cuprorivaite, the calcium copper silicate that is Egyptian blue, namely $BaCuSi_4O_{10}$, and was named after its appearance during the Qin Han Dynasty (221 BCE—220 CE).Its congener, Han purple, $BaCuSi_2O_6$, appeared in the decoration of the terracotta Xian warriors excavated in the mausoleum at Xian.

A wider range of blue mineral pigments was available historically and these are summarised with their alternative names Table 4.3.

4.3 The Mediaeval Blue Pigments and Transition into the Renaissance

Despite the greater variety of pigments that became available during the mediaeval period and thence into Renaissance which were adopted widely by artists [3, 11, 13, 15], which include minerals such as lead tin yellow and organic resins and dyes such as turnsole (heliotrope), orcein and orchil, none would have been suitable for use on high-temperature fired ceramics and the cobalt blues still reigned supreme in this context. The classic work of Agricola [1] is noteworthy for its report of the re-discovery of the production of Egyptian blue glass from cobalt containing alums in the mediaeval period, which had apparently been lost after the cessation of the Roman Empire. The advances in synthetic chemistry in the eighteenth and nineteenth centuries and a better understanding of the high-temperature reactions that could be undergone between the pigments and an applied glaze on a heated ceramic substrate facilitated the adoption of an even wider range of pigments that could be made available for artworks generally. Especially noteworthy in this respect were the aniline dyes based on mauveine following William Perkin's synthesis in 1858 which was then followed quickly by the aromatic organic azo-naphthol dyes which gave brighter colours. Inorganic congeners were materials such as chrome yellow, Scheele's green, Mars red, the cadmium yellows and oranges, and a synthetic ultramarine, which negated the use of lapis lazuli and its expensive acquisition from Afghanistan. These rapidly gained favour with post-Renaissance artists on account of their vibrant colours and small changes in chemical composition of derivatives which could be effected to give a much wider range of colour tones. Again, although

Table 4.3 Blue pigments used historically[a]

Name	Alternative names	Description	Chemical formula	Synthetic/ Natural	Date of synthesis/ First use	Reference
Azurite	Bergblau	Basic copper carbonate	$2CuCO_3.Cu(OH)_2$	Natural	2500 BCE	AP2
	Lapis armenius					
Blue verditer	Bremen blue	Basic copper carbonate	$2CuCO_3.Cu(OH)_2$	Synthetic	Seventeenth century CE	AP2
	Blue bice					
	Cendres bleu					
Ultramarine blue	Outremer	Sodium calcium aluminium silicate	$(Na,Ca)_8(AlSiO_4)_6(S,SO_4)$	Natural	Sixth century CE	AP2
	Azur d'Acre					
	Lapis lazuli					
	Persian blue					
Ultramarine artificial		Sodium calcium aluminium silicate	$Na_{6\text{-}10}Al_6Si_6O_{24}S_{2\text{-}4}$	Synthetic	1828 (Guimet, Gmelin)	AP2
	French ultramarine					
	Guimet's blue					
Smalt	Esmalti	Potassium cobalt glass	$(K_20)_{0.1}(SiO_2)_{0.6}(CoO)_{0.2}(As_2O_3)_{0.1}$	Synthetic	1540 (Schurer)	AP2
	Email					
	Kaiserblau					
Cobalt blue	Kobaltblau	Cobalt aluminate	$CoAl_2O_3$	Synthetic	1803 (Thenard)	AP4
	Thenard's blue					

Egyptian blue	Blue frit	Calcium copper silicate	$CaCuSi_4O_{10}$	Synthetic	2500 BCE	AP3
	Pompeiian blue					
	Alexandrian blue					
	Pozzuoli blue					
Cuprorivaite		Calcium copper silicate	$CaCuSi_4O_{10}$	Natural		AP3
Prussian blue	Brunswick blue	Potassium iron ferricyanide	$K_4Fe(CN)_6.xH_2O$	Synthetic	1704 (Diesbach)	AP3
	Iron blue					
	Turnbull's blue					
	bleu de Prussie					
Indigo	Woad	Plant extract: *Isatis tinctoria*	Footnote 1	Natural	100 BCE	AP3
	Indigotin					
	Anil					
	Guede					
Indigo				Synthetic	1878 (von Baeyer)	AP3
Posnjakite		Basic copper sulfate	$CuSO_4.3Cu(OH)_2.H_2O$	Natural/ Synthetic	1967	
Boothite		Copper sulfate heptahydrate	$CuSO_4.7H_2O$	Natural		
Cerulean blue	Caeruleum	Cobalt stannate	Co_2SnO_4	Synthetic	1789 (Hopfner)	
	Caerulean blue					
Leitsch's blue	Cyanine blue	Mixture of cobalt blue and Prussian blue		Synthetic	1900 (?)	
Turnsole	Heliotrope	Plant extract: *Chrozophora tinctoria*	Footnote 2	Natural	Fifteenth century	
	Folium					

(continued)

Table 4.3 (continued)

Name	Alternative names	Description	Chemical formula	Synthetic/ Natural	Date of synthesis/ First use	Reference
Litmus	Erythrolitmin	Plant extract: *Roccella tinctoria*	Footnote 3	Natural	Fifteenth century	
Maya blue		Mixture of indigo and palygorskite clay		Synthetic	800CE	
Han blue	China blue	Barium cuprosilicate	$BaCuSi_4O_{10}$	Synthetic	1000BCE	
	Effenbergite			Natural		
YInMn blue[b]	Mas Blue	Yttrium indium manganese oxide	$YIn_{0.8}\,Mn_{0.2}O_3$	Synthetic	2009	

Footnotes:

1. Indigo: $C_{16}H_{10}N_2O_2$, 2-(3-hydroxy-1H-indol-2-yl)indol-3-one
2. Turnsole: $C_{28}H_{24}N_2O_7$, 8-(2,4-dihydroxy-6-methylanilino)-2-(2,4-dihydroxy-6-methylphenyl)imino-7-hydroxy1,9-dimethyldibenzofuran-3-one
3. Litmus: $C_{12}H_7NO_3$, 7-hydroxyphenazone

AP1: Artists' Pigments: A Handbook of Their History and Characteristics, ed. Feller, R.L. National Gallery of Art: Washington and Oxford University Press: Oxford and New York, 1986. 10 pigments quoted but no blue!

AP2: Artists' Pigments: A Handbook of Their History and Characteristics, ed. Roy, A. National Gallery of Art: Washington and Oxford University Press: Oxford and New York, 1993

Gettens, R.J., Fitzhugh, E.W.: azurite and blue verditer, pp. 23–36

Plesters, J.: ultramarine blue, natural and artificial, pp. 37–66

Muhlethaler, B., Thissen, J.: smalt, pp. 113–130

AP3: Artists' Pigments: A Handbook of Their History and Characteristics, ed. Fitzhugh, E.W. National Gallery of Art:, Washington and Oxford University Press; Oxford and New York, 1997

Riederer, J.: Egyptian blue, pp. 23–46

Schweppe, H.: indigo/woad, pp. 81–108

Berrie, B.H.: Prussian blue, pp. 191–218

AP4: Artists' Pigments: A Handbook of Their History and Characteristics, ed. Berrie, B.H. National Gallery of Art: Washington and Archetype Publications Ltd.: London, 2007

Roy, A.: Cobalt blue, pp. 151–178

[a]The list of blue pigments in this table is comprehensive insofar as all have been recorded as having been used in paintings generically: however, because many of these are thermally unstable they would not have been seen as suitable for underglaze ceramics decoration which normally required a firing in a glost kiln at a temperature ranging between 800 and 1100 °C. Of course, the cold painting or overglaze decoration of ceramics would not be subject to this restriction, although overglaze decoration for utilitarian objects would not be protected from deterioration during commercial usage and would thereby suffer much damage. This latter statement would not apply to overglaze decorative ornamental artefacts. A classic example of this thermal instability would be provided by azurite, a basic copper carbonate, which decomposes at 350 °C into tenorite, CuO and cuprite, Cu_2O, which are both dark brown and black pigments. Other blue minerals are known such as bornite, Cu_5FeS_4 (also known as peacock ore or peacock copper) and covellite, CuS or better $Cu_3^{+}S^{2-}S_{2-}$, but there is no record of these ever having been used experimentally as artists' pigments

[b]YInMn blue is claimed to be the first synthetic blue pigment created for artists in the last 200 years, made in 2009 by heating yttrium, indium and manganese oxides to about 1100 °C and is now widely being promoted as a non-toxic version of blue pigment similar in tonal colour and depth to cobalt aluminate. This does not seem to recognise the synthesis of copper phthalocyanine (phthalocyanine blue), a vivid blue pigment that was synthesised in 1927

The chemical composition of a naturally occurring mineral by definition is accurately given but that of other naturally occurring rocks, such as lapis lazuli, can vary. The composition of synthetic materials such as glasses can vary and is only approximately rendered here

In ancient times it was quite common to only loosely describe a pigment and the Latin names of several pigments are still a matter for some debate e.g. the true identity of the blue Roman pigment *caeruleum* cited by Pliny and Vitruvius is still debated

The three blue plant extracts were much used in the Renaissance for paintings but the colours were labelled as "fugitive": they are not minerals but dyes and their chemical formulae are not strictly precise as their composition varied depending upon the methods of extraction and the specific botanical plants used. The formulae cited here represent the major chromophore but in most cases mixtures of up to 10 or more different combinations of dyes could come under the umbrella definition of the dye nomenclature undoubtedly available for the overglaze cold-decorative painting of ceramics, none of these were stable enough for use as a high-temperature enamel for an underglaze high-fired porcelain or stoneware decoration.

A comprehensive list of the blue pigments that became available to mediaeval, Renaissance and post-Renaissance and modern artists is given in Table 4.4: whereas some of these have been recounted already in Tables 4.1 and 4.3 as pigments used by the Romans and Greeks in antiquity and thereafter, the wider range of chemical compositions that were now made available by the advances in chemical synthesis in the nineteenth and twentieth centuries is noteworthy. It should be realised too that geological discoveries of small outcrops of naturally occurring blue coloured minerals were made during this time although their comparative rarity would have been

Table 4.4 Definitive list of Renaissance pigments, their chemical formulae and historically alternative names

Pigment	Alternative name(s)	Colour	Type	Chemical formula
A: Minerals and synthetic inorganics				
Lazurite	Natural ultramarine	Blue	Mineral	$Na_{8\text{-}10}Al_6Si_6O_{24}S_{2\text{-}4}$
	Lapis lazuli			
	Outremer			
Azurite	Basic copper carbonate	Blue	Mineral	$Cu(OH)_2.\ 2CuCO_3$
	Caerulium			
Haematite	Red oxide of iron	Red	Mineral	$Fe_2O_3.xH_2O$
	Red ochre			
	Red earth			
	Sinope			
	Sanguine			
Goethite	Yellow oxide of iron	Yellow	Mineral	$FeO(OH).xH_2O$
	Yellow ochre			
	Yellow earth			
Lead white	Basic lead carbonate	White	Mineral	$Pb(OH)_2.2PbCO_3$
	Cerussite			
	Ceruse			
Cinnabar	Vermilion	Red	Mineral	HgS
	Mercury sulfide			
Red lead	Lead(II) lead(IV) tetroxide	Red	Synthetic	Pb_3O_4
	Minium			
	Sandarach			
Burnt sienna	Brown ochre	Brown	Synthetic	$Fe_2O_3.FeO(OH).xH_2O$
	Burnt umber			

(continued)

Table 4.4 (continued)

Pigment	Alternative name(s)	Colour	Type	Chemical formula
Malachite	Basic copper carbonate	Green	Mineral	$Cu(OH)_2.CuCO_3$
	Chrysocolla			
Orpiment	Arsenic sulfide	Yellow	Mineral	As_2S_3
	Auripigmentum			
Realgar	Arsenic Sulfide	Red	Mineral	As_4S_4
	Sandaraca			
Pararealgar	Arsenic sulfide	Orange yellow	Synthetic	As_4S_4
Lead-tin-yellow I	Lead stannate	Yellow	Synthetic	Pb_2SnO_4
	Giallorino			
Lead-tin-yellow II	Lead tin silicate	Yellow	Synthetic	$Pb(Sn,Si)O_3$
Naples yellow	Lead antimonate	Yellow	Mineral/ Synthetic	$Pb_2Sb_2O_7$
	Pyrochlore			
Charcoal	Carbon black	Black	Synthetic	C
	Soot			
	Trigynium			
	Atramentium			
Bone Black	Ivory black	Black	Synthetic	$C + Ca_3(PO_4)_2$
	Elephantium			
Green earth	Verona green	Green	Mineral	$K(Al,Fe^{3+})(Fe,Mg^{2+})(AlSi_3, Si_4)O_{10}(OH)_2$
	Terre verde			
	Creta viridis			
	Appianum			
Verdigris	Basic copper acetate	Green	Synthetic	$Cu(CH_3CO_2)_2.xH_2O.Cu(OH)_2$
Calcite	Calcium carbonate	White	Mineral	$CaCO_3$
	White chalk			
	Melinium			
	Paretonium			
Litharge	Syricum	Red	Mineral	PbO
Gypsum	Calcium sulfate dihydrate	White	Mineral	$CaSO_4.2H_2O$
Egyptian blue	Cuprorivaite	Blue	Synthetic	$CaCuSi_4O_{10}$
	Pozzuoli blue			
	Smalto			
Crocoite	Lead chromate	Yellow	Mineral	$PbCrO_4$
Atacamite	Basic copper chloride	Blue	Mineral	$CuCl_2.Cu(OH)_2$

(continued)

Table 4.4 (continued)

Pigment	Alternative name(s)	Colour	Type	Chemical formula
Lead yellow	Massicot	Yellow	Mineral	PbO
Copper red	Cuprous oxide	Brown	Synthetic	Cu_2O
Tin white	Cassiterite	White	Mineral	SnO_2
Purple of Cassius		Purple	Synthetic	Au^0 in suspension
B: Organic extracts from plants, insects and shellfish				
Turnsole	Heliotrope	Blue	Plant extract	Chrozophora tinctoria/ chrozophoridine
	Folium			
	Solsequinium			
Indigo	Indigofera	Blue	Plant extract	Indigofera tinctoria/indigo
	Hysginium			
	Pigmentum indicum			
Woad	Isatis tinctoria	Blue	Plant extract	Isatis tinctoria/indigo
	Vitrium			
	Glastum			
Tyrian purple	Dibromoindigo	Purple	Shellfish extract	Murexide/Dibromoindigo
Madder	Purpurin/alizarin	Red	Plant extract	Rubia tinctorium/alizarin, purpurin
	Rubiae redux			
Orchil	Lichen dye	Red/blue	Plant extract	Roccella tinctoria/ hydroxyphenoxazone, orcein
	Litmus			
Dragon's blood	Dracaena cinnabari	Red	Plant extract	Calamus draco (Dracaena)/ anthocyanidin, dracoflavin
	Daemonorops draco			
	Sanguis draconis			
Brazilwood	Brasiliein	Red	Plant extract	Caesalpinia echinata/brasilein
	Haemotoxylin			
	Pernambuco wood			
Carmine	Carminic acid, Kermes	Red	Insect extract	Dactylopius coccus/carminic acid, anthraquinones
	Cochineal			
	Coccos			
	Vermiculum rubrum			
Lac	Shellac	Brown-red	Insect extract	Kerria lacca/aleutiric acid

(continued)

Table 4.4 (continued)

Pigment	Alternative name(s)	Colour	Type	Chemical formula
Indian yellow	Magnesium euxanthate	Yellow	Animal waste product	Medrecydon tinctorum/ euxanthic acid, Ca, Mg salts
	Purree			
Saffron	Crocetin	Yellow	Plant extract	Saffron/apocarotenoid carboxylic acid/crocetic acid
Berberine	Tree Turmeric	Yellow	Plant extract	Berberis vulgaris/ benzylisoquinoline
Maya blue	Indigo in palygorskite	Blue	Natural/ synthetic	Indigofera suffruticosa plus clay
Weld	Reseda luteola	Yellow	Plant extract	Reseda luteola/glucobarbarin/ luteolin

44 pigments and dyes: 29 Inorganic mineral and inorganic synthetic, 15 organic extracts from plants, insects and animal products

a negative factor for their adoption by Renaissance artists. Often, a small variation in chemical composition will define a different mineral source: an example is the blue mineral posnjakite, $Cu_4SO_4(OH)_6.H_2O$, a product found in the oxidised zones of copper sulfide mineral deposits in Karganda, Kazakhstan, and named after the geochemist Eugene Posnjak. Posnjakite has several closely related naturally occurring blue mineral derivatives such as langite, $Cu_4SO_4(OH)_4.2H_2O$, devilline, $CaCu_4(SO4)_2(OH)_6.3H_2O$, serpierite, $Ca(CuZn)_4(SO_4)_2(OH)_6.3H_2O$ and aurichalcite, $(CuZn)_5(CO_3)_2(OH)_6$. Brochantite, $Cu_4SO_4(OH)_6$, is often cited in pigment manuals as being a blue mineral of closely related formula to posnjakite but the essence of opinion seems to favour this now being classified better as a green or greenish-blue mineral as its colour varies with its substrate and it is frequently found associated with malachite and azurite. Easthaugh et al. [12] does list brochantite correctly as a green mineral used as a pigment but also puts posnjakite in the same colour category!

A survey of the comprehensive encyclopaedia of pigments by Easthaugh et al. [12] reveals that some 350 specific pigments are listed as blue, excluding generic chemical collections such as phthalocyanins or anthraquinones: several of the fundamental blue pigments are given in Table 4.4 but the situation is complicated on account of different names being given by artists to mixtures of pigments during the mediaeval and Renaissance. Examples of this practice are *academy blue,* a mixture of synthetic ultramarine and viridian, *Antwerp blue,* which is Prussian blue and alumina, *azurblau*, a mixture of Prussian blue, smalt and zinc white, *Brunswick blue*, Prussian blue deposited on barium sulfate, and *Monthiers blue*, an ammoniacal Prussian blue, whilst other names are given for lower grade pigments such as *azul baxo*, a poor quality azurite, and *biadetto*, a mixture of verdigris, sal ammoniac and tartar emetic. What is clear is that the pigments used during the Roman period lasted into mediaeval times and there generated others in an expanded nomenclature, which in turn reflected their sourcing and alternative methods of preparation. Towards the end of the mediaeval period and into the renaissance new blue pigments were discovered, such as the plant extracts orcein and turnsole, but the next stage needed the development of synthetic chemistry in the eighteenth century, which started with the provision of *Prussian blue* in 1704 and others such as Scheele's green, emerald green, leading into the major organic syntheses based on the aniline dyes, coal tar products, naphthols and anthraquinones in the mid—nineteenth century and onwards. The early twentieth century saw the production of many different dyes and pigments which often adopted their sourcing for nomenclature, especially the transition metal phthalocyanine dyes. However, most of these, although very usable by artists for textiles, oil paintings and watercolouring, did not immediately offer alternatives for ceramic underglaze high temperature decoration because of their instability towards the thermal treatment required for kiln firing processes. The blue pigments used following on from the mediaeval and during the Renaissance will be considered specifically and in more detail later.

Finally, the latest synthetic blue pigment that has appeared only recently is YInMn blue, alternatively known as Oregon Blue or Mas Blue (after its lead discoverer, Mas Subramanian), which was reported as the ferromagnetic/anti-ferromagnetic solid solution between $YInO_3$ and $YMnO_3$ synthesised at 1100 °C [18, 26, 27]: it has a range of blue colours dependent upon the In:Mn ratio and of a general formula $YIn_{1-x}Mn_xO_3$. The bluest colour, which is comparable with that of the best cobalt blue, $CoAl_2O_4$, is of the composition $YIn_{0.8}Mn_{0.2}O_3$. Unfortunately, the misconception has arisen that this is the first blue pigment to have been synthesised in the twentieth century whereas of course manganese blue (barium manganate/barium sulfate, $BaMnO_4.BaSO_4$) and copper phthalocyanine, $(C_8H_4N_2)_4Cu$, were both reported in 1907. YInMn blue, however, is extremely stable at high temperatures and although not yet reported it would certainly be deemed suitable for high temperature ceramics decoration applications in place of cobalt blue.

References

1. Agricola, G.: De Re Metallica (1556), translated by Hoover H.C. and Hoover, L.H. Mining Magazine Publishing, London (1912)
2. Aufrere, S.H., Johnson, C., Martelli, M., Beretta, M.: Introduction and Chapters 1, 3 and 4. In: Beretta M. (ed.) A Cultural History of Chemistry in Antiquity (2022) (see reference below), entitled Introduction, 1–22; Theory and Concepts: The Mythological Foundation of Chemical Theories and Ancient Civilisations, 23–50; Laboratories and Technology: From Temples to Workshops – Sites of Chemistry in Ancient Civilisations, 85–112; Culture and Science: Gods, Myths and Religion, 113–138
3. Ball, P.: Bright Earth: The Invention of Colour. Penguin/Viking, London (2001)
4. Beretta, M. (ed.): A Cultural History of Chemistry in Antiquity, Volume 1 General editors, Morris P.J.T., Rocke A.J. Bloomsbury Academic Press, London (2022)
5. Brazil, R.: Unwrapping ancient Egyptian chemistry. Chem. World. **19**, 24–29 (2022)

6. Caggiani, M.C., Colomban, P., Valotteau, C., Mangone, A., Cambon, P.: Mobile Raman microscopy analysis of ancient Enamelled glass masterpieces. Anal. Methods. **5**, 4345–4354 (2013)
7. Caggiani, M.C., Barone, G., de Ferri, L., Lavraio, R., Mangone, A., Mazzoleni, P.: Raman and SEM-EDS insights into technological aspects of mediaeval and Roman ceramics from southern Italy. J. Raman Spectrosc. **52**, 186–198 (2021)
8. Clark, R.J.H., Curri, M.L., Laganara, C.: The identification of Lapis Lazuli on mediaeval pottery fragments from the South of Italy. Spectrochim. Acta A. **53**, 597–603 (1997)
9. Colomban, P., Milande, V.: On-site Raman analysis of the earliest known Meissen porcelain and stoneware. J. Raman Spectrosc. **37**, 606–613 (2006)
10. Davy, H.: Some experiments and observations on the colours used in painting by the ancients. Philos. Trans. R. Soc. Lond. **105**, 97–124 (1815)
11. Delamere, F., Guineau, B.: Colour: Making and Using Dyes and Pigments. Thames & Hudson, London (2000)
12. Easthaugh, N., Walsh, V., Chaplin, T., Siddall, R.: Egyptian blue. In: The Pigment Compendium: A Dictionary of Historical Pigments, pp. 147–148. Elsevier/Butterworth Heinemann, Oxford (2004)
13. Gettens, R.J., Stout, G.L.: Painting Materials: A Short Encyclopaedia. Chapman & Hall Ltd/D. Van Nostrand, London/New York (1942)
14. Grissom, C.: Green earth. In: Feller, R.L. (ed.) Artists' Pigments: A Handbook of Their History and Characterisation, Volume 1, pp. 141–168. Oxford University Press, New York/Oxford (1986)
15. Guineau, B.: Pigments et Colorants de l'Antiquite et du Moyen Age (Pigments and Dyes from Antiquity and the Middle Ages). CNRS Editions, Paris (1990)
16. Harley, R.D.: Artists' Pigments c.1600–1835: A Study in English Documentary Sources, 2nd edn. Archetype Publications, London (2001)
17. Jay, W.H., Edwards, H.G.M., Cashion J.D., Hope, C.A.: Blue-painted pottery from the Egyptian new kingdom: a Raman spectroscopy and electron microscopy study of pigment sourcing and production. J. Raman Spectros., to be published (2023)
18. Jun, L., Subramanian, M.A.: Inorganic pigments with transition metal chromophores at trigonal Bipyramidal coordination $YInMnO_3$ blues and beyond. J. Solid State Chem. **272**, 9–20 (2019)
19. Kaczmarczyk, R.: The source of cobalt in ancient Egyptian pigments. In: Olin, J.S., Blackman, M.J. (eds.) Proceedings of the 24th International Archaeometry Symposium, pp. 369–376. Smithsonian Institution Press, Washington, DC (1986)
20. Muhlethaler, B., Thissen, J.: Smalt. In: Roy, A. (ed.) Artists' Pigments: A Handbook of Their History and Characterisation, Volume 2, pp. 113–130. Oxford University Press, New York/Oxford (1993)
21. Plesters, J.: Ultramarine blue: natural and artificial. In: Roy (ed.) Artists' Pigments: A Handbook of Their History and Characterisation, Volume 2, pp. 37–65. Oxford University Press, New York/Oxford (1993)
22. Pliny (Caius Plinius Secundus) 1949–1954: Naturalis Historiae, 10 Volumes, translated by Rackham H., (Vols. 1–5, 9), Jones, W.H.S., (Vols. 6–8), and Eichholz, D.E., (Volume 10). Loeb Classical Library. Harvard University Press/William Heinemann, Cambridge, Massachusetts, MA/London (1949)
23. Riederer, J.: Recently identified Egyptian source pigments. Archaeometry. **16**, 102–109 (1974)
24. Riederer, J.: Egyptian blue. In: Fitzhugh, E.W. (ed.) Artists' Pigments: A Handbook of Their History and Characterisation, Volume 3, pp. 23–46. Oxford University Press, New York/Oxford (1997)
25. Roy, A.: Cobalt blue. In: Berrie, B.W. (ed.) Artists' Pigments: A Handbook of Their History and Characterisation, Volume 4, pp. 150–177. National Gallery of Art/Washington and Archetype Publications, London (2007)
26. Smith, A.E., Mizoguchi, H., Delaney, K., Spaldin, N.A., Sleight, A.W., Subramanian, M.A.: Mn^{3+} in trigonal Bipyramidal coordination: a new blue chromophore. J. Am. Chem. Soc. **131**, 17084–17086 (2009)

27. Smith, A.E., Comstock, M.C., Subramanian, M.A.: Spectral properties of the UV absorbing and near-IR reflecting blue pigment $YIn_{1-x}Mn_xO_3$. Dyes Pigments. **133**, 214–221 (2016)
28. Thenard, L.J.: Sur les couleurs suivis d'un procede pour preparer une couleur aussi belle que l'outremer. Journal des Mines. 15, 128–136 (1803–1804)
29. Theophrastus: De Lapidibus, On Stones Greek Text translated with commentary by Caley, E.R., Richards, J.F.C. Graduate School Monographs (Contributions in Physical Sciences). Ohio State University Press, Columbus (1956)
30. Tite, M.S., Shortland, A.J.: Production technology for copper- and cobalt- blue vitreous materials from the New Kingdom Site of Amarna – a reappraisal. Archaeometry. **45**, 285–312 (2003)
31. Vitruvius: De Architectura Libri Decem, Ten Books on Architecture. Book VII (Finishing), Chapter XII (Blue) Translated by Rowland I.D. University Press, Cambridge (1999)
32. Winter, J., Fitzhugh, E.W.: Pigments based on carbon. In: Berrie, B.W. (ed.) Artists' Pigments: A Handbook of Their History and Characterisation, Volume 4, pp. 1–37. National Gallery of Art: Washington and Archetype Publications, London (2007)
33. Wyatt, J.: Notes on the use of cobalt in later Chinese ceramics. Ars Orientalis. **11**, 63–85 (1979)

Chapter 5
First Cobalt Blue Pottery

Abstract The chapter details studies concerning the first use (~1500 BCE) of cobalt ores for the preparation of a blue decoration and discusses the interpretations since the first studies and clearly illustrates the difficulties and inaccuracies of this type of work. Cobalt-decorated sherds from the 18th Dynasty Nile Valley occupation sites and from the Western Desert Oases together with cobalt-containing alum and kiln slag are examined. The cobalt frit applied to the sherds, often over a slip coating suggests that at least some of the cobalt, nickel and zinc ions have replaced magnesium ions within the diopside structure. Kiln slag would suggest that the kiln had been used for faience or cobalt frit production directly related to production of Egyptian blue. This is strongly indicated by the presence of calcium phosphate (from bone ash) which has been detected in both Egyptian blue and in the cobalt pigment.

Keywords Egyptian blue · Cobalt frit · Kiln slag · Western Desert · Calcium phosphate · Bone ash · Faience · 18th dynasty sherds

5.1 Aim

Egypt demonstrates the first use that has been identified to date, of cobalt ores to produce blue decorations using a synthetic pigment. This chapter is intended to: (i) present the evolution of the understanding of the technique of decoration in blue from the first works in the nineteenth century until today, (ii) characterise the alums considered to have been exploited at Dakhleh (Fig. 5.1) and Kharga Western Desert Oases, Egypt, for cobalt mineralisation; (iii) to examine a slag-like sample from possibly an undiscovered kiln at Mut al-Kharab, the capital city within the Dakhleh Oasis, and which was probably related to cobalt pigment or more likely to faience production; (iv) to discuss the chemistry and mineralogy of the conversion cobalt-containing alums into cobalt-based pigments and of these cobalt-based pigments applied to shards recovered at the Dakhleh Oasis and in the Nile Valley and

H. G. M. Edwards, P. Colomban, *Blue by Fire: A Marker of the Technical History of Glass and Ceramics*, Cultural Heritage Science,
https://doi.org/10.1007/978-3-031-77122-4_5

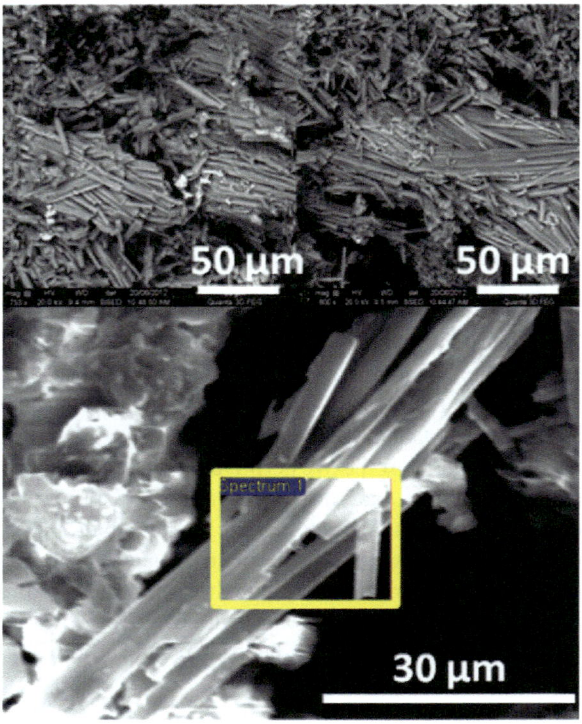

Fig. 5.1 SEM (BSE) images of Dakhleh Oasis alum sample (D1A) at different magnifications

Table 5.1 Periodisation of the Egyptian Dynasties (BCE)

Period	Dynasty	Start	End
Early Dynasty	I–II	3150	2686
Old Kingdom	III–VI	2687	2181
First Intermediate	VII–X	2181	2040
Middle Kingdom	XI–XIII	2130	1649
Second Intermediate	XIV–XVII	1725	1550
New Kingdom	XVIII–XX	1550	1077
Third Intermediate	XXI–XXV	1069	656
Late Period (including Persian)	XXVI–XXX	664	332
Graeco-Roman	–	332	30

manufactured during the 18th Dynasty (Table 5.1) and; (v) and finally to examine the potential for these cobalt-based pigments to have been used as the cobalt raw material in cobalt blue glass production [20, 122].

The Western Desert region, encompassing the Dakhleh and Kharga Oases, Egypt, has previously been discussed in some detail by Churcher and Mills [17] and in the reports contained therein. This region is considered to have provided the source of the cobalt mineral found in these alum deposits (representative of the tertiary deposit type, see Chap. 3).

5.2 Historical and Geological Context

A significant body of research has been devoted to the identification of the blue pigment employed on ceramics found throughout Egypt during the New Kingdom (Table 5.1), from the reign of Thutmose III (1479–1425 BCE) during the 18th Dynasty, with some reduction in the early 19th Dynasty; and following Ramesses II (1279–1213 BCE) there was a significant reduction when manufacture ceased during the reign of Ramesses IV (1153–1147 BCE) in the 20th Dynasty [5, 50, 112].

Attempts to identify the source of raw materials used to prepare blue glass are older. Richmond and Off [100] collected and analysed cobalt-containing, iron-rich alum samples from Egypt, but they did not report the location for these alums. Alums belong to evaporite minerals with the formulation $XAl(SO_4)_2 \cdot n\ H_2O$ where $X = Fe^{2+}$, Cr^{2+}, Co^{2+}, K^+, Na^+, NH_4^+ ions. Subsequently, Beadnell [7] provided analyses of iron-rich alums from Dakhleh but he was unable to determine the presence of cobalt. At Kharga, he observed alums which were often rose-tinted, and he therefore postulated the presence of cobalt.

The source of the cobalt in 18th Dynasty cobalt blue glass was initially proposed by Farnsworth and Ritchie [35] to possibly be the Egyptian alums. Lucas and Harris [73] stated that the only known cobalt compounds found in Egypt were present only as traces within the alums from Dakhleh (Fig. 5.1) and Kharga Oases and from nickel ore located on St. John's Island in the Red Sea. They considered that the extraction of the cobalt from the alums would have presented almost insuperable difficulties. Therefore, they hypothesised that any cobalt ore must have been imported into Egypt from Persia or the Caucasus region.

Noll and Hangst [90] noted however the slag material from (hydrothermal) copper deposits mined in the Eastern Desert contain some cobalt, nickel and zinc. Thus, they intimated that a mixed ore was the more likely source for the cobalt. They also reported that another possible cobalt source would include the Sinai Peninsula and the alum deposits at the Kharga and Dakhleh Oases.

Dayton [23–26] provocatively developed a highly controversial theory that rather than cobalt mineralisation being found and utilised as a pigment in Egypt, he proposed that the cobalt was first obtained as a by-product of silver smelting from ores being mined in Schneeberg, Saxony, and was transported to Egypt and subsequently incorporated into glass. It should be noted however, that the earliest written proof for cobalt extraction from the Saxony deposits at Schneeberg can be dated to the late fifteenth century CE [3] and these deposits contain a number of elements not present in the cobalt–containing glasses or pigments produced in Egypt during the 18th dynasty [20, 43, 44].

Furthermore, Dayton concluded that the appearance of cobalt- and copper-blue faience and glass is firmly dated to c. 1365 BCE and which usually contained significant concentrations of nickel, zinc and manganese. He suggested that it is doubtful that anyone in 1600 BCE could have mastered the chemical reactions by deliberately using pink cobalt bloom (erythrite, $Co_3(AsO_4)_2{\cdot}8H_2O$) to produce a cobalt coloured blue glass [25], noting that erythrite is an arsenical ore and arsenic has never been reported in any cobalt pigment used in Egypt. He then determined that cobalt is not found in Egypt, but rather in the Near East at Anarak in Persia or in the Caucasus at Laurion. And, provocatively, as silver is not found in Egypt, clearly the cobalt was sourced from the classic silver mines of Bohemia, namely, the geologically unique silver, cobalt and nickel ores of Schneeberg. Dayton then stated that it is no coincidence that the cobalt-blue pigment "smalt" was later re-invented at the silver mines of Schneeberg by Pierre Wiedenhammer in 1520 C.E. [69] proposed that there might have been a trade route during the late 18th Dynasty directly or via Syria, from the far side of the Balkans to Egypt.

An XRD analysis of the blue painted pottery from Amarna and from the palace of Amenophis II at Malkata conducted by Riederer [99] concluded that it was decorated using a cobalt aluminate. This is consistent with Dayton [24] who claimed that an XRD analysis indicated that the major phase was a spinel together with a mixture of ZnO, Co_2O_3 and $CoAl_2O_4$. He also reported that the zinc spinel, $ZnAl_2O_4$, gahnite is present, indicative of the blue pigment being a by-product of silver smelting. Indeed, the spinel structure can accommodate many substitutions of the two elements in the reference compound $MgAl_2O_4$. Importantly, Dayton [24] correctly observed that the pottery was decorated after firing and that the blue pigment was applied over a red ochre layer.

Moorey [77], Kaczmarczyk [62], and Nikita and Henderson [87] all disagreed with the Dayton hypothesis. Moorey stating that it is an illusion because he allows preconceived ideas of technological innovation and diffusion to override all other lines of inquiry. Kaczmarczyk [62] also disputed the suggestion from Dayton [24] that during the New Kingdom cobalt was imported into Egypt from the Mycenaean world. Kaczmarczyk noted that Riederer [99], Noll [88, 89], Bachmann et al. [6] all reported elevated aluminium and as a consequence the pigment was labelled as a cobalt aluminate.

Surprisingly, Wood and Hsu [130] have attempted to reinterpret the data provided by Abe et al. [1, 2] by revisiting the Dayton claim that the cobalt ore was obtained from Schneeberg, Saxony, rather than the Western Desert Oases as claimed by Abe and co-researchers. Wood and Hsu postulated that concentrated cobalt glass could alternatively be produced from the by-product or slag from silver extraction or from the cobaltiferous siliceous ores possibly obtained from central Iran. They hypothesised that this source for cobalt provided an explanation for the elevated levels of alumina and lower levels of potash reported in the cobalt blue glasses and, importantly, providing an explanation for the cessation of cobalt exploitation at the end of the Late Bronze Age (LBA). These researchers suggested that the exhaustion of the native silver and siliceous silver ore deposits during the Bronze Age, with argentiferous lead ores becoming the main source of silver, depleted the amount of cobalt available.

Wood and Hsu [130] adopted the Occam's razor principle to suggest that the New Kingdom cobalt may not have derived from the Western Desert. However, Occam's razor is considered to hold in the sense that with competing theories or explanations, then the simplest one, with fewer parameters, is to be preferred. This must surely mean that the Western Desert must be preferred over a cobalt source found in slags at Schneeberg on the far side of the Balkans and from a time when mining is not reported in this district. For additional detail in respect of Occam's razor refer to Edwards [33].

Wood and Hsu [130] noted that the Uluburun shipwreck, Trkiye, analysed by Jackson and Nicholson [57], established that coloured ingots were being transported from Egypt and the decorations on one wall of the Hall of the Annals at the Temple at Karnak in Thebes depicted the spoils from Thutmose III's successful campaigns in the Levant and Syria being transported back to Egypt. They stated that in their opinion, the Uluburun shipwreck indicates a common, but not necessarily an Egyptian origin.

Furthermore, Wood and Hsu [130] correctly agreed with Jackson and Nicholson [56, 85] and Tite and Shortland [123] that the cobalt pigment would have been introduced into the base glass via a frit with the addition of plant ash and quartz. They provided evidence in their Fig. 4 to show blue frit in fritting pans, glass rods and cobalt blue glass objects recovered at Amarna from the 18th Dynasty. However, referring to their Fig. 4, they provide as an example of this frit as sample UC24684 supplied by Weatherhead and Buckley from the Petrie Museum, London. Unfortunately, this sample is not of a cobalt frit but rather, Egyptian blue, thereby casting doubt over the interpretations being made by Wood and Hsu [130]. Sample UC24684 is clearly the copper complex, cuprorivaite, or Egyptian blue. A sample of this specific Egyptian blue "frit" has been analysed [58], its mineralogy has been determined by Raman spectroscopy and its composition by electron microscopy.

The presence of alumina and titanium in the blue glass analysis of Abe et al. [2] is ascribed by Wood and Hsu [130] to have been derived from granite and basalt grinding tools. The proximity of the Ramesside data to a plot of TiO_2 versus Al_2O_3 for basalt could indicate grinding tools of different composition. However, they concluded that the alumina concentration is high and difficult to be explained by grinding tools alone and concluded that it is apparent that whilst cobalt blue glass generally has a higher concentration when compared with glass not coloured by cobalt, contamination by the elements found in igneous rocks is sufficient to explain the presence of alumina.

Thus, they hypothesised that since Mesopotamian glass is less likely to have been derived from the alum deposits in Egypt, the linearity of their plot supports the alumina levels being related to contamination from igneous rocks. They proposed that both Egyptian and Mesopotamian glass would not have come from Egypt's Western desert oases, which are sedimentary deposits with much higher alumina and lower cobalt than found in cobalt blue glass and do not contain titania, when in fact, Shortland et al. reported that samples K1, K3 and K5 from Kharga oasis contained titanium. Jay [58] and Table 5.2 here also reported the presence of titanium in Dakhleh oasis alums.

Table 5.2 Average published analyses (or partial analyses) for alums recovered from Dakhleh and Kharga Oases by various researchers

	Na_2O	MgO	Al_2O_3	SiO_2	P_2O_5	SO_3	Cl	K_2O	CaO	TiO_2	MnO	FeO	CoO	NiO	CuO	ZnO	F
From Ref. [62]																	
Dakhla alum 1		6.1	14			46.9					0.88	0.37	**0.4**	0.33		0.8	
Dakhla alum 2		8.5	10.7			47.7					0.93	0.25	**0.43**	0.58		0.99	
Dakhla alum 3		6.5	16								0.84	0.32	**0.32**	0.18		0.6	
Dakhla alum 4		6	16.1								1.13	0.35	**0.31**	0.16		0.62	
From Ref. [113]												Fe2O3					
K1	0.06	0.3	2.8						0.81	0.01	0.09	24.1	**0.03**	0.04		1.4	
K3	0.01	3.5	9.9						0.01	0.78	0.52	0.4	**1.8**	0.09		2.6	
K5	0.02	0.3	3.3						0.78	0.09	0.17	22.6	**0.05**	0.07		2.2	
K6		4.1	10.9						0.09		0.68	0.3	**1.8**	1.1		3.3	
P3	0.01	2.3	12.4						1.49		2.99	0.3	**0.09**	0.05		1.3	
P4	1.31	3.8	12.1						0.09		0.66	0.2	**0.1**	0.09		1.3	
From Ref. [58]																	
D1A	1.2	8.1	22.9	0.3	0	61.1	0.1	0	0	0.1	2.6	1.2	**0.8**	0.3	0.9	0.7	0
D1A-2	5.3	4.9	14.7	23.6	0	33.4	0	1.1	0.2	0.4	0	12.1	**1.2**	1.6	0.6	1.1	0
w/p Spectrum 1	0.9	12.7	18.5	0.3	0.3	63.6	0	0	0.2	0	2.5	0.1	**0.3**	0.3	0	0.3	0
From Ref. [113]																	
Kharga P1	0.01	15.5	0.3						0.05		2.97	0.11	**0.1**	2.5		15	
Kharga KH6		20.5	0.1						0.03		0	0.08	**Tr**	Tr		Tr	
Kharga 17B	0.22	0.7	15						0.02		0.39	0.05	**Tr**	Tr		Tr	
Kharga 17CA		0.07	17						0.01		0.01	0.05	**Tr**	Tr		Tr	
Kharga 17F		3.5	11.6						0.02		1.82	0.03	**0.03**	0.02		0.04	
Kharga 17FA		3.4	11.5						0.02		1.8	0.04	**0.03**	0.02		0.04	
From Ref. [58]																	
Kharga K18b	0	3.7	24.9	31.2	0.2	31.4	0	0.3	0.2	1.3	1.6	5	**0.2**	<0.1	0	<0.1	0.1

Furthermore, Wood and Hsu [130] proposed that Ramesside glass is often presumed to contain cobalt from a different source, when remelting of 18th Dynasty glass and potential contamination from the furnace or crucible during recycling would increase both the alumina and titania concentration but not the ZnO, NiO, MnO, or CoO concentrations. Thus, Wood and Hsu concluded that more mixing results in more contamination and therefore more variation; this fits well with their narrative that glass was not produced in significant amounts after ca. 1250 BCE, with glass makers reducing the level of cobalt through dilution with a base glass to the point where the colour was still effective. These authors also hypothesised that the ZnO concentrations reported may be explained by remelting, remixing and reworking of the glass. Provocatively, they stated that it is no longer necessary to speculate on why such accessible sources as the Kharga and Dakhleh oases, with their proximity to both Amarna and Malkata and even Dahshur stopped being exploited by around 1250 BCE, the claim being made here is that they had not been used for colouring glass previously and they remain unused. Such a claim predicates that cobalt must have been transported from the Erzgebirge region of Saxony, then the cobalt extraction from this site was lost, to be rediscovered according to Wood and Hsu in 1520 CE. This surely is a fanciful proposal.

XRF analyses used by Abe et al. [1, 2] being a surface determination, correctly detects the surface depletion of sodium (and potassium) by leaching. That the analysis is consistent with the hypothesis that the initial glass is derived from mixing concentrated cobalt in the form of a frit with low soda, low lime and high silica to form a base glass. Thus, they suggested that the initial blue glass was diluted in the 18th Dynasty and further diluted in the Ramesside period. Importantly, they acknowledge that the concentrated cobalt frit may have been made elsewhere. This effectively supports the Tite and Shortland supposition [123], namely, that Egyptian cobalt blue glass contained both natron and plant ash, the former from the frit and the latter from the base glass into which the blue frit was mixed.

Strangely, Wood and Hsu [130] spend a considerable effort in discussing a cobaltiferous rock containing alumina, K_2O, TiO_2 and CaO. They suggested that copper entered the frit as an impurity of cobalt. Thus, there was a glass frit with low soda, low lime and high silica into which cobalt was added and that the frit was a natural by-product of a metallurgical smelting system in which a soda-based mineral acted as a flux to melt cobaltiferous rock in order to remove a desired metal, the cobalt in essence then being a component of the slag of the system.

Whilst Wood and Hsu [130] discussed the five element group of minerals [64] in an attempt to explain the source of the arsenical ores containing cobalt, they dismissed the lack of the presence of arsenic in cobalt decoration on pottery analyses being due to its volatility and to the difficulty to measure low level of arsenic. Whilst arsenic is certainly volatile, it would be expected that under roasting conditions where a reasonable depth of mineral in the crucible would exist, some arsenic may well have survived. Importantly, in discussing cobalt mineralisation from the Western Desert, Egypt, Jay [58] detected arsenic adsorbed on to bone ash in the cobalt frit, or in one instance, he found arsenic being spread over the entire shard surface in an effort to lighten the colour of the cobalt blue.

In conclusion, Wood and Hsu [130], postulated a number of prospective cobalt mineralisation sites, such that copper and lead objects recovered from 18th Dynasty Amarna also appear close in composition to Iranian ores, litharge and objects indicating that cobalt may have made a similar journey to Egypt, alternative to an origin in Schneeberg, Saxony.

For most researchers, following the work of Riederer [99] and subsequent 'serious' studies by Noll and Hangst [90], Bachmann et al. [6], Noll [89], Abe et al. [1, 2], there is a general agreement that the cobalt is most likely present as a cobalt-rich spinel of the general formula $(Co,Mn,Fe,Ni,Zn)Al_2O_4$ and that the source of the colourant was the alum deposits found in the Dakhleh and Kharga Oases of the Western Desert. In studying this pigment, Shortland et al. [112] indicated that there are some tenuous differences between the composition of the pigment as found on shards from Malkata and Amarna, and that this may relate to the specific sources of alum exploited during the reigns of Amenhotep III and Akhenaten to obtain the pigment. In addition, there remains some difference in scholarly opinion in respect of the exact way in which the cobalt blue pigment was produced and this will be further investigated.

Kaczmarczyk [62] was the first to conclusively make the connection between these oasis-derived cobalt-containing alums and their use in pigment and glass manufacture. In this paper he drew a clear connection between a specific suite of transition metals present and their similar presence in Egyptian faience. Rehren [97] supported the Kaczmarczyk concept for the cobalt being obtained from the Western Desert alums.

In dismissing Central Europe as the cobalt source, Kaczmarczyk and Hedges [63] stated that the preferred source of this raw material would be the Dakhleh and Kharga Oases or the Iranian plateau. They provided an analysis for the cobalt-containing alums that give after thermal treatment a mixture with the following typical composition (% wt): 10.6% Al_2O_3, 1.65% Fe_2O_3, 4.25% FeO, 2.55% MnO and 1.0% CoO. They reserved their opinion on these oases as being the source of the cobalt until the presence of NiO and ZnO can be determined. The level of magnesia and alumina present in Egyptian faience was proposed as having been derived from the local sand. Importantly, Kaczmarczyk reported that from the twelfth to the seventh century BCE, cobalt was not employed as a colourant in Egypt and that the cobalt when it reappeared in the seventh century BCE (26th Dynasty) was from a different type of mineral deposit Kaczmarczyk and Hedges [63].

Kaczmarczyk [62] disputed the suggestion of Dayton [24, 25] that during the period of the New Kingdom that cobalt was imported into Egypt from the Mycenean world. Furthermore, Kaczmarczyk reported that it was highly unlikely that another cobalt deposit would be discovered with a similar sequential combination of five transition metals as found in the Dakhleh Oasis alums. Therefore, he considered that it was evident that if the New Kingdom colourants containing cobalt also contain similar transition metal constituents, then the cobalt source was from these cobalt-containing alums. Kaczmarczyk then proposed that if the alums are dissolved in water and the solution is made mildly alkaline using say natron, plant ash or ammonia, then the transition metals would precipitate as their hydroxides (mixed hydroxide precipitation—MHP) leaving the sulphate behind. Drying and heating the MHP precipitate would then yield a blue mixture containing aluminates and oxides with a much higher cobalt content. It should be noted that the distinctive blue colour developing when the MHP is incorporated into a glassy structure forms Co^{2+} ions or that it forms a spinel or an olivine structure.

Abe et al. [1, 2] reported that in Egypt, the cobalt blue pigment derived from the cobalt-containing alums was continually produced until the 20th Dynasty. However, whilst the source of the alum had not been depleted, the use of the cobalt-blue colourant decreased during the 19th Dynasty, eventually disappearing at the end of the New Kingdom. After the end of the New Kingdom, blue colourants did not reappear until the late period (mid-seventh century BCE). And, in agreement with Kaczmarczyk, these researchers reported that the composition of this later cobalt-blue colour was significantly different in mineralogical composition to the earlier cobalt pigment.

5.3 Phase Identification and Transition Metal Content in Alum Samples

Samples (numbered D1-D4, D1A and D1A-2, see an example in Fig. 5.1) were sourced from an artesian well located at Bir el-Hurritiya el-Gubli within the Dakhleh Oasis region and one sample (K18b) from the extensive sampling conducted at the Kharga Oasis in 2002 by Shortland and co-workers in collaboration with the Egyptian Geological Survey [113]. The Dakhleh Oasis samples D1 to D4 were found to be clay/carbonate minerals and as such will not be discussed further. Whilst single samples cannot be considered to be representative of the entire mineralisation at either site, they do provide an indication into the likely composition of these minerals and the probable geographic spread of the cobalt-containing pigments and glasses in Egypt and possibly beyond. Thus, the analytical findings must be considered to be tentative until additional samples can be sourced by either hand Auger or air core (Rotary Air Blast) drilling. The elemental analyses are provided in Table 5.2 and the Raman spectroscopy results are given in Fig. 5.2 and in Table 5.3.

SEM-EDS analyses are provided in Table 5.2. The Raman spectral stacked plots for these alums are given in Fig. 5.2. The assignments of the Raman spectra are provided in Table 5.3. This table also lists the Raman spectra for additional minerals which have been reported to be present in alum analyses conducted by other workers using XRD.

Comparison of the alum samples from Dakhleh and Kharga Oases shows some similarities in their chemical composition as given in Table 5.2. Importantly, Dakhleh sample D1A is high in magnesium and sulphate, low in alumina and almost devoid of silica; whereas, the Kharga sample K18b is high in alumina and sulphate, variable but higher in silica and generally lower in magnesia. The Kharga sample also has a much higher concentration of manganese. This is reflected in its mineralogy where Raman spectroscopy has shown that for the samples examined in which

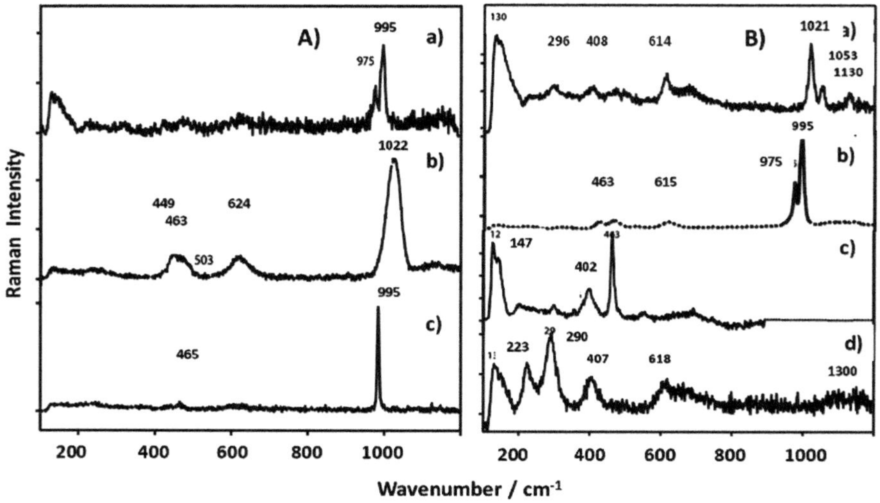

Fig. 5.2 Representative Raman spectra of Dakhleh Oasis alum (λ = 514.5 nm): (**A**) D1A sample; (**B**) K18b sample

alums are present, the transition element-based minerals are related to the trichite family.

In Table 5.2, the elemental analysis marked 'w/p Spectrum 1' provides the analysis of the acicular minerals present in the Dakhleh alum D1A and marked Spectrum 1 in Fig. 5.1. It is high in sulphate, alumina and in magnesia. Furthermore, manganese is concentrated within this mineral as shown by its analysis relative to the manganese analysis in D1A. In this sample, the weight percent for CoO, NiO and ZnO are approximately equal and copper was in the order of approximately 50% of the concentration of each of the other transition metals. However, copper was not reported in w/p Spectrum 1, indicating that the copper present in D1A may not be directly related to the alum itself.

A large sample of the Kharga K18b ore was analysed under low vacuum conditions to preserve hydrated minerals and as shown in Table 5.2, no copper was detected in any of the area or spot analyses conducted. In the report of [113] the presence of copper was not reported. One spot analysis did indicate the presence of a titanium-based mineral such as anatase, brookite, rutile or more likely, ilmenite.

As shown by their respective Raman signatures given in Table 5.3 for samples D1A and K18b, pickeringite is presently favoured over halotrichite or other members of the trichite family. This is supported by the Raman signature at 423, 463, 975 and 995 cm^{-1} in the Kharga sample and at 425, 975 and 995 cm^{-1} for the Dakhleh sample. The presence of apjohnite (a Mn^{2+} trichite member), dietrichite (a Zn^{2+} trichite member), (and the zinc-containing mineral gunningite [$ZnSO_4.H_2O$]), halotrichite (Fe^{2+} trichite member) and wupatkiite (Co trichite member) epsomite (and possibly hexahydrite) are all indicated by their respective Raman signatures. Also

Table 5.3 Raman characteristic peak wavenumbers of minerals identified in Dakhleh and Kharga Oases cobalt-containing alum mineralisation. Raman spectroscopy band assignments are from references [10, 67, 116] and the RRUFF database unless otherwise given in the footnote. Stronger peak in bold (# Ref. [94])

Sample mineral	Formula	Wavenumber/cm^{-1}	Dakhleh Oasis D1A	Kharga Oasis K18b
Kieserite	$MgSO_4.H_2O$	433, 499, **1042/1049**		
Starkeyite	$MgSO_4.4H_2O$	238, 464, 613, **1001**, 1047, 1155		
Pentahydrite	$MgSO_4.5H_2O$	446, 597, **1005**		
Hexahydrite	$MgSO_4.6H_2O$	485, 617, **998**		
Epsomite	$MgSO_4.7H_2O$	227, 445, 467, 610, **983**, 1087, 1149	465, **984**	
Apjohnite	$Mn^{2+}Al_2(SO_4)_4.22H_2O$	209, 309, 420, 454, 612, **974, 991**	425, 612, **975, 995**	423, 463, **975, 995**
Boussingaultite	$(NH_4)_2Mg(SO_4)_2.6H_2O$	451, 622, 829, **983**		
Dietrichite	$ZnAl_2(SO_4)_4.22H_2O$	321, 416, 465, 618, **973, 990**	320, 617, **975, 995**	463, 619, **975, 995**
Halotrichite	$Fe^{2+}Al_2(SO_4)_4.22H_2O$	311, 421, 465, 613, **974, 994**	425, 617, **975, 995**	423, 463, 619, **975, 995**
Pickeringite	$MgAl_2(SO_4)_4.22H_2O$	425, 460, 609, **975, 993**	425, **975, 995**	423, 463, **975, 995**
Wupatkiite	$CoAl_2(SO_4)_4.22H_2O$	368, 428, 471, 619, **974, 994**, 1074	425, 472, 617, **975, 995**	423, **975, 995**
Bieberite	$CoSO_4.7H_2O$	211, 275, 423, 504, 619, **984, 1025**	425, 617, **984**	
blödite	$Na_2Mg(SO_4)_2.4H_2O$	**992**		
Geikietite	$MgTiO_3$	**209, 267, 334**, 386, 471, **699**		
Ilmenite	$FeTiO_3$	223, 368, **678**		
Rutile	TiO_2	233, 445, **608**		
Brookite	TiO_2	**151,** 316, 407, 639		
Wairakite	$Ca(Si_4Al_2)O_{12}.2H_2O$	**131, 146**, 326, 440, 912	**129, 147**	**130, 148**
Muscovite	$KAl_2(Si_3Al)O_{10}(OH)_2$	**263**, 420, **703**		
Dwornikite	$NiSO_4.H_2O$	**251,** 307, 353, 373, 447, **526**		

(continued)

Table 5.3 (continued)

Sample mineral	Formula	Wavenumber/cm^{-1}	Dakhleh Oasis D1A	Kharga Oasis K18b
Nickelhexahydrite	$NiSO_4.6H_2O$	**229, 377,** 462, 524, 682**, 984,** 1043		
Retgersite	$NiSO_4.6H_2O$	208, 369, 465, 611, **986**		
Gunningite	$ZnSO_4.H_2O$	219, 275, 428, 503, 627, **1029**, 1083	503, 624, **1022**	
Gypsum#	$CaSO_4.2H_2O$	420, 491, 623, **1008**, 1141		
Bassanite#	$CaSO_4.0.5H_2O$	421, 490, 630, 680, **1014**, 1152, 1174		
γ-Anhydrite#	$CaSO_4$	424, 490, 628, 668, **1026**, 1174		
α-Anhydrite#	$CaSO_4$	490, 670**, 1017**		
Quartz	SiO_2	128, 205**, 463**		129, 402, **463**
Burkite	$Na_4(SO_4)(CO_3)$	155, 452, 620, **993, 1063**		
Alunogen	$Al_2(SO_4)_3(H_2O)_{12}.5H_2O$	300, 470, 607, **992**		
Natroalunite	$NaAl_3(SO_4)_2(OH)_6$	**207**, 380, 458, 652, **1024**, 1048		**1021**, 1053
Tamarugite	$NaAl(SO_4)_2.6H_2O$	460, 601, 624, **991,** 1066		
Natrojarosite	$NaFe^{3+}{}_3(SO_4)_2(OH)_6$	**138, 224, 442,** 561, 622, **1010, 1108**		
Ferricopiapite	$Fe^{3+}{}_{0.67}Fe^{3+}{}_4(SO_4)_6(OH)_2.20H_2O$	193, 268, 301, 451, **479**, 598, **993, 1019,** 1147		
Coquimbite	$Fe^{3+}{}_2(SO_4)_3.9H_2O$	211, 284, 505, **1025**, 1202		
Hematite	Fe_2O_3	226, **293, 411**, 613		223, **291, 407,** 618

present in both D1A and K18b is the mineral wairakite [$Ca(Si_4Al_2)O_{12}.2H_2O$] shown by the strong Raman bands at 131 and 146 cm^{-1}. As shown in Table 5.3, the Kharga Oasis sample K18b also contained natroalunite [$NaAl_3(SO_4)_2(OH)_6$], quartz [SiO_2], and hematite [Fe_2O_3].

As shown in Table 5.2, the transition metal concentrations of Dakhleh D1A and Kharga K18b (Mn, Co, Ni, Zn) specimens are generally of the same order of magnitude as those provided by [62] (in Table 34.3) and reported in Table 5.2. However, in the present study, copper was detected in all Dakhleh Oasis samples in which these transition metals were also present. Riederer [99] also reported traces of both copper and beryllium; beryllium cannot be detected by the electron microscope used in the data used presented as illustration.

The variability in composition of these samples suggests that additional minerals may be detected if in the future a larger suite of minerals becomes available. Furthermore, long-term partial ageing by hydrolysis of these samples could have modified the alum samples. Indeed, the subterranean waters feeding the mineral springs at Dakhleh and Kharga are anaerobic, acidic and rich in both hydrogen sulphide and carbon dioxide [18]. Furthermore, Clarke postulated that based on the carbon dioxide, iron, and calcium carbonate saturation values that the Dakhleh Oasis waters are more corrosive than those located at Kharga. Thus, these acidic and corrosive waters have the ability to dissolve iron and other components and re-precipitate them at the air/water table interface.

These alums have previously been described by Segnit and Shortland et al. [107, 112, 113]. Bachmann et al. [6] reported that the alums were based on potash as alunite ($KAl_2(SO_4)_2(OH)_6$), which when heated to 800 °C would form alumina. This would then react with the cobalt oxides to form the spinel phase ($CoO.Al_2O_3$). However, as shown in Table 5.2, potassium was not present in the alum. Furthermore, as shown by [131] aluminium sulphate decomposes at 770 ° C whilst magnesium sulphate decomposes at 1124 ° C. Magnesium sulphate will react with silica, particularly in the presence of an alkali flux at approximately 1000 °C. Hope et al. [54] reported that the Dakhleh Oasis alums contained alunogen, together with apjohnite ($Mn^{2+}Al_2(SO_4)_4.22H_2O$), wupatkiite ($CoAl_2(SO_4)_4.22H_2O$) and dietrichite ($ZnAl_2(SO_4)_4.22H_2O$). The presence of these minerals is confirmed by Raman spectroscopy as indicated in Table 5.3. In their report, Shortland et al. [113] indicated that a field study conducted at the Dakhleh Oasis in September 2002 failed to detect cobalt in any of the samples analysed! Shortland et al. [113] then re-analysed the original samples obtained by Kaczmarczyk and by Picon group. They proposed that from the XRD analyses they conducted on 6 samples from the Dakhleh Oasis that, together with pickeringite, the samples variously contained coquimbite (aluminocopiapite), ferricopiapite, alunogen, starkeyite, tamarugite and gypsum [116]. Segnit [107] analysed a separate suite of samples from the Dakhleh Oasis and reported that an XRD analysis showed the presence of coquimbite (which would undergo gradual hydrolysis) together with copiapite, natrojarosite, pickeringite, hexahydrite and epsomite. As previously noted, these researchers did not detect copper other than at trace concentrations. In examining six samples from Kharga Oasis using XRD Shortland et al. [113] reported that the principal minerals determined were pickeringite, alunogen ($Al_2(SO_4)_3(H_2O)_{12}.5H_2O$), halotrichite ($Fe^{2+}Al_2(SO_4)_4.22H_2O$) and pentahydrite (kieserite). The principal Raman signatures for these minerals are given in Table 5.3; the Kharga sample 18b does not support the presence of all of these minerals, but rather a different suite of minerals is suggested to be present in this sample. Traces of cobalt were detected in Kharga Oasis sample K18b.

Zinc is present in similar or greater concentrations than cobalt in the Dakhleh and Kharga Oases alums. It has been identified as dietrichite and as gunningite ($ZnSO_4$) at Dakhleh, but not at Kharga. Zinc oxide has been widely used as a white pigment for millennia. Cobalt when doped in oxide matrices will produce various colours, particularly blue, but also green [9, 28, 37, 68]. As demonstrated by Florés et al. [37], the presence of very low concentrations of Co substituting for zinc in a willemite structure provides a lighter shade of blue. They indicated that the vibrations of the ZnO_4 group decrease, confirming the preferential Co-occupation of the tetrahedral sites.

In the Dakhleh Oasis sample D1A, other than pickeringite, epsomite (and possibly hexahydrite) and the zinc-containing mineral gunningite [$ZnSO_4.H_2O$], Raman spectroscopy failed to identify the presence of any other minerals in this sample whereas, as shown in Table 5.3, the Kharga Oasis sample K18b also contained natroalunite [$NaAl_3(SO_4)_2(OH)_6$], anhydrite [$CaSO_4$], quartz [SiO_2], and hematite [Fe_2O_3].

5.4 Phase Identification and Transition Metal Content in Blue Decoration

A range of blue painted pottery particularly from Amarna, Malkata, and North Karnak, with a more limited number of samples from Memphis, Deir al-Medineh, North Sinai, Dendera and Mut al-Kharab are presented in Figs. 5.3 and 5.4. Their elemental analyses are provided in Tables 5.4a and 5.4b and the Raman spectra of the blue decoration are given in Fig. 5.5 (Makata and Armana samples) and Fig. 5.6 (Karnak and other samples) and listed in Table 5.5. A number of these samples are surface samples and despite the abrasion of a significant proportion of their cobalt decoration, their Raman signatures were obtained by interrogating specific crystals and their mineralogy is given in Table 5.5. Specific samples suggestive of their being slag were analysed by Raman spectroscopy (Fig. 5.7), Mossbauer spectroscopy (Fig. 5.8, Table 5.6) and SEM-EDS (Table 5.8, Fig. 5.9). Elemental mineralisation maps were developed for a number of the shards using the XFM synchroton beamline and are provided in Figs. 5.10 and 5.11.

Pottery kilns have been excavated at a number of sites throughout the Dakhleh Oasis region. Several kilns have been located in close proximity to each other, potentially suggesting the existence of one or more high temperature workshops [32]. At site 31/405-F9–1 a large Roman period kiln has been excavated and based upon the clinker recovered, it would appear to have been employed for the burning of limestone, the lime then being used as wall plaster.

Fig. 5.3 Malkata and Amarna shard samples (Nile silt with cream burnished surface; MA and MC, Amarna 2 and 3: white slip and beeswax).

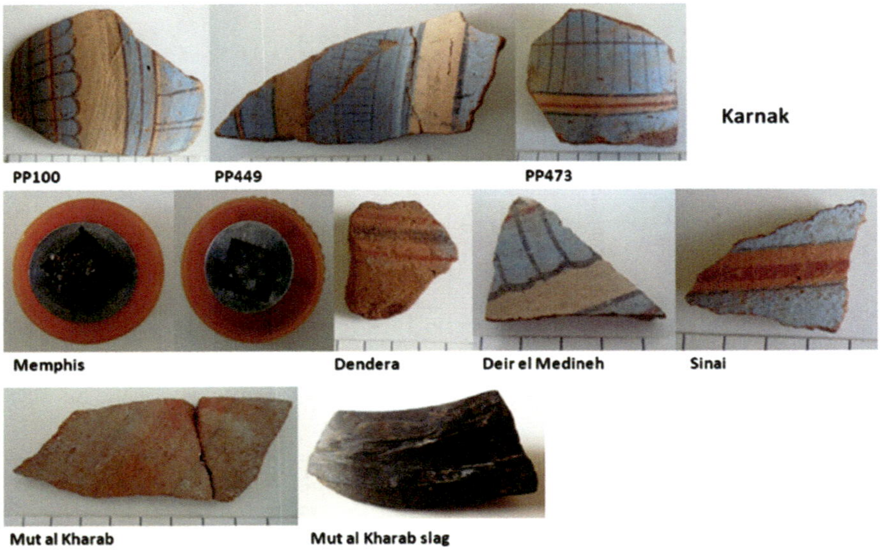

Fig. 5.4 Karnak, Dendera, Mut al Kharab and other shard samples (Nile silt with cream burnished surface; MA and MC, Amarna 2 and 3: white slip and beeswax)

Table 5.4a Average analyses of blue pigment applied to Egyptian pottery from various locations and sites (after [58])

	Na_2O	MgO	Al_2O_3	SiO_2	P_2O_5	SO_3	Cl	K_2O	CaO	TiO2	MnO	FeO	CoO	NiO	CuO	ZnO	PbO	As_2O_3	F
Malkata	1.8	4.3	29.5	21.4	0.8	7.8	1.4	1.4	15	0.6	0.9	5.1	2.8	1.7	1	3.8	0.3	0	0
Amarna	2.4	3.6	30.4	21.8	1.1	7.8	2.5	1.7	15.5	0.6	0.6	6	1.5	1	1.2	2.3	0	0	0
Karnak	1.2	4	33.9	25.8	0.4	4.5	Tr	2.1	5.6	0.9	1.3	13.9	1.8	1	1.3	1.6	0	0	0
Memphis	1.8	5.4	37.3	30	2.8	0.4	0	1.5	8.3	0.5	1.2	4.5	1.6	1.4	0.7	3.4	0	0	0
Deir el Medineh	1.8	2.6	42.3	18.8	0.9	6.1	0	1	12	0	0.7	3.5	2.7	2.4	0	5.2	0	0	0
Sinai	0.4	5.3	39.4	30.1	0.2	6.3	0.4	2.5	3.9	0.6	0.8	5	1.5	1.1	0.7	1.7	0	0	0
MUT	1.9	4.3	27.4	36.6	1.3	9.3	0.2	2.4	8.1	0.7	0.2	6.2	0.5	0.1	0.6	0.3	0	Tr	0
MUTB_3A1	1.8	2.2	17.9	9.7	20.8	9.0	0.5	1.0	31.5	0.4	0.3	2.6	0.3	0.0	0.7	0.0	0.0	0.0	1.2

MUT Hydroxy apatite (bone ash) particle

Table 5.4b Analyses of cobalt glass

	Na_2O	MgO	Al_2O_3	SiO_2	P_2O5	SO3	Cl	K2O	CaO	TiO2	MnO	FeO	CoO	NiO	CuO	ZnO	As2O3	F
Malkata glass	20.7	3.98	1.86	63.2				1.57	7.69			0.92						
Amarna glass	18.6	4.08	2.09	64.8				1.1	8.53			0.81						

From Ref. [70] (p.41, their Table 3)

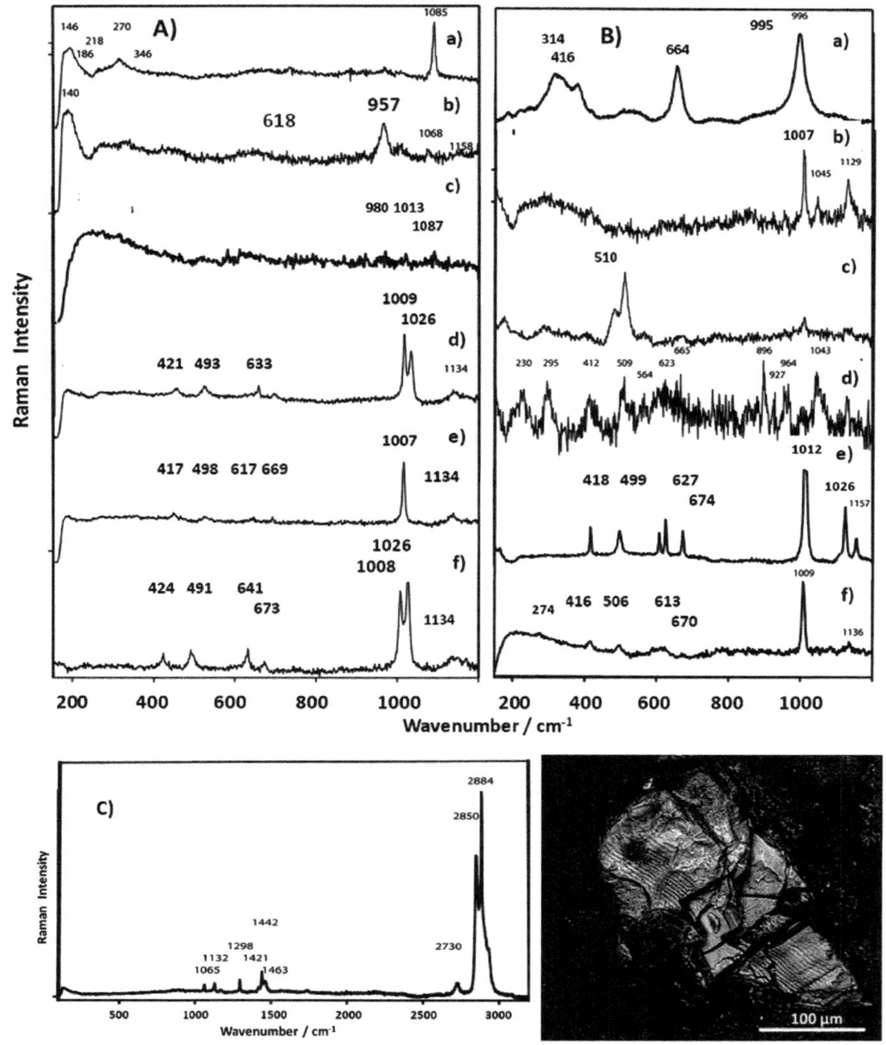

Fig. 5.5 Representative Raman spectra of the blue pigment applied to Malkata samples (**A**: **a**–**c**) Bachmann, ($\lambda = 413.2$ nm), (**d**) MA, (**e**) MC and (**f**) G3989 ($\lambda = 514.5$ nm)) and Amarna samples (**B**: **a**, **b**) Amarna 1, (**c**) Amarna 2, (**d**) Amarna 4, (**e**) Amarna 362, and (**f**) Amarna 423). (**C**), left: the Raman spectrum recorded from a Malkata sample: this spectrum precisely matches the beeswax spectrum provided by Burgio and Clark [13]; right: Particle of degraded hydroxyapatite (bone ash) in MUT cobalt blue pigment (all data with $\lambda = 514.5$ nm). (Courtesy Jay)

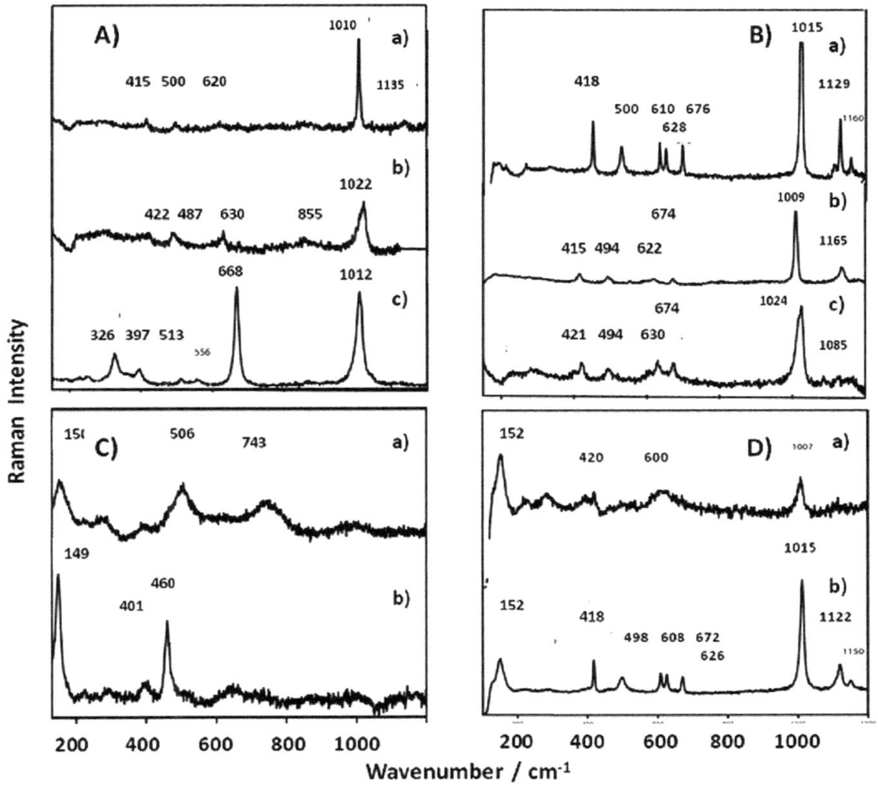

Fig. 5.6 Representative Raman stackplots of Karnak (**A**: **a**) and (**b**) PP 100, (**c**) PP473), Memphis (**B**: **a**) RAT 10814812, (**b**) RAT 1054199), Dendera, (**C**, **a**), Sinai BEA (**C**, **b**), Deir al Medineh (**C**, **d**) and Mut al Kharab shard (**D**): (**a**) slip coating, (**b**) blue pigment) ($\lambda = 514.5$ nm)

5.4.1 Mut al Kharab Slag: Mössbauer Spectroscopy

An image of a stalactite-like sample showing a conchoidal fracture is provided in Fig. 5.4. The sample was recovered in an excavation below the floor of the probable 25th Dynasty (c. 747–656 BCE) temple erected to Seth ("Lord of the Oasis") at Mut al Kharab. The visual appearance would suggest that this is slag which was removed from the internal walls of a kiln. A number of visually similar lumps of slag were excavated and remain at the site awaiting approval for their removal. However, to date, no kiln has been located and excavated within the immediate location of this sample. This sample is presently tentatively dated to an 18th Dynasty production. It was recovered from a highly disturbed site and therefore this tentative dating may well be revised in the light of further studies.

Table 5.5 Raman spectral tentative assignment for the cobalt-based pigments used to decorate shards from various locations and Dynasties (after Ref. [58, 59])

		Wavenumber/cm^{-1}			
Mineral	Formula	Malkata	Amarna	Karnak	Memphis
Quartz	SiO_2	466	469	464	464
Tridymite	SiO_2	293, 310, 354			
Apatite. calcium phosphate	$Ca_5(PO_4)_3(OH,F,Cl)/Ca_3(PO_4)_2$	960, 1059	924, 966		
Diopside	$Ca(Mg,\square)Si_2O_6$	324, 395, ~670, 1008, 1026	~670, 1009,	668, 1011	667, 1006
Wollastonite	$CaSiO_3$		963, 1045		401, 506, 973
Anhydrite	$CaSO_4$	420, 503, 1016, 1129	416, 1015, 1127	502, 609, 1015	1017
Bassanite					
Gypsum	$CaSO_4.2H_2O$	491, 673, 1134, 1007, 1143	414, 496, 626, 1009, 1131		
Aragonite	$CaCO_3$	278, 713, 1085			
Calcite	$CaCO_3$	1087			
Coquimbite	$Fe^{3+}{}_2(SO)_3.9H_2O$			291, 496, 1025	281, 506, 1028
Gunningite	$ZnSO_4.H_2O$				506, 621, 1028
Albite	$Na(Si_3Al)O_8$	490, 510			
Plagioclase feldspar	$Na_xCa_{1-x}Al_{1+x}Si_{3-x}O_8$		176, 287, 481, 510		
Omphasite	$(Ca,Na)(Mg,Fe^{2+}Fe^{3+},Al(Si_2O_6)$	504, 905, 1015			
Gunningite	$ZnSO_4.H_2O$				506, 621, 1028
	$Zn_{1-x}Co_xO$ (x = 0.2)	490, 628	627, 717, 992		
Heterogenite	$Co^{3+}O(OH)$	496, 621	495, 624	502, 628	506, 621
Hypersthene	$(Mg,Fe^{2+})_2Si_2O_6$				392, 661, 871, 936, 1003
Spinel	$(Mg,Zn,Mn,Fe)Al_2O_4$	~630	~630	~630	287, 507, 643
Cobalt aluminate	$CoAl_2O_4$	200, 512			
Alunogen	$Al_2(SO_4)_3(H_2O)_{12}.5H_2O$		610, 996	992	

(continued)

Table 5.5 (continued)

Mineral	Formula	Wavenumber/cm^{-1}			
		Malkata	Amarna	Karnak	Memphis
Aluminite	$Al_2(SO_4)(OH)_4.7H_2O$			992	
Mooreite	$Mg_{15}(SO_4)_2(OH)_{26}.8H_2O$		129, 457, 982		
Anatase	TiO_2		143, 399, 514		
Hematite	α-Fe_2O_3		222, 292, 412, 609	225, 294, 407, 502	
Mullite	$3Al_2O_32SiO_2$		610, 1127, 1161		
Number of samples analysed		9	8	3	2

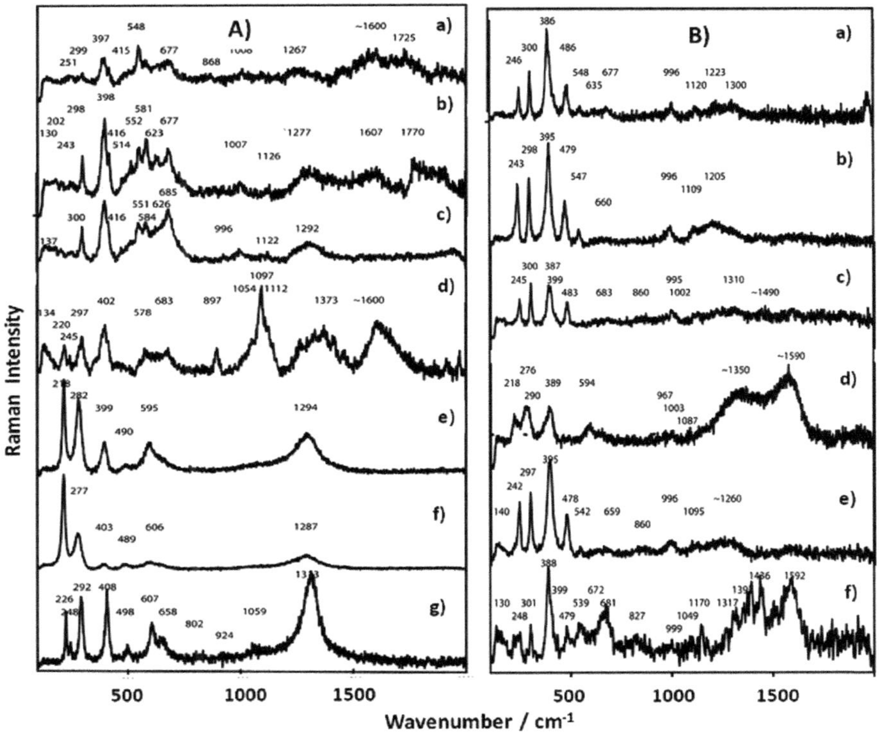

Fig. 5.7 Raman stackplots of Mut slag core (**A**) and outer surface (**B**) ($\lambda = 514.5$ nm)

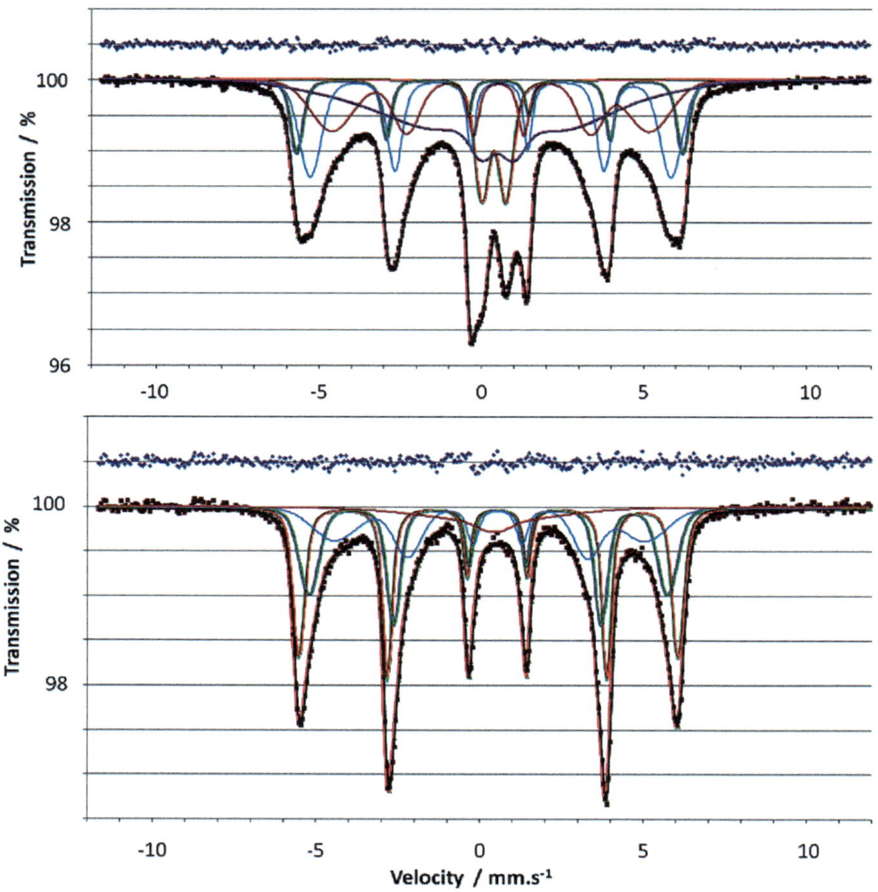

Fig. 5.8 ^{57}Fe Mössbauer spectrum of powder scraped from the core (**A**, top) and mixed powder and needles scraped from the outer surface (**B**, bottom). It has been fitted using a Voigt profile to four magnetically ordered sextets. The blue line at the top is the difference between the data and the fit. (Courtesy of Prof. J.D. Cashion, Monash University)

The term "slag" applied to this sample is in accordance with the discussion provided by Nicholson [82, p. 72]. If the stalactite-like runs illustrated and photographed in the internal structure of kiln HAC3 [82, p. 64] were to be broken away, their visual appearance would generally relate to the present sample. Nicholson [82] noted on page 78 that the slag recovered from kiln HAC3 at Kom Helul has not resulted from either metallurgical processes nor derived from faience glaze raw materials. The present sample from Mut al Kharab may indicate a slightly different kiln usage and therefore a different slag formation. It does show the presence of transition elements, including cobalt, nickel and zinc and both magnesium and

Table 5.6 Fitted parameters of the Mössbauer spectra in Fig. 5.8 together with the literature values for goethite and hematite

Sample	IS (δ)	QS (ε or Δ)	HFF (B)	σ(B)	A_2/A_3	Area
	mm/s	mm/s	T	T		%
Core	0.38(1)	−0.13(1)	36.9(1)	0.7	2*	10.2
	0.39(1)	−0.14(1)	34.5(4)	1.7	2*	11.1
	0.39(1)	−0.12(1)	30.3(10)	3.7	2*	23.8
	0.37(4)	−0.10(4)	18.8(11)	12.8	2*	22.8
	0.38(1)	0.74(1)	n/a	n/a	n/a	32.0
Surface	0.39(1)	−0.14(1)	36.0(1)	0.7	2.86(9)	36.0
	0.39(1)	−0.14(1)	34.0(8)	1.5	2.93(13)	32.0
	0.40(1)	−0.12(2)	29.5(13)	3.9	2.87(20)	24.7
	0.39*	−0.13(25)	10.1(42)	12	2*	8.1
Goethite	0.37	−0.13	38.0	n/a	2*	100.0
Hematite	0.37	−0.10	51.8	n/a	2*	100.0

IS = isomer shift, QS = quadrupole splitting, ε = value of the QS associated with a sextet, Δ = QS of a doublet, HFF = magnetic hyperfine field, σ(B) = the standard deviation of the hyperfine field distribution, A_2/A_3 = ratio of lines 2 and 3 in the sextet, * indicates that the value was fixed in the fitting

Fig. 5.9 Mut Image of high iron particle within the core of the slag sample

alumina. However, within this report, alternative possible uses or requirements for this undiscovered kiln are postulated.

Figure 5.7A shows the Raman spectra obtained from the conchoidal fractured surface (core), and Fig. 5.7B the Raman spectra obtained from the external surfaces. The elemental analysis is given in Table 5.4a and the Raman spectral interpretations are provided in Table 5.8. Figure 5.8 shows the Mössbauer spectra of the iron phases

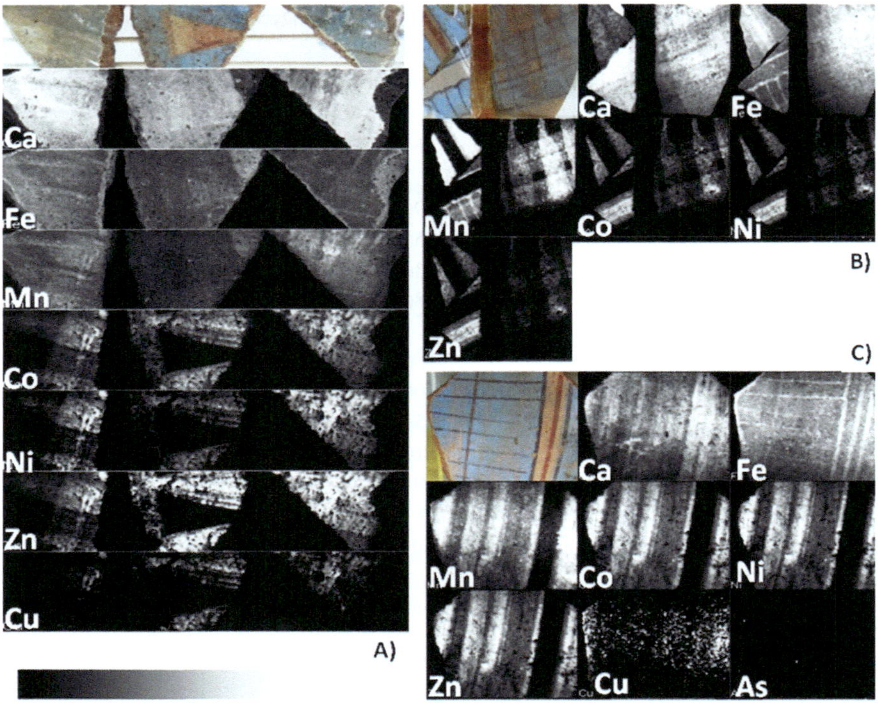

Fig. 5.10 Synchrotron elemental maps generated using the XFM beamline for shards: (**A**) (Amarna A2, Malkata M1 and Malkata M2), (**B**) (North Sinai (NS 1, upper left), Deir al Medineh (DM 1, lower left), Amarna A4 (right)) and (**C**) for North Karnak shard NK3

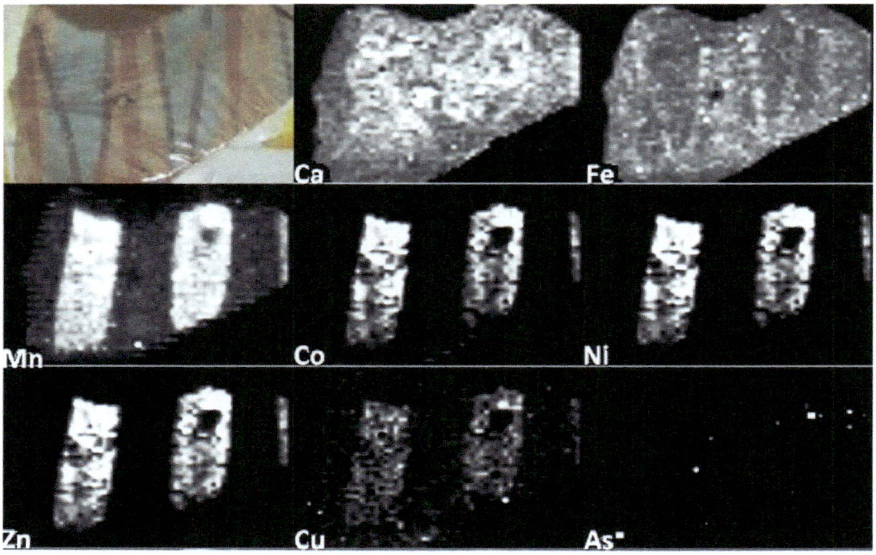

Fig. 5.11 Synchrotron elemental maps generated using the XFM beamline for Malkata shard M4

Table 5.7 SEM-EDS analyses of the slag sample marked MUT 31/405 G10 1/6 (104) S2002.58 10.1.02 at the external surface and at the conchoidal fractured surface core

External Surface

	Na_2O	MgO	Al_2O_3	SiO_2	P_2O_5	SO_3	Cl	K_2O	CaO	TiO_2	MnO	FeO	CoO	NiO	CuO	ZnO	Cr_2O_3	Sb_2O_5
G101_A	0.8	5.6	20.7	36.2	1.7	3.0	0.6	1.0	20.6	0.6	0.7	7.4	**0.5**	0.3	0.5	0.0	0.0	0.0
G101_A1	1.0	4.5	30.0	16.7	1.4	1.8	1.0	0.8	24.6	0.6	2.0	6.2	**3.1**	2.2	1.2	2.8	0.0	0.0
G101_S2	10.0	3.9	25.8	7.1	1.3	2.2	21.2	0.7	8.3	0.7	2.2	4.3	**3.8**	3.1	1.5	3.7	0.0	0.0
G101_S4	1.8	8.6	48.8	12.6	2.1	1.6	1.2	0.6	8.0	0.0	2.0	3.9	**3.2**	2.4	0.0	3.1	0.0	0.0
G101_S5	26.0	3.6	20.2	7.2	1.1	1.9	27.2	0.5	5.2	0.0	0.8	1.7	**1.3**	1.1	0.9	1.4	0.0	0.0
G101_S6	1.2	9.7	19.9	34.0	0.3	4.3	0.5	4.8	6.9	1.6	0.7	14.3	**0.6**	0.7	0.0	0.0	0.3	0.0
G101_S7	0.9	5.5	24.6	32.7	2.1	1.7	0.5	0.9	18.9	0.7	1.1	6.6	**1.3**	1.2	1.2	0.0	0.2	0.0

core

	Na_2O	MgO	Al_2O_3	SiO_2	P_2O_5	SO_3	Cl	K_2O	CaO	TiO_2	MnO	FeO	CoO	NiO	CuO	ZnO	Cr_2O_3	BaO	PbO
Slag 1	0.2	0.5	30.8	39.7	0.2	0.3	0.0	2.9	0.0	0.5	0.1	12.6	**0.0**	0.1	0.7	0.3	0.0	0.0	11.2
Slag 2	1.4	0.7	23.4	45.9	1.3	5.7	0.0	3.0	1.0	0.3	0.0	7.1	**0.0**	0.4	1.4	0.3	0.1	0.0	8.0
Slag 3	0.8	0.0	0.5	2.9	0.0	0.4	0.3	0.2	0.4	0.2	5.2	87.5	**1.3**	0.0	0.3	0.0	0.0	0.0	0.0
Slag 4_rod	2.4	0.7	0.5	1.4	0.7	0.0	0.3	0.0	0.0	0.0	1.6	92.4	**0.0**	0.0	0.0	0.0	0.0	0.0	0.0
Slag 5	0.5	0.6	1.4	4.2	0.0	0.5	0.3	0.3	0.5	0.4	4.7	86.6	**0.0**	0.0	0.0	0.0	0.0	0.0	0.0
Slag 6	0.3	0.0	0.0	0.2	0.0	33.1	0.0	0.0	0.0	0.0	0.0	0.5	**0.0**	0.0	0.6	0.0	0.0	65.4	8.4

present in, respectively, the core at the broken surface and at the external surface of the slag sample as described below.

Samples for ^{57}Fe Mössbauer spectroscopy were prepared by scraping material from the curved outer surface of the slag (the top surface in Fig. 5.4 and from the fractured surface (left hand end in Fig. 5.4 with a diamond tipped tool. The material from the curved surface came away both as powder and in the form of long needles, some up to 5 mm in length, which had formed in the flutes. Scraping the fractured surface produced fine powder as expected. Both samples were made up as absorbers at approximately 20 mg/cm^2 and mixed with boron nitride to ensure a uniform distribution. In contrast to usual practice, the material from the outer surface was not crushed, with the powder and needles used together. This allowed for the needles of various lengths to lie approximately flat in the absorber holder.

The room temperature Mossbauer spectrum of the core is shown in Fig. 5.8A (top). The majority of the spectrum shows strong magnetic ordering and this part has been fitted to four sextets using Voigtian lineshapes. The parameters are given in Table 5.6. From the similarity of the isomer shift and quadrupole splitting, we can conclude that they are all due to the same species with the parameters of the best formed one being very close to that of goethite, α-FeOOH, the literature values of which are also given in Table 5.6, together with the only other possibility, hematite. Natural mineral samples of goethite very rarely have hyperfine fields of 38 T due to poor crystallinity or the presence of impurities such as Al^{3+}, other transition elements, vacancies or embodied OH^- or H_2O. These all have the effect of lowering both the Néel temperature from its usually accepted value of 400 K and the saturation hyperfine field. Consequently the spectra are commonly fitted with a distribution of hyperfine fields to account for the range of local environments of the iron ions [126]. This provides rational and reproducible fits over a range of temperatures as reviewed, for example, by Murad and Cashion [79]. In the present case, we have sought to minimize the number of parameters and have used four sextets to approximate this distribution to sufficient accuracy. The largest fitted hyperfine field in this spectrum is 97% of the maximum literature value, attesting to the purity of this best component of the iron phase.

There is one additional component in Fig. 5.8A which is a doublet at ~1 mm/s due to a paramagnetic ferric species. The quadrupole splitting of 0.74 mm/s is higher than that usually recorded for superparamagnetic goethite of approximately 0.5 mm/s, although the isomer shift value is correct. Consequently, it is likely that the Fe^{3+} is in an aluminosilicate or other mineral approaching that of a high iron-containing oxide whose structure may possibly have been changed by firing in the kiln.

The room temperature spectrum of the scrapings from the outer surface is shown in Fig. 5.8B. It has been fitted to four magnetic sextets as for Fig. 5.8A, and is similarly an approximation to a relatively continuous distribution of hyperfine fields. The outer field of 36 T is again a remarkably high value for goethite which has been exposed to the atmosphere. It is noticeable that there is no paramagnetic doublet present in this spectrum, suggesting that the material has been recrystallized on the outer surface in a process which was not available to the mineral responsible for the doublet. This conjecture is supported by the unusual intensity of the lines in the sextets. In a powder sample, the six lines in a magnetic sextet normally have the relative intensities 3:2:1:1:2:3 and this constraint was used in fitting Fig. 5.8A. However, in a single crystal, these intensities change to 3:4:1:1:4:3 if viewed parallel to the magnetization direction, or 3:0:1:1:0:3 if viewed perpendicular to the magnetization direction. The suspicion that the needles which flaked out of the flutes were single crystals is confirmed by the intensities of the lines in Fig. 5.8B which are, on average, 3:2.9:1:1:2.9:3, a conclusive result given that no attempt was made to separate the needles from the powder, nor to orient the needles. Thus the outer surface has been modified by some process of dissolution and recrystallization during burial since the original firing as it is most unlikely that single crystals of many mm in length would be formed during the firing, even for an extended firing. The density of the sample was measured to be 3.3 g/cm^3. It is noted that goethite has a density variously quoted in the range 3.4–4.4 g/cm^3 while diopside has a density of 3.39 g/cm^3 and wollastonite 2.8 g/cm^3 [4].

The sample can be described as being high in alumina (clay), silica and iron and containing variable concentrations of magnesium together with low concentrations of transition elements. The variably in its high iron concentration is indicated in Table 5.7 and the Raman spectra provided in Fig. 5.7A and B indicate the presence of goethite (α-FeOOH) and the Raman spectrum shown in Fig. 5.7g supports the presence of hematite. The surface also shows the Raman signature for diopside with bands at 667–675 cm^{-1} and at 1002–1008 cm^{-1}. These bands are slightly elevated from that of diopside itself due to the partial replacement of magnesium by transition elements. As shown in Fig. 5.7A, the core of the sample contains a particle of what would relate to bloomery iron; the elemental analysis of which provided an analysis of 92.4% FeO. Within the sample are several particles of baryte, the analysis of which is provided in Table 5.7. The external surface of the slag contained the transition elements Mn, Co, Ni, Zn in approximate equal proportions. Variable concentrations of these transition elements exist within the slag particle. Lead oxide at an approximate concentration of 8–11% was found to be present in two core sample analyses. Tin was not detected. The core contained only minor calcium; however, the surface was variably high in CaO.

This siliceous lump of slag would have almost certainly been deliberately broken away from the internal surface of a kiln walls when cold. The kiln would have been operated under highly reducing conditions and at a sufficiently high temperature to volatilise various elements which have then condensed on the cooler internal wall surfaces of the kiln. As noted above, the biomass fuel used would have contributed significantly to slag formation. As shown in Fig. 5.9, some of the iron oxide has been reduced to metal (see the needle ##3), which indicates the redox firing conditions present in the kiln. Almost certainly, the aggressive siliceous slag has attacked the iron oxides and alumina present in the clay bricks forming the kiln walls. This would account for the high silica and alumina recorded in the slag. And, the alkali metals present in the biomass fuel would have lowered the melting point of the slag.

The surface of the sample as shown by its Raman signature contains goethite, which is indicative of oxidative conditions resulting from the long burial period in a variably damp environment. Mössbauer spectroscopy indicates that the iron phases present at the various surfaces are complex and are more likely to consist of a mixture of iron oxides and iron-containing aluminosilicates. Raman spectroscopy supports the presence of hypersthene as one such aluminosilicate and also the presence of ankerite, a calcium-iron carbonate. The presence of hematite as shown by its Raman signature (Fig. 5.7A, spectra (e and g)) obtained from well below the sample surface and within the core of the sample would suggest that at the time of formation of the slag reducing conditions would have been present in the kiln. Iron smelting or iron working technology was probably introduced into Egypt during the Assyrian occupation in about the seventh century BCE. The area around Meroe in Nubia contains rich deposits of iron ore and which have been mined for about 2000 years. At Mut there are prominent iron-rich ore and ochre deposits which could have been smelted for iron production [32].

Eccleston [32] reported that at site C/2/6 at Kellis, debris from various high temperature industrial processes were recovered. This included a large number of pieces of slag and vitrified kiln linings. Iron bloom production would require a kiln temperature of 1150 °C. At site C/2/7 Room 1, Kellis, fayalitic slag containing wüstite and several small iron prills in a glass-like phase were recovered. SEM analyses indicated the presence of Fe-Ca-Si and Al-Ca-Si phases containing some K, Na and Mg. Fuel ash slag containing high quantities of Ca, Mg and Al were also recovered at Kellis. During the Roman period the biofuel used was reported to be acacia wood, palm fronds and straw [32].

Whilst this slag sample has many characteristics indicative of iron production, this is presently rejected. Firstly, the indicative age of the sample would preclude iron production. More importantly, the presence of cobalt and its accompanying transition elements in the sample would suggest that it has emanated from an alternative kiln use.

The temperature within the kiln would have been of the order of 1000° C as shown by the reaction of hematite, or other iron source with magnesium to form hypersthene [$(Fe,Mg)_2Si_2O_6$] and the reaction of the transition elements also with magnesium to form the doped diopside as shown by their respective Raman signatures provided in Table 5.8. Several of the Raman spectra provided in Fig. 5.7 show the presence of amorphous carbon (from very broad bands at ~1350 and 1590 cm^{-1}) likely indicating that the biofuel during combustion was generating some soot. The presence of this amorphous carbon supports the kiln being operated under reducing conditions.

The cobalt and related transition metals found in the Dakhleh Oasis alums together with magnesium oxides found in this slag sample suggests that this slag could have resulted from the preparation of the cobalt-containing silica frit. The composition of the external surface material of the slag particle is remarkably similar to that of the cobalt used to decorate the vessels at Mut and within the Nile Valley. As will be discussed below, the final production step of the cobalt-based product involved a high temperature reaction step.

A high calcium concentration was found on the surface of the Mut slag sample but not within its core. Nicholson [82] reported that the Kom Helul HAC3 kiln slag

Table 5.8 Raman spectral analyses for the cobalt-based pigments used to decorate a shard from MUT and a slag from the same context.

		Wavenumber/cm^{-1}			
Mineral	Formula	RRUFF	MUT blue pigment	MUT slip	MUT slag
Quartz	SiO_2	266, 357, 402, **467**, 699, 810, 1162	464		
Apatite/Ca phosphate	Degraded $Ca_5(PO_4)_3(OH,F,Cl)$ $Ca_3(PO_4)_2$	960, 1059			
Diopside	$Ca(Mg,\square)Si_2O_6$	323, 356, 391, **666, 1012**, 1026	671, 1012		677, 1008
Wollastonite	$CaSiO_3$	338, 412, 502, 581, 636, 972			
Anhydrite	$CaSO_4$	417, 499, 601, 676, **1017**, 1129			
Gypsum	$CaSO_4.2H_2O$	414, 493, 670, **1007,** 1116		493, 1007, 1115	
Aragonite	$CaCO_3$	155, 208, **1085/1088**			
Calcite	$CaCO_3$	149, 276, 708, 1012, **1087**			
Coquimbite	$Fe^{3+}{}_2(SO)_3.9H_2O$	211, **284**, 505, **1025**, 1202			
Albite	$Na(Si_3Al)O_8$	290, 480, 508			
Plagioclase feldspar	$Na_xCa_{1-x}Al_{1+x}Si_{3-x}O_8$	290, 486, **510**			
Omphasite	(Ca,Na) $(Mg,Fe^{2+}Fe^{3+},Al(Si_2O_6)$	335, 377, 407, 556, **679**, 915, 1022			
Heterogenite	$Co^{3+}O(OH)$	496, 621			
Hypersthene	$(Mg,Fe^{2+})_2Si_2O_6$	656, 674, 1003			655, 677, 1007
Spinel	$(Mg,Zn,Mn,Fe)Al_2O_4$	~630	626		
Cobalt aluminate	$CoAl_2O_4$	~203, ~512			
Cobalt olivine	Co_2SiO_4	**824/826**			
Magnesium olivine	Mg_2SiO_4	**820, 853**			
Alunogen	$Al_2(SO_4)_3(H_2O)_{12}.5H_2O$	474, 607, **995**			
Aluminite	$Al_2(SO_4)(OH)_4.7H_2O$	437, 603, **992**			

(continued)

Table 5.8 (continued)

Mineral	Formula	Wavenumber/cm^{-1} RRUFF	MUT blue pigment	MUT slip	MUT slag
Mooreite	$Mg_{15}(SO_4)_2(OH)_{26}.8H_2O$	437, 457, **982**			
Amorphous carbon	C	~1350, ~1590			1350, 1590
Anatase	TiO_2	**142/144**, 198, 392–398, 514–517, 637/			
Ilmenite	$Fe^{2+}Ti^{4+})_3$	223, 368, **678**			677
Actinolite	$Ca_2(Mg,Fe^{2+})_5Si_8O_{22}(OH)_2$	121, 178, 224, 393, **674**			178, 224, 388, 672
Ankerite	$Ca_2Fe^{2+}(CO_3)_2$	287, **1094/1102**			290, 1087
Copiapite	$Fe^{2+}Fe^{3+}{}_4(SO_4)_6(OH)_2.20H_2O$	242, 270, 302, **476, 985, 1027, 1114**			
Goethite	FeO(OH)	245, **299, 396**, 478, 549			299, 395, 474, 554
Hematite	α-Fe_2O_3	225, 245, 292, 410, 496, 610, 1310			248, 292, 408, 498, 607, 1313
kôzulite	$NaNa_2[Mn^{2+}{}_4(Fe^{3+},Al)]S_{i8}O_{22}(OH)_2$	**386, 414**, 512, **597, 682, 904**, 1008,			388, 681, 900
Mullite	$3Al_2O_32SiO_2$	480, 960, 1130, 1160			
Number of samples analysed			1	1	1

visually contained small particles of lime on its surface. He suggested that handfuls of lime were presumably thrown into the Kom Helul kiln to help flux the slag and thereby allow it to run down the kiln walls. In Fig. 5.5 (spectrum a) the Raman band at ~1085 cm^{-1} is indicative of the presence of calcite. At the kiln firing temperature, it would be expected that calcite would decompose to liberate carbon dioxide (Table 5.8) and form lime. Such lime would have been highly reactive and would have assisted in the formation of diopside. Therefore, the presence of residual calcite and the formation of ankerite would support the proposition of Nicholson [82] for the use of lime to flux the slag on the kiln internal surface to facilitate its later removal. The dissolution and recrystallization of the various iron phases present at the exposed surfaces of this slag sample would lead to further chemical modification of the surface of the sample during burial.

The presence of lead in several analyses together with the presence of manganese, cobalt, nickel and zinc suggests that the kiln may have been used for the production of faience glazes with a blue appearance. The relatively low copper analyses would suggest that if faience decoration was conducted it was for the production of blues or reds. This hypothesis is supported by Eccleston [32] who reported that a range of vessel types, including plates and bowls were recovered throughout the Dakhleh region. They were decorated using the faience application method (the faience slip is applied to the vessel, air dried and then fired). Importantly, most objects were said to have been blue-decorated. He also reported that only limited evidence exists for the actual production of faience at Kellis or elsewhere in Dakhleh, as indicated by the discovery of overfired shards amongst the excavated material. Furthermore, Nicholson and Henderson [83] stated that Egyptian glass used a local lead source, distinct from that obtainable from Mesopotamia.

5.4.2 Cobalt-Based Phases

The pottery used in the construction of the vessels from the various sites analysed in this document has been previously described by Hope [50, 51]. In the Nile Valley, either Nile silt or marl clays were used to manufacture pottery. According to Soukiassian et al. [119], the potters at 'Ayn-Asil in the Dakhleh Oasis added plant ash to their kaolinitic clays when they prepared their pottery vessels.

As suggested by Nicholson and Henderson [83], cobalt would have been in limited supply. Therefore, it is likely that clusters of workshops using cobalt would have been established. These workshops would most likely have been subject to direct state control.

The presence of significant concentrations of both aluminium and silica in the analysed pigment samples, together with the presence of various transition metals, provides the opportunity for either or both olivine and spinel formation together with cobalt-doped willemite ($Co_xZn_{2-x}SiO_4$) or gahnite ($Co_xZn_{1-x}Al_2O_4$) type structures [8, 11, 72, 76, 103]. Magnesium has reacted with calcium and silica to form diopside, which in turn can form 'copside' and 'niopside'. The presence of magnesium-doped Co-Al spinels may also form. Furthermore, the presence of zinc at higher levels then cobalt and nickel in the Oases alums provides the opportunity for these elements to react at elevated temperature and then enter the melt phase. Thus, the mineralogical conditions are conducive for cobalt to be present as an olivine, a spinel, a pyroxene or as a zinc complex, all of which exhibit a different colouring efficiency. These various cobalt mineral forms are discussed herein below in some detail.

Crystallinity and mineral chemistry provides the likely method(s) by which cobalt is concentrated and then introduced into a glass or formed into a blue pigment. The potential for this pigment to be used as a raw material in the production of cobalt glass requires further investigation and will be discussed below in Sect. 5.5. As discussed by Bouchard and Gambardella [11], Raman spectroscopy is a good tool for studying cobalt-based phases whereas at low concentrations of doping agents, XRD is almost unable to distinguish between doped and non-doped spinels traces. The Raman signatures for the various cobalt minerals are provided in Table 5.8.

Production of stable blue pigments is normally based upon the use of phases which are thermally stable at the glass or glaze melting temperatures and which exhibit a low reactivity with the glass component. Reaction with and/or dissolution of the pigment into the glass cannot be completely prevented but Co^{2+} ions dissolved in glassy silicate are very efficient colouring agents: 0.5 wt% equivalent CoO is sufficient to achieve a vivid blue colour. The potential phases are the cobalt spinels including $CoAl_2O_4$, Co_2SnO_4, $(Mg,Co)Al_2O_4$, $(Co,Zn)Al_2O_4$ and $Co(Al,Cr)_2O_4$, $CoCr_2O_4$ and Co_2TiO_4 together with the blue cobalt orthosilicate or olivine (Co_2SiO_4) structures [11, 31, 36, 46, 103, 104, 108, 128].

The Kharga alum analysed in this research contained approximately 4% MgO and the Dakhleh alum approximately 22% MgO. Therefore, the presence of magnesium would be expected to materially affect the spinel formation. Warachim et al. [129] showed that co-precipitating aluminium, cobalt and magnesium as a mixed hydroxide and then calcining at temperatures between 400–1200 °C for 0.5 to 2 hours produced a series of spinels of the general formula $Mg_{1-x}Co_xAl_2O_4$. Above 900 °C the colour varied from white to blue. They proposed the following reactions:

- $Co(OH)_2 \xrightarrow{\Delta} CoO + H_2O$ (reaction temp. >100 °C)
- $2\,CoO + 0.5\,O_2 \xrightarrow{\Delta} Co_2O_3$ (reaction temp. >300 °C)
- $3\,Co_2O_3 \xrightarrow{\Delta} 2CoCo_2O_4 + 0.5\,O_2$ (reaction temp. ~500 °C)
- $MgO + Co_2O_3 \xrightarrow{\Delta} MgCo_2O_4$ (reaction temp, 500–700 °C)
- $CoCo_2O_4 \overset{\Delta}{\leftrightarrow} 3CoO + 0.5\,O_2$ (reaction temp ~900 °C)
- $MgCo_2O_4 \xrightarrow{\Delta} MgO + 2\,CoO + 0.5\,O_2$ (reaction temp. >900 °C)

Furthermore, as indicated by Warachim et al. [129], Co^{2+}-Co^{3+} and Mg-Co^{3+} spinels may react with each other and with magnesium-aluminium or Co^{2+}-aluminium spinels to form a range of variously coloured solid solutions. They proposed the following colour range:

- CoO blue
- Co_2O_3 black
- $CoCo_2O_4$ black
- $MgCo_2O_4$ green

This colour range confirms that mixed hydroxides of aluminium, cobalt(II) and magnesium when heated to 500–900 °C produce dark grey to black pigments, the blue colour only developing when the sintering temperature is above 900 °C. In the temperature range of 1000–1200 °C and at a cobalt(II) oxide concentration of up to 10%, the colour is independent of the actual calcining temperature.

Analysis of the Raman spectra of the $MgAl_2O_4$ spinel and the many other spinels is complicated by disordering of the cations and the sample history [19, 20, 22, 65, 106, 109, 110, 115, 127] and by operation of the Resonance Raman Effect that modifies relative peak intensity and shifts slightly the peak wavenumbers [117].

The chemistry and mineralogy of the olivine group and the spinel group have been discussed in detail elsewhere [27].

In the olivine structures, $(A,B)_2SiO_4$, the A and B cations are distributed over two non-equivalent sites of octahedral coordination, M1 and M2. Mutke et al. [80] studied the kinetics of cation redistribution between the two sub-lattices for the olivine $(Co_{0.6}Mg_{0.4})_2SiO_4$.

Stola et al. [120] reported that CoO will react with silica leading to the formation of the olivine structure Co_2SiO_4 which will crystallise from a melt at 700 °C. However, the conventional solid state synthesis of the cobalt olivine from SiO_2 and Co_3O_4 requires temperatures of about 950 °C to reduce the Co_3O_4 to CoO to allow the reaction with silica to proceed. Jonynaite et al. [61] have reported that the formation temperature of the blue pigment known as smalt ($CoO.nSiO_2$) is 850–900 °C.

All olivines provide an intense Raman doublet near 858–837 cm^{-1} (Si-O asymmetric stretching bands) and 825–808 cm^{-1} (Si-O symmetric stretching bands) [31]. As observed by Mouri and Enami [78], olivines can be grouped as follows:

- forsterite, monticellite and kirschsteinite at 858–847 cm^{-1} and 825–815 cm^{-1} and wavenumbers decrease in the order listed;
- nickel olivine, fayalite, cobalt olivine and tephroite display decreasing wavenumber bands at 819–808 cm^{-1} in the order given; all provide a similar band at 837–839 cm^{-1}.

Cobalt olivines provide a Raman signature at 328, 387, 516, 552, **811, 837**, 885, 928 cm^{-1} and nickel olivines provide a Raman signature an order of magnitude more intense than for cobalt olivines. Their Raman signature is found at 298, 414, 521, **560, 593, 819, 839**, 868, 889, 952 cm^{-1} [71].

Diopside-like crystals are clinopyroxenesPyroxenes are metasilicates composed of $[SiO^{2+}{}_3]_\infty$ chains formed by sharing SiO_4 tetrahedron. As previously noted, diopside ($CaMgSi_2O_6$) has been found present in all shards containing cobalt blue pigmentation. In ternary systems with SiO_2, CaO and a divalent transition metal (Mn^{2+}, Fe^{2+}, Co^{2+}, Ni^{2+}, Cu^{2+}, or Zn^{2+}), the diopside structure is encountered when the divalent cation can substitute for Mg in octahedral sites. Thus, the cation radius must be close to that of $Mg^{2+}$and the crystalline field must favour octahedral coordination [75]. It has been shown by Schairer and Bowen [105] that even with <2% diopside in a ternary silicate mixture containing potash that diopside may be the first crystalline phase to appear on cooling.

As reported by Buzatu and Buzgar [14], the wavenumbers of Raman bands decrease with the increase of the Fe concentration (from enstatite to ferrosilite).

'Copside' is the cobalt analog of diopside and has been investigated by Nicolini and Porter [86]. Thus, when $CaO.CoO.2SiO_2$ is heated to about 1000 °C 'copside' ($CaCoSi_2O_6$), cobalt akermanite ($Ca_2CoSi_2O_7$) and cobalt silicate (Co_2SiO_4) can develop. In a similar manner, NiO crystallised as 'niopside' ($CaNiSi_2O_6$) with wollastonite ($CaSiO_3$) and larnite (Ca_2SiO_4) as minor phases. Niopside begins to form at about 1000 °C at the expense of the other phases [75].

Forsterite [Mg_2SiO_6] has been reported in the literature to be doped with cobalt to form $(MgCo)_2SiO_4$ [41]. When MgO was equimolecularly replaced by CoO and heated with SiO_2 to 1250 °C the crystal phase so formed was shown by XRD to consist of Co-containing forsterite [$(MgCo)_2SiO_4$] in which the lattice parameters with the increased CoO concentration altered slightly.

As provided in the RRUFF database (Table 5.8), the principal Raman bands are located at 659–669 cm^{-1} and at 1007–1015 cm^{-1}. The partial replacement of magnesium by cobalt or nickel in the mineral structure will normally result in a movement to higher wavenumbers, particularly by the band observed at ~666 cm^{-1}.

5.4.3 Pottery Shard

The fabric of the Malkata shard samples examined were either marl clays on to which the pigment has been directly applied or Nile silt in which beeswax (Fig. 5.5) was used to apply the pigment to the body. The Amarna and the Sinai samples were also Nile silt fabrics to which the cobalt pigment has been applied. The Deir el Medineh sample was a marl clay fabric and the Mut el Kharab shard was a Dakhleh fabric. Their body compositions have been described in detail [50, 52, 53].

Malkata was the palace of Amenhotep III (c. 1479–1425 BCE) at Thebes and predates the establishment of Amarna. According to Nicholson and Peltenburg [84] no faience kilns or furnaces were discovered at Malkata. However, from examination of the industrial debris, Malkata is credited with faience making and glass production. Maas et al. [74] suggest that the Malkata glassworks were operated by foreign workers. They considered this to be the case based upon the superiority of the quality of the production when compared with that produced during the late New Kingdom period at Lisht; it is not certain whether Malkata was a true glass-making site or only a glass working operation.

Amarna was established in approximately year 5 of the 18th Dynasty reign of Amenhotep IV/Akhenaten (c. 1352–1336 BCE). At site O45.1 at Amarna is a workshop showing evidence for the production of glass, faience and pigments [81]. The interaction between the Dakhleh and Kharga Oases desert dwellers and the Nile Valley people by way of the cobalt pigment supply will be demonstrated.

The Raman spectra for the cobalt pigment applied to the surface of the various shards are provided in Figs. 5.5, 5.6, and 5.7 and their interpretation is given in Table 5.8. It should be noted that not every shard provided bands indicating the presence of all minerals, indicative of the heterogeneous nature of the samples. Rather, a number of the listed minerals as shown are present in individual cases. However, diopside was observed in almost every instance and from the position of the Raman bands it is evident that in this mineral a partial replacement of magnesium by transition elements has occurred. The shift to higher wavenumbers is indicative of the ionic radii of these transition elements being less than that of magnesium. Thus, this shift clearly indicates that at least one of the cobalt, nickel and zinc ions are bound into this diopside structure.

5.4.3.1 Slip

The Raman spectra provide signatures for aragonite ($CaCO_3$) and gypsum ($CaSO_4.2H_2O$) [93, 94, 95, 101] present on the surface various shards, either within the pigment or as a slip coating applied directly to the pottery surface. The presence of aragonite as a surface deposit occurs during burial by rehydration and carbonation of of lime or by surface deposition from the surrounding soils.

Phosphorus is a common pollutant of archaeological samples. However, Nicholson has shown that a kiln at site O45.1 at Amarna and also a kiln HAC3 at Kom Helul when excavated contained bone fragments [81, 82]. Phosphorus has also been detected in several samples of Egyptian blue from Amarna and in cobalt pigments applied to pottery from Amarna.

The desert sand at Amarna is known to contain significant lime. Turner [125] reported the presence of 18.85% CaO in this sand. As proposed by Lucas and Harris [73] and by Lilyquist and Brill [70] this sand when combined with natron produces a glass containing 15–20% Na_2O. The high potash concentration in this glass was considered by Lucas and Harris to result from contamination of the frit by fuel ash. Caley [15] however suggested that the ash from the fuel employed to melt the glass was often incorporated as one of the glass-making raw materials. Such an inclusion would provide the opportunity for phosphorus to be incorporated into the glass. Thus, as discussed below, it is feasible that the fuel ash, incorporating bone material, was used in the production of the cobalt blue pigment by one of the methods discussed below.

5.4.3.2 Transition Element Data Comparison

One of the great strengths of the synchrotron XFM beamline is its ability to scan and elementally map (Figs. 5.10 and 5.11) much larger areas of an object than is able to be achieved by other experimental methods. As shown in Fig. 5.12, the transition element weight percent ratios found in various Nile Valley shards manufactured during the period between the 18th and 20th Dynasties has been plotted. The results obtained clearly identify a very strong and direct relationship between the various transition elements mapped. Zinc appears to be normally in a higher concentration and manganese in a lower concentration when compared with cobalt; nickel closely matches the cobalt and zinc concentrations. This finding is in accord with the alums analysed from the Dakhleh Oasis and support the proposition that the Western Desert oases provided the source for all, or most, of the cobalt-containing alums mined. Normal variation within various parts of the alum ore body, together with minor adjustment in production methods and normal batch-to-batch variation in production over this extended period would be expected. However, the results do show a very uniform quality of blue pigment produced over this extended period, supporting the proposal that cobalt-blue frit production was conducted in very few locations. The very close relationship, particularly between NiO and CoO would suggest a single alum deposit was mined and this raw material could support a

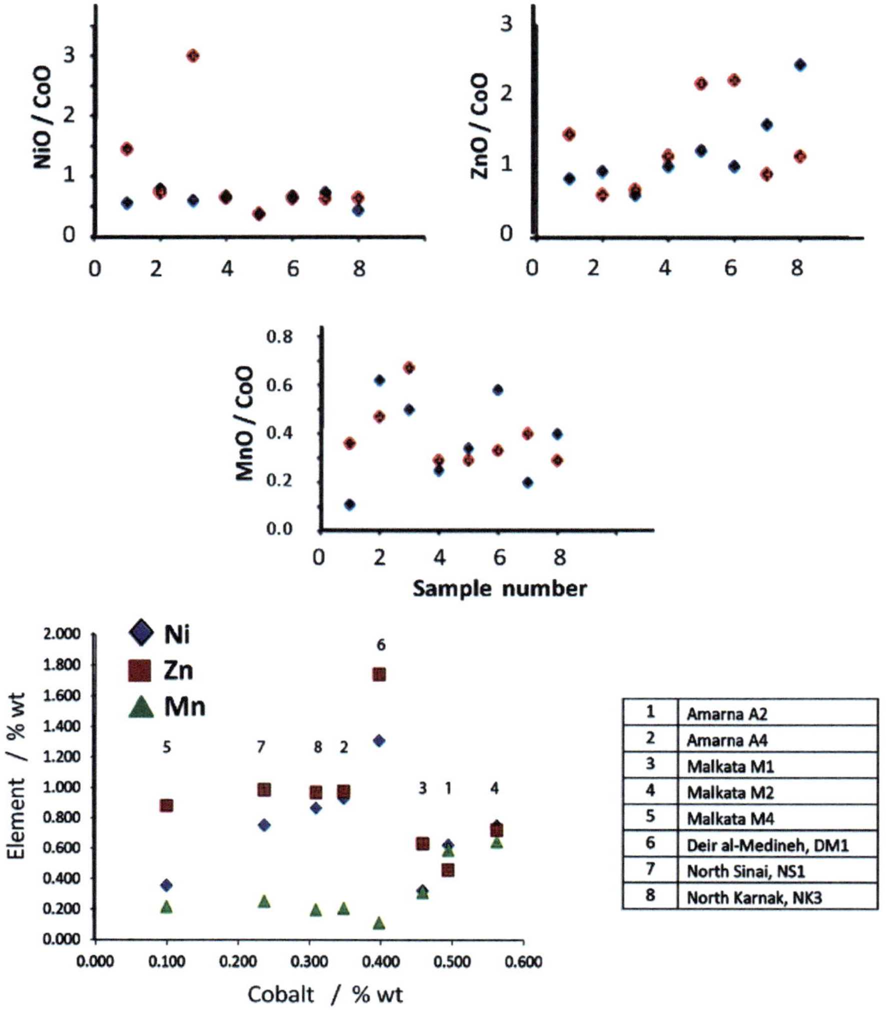

Fig. 5.12 (Top) Plots of eight Amarna shards and eight Malkata shards providing the NiO to CoO, ZnO to CoO and MnO_2 to CoO ratios for the blue pigment applied to each of the shards examined and listed in Fig. 5.2. The **blue** data points are for the Malkata shards and the red data points are for the Amarna shards. Several points show an overlap of ratio data. (Bottom) Synchrotron XFM beamline transition elements compared with cobalt, for various cobalt-blue pigment-decorated sherds recovered at sites located throughout the Nile Valley and shown in Fig. 5.2. All element results are expressed as weight percent, (◆ Ni, ■ Zn, ▲ Mn)

limited number of manufacturing sites. As discussed above, all of the necessary raw materials to produce the cobalt-blue pigment were available at Dakhleh oasis making it an ideal production site. All of the necessary raw materials to produce the cobalt-blue pigment were available at Dakhleh oasis, making it an ideal production site.

Shortland et al. [112] suggested that there were probable differences in the quality of the cobalt blue pigment used at Malkata and Amarna. In Fig. 5.12, the EDS weight percent ratios of manganese dioxide, nickel oxide and zinc oxide compared to that of cobalt oxide are provided. Nickel and cobalt ratios show almost no differences for the various shards analysed. There is slightly more variance in the zinc and cobalt results, but both of these elements also support the clear indication that both Malkata and Amarna obtained their cobalt-blue pigment frit from a similar specialist production site. It is now proposed that this site is located within the Dakhleh Oasis.

Shortland et al. [112] postulated that the ratios of Mg, Mn, Fe, Ni and Al to Co for the blue pigment found at Malkata tend to be lower than those for the same pigment ratios at Amarna. This data is considered to be somewhat tenuous. These element oxide ratios in the present study show rather similar ratios as provided in Tables 5.4a and 5.4b. Figure 5.12 would suggest that no substantive changes in the ratios of the various transition elements to cobalt occurred either at Amarna, Malkata, or more broadly throughout the Nile Valley. The data generated in Fig. 5.10 indicates that the cobalt blue pigment used between the 18th and 20th Dynasties did not substantially alter in its composition. Minimal variability is shown in Fig. 5.12 for the ratios of MnO_2, NiO and ZnO to CoO for eight shards from Amarna and eight shards from Malkata shown in Fig. 5.4.

As shown in Figs. 5.10 and 5.11, it is evident that different concentrations of manganese are present in the Malkata shards A2 and A4. The Deir al-Medineh shard also contains a lower manganese concentration. This would support the above suggestion that element oxide ratios may not provide a suitable method for differentiation of the actual sources of the cobalt-rich alums exploited.

El Goresy [34] reported that the black pigment used for decoration requires scientific investigation and was generally soot or pyrolusite (MnO_2). However, Fig. 5.10 clearly identifies the black pigment decoration to be the black iron oxide spinel, magnetite ($Fe^{2+}Fe^{3+}{}_2O_4$). Manganese is associated with some of the iron-black lines, but not in every case.

The Malkata shard M3 (MA) shown in Fig. 5.3 has an area in which the blue had been rendered a paler shade. This was shown by EDS analysis to be achieved by overpainting with white lead, possibly as the mineral cerussite ($PbCO_3$) although the Raman signature for this mineral was not observed. Lead mineralisation was also detected in the blue pigment applied to the Amarna shard A1 and more particularly on the Amarna shard A5 (362), where again white lead has been used to deliberately overcoat the blue to obtain a paler shade of blue. Alterations of the painted areas on pottery have been observed at Amarna by Rose [102]. In discussing overpainting, Rose described an additional form of alteration; namely the application of a slip coating over the blue decoration to obscure the original design. In one example, the uppermost slip coating was decorated in black and blue, below this is a middle layer decorated in blue, black and red again on a cream slip and under both of these there is, at least in some areas, a blue coating. How this could relate to the section heading "Pre-firing painted decoration" and whether this overpainting was again fired is not discussed. The application of the lead pigment is different insofar as it does not cover the blue, but is intended to apparently change the shade of blue within the various decorative areas.

Hope [49, 50] discussed the order in which the decorators applied the various pigments being used. In some instances black outlined patterns were in-filled with blue but generally coloured areas were later framed with black lines as shown in Fig. 5.4. This latter sequence was proposed by Rose [102] in her study of the blue-painted material from Amarna and it has been confirmed by detailed analysis of photographs of the shards analysed in this study. In Fig. 5.3a–e, the pigment order of application on the shards analysed is provided.

5.4.3.3 Cobalt Pigment Discussion

Kaczmarczyk [62] provided in his Table 34.3 (reproduced herein in Table 5.2) the analysis of four samples of alum from the Mut al Kharab region of the Dakhleh Oasis. These analyses showed MgO varying from 6.0–8.5%, MnO 0.84–1.13%, Fe_2O_3 0.25–0.35%, CoO 0.31–0.45%, NiO 0.15–0.6%, ZnO 0.6–1% and in two samples 44.7 and 46.9% SO_3 was present. No silica was reported and therefore it is unclear if this mineral was found or ignored in the analyses. Kaczmarczyk argued that cobalt could not have been imported into Egypt. He supported the proposal of Noll that the cobalt and the associated oxides were precipitated from an aqueous solution by rendering the solution mildly alkaline using natron, plant ash or ammonia leaving the sulphates behind. The (MHP) precipitate is then dried and heated to produce the aluminates with much higher transition element content in the pigment.

It is well established that transition metal ions are efficient chromophores for colouring glasses because of the mobility of their 3d electrons. Thus, the colouring performance of cobalt minerals is largely dependent on their thermal stability, chemical reactivity towards glass and glaze components and on the coordination of the Co^{2+} ions, the tetrahedral coordination being preferred to the octahedral. de Waal has applied Raman spectroscopy to a range of ceramics and has demonstrated that the cobalt signature may be enhanced by focussing the microscope to below the actual mineral surface [29, 30]. However, when cobalt is in an ionic form, it cannot provide a specific Raman signal [20].

Riederer [99] examined the blue decoration on several vases and concluded that the cobalt pigment was cobalt aluminate. Importantly, he reported that copper, manganese, zinc and beryllium were present as trace elements. He was unsure as to whether the cobalt was derived from small cobalt deposits in the Eastern Desert or was imported from cobalt deposits near Isfahan in Persia. Furthermore, he postulated that aluminium hydrate may have been mixed with a cobalt mineral, applied to the vessel and then fired, forming the blue cobalt aluminate (a spinel).

Spinel formation ($CoAl_2O_4$) was also reported by Noll and Hangst [90] and Noll [89] who disputed Riederer's suggestion. These workers did not physically determine the presence of a cobalt spinel and relied upon the EDX elemental analysis to propose the presence of the cobalt aluminate. Rather than a solid-state, direct firing of the pigment composition to form the colour on the vessel they proposed that the pigment must have been produced by precipitation from an aqueous solution of cobalt salts and alum by pH adjustment using sodium or ammonium carbonates.

The precipitate would then have been heated to between 800 and 1000 °C to produce the Al-spinel. They also reported that a low cobalt concentration would facilitate the spinel formation; that the presence of chlorides accelerates the reaction and zinc stabilises the pigment. Importantly, these researchers reported the presence of diopside in the body of the shards examined and therefore this would indicate a firing temperature above 850 °C. The pigments were applied to the wares either using a binding medium such as gypsum or an organic material ("post-firing painting") or the pigment was applied to the unfired body and both were fired simultaneously ("ceramic painting"). They concluded that the cobalt aluminate was most likely applied by a post-firing technology, but because of its heat stability it could well have been applied by direct firing.

Bachmann and co-workers [6] agreed with the results published by Noll and Hangst [90] that the pigment was prepared prior to its application to the vessel. They suggested that the pigment may have been prepared by a solid state reaction between a cobalt-zinc ore and aluminium-based minerals, or by calcination of the cobalt-containing alum (they incorrectly assumed it to be a potash alum) at above 800 °C to transform the cobalt to its oxides and then combine with alumina to form the blue cobalt spinel. Furthermore, they agreed with Hope [49] that the pigment was applied before firing and the possible presence of gypsum could have resulted from burial conditions rather than from the proposal of Noll and Hangst [90] that the cobalt pigment was applied as a slip mixture containing gypsum.

Eight shards, four from Malkata and four from Amarna were sectioned and polished and examined by SEM-EDS analysis and XRD by Shortland et al. [112]. They did not report the presence of sulphur, phosphorus or copper in their analytical tables, thus no conclusion can be reached as to whether or not these elements were present. Surprisingly, the actual concentrations of Co, Ni and Zn determined by SEM-EDS were significantly higher than those reported by all earlier workers. They noted that the sample size available for XRD analysis was small and therefore individual spinels could not be determined with any certainty. Thus, based upon the bulk chemical analyses, they conjectured that Al-spinels incorporating cobalt, zinc, magnesium, iron and nickel were present. They also postulated that the variation in the magnesium content may have resulted from incomplete precipitation of this element, it requiring the highest solution pH to proceed to completion. In discussing the slip layer, they considered that gypsum was the most likely mineral used. However, based upon the sulphate content, they suggested that calcareous clay was also present. The addition of calcite or huntite to the gypsum in preparing the slip was not considered. Unfortunately, this and all other comparative studies have relied upon the four samples of alum from the Dakhleh Oasis analysed by Kaczmarczyk.

Tite and Shortland [123] discussed the work of Rehren [97] and reported that, based upon the variation in the decomposition temperatures of aluminium sulphate and magnesium sulphate they suggested, that prior to decomposition, there would be a preferential partitioning of alumina over the magnesia depending upon whether or not the two sulphates decomposed separately or at the same temperature. Bretsznajder and Pysiak [12] demonstrated that ammonium aluminium sulphate undergoes a step-wise decomposition in which water, ammonia and sulphur trioxide are sequentially evolved with increasing temperature. Evolution of residual water and ammonia commenced at 400–470 °C and is diffusion controlled. Above 470 °C the reaction increases rapidly and above 860 °C decomposition of anhydrous $3Al_2O_3.4SO_3$ commences. Below 860 °C the reaction is retarded by the diffusion of gaseous decomposition products through the grains of the solid Al_2O_3 produced. As shown by Ingo et al. [55] the thermal decomposition of calcium sulphate commenced at 1240 °C. However, in the presence of silica, thermal decomposition commenced at about 990 °C and is practically complete at 1260 °C.

A blue-decorated shard from a surface find at Malkata was examined by Bachmann et al. [6] and from their analyses they stated that these agreed exactly with those of Noll and Hangst [90] and in particular with regard to the Co/Zn ratio and the admixture of nickel. They reported that three cubic crystals on the surface when analysed by XRF indicated the presence of a spinel of the general formula $(Zn,Co,Ni)Al_2O_4$. The XRF diagram presented indicates a very small peak slightly above a voltage of 8 keV where copper would be anticipated. It is assumed therefore that if copper was present, it was below the limit of detection for the instrument employed. Whilst these researchers noted that the normal habit for a spinel is octahedral in shape, dodecahedral and cubic crystal forms have been observed. They reported that the sample was washed thoroughly in water prior to analysis, thereby eliminating sodium chloride as a source for these cubic crystals. Caracas and Banigan [16] confirmed that at room temperature and pressure, $MgAl_2O_4$ is a cubic crystal. Bachmann et al. [6] provided their XRF spectrum for the analysis of the cobalt mineral as their Fig. 1. It is evident that they were unable to detect the presence of any elements with $Z < 13$. Thus, they were unable to detect magnesium and therefore their proposed spinel with a general formula of $(Zn,Co,Ni)Al_2O_4$ is open to question.

Importantly, examination of the published literature reveals that Bachmann et al. [6] relied on an XRF analysis to determine the cobalt complexation in the Malkata shard they examined. This same Malkata shard is identified in the present work as "Bachmann". The XRF instrument used was unable to detect magnesium or lighter elements, and therefore they presumed that the aluminium determined had complexed with the cobalt. Noll [89] and Noll and Hangst [90] relied upon an EDX analysis which indicated the presence of aluminium to also assume that a cobalt spinel had formed. Shortland et al. [112] relied upon XRD analysis to determine the cobalt homologues formed; however their small sample size prevented them from reliably determining specific spinels. Riederer [99] examined four vases exhibiting a cobalt blue decoration and found all lines to support cobalt aluminate formation in amongst a considerable number of additional lines due to the presence of impurities. As noted previously, XRD is not able to clearly distinguish between the various cobalt homologues.

In respect of this Malkata sample, previously analysed by Bachmann and coworkers [6], the presence of a spinel, is confirmed by its Raman signature with bands at 200 and 512 cm^{-1}. Other spinel-based minerals, including magnesium spinels were also present in all other Malkata samples. However, as provided in Fig. 5.6A(c) this was the only sample found to contain traces of cobalt aluminate spinel. Because of the stated concern over the Bachmann work, this specific shard was examined using both the blue and green laser lines to ensure that the maximum data could be ascertained by Raman spectroscopy. The present research would strongly support a cobalt-diopside complex as shown by the Raman bands at 669–673 cm^{-1} and 1007–1013 cm^{-1}. Thus, as provided in Table 5.8, the most likely cobalt mineral present in this and other 18th Dynasty pigments is based upon the pyroxene (diopside) structure [66].

As shown by Cultrone et al. [21], calcium oxide will react with silica to form wollastonite at 800 °C. Diopside will also commence to form in very low concentrations at this temperature. Noll [89] reported that the formation temperature for diopside is 850 °C.

The other cobalt minerals determined by their Raman signatures are omphasite, heterogenite and a cobalt doped zinc oxide which would have formed within the glass. Co^{2+} ions do not provide a Raman signature, if cobalt is dissolved either in a glass or in another structure [20]. In several of the sherds the cobalt-based pigment is present only as a superficial, surface remnant, enabling the minerals in the substrate to become visible. Thus, the feldspars and the calcium minerals are most likely related to the substrate on which the pigment has been applied. The fact that degraded apatite has been detected indicates that at least some of the reaction vessels were partially lined with a hydroxyapatite (bone ash) and gypsum composition, or that kiln ash was employed as part or all of the alkali flux. Thus, these minerals could also be present as contamination within the cobalt pigment.

Raman spectroscopy revealed the presence of traces of spinels and, on only one occasion was the cobalt spinel determined (Table 5.8). In the case of this sample, and all other samples examined, all indicated that the cobalt was bound into a diopside lattice by the partial replacement of a magnesium ion by a cobalt ion. This was true for the sample from the desert dwellers at Mut al Kharab and for all sites examined from the Nile Valley.

The commonality in composition of the cobalt pigment used at Mut al Kharab and in the Nile Valley would indicate that they shared a common genesis. Therefore, it is proposed that the cobalt pigment was produced within the Dakhleh Oasis using the locally available cobalt alum and then transported to the Nile Valley where it was then used to decorate important vessels. It is hypothesised that this cobalt pigment was also used as a raw material in the production of cobalt coloured glass.

5.5 Blue Decoration: Expected 'chaîne opératoire'

5.5.1 Cobalt Extraction from Dakhleh and Kharga Oasis Alums: Expected Procedure

Kaczmarczyk proposed that if the alums are dissolved in water and made mildly alkaline using say natron, plant ash or ammonia, then the transition metals would precipitate as their hydroxides(MHP) and oxides leaving the sulphate behind. Drying and heating the precipitate would then yield a blue mixture containing aluminates and oxides with a much higher cobalt content. It should be noted that cobalt oxides vary in colour between dark green and black, the distinctive blue colour developing when the oxide is incorporated into a glassy structure and dissolves to form Co^{2+} ions or reacts to form a spinel or olivine structure.

Shortland et al. [112] summarised the previously published data and noted that the cobalt blue pigment was in the form of a CoAl-spinel containing significant amounts of zinc and nickel and, that the pigment was applied over a thin slip containing gypsum. These researchers noted that Shortland et al. [113] had previously reported that when aqueous solutions of the alums are rendered alkaline using sodium carbonate that precipitation proceeds in the order as suggested by Rehren [97]; namely, aluminium hydroxide, followed by cobalt and manganese hydroxides and finally magnesium hydroxide. Shortland and his co-workers did not report the solution pHs at which precipitation proceeded. Nor did they consider the varying solubility between sodium, ammonium and potassium double alums as provided in Table 5.9 or the effect of temperature on solubility or the addition of urine to precipitate the ammonium double salt as noted in Table 5.9.

Nicholson and Henderson [83] suggested that the alum was mixed with water and natron, then plant ash or ammonia were added to make it slightly alkaline. The MHP precipitate was then heated at between 800 and 1000 °C to produce the blue pigment. This proposal does not explain the high silica content found in all pigment analyses as provided in Table 5.3, nor the source of the ammonia. Pourbaix (Eh-pH) diagrams for the equilibrium potential of electrochemical reactions against pH for the transitional metals and of the Monhemius precipitation diagram are provided in Jay [58].

Agricola, [3] described various methods for the European manufacture of alum from various rock types including alum rock "alunite", and certain types of shale. Hoover, the translator of the Agricola document noted that the general methods described had been known for centuries, the Roman alum of commerce being a double basic potassium alum. In the various processes described by Agricola, the rock mineral was allowed to decompose over a period of many months by being kept wet. This caused the lixiviant to form aluminium sulphate. To this solution, urine was added to convert the salt to its ammonium form [3]. The ammonium alum would then crystallise normally in four days in winter or six days in summer. Crystallisation was encouraged to form by placing horizontal and vertical twigs in the solution on to which the crystals formed. And, by the 1600s, alum purification using urine was widely practiced in England. Ammonium and potash alums offer

Table 5.9 Solubilities of various alums in water (data extracted from [92])

Alum	Formula	Melting Point °C	Solubility in cold water	Solubility in hot water
Ammonium	$Al_2(SO_4)_3.(NH_4)_2SO_4.24H_2O$	93.5	3.9 (0 °C)	∞ (100 °C)
Ammonium iron	$Fe_2(SO_4)_3.(NH_4)_2(SO_4).24H_2O$	40	124 (25 °C)	
Potassium	$Al_2(SO_4)_3.K_2SO_4.24H_2O$	92	5.7 (0 °C)	∞ (93 °C)
Sodium	$Al_2(SO_4)_3.Na_2SO_4.24H_2O$	61	106.4 (0 °C)	121.7 (45 °C)

much lower solubility in cold water and therefore more readily crystallise from solution [92; Table 9].

Urine did have other known uses in Egypt in the 18th Dynasty, particularly in gold purification and in the preparation of animal skins. Alum was also widely used in medicines, mummification, as a mordant in dyeing, etc. Therefore, it is conceivable that the use of urine to form ammonium alum was a likely method for its purification in Egypt during the 18th Dynasty.

As suggested by Warachim et al. [129], ammonia precipitation would enable aluminium, magnesium and the transition metals to be recovered. However, such a quantitative precipitation was not favoured by Rehren [97] as this would have precipitated all of the metals. Whereas, this quantitative precipitation (MHP) would have approximated the composition of the magnesia, alumina and transition metals found in the cobalt-based pigments as indicated in Table 5.4a and 5.4b. This precipitate could then have been combined with desert sand, (and thus calcium) and plant ash (probably kiln ash) as the alkali flux and possibly additional lime, and when heated to about 1000 °C would have formed a cobalt based glass as a frit with a composition similar to the cobalt pigment analysis provided in Tables 5.4a, 5.4b, and 5.7. The process would therefore have emulated the process for the production of Egyptian blue with the copper being replaced by the cobalt alum precipitate.

Therefore, based on the significantly lower cold water solubility of the potash and ammonium double salts over sodium as shown in Table 5.9, it is more likely that these would have been the preferred alum double salts. The potash and ammonium salts would have more readily crystallised and recovered from aqueous solution and therefore they are more suited for the purpose of cobalt pigment production. The double salt containing the co-precipitated transition metals would have only required the addition of calcium-containing quartz (desert sand) and the mixturebeing heated to 800–1000 °C. The use of desert sand containing calcium was a well established technology for Egyptian blue production and this cobalt pigment firing would have proceeded in a like manner.

The use of kiln ash would have provided the source for the potassium derived from combustion of the plant biomass used to fire the kilns. Potash would therefore have been available through the use of this residual ash from various kilns [48]. Halfa grass was evidently used to fire the kilns at Kom Helul, Memphis, during the Roman period for faience production [82]; and the salt-resistant halophytic plants of the genus such as *Salsola* [124] and *Salicornia* [48] have also been proposed as a suitable source for plant ash. Nicholson and Henderson [83] also noted the use of plant ash at Amarna. In respect of the use of these plant ashes consideration must be given to their method of application. Namely, were they added without modification (other than possible sieving), or were they dissolved in water and the soluble salts evaporated to dryness and used? Caley [15], on the basis of previous work conducted by Turner [125], suggested that possibly the fuel ash employed to melt the glass was incorporated into successive batches as one of the raw materials.

As reported by Rehren [97] the precipitate, or concentrate, was probably similar to, and therefore related to the dark blue cobalt aluminium spinel found on the blue painted pottery from the Amarna period and would have been an almost ideal colourant for plant ash based glass. Rehren favoured the use of natron as the precipitant as favoured by Noll [89] as this would lower the solution pH allowing the transition elements and some of the aluminium but probably none of the magnesium to precipitate from solution. The analyses for the cobalt blue pigments given in Table 5.4b require the presence of magnesium in concentrations greater than that obtainable from plant ash. Therefore, the present work would suggest that natron is not the preferred precipitant. However, the glass into which the pigment-based cobalt-containing frit, or the calcium-containing sand-alkali-precipitate mixture could have contained some natron.

5.5.2 Cobalt Glass Production, Current Understanding

Comparisons have been made by Shortland et al. [112, 113] with respect to the similarities and differences between the blue pigment applied to decorate vessels and the contemporaneously produced cobalt-containing glasses. Rehren [96] speculated that this cobalt pigment may have supplied the colourant necessary for cobalt blue glass production.

Shortland and Tite [111] undertook a study into copper- and cobalt-blue frit, glass and faience from the 18th Dynasty New Kingdom site at Amarna. In discussing the glass samples they found at Amarna they suggested that different alkali raw materials were used for the production of cobalt-containing glass and copper-containing glass. They noted that in the copper- and the cobalt-blue vitreous materials the copper-blue is significantly lower in magnesium and alumina than found in the cobalt blue glasses. The cobalt glass is also associated with nickel, zinc, and manganese and the presence of these metals could serve as a tentative provenance for the colourant source. These authors suggested that the copper-containing glass was manufactured using plant ash and the cobalt-containing glass used natron. They did not comment upon the method for concentrating the cobalt and/or its conversion into an olivine, pyroxene or an Al- or Mg-spinel [40, 42, 45, 60, 91]. Furthermore, Shortland and Tite [111] proposed that the cobalt-bearing alum was directly added to a raw glass of a unique composition, different to that used for all other coloured glasses. The alkali source for Late Bronze Age (LBA) Egyptian and Mesopotamian glasses is generally considered to be plant ash-based and which introduces a few per cent of elements such as K and Mg into the glass. By way of contrast, soda-rich glass with less than 1% of these components is generally considered to have been derived using the mineral natron as the alkali source [38, 39, 47].

In examining the chemistry of the various coloured glasses produced in the Near and Middle Eastern Late Bronze Age, Rehren [97] observed that the cobalt glasses were unique in their composition. These cobalt glasses are characterised by significantly higher concentrations of Al_2O_3 and lower concentrations of K_2O when compared with other glasses. As shown in Table 5.2, both the Dakhleh Oasis sample D1A and the Kharga Oasis sample K18b contain little to no K_2O but of the order of 20% alumina. The fact that the Dakhleh Oasis alums contain the same suite of trace elements known to be also present in the cobalt-blue glasses has been considered to indicate that these alums were the source for the cobalt. Rehren also observed that the glasses containing this cobalt colourant were higher in aluminium but with a greatly reduced potash content. He therefore questioned how the direct addition of the cobalt colourant could have reduced the potash content in the glass.

This inconsistency led him to further investigate the methods by which the cobalt was extracted from the alum. He suggested that the methods previously proposed fail to explain how the colourant was actually added to the glass and in particular, to address the nature of the raw materials used, and the problems of salt melt and glass melt interaction if the cobalt-containing alum was directly added to the glass. With a direct addition of alum, then the partitioning of the pigments between the two phases created would require further investigation.

The diagnostic criteria for the separation of glasses based upon plant ash and soda-based glasses is the approximate 2–3% higher content of K_2O and MgO in the plant ash glasses. Shugar and Rehren [114] reported that in considering soda-lime-silica glass production over a period of some 2000 years that two compositional groups prevailed. The first type of glass was high in magnesium and potash, plant ash based and described as Late Bronze Age (LBA) Egyptian and Mesopotamian glass. The second type of glass identified was low in magnesium and potash, natron-based Hellenistic and Roman glass. To further understand the various relationships, these workers investigated a smelting model that included the formation of a cotectic glass melt in the presence of a considerable quantity of crystalline material to act as a buffer. They investigated chlorine and determined that it did not act as a thermometer as the concentration of calcium oxide controls the chlorine uptake and not temperature or the soda content.

Rehren [97] reported that from his perspective, the various alternative models proposed at that time are all lacking in essential detail. In particular, the thermal process lacks an explanation for the lower potash content in these glasses when compared with their counter-parts coloured using alternative colouring agents. The dissolution, pH adjustment and subsequent precipitation method previously proposed by Noll [89] and again considered by Shortland and Tite [111], Shortland et al. [[112, 113]) was seen to overcome this potash "problem". Thus, by starting with natron-based glass that is initially poor in potash and magnesia, the final glass composition could have its magnesia and alumina concentrations raised through the addition of the cobalt-bearing alum. However, this concept fails to explain why this natron-based glass production remained restricted to one colour only. Furthermore, it requires the addition of unprocessed salt in order to adjust the ratio of MgO to CoO; and it does not discuss the implications of mixing these sulfate salts and glass at melting temperatures, the subsequent decomposition of such salts at glass melting temperatures, and the different ratios of MgO to Al_2O_3 in the alum and the glass respectively.

Smirniou and Rehren [118] stated that it is still uncertain whether the colourant was mixed with the initial raw materials or added to the crushed semi-finished glass during a second stage, or whether a third stage heating was required in which the colourant was added. To further elucidate the likelihood for the use during the 18th

Dynasty of a high temperature approach, Rehren and co-workers have investigated glass production at temperatures of up to 1050 °C, this being the maximum temperature that the pottery vessels used for glass manufacture during the 18th Dynasty could withstand. Aspects such as the alkali metal and alkaline earth partitioning and the transfer of these fluxes into the glass were investigated [98, 118, 121].

Tite and Shortland [123] subsequently considered the various proposed technologies, namely the direct high temperature addition as discussed by Rehren, alternatively a hydroxide precipitation process and finally a two stage process as proposed by Turner [125] in which a cobalt frit was formed and that this was then added to the glass. They concluded that probably four different plant ashes were employed for the production of copper- and cobalt-blue glasses at Amarna. Furthermore, that natron was the most likely alkali used to precipitate the cobalt and therefore that natron would have also been used in the production of a cobalt-containing frit. They stated that depending upon whether the supernatant liquid was decanted or allowed to evaporate, that this would have influenced the concentration of magnesium hydroxide in the precipitate and also the incorporation of natron into the frit. This frit would have then been formed into a glass using additional plant ash and quartz. These researchers then proposed further work should be conducted in an effort to increase the understanding of the raw materials used and the production techniques adopted.

5.5.3 Cobalt Glass and/or Frit Production, a New Hypothesis

The present research summarized in Table 5.10 confirms that the Dakhleh and Kharga Oases were the sources for the alum raw material used in the production of the cobalt pigment. Although no evidence has been found to date to indicate any site at either oasis where such pigment conversion could have been conducted. It would have been a logical step to undertake the preparation of the pigment at, or close to the alum site and thereby only require minimal transport to convey the cobalt pigment to decorating sites as required throughout Egypt.

As discussed Sect. 5.7, a modified precipitation and firing regime has now been proposed for the conversion of the alum into a pigment. The cobalt extraction from the alum was probably conducted by either the direct addition of plant ash or alternatively and preferably, the addition pf urine to the alum solution to form the potash and/or ammonium double salt. Precipitation and crystallisation of these salts together with the precipitated transition metals present in the alum would have been favoured by lower temperatures. The product obtained would then have been combined with desert sand and a probable flux and then fired at 800 to 1000 °C to form the final fritted cobalt blue pigment.

Based upon the mineralogy and the elemental analysis of these cobalt pigments it is feasible that their admixture with a Late Bronze Age glass frit and subsequent firing could have provided the conditions discussed by Shugar and Rehren [114] for cobalt glass production. The major portion of the aluminium and iron would have partitioned into a second phase.

Table 5.10 Cobalt pigment mineralisation determined by Raman spectroscopy and EDS elemental analysis

Sample	MgAl spinel	CoAl spinel	(Co,Zn)Al spinel	ZnAl gahnite	CoSi olivine	Co Pyroxene#	CoO	CoOOH	Co hydrox*	$CoSO_4$ bieberite
Amarna										
A1	•		•	•		•				
A2					**NiSi**	•	•		•	
A3		?##	•			•	?	?		
A4						•	•	•		
A5						•	•	•		
A6						•		•		•
A7	•	•				•	•			
A8	•		•			•				•
Malkata										
M3	?	?	•			•		•		?
M5		?				•			•	**
M6			•			•		•	•	
M9		?		•		•			•	•
M12		?		**NiAl**		•		•	•	
M13					**•/Ni**	•	•	•		•
M14		•		•		•				•
M15	•					•	•			
M16		?			•	•	?			

(continued)

Table 5.10 (continued)

Sample	MgAl spinel	CoAl spinel	(Co,Zn)Al spinel	ZnAl gahnite	CoSi olivine	Co Pyroxene#	CoO	CoOOH	Co hydrox*	$CoSO_4$ bieberite
Mut al-Kharab										
D7						●	●	●		
Karnak										
K1		?	●			●	●	●		●
K2	●			●		●	●	●		●
K3		?				●			?	●
Memphis										
R1						●			●	●
R2						●	●			●
Dendera										
DE1						●		●	●	●
Deir al Medineh										
DM1	●		●			●		●	●	●
Bir al-Abd, North Sinai										
S1			●			●		●	●	●

#Co pyroxene in this context may be considered as Cobalt substituting for Mg in diopside
##**?** Indicates that additional bands to support the mineral assignment are missing and further, that the specific band identified is shared by other minerals
*Co hydrox is $Co(OH)_2$
**Possibly spherocobaltite [$CoCO_3$]

5.6 Conclusions

The sample of slag, tentatively dated to 18th Dynasty and found in a highly disturbed site at Mut al Kharab, is considered to have been removed from the internal surface of an as-yet undiscovered kiln during its cleaning. The presence of the cobalt, nickel and zinc elements within this slag sample would suggest that the kiln had been used for faience production.

The Raman signature for the pyroxene (diopside) and its principal bands observed at slightly higher wavenumbers strongly suggests that at least some of the cobalt, nickel and zinc ions have replaced magnesium ions within the diopside structure. Cobalt is also present in the pigment as heterogenite and it may also have complexed with zinc to enter the glass structure. Al- and Mg-spinels are widely disseminated throughout the various sherds exhibiting a cobalt mineral decoration.

The strong likelihood is that the desert dwellers at Dakhleh Oasis prepared the cobalt pigment as a glass frit using techniques directly related to those employed in the production of Egyptian blue. This is strongly indicated by the presence of calcium phosphate (as bone ash) which has been detected in both Egyptian blue and in the cobalt pigment; the bone ash probably formed a proportion of the high calcium slip lining to the high temperature reaction vessels and/or entered the glass as part, or all, of the alkali kiln ash flux. It would then only require transportation of small volumes of this precious colourant to the Nile Valley workshops for vessel decoration. As demonstrated, decorators at Mut al Kharab also used this pigment for their own decorative purposes.

References

1. Abe, Y., Nakai, I., Takahachi, K., Kawai, N., Yoshimura, S.: On-site analysis of archaeological artifacts excavated from the site on the outcrop at Northwest Saqqara, Egypt, by using a newly developed portable fluorescence spectrometer and diffractometer. Anal. Bioanal. Chem. **395**, 1987–1996 (2009). https://doi.org/10.1007/s00216-009-3141-x
2. Abe, Y., Harimoto, R., Kikugawa, T., Yazawa, K., Nishisaka, A., Kawai, N., Yoshimura, S., Nakai, I.: Transition in the use of cobalt-blue colorant in the New Kingdom of Egypt. J. Archaeol. Sci. **39**, 1793–1808 (2012). https://doi.org/10.1016/j.jas.2012.01.021
3. Agricola, G. (G. Bauer alias), (1556) "De Re Metallica". Translated by H.C. Hoover and L.H. Hoover, Pub. by The Mining magazine, London (1912)
4. Alizadeh, P., Murghussian, V.K.: The effect of compositional changes on the crystallisation behaviour and mechanical properties of diopside-wollastonite glass ceramics in the SiO_2-CaO-MgO(Na_2O) system. J. Eur. Ceram. Soc. **20**, 765–773 (2000). https://doi.org/10.1016/S0955-2219(99)00135-1
5. Arnold, D., Bourriau, J.: An Introduction to Ancient Egyptian Pottery. Phillip von Zabern, Mainz am Rhein (1993)
6. Bachmann, H.G., Everts, H., Hope, C.: Cobalt-Blue pigment on 18th Dynasty Egyptian pottery. In: Mitteilungen des Deutschen Archäologischen Instituts Abeilung Kairo, vol. Band 36, pp. 33–38. Pub. Verlag, Mainz/Rhein (1980)
7. Beadnell, H.J.L.: *Dakhla Oasis: Its Topography and Geology* Geological Survey Report 1899, Part IV. National Printing Department, Cairo (1899)., 1901

8. Bell, I.M., Clark, R.J.H., Gibbs, P.J.: Raman spectroscopic library of natural and synthetic pigments (pre- ~1850 AD). Spectrochim Acta A. **53**, 2159–2179 (1997). https://doi.org/10.1016/S1386-1425(97)00140-6
9. Bhargava, R., Sharma, P.K., Kumar, S., Pandey, A.C., Kumar, N.: Raman investigations of $Zn_{1-x}Co_xO$ nanocrystal: role of starting precursors on vibrational properties. J. Raman Spectrosc. **42**(9), 1802–1807 (2011). https://doi.org/10.1002/jrs.2924
10. Bishop, J., Lane, M.D., Darby Dyer, M., King, S.J., Brown, A.J., Swayze, G.: What lurks in the martian rocks and soil? Investigations of sulfates, phosphates, and perchlorates. Spectral properties of Ca-sulfates: gypsum, bassanite, and anhydrite. Am. Mineral. **99**(10), 2105–2115 (2014). https://doi.org/10.2138/am-2014-4756
11. Bouchard, M., Gambardella, A.: Raman microscopy study of synthetic cobalt blue spinels used in the field of art. J. Raman Spectrosc. **41**, 1487–1495 (2010). https://doi.org/10.1002/jrs.2645
12. Bretsznajder, S., Pysiak, J.: Thermal dissociation of basic ammonium aluminium sulphate II. Measurements of the decomposition rate and attempt to explain the reaction mechanism. *NASA Technical Translations TT F-11.* ***479***(Feb.), 1–15 (1965)
13. Burgio, L., Clark, R.J.: Library of FT-Raman spectra of pigments, minerals, pigment media and varnishes, and supplement to existing library of Raman spectra of pigments with visible excitation. Spectrochim. Acta A Mol. Biomol. Spectrosc. **57**(7), 1491–1521 (2001). https://doi.org/10.1016/S1386-1425(00)00495-9
14. Buzatu, A., Buzgar, N.: The Raman study of single-chain silicates. Analele Ştiinţifice Ale Universităţii,,Al. I. Cuza" Iaşi Geologie Tomul **LVI**, nr.1 (2010). http://geology.uaic.ro/auig/articole/2010%20no1/1_L08-Buzatu%20-%20pag%20XX-XX.pdf
15. Caley, E.R.: Analyses of Ancient Glasses 1790–1957. Corning Museum of Glass, Corning (1962)
16. Caracas, R., Banigan, E.J.: Elasticity and Raman and infrared spectroscopy of $MgAl_2O_4$ spinel from density functional perturbation theory. Phys. Earth Planet. Inter. **174**, 113–121 (2009). https://doi.org/10.1016/j.pepi.2009.01.001
17. Churcher, C.S., Mills, A.J.: Reports from the Survey of the Dakhleh Oasis Western Desert of Egypt 1977–1987 Oxbow Monograph 99. Oxbow Books, London (1999)
18. Clarke, F.E.: The Corrosive Well Waters of Egypt's Western Desert Geological Survey Water-Supply Paper 1757-O. Pub. U.S.A, Govt. Printing Office, Washington, DC (1979)
19. Colomban, P., Jullian, S., Parlier, M., Monge-Cadet, P.: Identification of the high-temperature impact/friction of aeroengine blades and cases by micro Raman spectroscopy. Aerosp. Sci. Technol. **3**, 447–459 (1999). https://doi.org/10.1016/S1270-9638(99)00102-9
20. Colomban, P., Kırmızı, B., Simsek Franci, G.: Cobalt and associated impurities in blue (and Green) glass, glaze and enamel: relationships between raw materials, processing, composition, phases and international trade. Minerals. **11**, 633 (2021). https://doi.org/10.3390/min11060633
21. Cultrone, G., Rodriguez-Navaro, C., Sebastian, E., Cazalla, O., de La Torre, M.J.: Carbonate and silicate phase reactions during ceramic firing. Eur. J. Mineral. **13**, 621–634 (2001). https://citeseerx.ist.psu.edu/document?repid=rep1&type=pdf&doi=c00c990e77c50415cc3a653705893ded0d1a63f0
22. Cvejic, Z., Rakic, S., Kremenovic, A., Antic, B., Jovalekic, C., Colomban, P.: Nanosize ferrites obtained by ball milling: crystal structure, cation distribution, size-strain analysis and Raman investigation. Solid State Sci. **8**, 908–915 (2006). https://doi.org/10.1016/j.solidstatesciences.2006.02.041
23. Dayton, J.E.: Minerals Metals Glazing and Man, or Who Was Sesostris 1? George G. Harrup & Co, London (1978)
24. Dayton, J.E.: Cobalt, silver and nickel in Late Bronze Age glazes, pigments and bronzes, and the identification of silver sources for the Aegean and near east by lead isotope and trace element analysis. In: Hughes, M.J. (ed.) Scientific Studies in ancient Ceramics British Museum Occasional paper, 19, pp. 129–142 (1981a)

25. Dayton, J.E.: Geological evidence for the discovery of cobalt blue glass in mycenaean times as a by-product of silver smelting in the Schneeberg area of the bohemian Erzgebirge, *Revue d'Archéométrie*, n°1, Suppl. 57–61 (1981b). https://doi.org/10.3406/arsci.1981.1131
26. Dayton, J.E., Lamberg-Karlovsky, C.C.: The Discovery of Glass: Experiments in the Smelting of Rich, Dry Silver Ores, and the Reproduction of Bronze Age-Type Cobalt Blue Glass as a Slag. Peabody Museum of Archaeology and Ethnology, Harvard University (1993)
27. Deer, W.A., Howie, R.A., Zussman, J.: An Introduction to Rock-Forming Minerals. Pearson, Prentice Hall/London (1992)
28. de Souza, L.K.C., Zamian, da Rocha Filo, G.N., Soledade, L.E.B., dos Santos, I.M.G., Souza, A.G., Scheller, T., Angélica, R.S., da Costa, C.E.F.: Blue pigments based on $Co_xZn_{1-x}Al_2O_4$ spinels synthesised by the polymeric precursor method. Dyes Pigments. **81**, 187–192 (2009). https://doi.org/10.1016/j.dyepig.2008.09.017
29. de Waal, D.: Raman investigation of ceramics from 16th and 17th century Portuguese shipwrecks. J. Raman Spectrosc. **35**(8–9), 646–649 (2004). https://doi.org/10.1002/jrs.1210
30. de Waal, D.: Micro-Raman and portable Raman spectroscopic investigation of blue pigments in selected Delft plates (17th–20th century). J. Raman Spectrosc. **40**(12), 2162–2170 (2009). https://doi.org/10.1002/jrs.2389
31. do Nascimento-Dias, L.: Overview about Raman spectroscopy of types of olivine group minerals: a brief review. J. Raman Spectrosc. **553**(11), 1942–1946 (2022). https://doi.org/10.1002/jrs.6412
32. Eccleston, M.: Technological and Social Aspects of High-Temperature Industries in the Dakhleh Oasis, Egypt, During the Ptolemaic and Raman Periods. PhD Thesis, Monash University (2006)
33. Edwards, H.G.M.: Porcelain Analysis and Its Role in the Forensic Attribution of Ceramic Specimens. Springer Nature, Cham (2022)
34. El Goresy, A., (2000), "Polychromatic Wall Painting Decoration in Monuments of Pharaonic Egypt: Composition, Chronology and Painting Technology", in: Proc. Of the First International Symp., *The Wall paintings of Thera,* (Ed. S. Sherratt), Thera Foundation, Piraeus, pp. 49-70
35. Farnsworth, M., Ritchie, P.D.: Spectrographic studies on ancient glass. Egyptian glass mainly of 18th dynasty, with special reference to its cobalt content. Spectroscopic studies in the Field of the Fine Arts. **6**(3), 155–173 (1938)
36. Fernández, A.L., de Pablo, L.: Formation and the colour development in cobalt spinel pigments. Pigment Resin Technol. **31**(6), 350–356 (2002). https://doi.org/10.1108/03699420210449043
37. Florés, A., Llusar, M., Badenes, J.A., Calbo, J., Tena, M.A., Monrós, G.: Cobalt minimisation in willemite ($Co_xZn_{1-x}SiO_4$) ceramic. Green Chem. **2**, 93–100 (2000). https://doi.org/10.1039/B000748J
38. Freestone, I.C.: Composition and microstructure of early opaque red glasses. In: Bimson, M., Freestone, I. (eds.) Early vitreous material British Museum Occasional Paper 56, pp. 173–191. British Museum Publications, London (1987)
39. Freestone, I., Gortin-Rosen, Y.: The great glass slab from Bet She'arim, Israel: an early Islamic glassmaking experiment? J. Glass Stud. **41**, 105–116 (1999). https://www.jstor.org/stable/24190848
40. Giannakas, A.E., Ladavos, A.K., Armatas, G.S., Pomonis, P.J.: Surface properties, structural features and catalytic performance for NO + CO abatement of spinels MAl_2O_4 (M = Mg, Co and Zn) developed by reverse and bicontinuous microemulsion method. Appl. Surf. Sci. **253**, 6969–6972 (2007). https://doi.org/10.1016/j.apsusc.2007.02.031
41. Glebycheva, A.I., KJazakova, N.N., Vizir, L.A., Bystrikov, A.S., Gorbachev, V.V.: Ceramic pigments based on cobalt-containing forsterite. Steklo I Keramika. **2**, 24 (1983)
42. Goodgame, M., Cotton, F.A.: Magnetic investigations of spin-free cobaltous complexes. IV. Magnetic properties and spectrum of cobalt(II) orthosilicate. J. Phys. Chem. **65**(5), 791–792 (1961). https://doi.org/10.1021/j100823a020

43. Gratuze, B., Soulier, I., Barrandon, J.N., Foy, D.: De l'origine du cobalt dans les verres. Revue d'Archéométrie. **16**, 97–108 (1992)
44. Gratuze, B., Soulier, I., Blet, M., Vallauri, L.: De l'origine du cobalt: Du verre à la céramique. Revue d'Archéométrie. **20**, 77–94 (1996). https://www.persee.fr/doc/arsci_0399-1237_1996_num_20_1_939
45. Gratuze, B., Pactat, I., Schibile, N.: Changes in the signature of cobalt colorants in late antique and early Islamic glass production. Fortschr. Mineral. **8**, 225 (2018). https://doi.org/10.3390/min8060225
46. Hagen, A., Östby, J.: Oxidation States of Mn, Cr, and Co in Mixed Spinels Studied by XANES, pp. 1–10. Risø National Laboratory, Roskilde, 130.226.56.153 (2006)
47. Henderson, J.: The raw materials of early glass production. Oxford J. Archaeol. **4**(3), 267–291 (1985). https://doi.org/10.1111/j.1468-0092.1985.tb00248.x
48. Henderson, J.: The scientific analysis of ancient glass and its archaeological interpretation. In: Henderson, J. (ed.) Scientific Analysis in Archaeology Monograph 19, pp. 30–62. Oxford University Committee for Archaeology, Oxford (1989)
49. Hope, C.: Two examples of Egyptian blue-painted Pottery in the Medelhavsmuseet. Medelhavsmuseet Bull. **12**, 7–11 (1977)
50. Hope, C.A.: Pottery of the Egyptian New Kingdom: Three Studies Occasional Paper No. 2. Victoria College, Melbourne (1989)
51. Hope, C.A.: Pottery manufacture in the Dakhleh Oasius. In: Churcher, C.S., Mills, A.J. (eds.) Reports from the Survey of the Dakhleh Oasis Western Desert Egypt 1977–1987 Oxbow Monograph 99, pp. 215–250. Oxbow Books, Oxford (1999)
52. Hope, C.A.: Early and mid-holocene ceramics from the Dakhleh Oasis: Traditions and influences. In: Friedman, R. (ed.) Egypt and Nubia – Gifts of the Desert, pp. 39–61. The British Museum Press, London (2002)
53. Hope, C.A., Blauer, M., Riederer, J.: Studien zur altägyptischen Keramik, pp. 139–166. Verlag Philipp Von Zabern, Mainz am Rhein (1981)
54. Hope, C.A., Kucera, P., Smith, J.: Alum exploitation at Qasr el-Dakhleh in the Dakhleh Oasis. In: Ikram, S., Dodson, A. (eds.) Beyond the Horizon, vol. 1, pp. 165–179. Supreme Council of Antiquities Press, Cairo (2009)
55. Ingo, G.M., Chiozzini, G., Faccenda, V., Bemporad, E., Riccucci, C.: Thermal microchemical Characterisations of $CaSO_4$-SiO_2 investment materials for casting jewellery alloys. Thermochim. Acta. **321**(1–2), 175–183 (1998). https://doi.org/10.1016/S0040-6031(98)00457-2
56. Jackson, C.M., Nicholson, P.T.: Compositional Analysis of the Vitreous Materials Found at Amarna, in: Brilliant Thins for Akhenaten: the Production of Glass, Vitreous Materials and Pottery at Amarna Site O45.1. Egypt Exploration Society, London (2007)
57. Jackson, C.M., Nicholson, P.T.: The provenance of some glass ingots from the Uluburun shipwreck. J. Archaeol. Sci. **37**(2), 295–301 (2010). https://doi.org/10.1016/j.jas.2009.09.040
58. Jay, W.H.: Non-destructive characterisation of pigments from Dakhleh Oasis, and the Nile Valley, Egypt from the New Kingdom, Ptolemaic and Roman periods. PhD Thesis, Monash University, Australia (2017)
59. Jay, W.H., Cashion, J.D.: Raman spectroscopy of Limehouse porcelain sherds supported by Mössbauer spectroscopy and electron microscopy. J. Raman Spectrosc. **44**(12), 1718–1732 (2013). https://doi.org/10.1002/jrs.4404
60. Jongsomjit, B., Panpranot, J., Goodwin Jr., J.G.: Co-support compound formation in alumina-supported cobalt catalysts. J. Catalysis. **204**(1), 98–109 (2001). https://doi.org/10.1006/jcat.2001.3387
61. Jonynaite, D., Senvaitiene, J., Beganskiene, A., Karieva, A.: Spectroscopic analysis of blue cobalt smalt pigment. Vib. Spectrosc. **52**(2), 158–162 (2010). https://doi.org/10.1016/j.vibspec.2009.12.005
62. Kaczmarczyk, A.: The source of cobalt in ancient Egyptian pigments. In: Olin, J.S., Blackman, M.J. (eds.) Proceeding of the 24th International. Archaeometry Symp, pp. 369–376. Smithsonian Institution Press, Washington, DC (1986)

63. Kaczmarczyk, A., Hedges, R.E.M.: Ancient Egyptian Faience: an Analytical Survey of Egyptian Faience from Predynastic to Roman Times. Aris and Phillips, Warminster (1983)
64. Kissin, S.A.: Five-element (Ni-Co-As-Ag-Bi) veins. Geosci. Can. **19**, 113–124 (1992) Available online: https://journals.lib.unb.ca/index.php/gc/article/view/3768/4282/. Accessed on 15 Dec 2019
65. Koch, L.D., de Waal, D.: Raman studies of the underglaze blue pigment on ceramic artefacts of the Ming Dynasty and of unknown origins. J. Raman Spectrosc. **38**(11), 1480–1487 (2007). https://doi.org/10.1002/jrs.1805
66. Krzemnicki, M.S.: Diopside needles as inclusions in Demantoid garnet from Russia: a Raman microspectrometric study. Gems Gemol. **35**(4), 192–195 (1999). https://www.gia.edu/doc/Diopside-Needles-as-Inclusions-in-Demantoid-Garnet-from-Russia-A-Raman-Microspectrometric-Study.pdf
67. Lager, G.A., Armbruster, T., Rotella, F.J.: A crystallographic study of the low-temperature dehydration products of gypsum, $CaSO_4.2H_2O$; hemihydrate $CaSO_4.0.5H_2O$ and γ-$CaSO_4$. Am. Mineral. **69**(9–10), 910–918 (1984). https://pubs.geoscienceworld.org/msa/ammin/article-abstract/69/9-10/910/41637/A-crystallographic-study-of-the-low-temperature
68. Lavat, A.E., Wagner, C.C., Tasca, J.E.: Interaction of Co-ZnO pigments with ceramic frits: a combined study by XRD, FTIR and UV-visible. Ceram. Int. **34**(8), 2147–2153 (2008). https://doi.org/10.1016/j.ceramint.2007.09.003
69. Lee, L., Quirke, S.: Painting materials. In: Nicholson, P.T., Shaw, I. (eds.) Ancient Egyptian Materials and Technology, pp.104–120. Cambridge University Press, Cambridge, UK (2000)
70. Lilyquist, C., Brill, R.H.: Studies in Ancient Egyptian Glass. Metropolitan Museum of Art, New York (1993)
71. Lin, C.-C.: Vibrational spectroscopic study of the system α-Co_2SiO_4-α-Ni_2SiO_4. J. Solid State Chem. **157**(1), 102–109 (2001). https://doi.org/10.1006/jssc.2000.9044
72. Llusar, M., Forés, A., Badenes, J.A., Calbo, J., Tena, M.A., Monrós, G.: Colour analysis of some cobalt-based blue pigments. J. Eur. Ceramic Soc. **21**(8), 1121–1130 (2001). https://doi.org/10.1016/S0955-2219(00)00295-8
73. Lucas, A., Harris, J.R.: Ancient Egyptian Materials and Industries, 4th edn, London (1962)
74. Maas, J.L., Wypski, M.T., Stone, R.E.: Malkata and Lisht glassmaking technologies: towards a specific link between Second Millenium BC metallurgists and glassmakers. Archaeometry. **44**(1), 67–82 (2002). https://doi.org/10.1111/1475-4754.00043
75. Masse, S., Boch, P., Vaissière, N.: Trapping of nickel and cobalt in $CaNiSi_2O_6$ and $CaCoSi_2O_6$ Diopside-like phases in heat-treated cement. J. Eur. Ceram. Soc. **19**(1), 93–98 (1999). https://doi.org/10.1016/S0955-2219(98)00114-9
76. Monari, G., Manfredini, T.: Colouring effects of synthetic inorganic cobalt pigments in fast-fired porcelainized tiles. Ceram. Eng. Sci. Proc. **17**, 167–172 (1996)
77. Moorey, P.R.S.: Antiquity, **53**, p 104, Gloucester, UK (1978)
78. Mouri, T., Enami, M.: Raman spectroscopic study of olivine-group minerals. J. Mineral. Petrol. Sci. **103**(2), 100–104 (2008). https://doi.org/10.2465/jmps.071015
79. Murad, E., Cashion, J.: Mössbauer Spectroscopy of Environmental Materials and their Industrial Utilisation. Kluwer Academic publishers, Boston (2004)
80. Mutke, M., Kreye, M., Shi, J., Becker, K.D.: Kinetics of cation distribution in cobalt-containing olivine, $(Co_{0.6}Mg_{0.4})_2SiO_4$. Phys. Chem. Chem. Phys. **10**, 3895–3902 (2008). https://doi.org/10.1039/B718182E
81. Nicholson, P.T.: Brilliant Things for Akhenaten. In: The Production of Glass, Vitreous materials and Pottery at Amarna Site O45.1 Excavation Memoir 80. Egypt Exploration Society, London (2007)
82. Nicholson, P.T.: Working in Memphis: The Production of Faience at Roman Period Kom Helul Excavation Memoir 105. Egypt Exploration Society, London (2013)
83. Nicholson, P.T., Henderson, J.: Glass. In: Ancient Egyptian Materials and Technology, pp. 195–226. Cambridge University Press, Cambridge (2000)

84. Nicholson, P.T., Peltenburg, E.: Egyptian faience. In: Ancient Egyptian Materials and Technology. Cambridge University Press, Cambridge (2000)
85. Nicholson, P.T., Shaw, I.: Ancient Egyptian Materials and Technology. Cambridge University Press, Cambridge (2000)
86. Nicolini, L., Porter, P.: Preparation, X-ray and magnetic investigations of some silicates containing transition metal ions. Gazz. Chim. Ital. **100**, 923–930 (1970)
87. Nikita, K., Henderson, J.: Glass analyses from Mycenaean Thebes and Elateia: compositional evidence for a Mycenaean glass industry. J. Glass Stud. **48**, 71–120 (2006). https://www.jstor.org/stable/24191147
88. Noll, V.W.: Mineralogie und Technik der bemalten Keramiken Altägyptens. N. Jb. Miner. Mh. **133**, 227–290 (1978)
89. Noll, W.: Mineralogy and technology of the painted ceramics of ancient Egypt. In: Hughes, M. (ed.) Scientific Studies in Ceramics British Museum Occasional Paper 19, pp. 143–154. British Museum Publication, London (1981)
90. Noll, W., Hangst, K.: Zuz Kenntnis der altägyptischen Blaupigmente. N. Jb. Miner. Mh. H. **12**, 309–314 (1975)
91. O'Neill, H., Navrotsky, A.: Simple spinels: crystallographic parameters, cation radii, lattice energies, and cation distribution. Am. Mineral. **68**(1–2), 181–194 (1983). https://pubs.geoscienceworld.org/msa/ammin/article-abstract/68/1-2/181/41446/Simple-spinels-crystallographic-parameters-cation
92. Perry, R.H., Green, D.: Perry's Chemical Engineers' Handbook, 6th edn. McGraw-Hill International (1984)
93. Prasad, P.S.R.: Raman intensities near gypsum-bassanite transition in natural gypsum. J. Raman Spectrosc. **30**(8), 693–696 (1999). https://doi.org/10.1002/(SICI)1097-4555(199908)30:8%3C693::AID-JRS434%3E3.0.CO;2-8
94. Prasad, P.S.R., Pradham, A., Gowd, T.N.: *In situ* micro-Raman investigation of dehydration mechanism in natural gypsum. Curr. Sci. **80**(9), 1202–1207 (2001)
95. Ramsdell, L.S., Partridge, E.P.: The crystal forms of calcium sulphate. Am. Mineral. **14**(2), 59–74 (1929). https://pubs.geoscienceworld.org/msa/ammin/article-abstract/14/2/59/535480/The-crystal-forms-of-calcium-sulphate
96. Rehren, T.: Rationales in Old World base glass compositions. J. Archaeol. Sci. **27**(12), 1225–1234 (2000). https://doi.org/10.1006/jasc.1999.0620
97. Rehren, T.: Aspects of the production of cobalt-blue glass in Egypt. Archaeometry. **43**(4), 483–489 (2001). https://doi.org/10.1111/1475-4754.00031
98. Rehren, T.: A review of the factors affecting the composition of early Egyptian glasses and faience: alkali and alkaline earth oxides. J. Archaeol. Sci. **35**(5), 1345–1354 (2008). https://doi.org/10.1016/j.jas.2007.09.005
99. Riederer, J.: Recently identified Egyptian pigments. Archaeometry. **16**(1), 102–109 (1974). https://doi.org/10.1111/j.1475-4754.1974.tb01098.x
100. Richmond, H.D., Off, H.: Indications of a possible new element in an Egyptian mineral. J. Chem. Soc Transactions. **61**, 491–495 (1892). https://pubs.rsc.org/en/content/articlelanding/1892/ct/ct8926100491/unauth
101. Robertson, K., Bish, D.: Constraints on the distribution of $CaSO_4 \cdot nH_2O$ phases on Mars and implications for their contribution to the hydrological cycle. Icarus. **223**(1), 407–417 (2013). https://doi.org/10.1016/j.icarus.2012.10.028
102. Rose, P.J.: The Eighteenth Dynasty Pottery Corpus from Amarna 83rd Excavation Memoir, (Ed. J. Bourriau). Egypt Exploration Society, London (2007)
103. Rosi, F., Miliani, C., Borgia, I., Brunetti, B., Sgamellotti, A.: Identification of nineteenth century blue and green pigments by *in situ* x-ray fluorescence and micro-Raman spectroscopy. J. Raman Spectrosc. **35**(8–9), 610–615 (2004). https://doi.org/10.1002/jrs.1180
104. Salem, S.: Effect of calcination temperature on colorant behaviour of cobalt-aluminate nanoparticles synthesised by combustion technique. J. Industrial Eng. Chem. **20**(3), 818–823 (2014). https://doi.org/10.1016/j.jiec.2013.06.011

105. Schairer, J.F., Bowen, N.L.: The system leucite-Diopside-silica. Am. J. Science. **5**, 289–309 (1938). https://ajsonline.org/article/120541.pdf
106. Sedel'nikova, M.B., Pogrebenkov, V.M.: The effect of mineralising additives on synthesis of ceramic pigments based on natural wollastonite. Glas. Ceram. **63**, 22–225 (2006)
107. Segnit, E.R.: Evaporite minerals from the Dakhleh Oasis. In: Edwards, W.I., Hope, C.A., Segnit, E.R.T. (eds.) Ceramics from the Dakhleh Oasis Occasional Paper 1, pp. 97–102. Victoria College Press, Burwood (1987)
108. Segnit, E.R., Holland, A.E.: The system MgO-ZnO-SiO_2. J. Am. Ceram. Soc. **48**(8), 409–413 (1965). https://doi.org/10.1111/j.1151-2916.1965.tb14778.x
109. Sepelak, V., Becker, K.D.: Comparison of the cation inversion parameter of the nanoscale milled spinel ferrites with that of the quenched bulk materials. Mater. Sci. Eng. A. **375–377**, 861–864 (2004). https://doi.org/10.1016/j.msea.2003.10.178
110. Shirai, H., Morioka, Y., Nakagawa, I.: Infrared and Raman spectra and lattice vibrations of some oxide spinels. J. Phys. Soc. Jpn. **51**, 592–597 (1982). https://doi.org/10.1143/JPSJ.51.592
111. Shortland, A.J., Tite, M.S.: Raw materials of glass from Amarna and implications for the origins of Egyptian glass. Archaeometry. **42**(1), 141–151 (2000). https://doi.org/10.1111/j.1475-4754.2000.tb00872.x
112. Shortland, A.J., Hope, C.A., Tite, M.S.: Cobalt blue painted pottery from the 18th Dynasty Egypt. In: Maggetti, M., Messiga, B. (eds.) Geomaterials in Cultural Heritage Special Publications, 257, pp. 91–99. Geological Society, London (2006a)
113. Shortland, A.J., Tite, M.S., Ewart, I.: Ancient exploitation and use of cobalt alums from the Western Oases of Egypt. Archaeometry. **48**(1), 153–168 (2006b). https://doi.org/10.1111/j.1475-4754.2006.00248.x
114. Shugar, A., Rehren, T.: Formation and composition of glass as a function of firing temperature. Glass Technol. **43**, 145–150 (2002). https://discovery.ucl.ac.uk/id/eprint/93659/1/Rehren_ShugarRehren%202002%20Glass%20firing%20temperature%20Glass%20Technol%2043C.pdf
115. Sickafus, K.E., Wills, J.M.: Structure of spinel. J. Am. Ceram. Soc. **82**(12), 3279–3292 (1999). https://doi.org/10.1111/j.1151-2916.1999.tb02241.x
116. Sievert, T., Wolter, A., Singh, N.B.: Hydration of anhydrite of gypsum ($CaSO_4$.II) in a ball mill. Cem. Concr. Res. **35**(4), 623–630 (2005). https://doi.org/10.1016/j.cemconres.2004.02.010
117. Simon, G., Costanzo, S., Lisiecki, I., Colomban, P.: Multiscale identification of the inorganic shell of core (Co)/shell-assembled nanoparticles, J. Raman Spectrosc., 55 (2024 in press). https://doi.org/10.1002/jrs.668
118. Smirniou, M., Rehren, T.: Direct evidence of primary glass production in late Bronze Age Amarna, Egypt. Archaeometry. **53**(1), 58–80 (2011). https://doi.org/10.1111/j.1475-4754.2010.00521.x
119. Soukiassian, G., Wuttmann, M., Pantalacci, I., Ballet, P., Picon, M.: Balat III. Les Ateliers de Potiers d'Ayn-Asil. Institut Français d'Archéologie Orientale, Cairo (1990)
120. Stola, M., Stefanescu, M., Dippong, T., Stefanescu, O., Barvinschi, P.: Low temperature synthesis of Co_2SiO_4/SiO_2 nanocomposite using a modified sol-gel method. J. Sol Gel Technol. **54**, 49–56 (2010). https://doi.org/10.1007/s10971-010-2156-2
121. Tanimoto, S., Rehren, T.: Interactions between silicate and salt melts in LBA glassmaking. J. Archaeol. Sci. **35**(9), 2566–2573 (2008). https://doi.org/10.1016/j.jas.2008.04.016
122. Tite, M.S.: Characterisation of early vitreous materials. Archaeometry. **29**(1), 21–34 (1987). https://doi.org/10.1111/j.1475-4754.1987.tb00394.x
123. Tite, M.S., Shortland, A.J.: Production technology for copper- and cobalt-blue vitreous materials from the New Kingdom site of Amarna – a reappraisal. Archaeometry. **45**(2), 285–312 (2003). https://doi.org/10.1111/1475-4754.00109
124. Tite, M.S., Shortland, A., Maniatis, Y., Kavoussanaki, D., Harris, S.A.: The composition of the soda-rich and mixed alkali plant ashes used in the production of glass. J. Archaeol. Sci. **33**(9), 1284–1292 (2006). https://doi.org/10.1016/j.jas.2006.01.004

125. Turner, W.E.S.: Studies of ancient glass and glassmaking processes, Part V. Raw materials and melting processes. J. Soc. Glass Technol. **40**, 277–300 (1956). http://sgthome.co.uk/Downloads/JSGT_1919_P001.pdf
126. Vandenberghe, R.E., De Grave, E.: Mössbauer effect studies of Oxidic Spinels. In: Long, G.J., Grandjean, F. (eds.) Mössbauer Spectroscopy Applied to Inorganic Chemistry, vol. 3. Plenum Press, New York (1989) Ser. Title: Modern Inorganic Chemistry, Ed. J.P. Fackler, Jr., 59–182
127. Van Minh, N., Yang, I.S.: A Raman study of cation-disorder transition temperature of natural $MgAl_2O_4$ spinel. Vib. Spectrosc. **35**(1–2), 93–96 (2004). https://doi.org/10.1016/j.vibspec.2003.12.013
128. Vieira Ferreira, L.F., Casimiro, T.M., Colomban, P.: Portuguese tin-glazed earthenware from the 17th century, Part 1: pigments and glazes characterisation. Spectrochim Acta A. **104**, 437–444 (2013). https://doi.org/10.1016/j.saa.2012.11.069
129. Warachim, H., Rzechula, J., Pilak, A.: Magnesium-cobalt-aluminium spinels for pigments. Ceram. Int. **11**(3), 103–106 (1985). https://doi.org/10.1016/0272-8842(85)90005-7
130. Wood, J.R., Hsu, Y.-T.: An archaeometallurgical explanation for the disappearance of Egyptian and near Eastern Cobalt-blue Glass at the end of the Late Bronze Age. Internet Archaeology, 52 (2019). https://doi.org/10.11141/ia.52.3
131. Weast, R.: RC Handbook of Chemistry and Physics. CRC Press, Boca Raton (1978)

Chapter 6
Egyptian Green

Abstract The ancient Egyptians attached a special significance to the colour green as it was considered to be associated with life itself. Egyptian green is shown to be produced by variation of the CaO/CuO ratio, kiln temperature and kiln atmosphere used in the production of Egyptian blue. The major minerals synthesised are para-wollastonite and tridymite, indicating a firing temperature of ~1000 °C. Egyptian green was often applied to pottery using beeswax, often over either huntite or a silica slip and, what are called generically "green earths" were applied to cartonnage in a tomb at Kellis cemetery. Alkali-lime glazes, later with added lead to lower the fusion temperature, were applied to pottery surfaces. During the Roman period, a lustre glaze was introduced by the firing under reducing conditions of silver or copper dissolved in vinegar on to a lead-free aluminosilicate glaze.

Keywords Egyptian blue · Egyptian green · Kellis · Green earths · Lustre glaze · Calcium to copper ratios · Alkali lime glazes · Reducing conditions

6.1 Historical Context

In ancient cultures, art and technique were a single entity due mainly to their early stage of development. They were defined by the necessities of life and their technical possibilities [7]. Thus, Egyptian blue, closely followed by Egyptian green (green frit) were first developed in Egypt during the Old Kingdom. They are considered by many to be the very first chemical synthetic processes to be developed by human endeavour.

The Egyptians attached a special significance to the colour green [58]. It is believed that the colour was most likely associated with life itself because of the association of green with the propagation of plants and crops. It was also seen as a sign of resurrection and early Egyptian texts refer to the afterlife as the "field of malachite". The colour is also closely associated with water [115]. Consequently the colour green was highly prized in both natural and manufactured forms.

H. G. M. Edwards, P. Colomban, *Blue by Fire: A Marker of the Technical History of Glass and Ceramics*, Cultural Heritage Science,
https://doi.org/10.1007/978-3-031-77122-4_6

The use of the terms Egyptian green, or "green frit", to describe an early green pigment has become a matter of some conjecture with Schiegl and El Goresy [90] stating in response to a paper by Pagès-Camagna and Colinart [78,79] that the correct term is green frit "in accord with the previously established international nomenclature by Lucas" [56]. In a further response, Pagès-Camagna and Colinart stated that Lucas and Harris [57] did not define "frit" and stated that this term describes the process to obtain a solid from powders, such as blue frit for Egyptian blue. According to Ullrich [108], because it appears as a glassy mass, therefore the term "frit" is justified. Furthermore, he claimed that ground glass itself was used as a pigment. Because glass is isotropic, and Egyptian green has major glass phases when examined by XRD, Ullrich was unable to provide adequate additional information. The main phases he detected were wollastonite and quartz. The term Egyptian green will be used throughout this text.

Commencing in the Late Bronze Age, Egypt was a leading glass production site. The primary production of Roman and late antique glasses was located close to suitable silica deposits either on the Palestinian coast or in Egypt itself and relied on trona from the Wadi el-Natron as the alkali flux prior to the change in the Levant and Egypt to soda-rich plant ash fluxes [92].

A number of authors have reviewed glass manufacture in the Near East [5, 68, 77, 84]. Glass beads have been recovered from archaeological contexts dated to the early third millennium BCE onwards in Mesopotamia and by the Fifth Dynasty in Egypt [68, 70, 94]. Rehren [85] reviewed glass production in ancient Egypt, proposing that its manufacture first commenced in Mesopotamia in around 1600 BCE and in Egypt in about 1550 BCE. The study of Roman and early Medieval glass has been reviewed by Smith and Clark [95] and glass production dated between the first and seventh century CE from Roman and Byzantine glass industries by Maltoni et al. [60]. Rehren proposed that glass was produced using only two raw materials, namely crushed quartz pebbles and plant ash, such as the unrefined ash of halophytic plants. Freestone [33] noted that if the flux was ash, then a lime-free silica was used, however, if a soda-based flux was used, then a calcareous quartz sand was required. Wedepohl et al. [114] suggested that the term plant ash should be replaced with soda ash glass as wood ash can also be used (wood ash glass); and trona [$Na_3H(CO_3)_2.2H_2O$] rather than natron is the major sodium mineral found in low-Mg glass (soda-lime glass). Caley [13] suggested that perhaps the ash of the fuel employed to melt the glass was often incorporated into successive batches as one of the glassy raw materials.

6.2 Synthesis and Phases Characterisation

Egyptian green has received significantly less reporting than Egyptian blue [37] in the published literature. A green tone in Egyptian blue was alternatively said to be introduced by the presence of iron in the raw materials [53, 108, 113: 208] and Pagès-Camagna et al. [78, 79, 80, 81] suggest that earlier researchers considered that Egyptian green was merely a variation of Egyptian blue, the green alternatively being introduced by varying the CaO/CuO ratio and the firing conditions, particularly temperature and kiln atmosphere.

It has been postulated that Egyptian green was synthesised soon after the discovery of Egyptian blue [78]. These researchers reported that Egyptian green is characterised by the presence of para-wollastonite ($CaSiO_3$) crystals smaller than 10 μm and residual silica (quartz and/or tridymite or cristobalite) embedded in an amorphous, silica-rich glassy phase. They also observed firing residues of tenorite (CuO), which they proposed indicated that this pigment was synthesised under oxidising conditions, because under reducing conditions red cuprite (Cu_2O) or Cu° metal would have formed. Tin was also found but the form was unable to be identified by these workers; they suggested that it was present either as cassiterite or as an unidentified Ca-Si-Sn complex (60% SnO_2, 20% SiO_2 and 19.6% CaO), which is structurally different from malayaite ($CaSiSnO_5$) as discussed by Groat et al. [35].

Egyptian green is obtained either by modification of the proportions of the initial compounds [53, 91, 108] or by variation in the firing conditions, namely temperature and atmosphere in the furnace [43, 75, 76, 108]. Furthermore, the colour of the synthetic product can be influenced by the heating temperature and the quantity of the added flux [78], and the presence of nucleation agent(s).

Egyptian green was never mentioned in ancient texts and early studies have suggested that it was perhaps an alteration product induced by the presence of iron. In the case of Egyptian green, the historical usage of this pigment is disputed by several writers. Noll and Hangst ([75, 76]] and Noll [73, 74] suggest that a green pigment containing wollastonite (Egyptian green) was possibly an accidental by-product obtained from the synthesis of Egyptian blue. However, a similar wollastonite-containing pigment, called "green frit" was intentionally produced according to Jaksch et al. [43]. Wollastonite ($CaSiO_3$), together with glass, are major phases in numerous Egyptian green pigments [43], who also stated that the manufacture of Egyptian blue under reducing conditions will lead to the formation of Egyptian green, the major phase consisting of copper wollastonite.

According to Ullrich [108] Egyptian green was developed during the 6th Dynasty. Jaksch et al. [43] claimed that it has been found in many 5th and 6th Dynasty sites. And, as discussed by Noll and Hangst [76], Egyptian green production was achieved by the ancient Egyptians prior to, or during the Fifth Dynasty of the Old Kingdom (2480–2320 BCE). Lucas and Harris [57] observed that the green pigments in paintings from the Fourth Dynasty to the New Kingdom were malachite or chrysocolla, the former dominating. According to Lucas [56], Spurrell recorded the use of malachite, or malachite and gypsum and/or malachite over yellow, in tomb paintings of the Fourth Dynasty. In paintings from the Middle and New Kingdoms, both malachite and chrysocolla were employed. The green used in decorating the tomb of Perneb (5th Dynasty) was reported to be malachite. The use of green frit or Egyptian green was recorded by Lucas [57] in an unnamed 6th Dynasty tomb, together with six specimens from 18th Dynasty tombs and, one from a 26th Dynasty tomb.

El Goresy et al. [32] initially suggested that the earliest use of this pigment was during the 5th and 6th Dynasties in tomb paintings at Saqqara.

However, El Goresy [31] revised this time period and stated that these early green pigments consisted of a copper-blue glass with a low calcium concentration which has weathered to form atacamite ($Cu_2(OH)_3Cl$), a basic copper chloride or copper carbonate (such as malachite). El Goresy et al. [32] proposed that the copper-bearing wollastonite, which he termed "green frit" was first used at the beginning of the 18th Dynasty on tomb paintings at Edfu and El Kab. Without providing supporting analytical data, Pagès-Camagna and Colinart [78] proposed a much earlier date in tomb paintings and on stone statues which occurred during the First Intermediate period (ca. 2200 BCE).

Hatton et al. [38] suggested that Egyptian green seems to have been confined to Egypt, and Pagès-Camagna and Colinart [78], Pagès-Camagna et al. [81] and El Goresy [31] concurred with this, suggesting that this pigment was only used during the Middle and New Kingdoms (2100–1069 BCE) and was confined to Egyptian territory (Nubia, Sinai, Egypt). Hatton [36] agreed with these workers and stated that the use of Egyptian green commenced sometime between the second half of the 3rd and the mid-2nd millennium BCE.

Lau et al. [52] noted that in the literature it is suggested that Egyptian blue degrades to green by either of two mechanisms. In the first mechanism, the original blue colour may change to a faint green as copper ions are leached out and combine with chloride ions and calcium carbonate forming the basic copper chlorides, atacamite and para-atacamite. The second mechanism described instances where the organic binding medium around Egyptian blue had darkened or yellowed sufficiently altering the appearance of the blue to a green colour. Lau artificially aged Egyptian blue using HCl and then examined the degraded product by XRD and showed that atacamite, copper chloride hydroxide hydrate ($Cu_xCl_4(OH)_z.2H_2O$) and eriochalcite ($CuCl_2(H_2O)_2$) were formed. Baraldi et al. [3] suggested that a weak Raman band observed at 222 cm^{-1} is attributable to $CuCl_2$. Other suggested minerals present include lithiodionite ($KNaCuSi_4O_{10}$), effenbergite [$BaCuSi_4O_{10}$. $CaCuSi_2O_6$] and dioptase ($Cu_6S_{i6}O_{18}$).

Note, kohls, which are a cosmetic prepared from Cu-rich mineral (malachite), when applied to the eyes in the form of eye shadow were also used to treat eye problems. The most common type of kohl used was black [57, 88]. But, the term 'green kohl' as described, can be found in 18 medical paragraphs. Again the use of 'green kohl' (Egyptian green ?) in the treatment of the eyes is well attested with 88% (16 out of 18 medical paragraphs) requiring it to be administered to the eye in some particular manner [88]. Cataracts, poor eyesight and being bleary eyed are all reasons given for the use of 'green kohl' [88].

6.3 History of Glazing in Egypt

Alkali-based glazes were developed in Egypt from the 4th millennium BCE onwards. Toffolo et al. [107] and Tite and Bimson [100–102], state that in Predynastic Egypt, quartz pebbles were coated with a mixture of natron and malachite and fired.

- The glaze was subsequently removed from the pebbles, powdered and then applied to a prepared quartz body and again fired. Such a method was termed the application method;
- Wulff et al. [116] proposed an alternative approach adopted in Qom, Iran, in which the quartz core is embedded within the glazing mix and then fired. In this method, the glaze consisted of a mixture of quartz powder, copper oxide, hydrated lime, plant ash and charcoal. This method has been described as cementation glazing.
- A third method has been described by Noble [72] in which a mixture of powdered quartz, natron, copper oxide, bentonite clay, whiting (calcium carbonate) and water is applied to the surface of the object. On drying, the alkali salts will effloresce and, by capillary action precipitate at the surface of the ware to provide a coating. When fired to ~95 °C, a glaze forms (termed efflorescence glazing) [93].
- These three methods have been discussed in detail by Nicholson [71] and confirmed experimentally in studies conducted by Vandiver [109–111] and by Tite et al. [103, 104, 105]. Replicas were also made by Kieffer at the Sèvres National Factory (France) in the 1960s.

During the fourteenth century BCE the use of glazed clay ceramics became widespread in the Near East from Iran in the east to Syria in the west. Tite and Shortland [106] proposed that the first glazed clay objects were produced in Mesopotamia at around 1500 BCE. Furthermore, glazed pottery and bricks with quartz-based bodies were produced during the Neo-Elamite (c. 725–539 BCE) and Achaemenid (c. 539–330 BCE) periods. McCarthy and Paynter [64] concluded that clay-based ceramics which were glazed using a lime-soda composition first appeared in Mesopotamia in the 2nd millennium BCE and their production continued into the Islamic period (650 CE) onwards. Greene [34] proposed that glazed clay-based ceramics were not produced in Egypt until the introduction of lead glazed pottery during the Roman period. Until this introduction, the body fabric in Egypt was always high in silica. Nicholson [71] proposed that the first glazes on clay-based ceramics did not appear in Egypt until the Late Bronze Age, in around 1600 BCE. These glazes were all alkali-lime based. McCarthy and Paynter [64] and Paynter and Tite [82] proposed that the delay in the introduction of lead-glazed clay ceramics into Egypt was probably related to cultural factors; building stone was widely used in Egypt but rarely available in Mesopotamia, leading to them developing and extensively using both fired and unfired clay to construct buildings and related structures.

Hedges [39] reported the sporadic presence of lead and tin in glazes rich in copper. He was uncertain if these two elements entered the glaze from the use of bronze. He also reported that tin and lead only occur at Kish in Parthian and later times and, also that the glazes from Nineveh normally contained zinc. This could indicate the use of brass rather than bronze as the copper source. He also suggested that in Egypt, brown glazes result from high Fe containing Mn ores.

An alkali glaze was initially used in Egypt and the Near East up to as late as the fourteenth or fifteenth century CE, this glaze being a mixture of silica and plant ash [40]. Pre-Islamic faience from Nippur, Iraq which spans the time from late Kassite (c. 1300 BCE) to Parthian (250 CE) are all alkali-lime-based, lead glazes not being detected [39]. Some slight differences were observed by Hedges in glazes from Kish and Nineveh, Iraq.

Hedges and Moorey [40] considered that around the time of the 22nd Dynasty a lead-based glaze was introduced, into Egypt, both types of glazes continuing to be used concurrently. However, Tite and Shortland [106] proposed that the first appearance of transparent high-lead glazes was in Anatolia sometime during the first century BCE. Lead glazes were found to melt at lower temperatures, to offer a thermal expansion similar to clay bodies and to fuse more uniformly to it. The systematic use of a fine, white slip in conjunction with a colourless lead glaze became a major characteristic of the eastern Islamic world during the eighth to ninth century CE [61]. In Mesopotamia, alkaline glazed pottery, often based on Hellenistic forms was current from the first century BCE to third century CE and was more limited in its application than the lead-glazed materials [71]. As discussed by Mason and Tite [63], tin-opacified glazes were then introduced into Abbasid Iraq during the eighth or ninth centuries CE, possibly in an attempt to replicate Chinese porcelain imports. Tin-opacified glazes were introduced into Fatimid Egypt in the tenth or eleventh century CE and from thence spread to Islamic Spain and subsequently to Italy. Lucas and Harris [57] acknowledge that there is considerable dispute in the literature over the dates at which lead-based glazing commenced in Egypt and in the Middle-East generally.

Glazing spread to China during the Han Dynasty [2], although Kerr and Wood [48] suggest that rare examples of lead glazing can be found during the Warring States and as early as the fourth century BCE. The Romans spread the method of glazing and in particular, the use of lead glazing on pottery throughout the Roman Empire. The technology, developed in Islamic countries and then Europe, was then adopted into England in the form of tin-glazed pottery or delftware [44, 54, 98].

6.4 The Body: 'Egyptian Faience'

In Europe, faience is considered to refer to earthenware made of clay (and/or marl) and sand onto which a lime-alkali or a lead-alkali (tin-opacified) glaze has been applied. The term 'faience' was also adopted by archaeologists working in Egypt for its relatively similar appearance to tin-glazed earthenware originally made in the thirteenth century CE in Faenza, Italy (also variously described as maiolica, majolica, delftware) [54, 96, 98]. The term Egyptian 'faience' was then adopted to describe the relatively high temperature fired, silica-based ware found on a range of objects produced throughout Egypt and the Near East. The body consists of a ground quartz body containing varying amounts of interstitial glass and clay, in other words it is a stonepaste or a fritware, not very different from soft-paste porcelain. Faience can best be categorised on the basis of the morphology of the surface glaze layer, the body fabric and the glaze/core interaction layer [55].

Egyptian faience has also been referred to as the 'first high-tech ceramic' the production of which spans up to 5000 years of ancient Egyptian history, with extant examples dating from the Predynastic period [15, 111] to at least the first century Roman period. There is strong evidence that the glazing of stones and steatite originated in the Near East in about 4000 BCE [65] and with time developed into the glazing being applied to clay and silica-based objects.

The most comprehensive early surveys of Egyptian faience were provided by Kühne [50] and by Kaczmarczyk and Hedges [46] (they set out to research the origin of raw materials) and later by Lucas and Harris [57]. These authors produced elemental analytical analyses using air path XRF instrumentation in order to develop a limited understanding of this material.

Lucas and Harris [57] stated that typical Egyptian faience consisted of a body material (core) coated with a vitreous alkaline glaze. They stated that glazed wares in Egypt commenced in the Badarian civilisation with the production of glazed natural steatite. Then, glazed powdered quartz fabric was introduced during the Predynastic period and finally, glazed pottery from the Islamic period onwards. These authors discussed in some detail variations of the basic production methods adopted, the body fabric construction and the changes in glaze composition.

In an effort to differentiate each of the application methods, Tite et al. [103] and Tite and Bimson [100] applied SEM-EDS techniques to conduct a series of microstructure examinations. They commented that alkali may have been added to the quartz core when the direct application method was adopted. This alkali would have resulted in interstitial glass formation and could therefore confound attempts to identify the method of application of the glaze on the basis of its microstructure. Subsequent burial, leading to leaching of the alkalis present, or other weathering processes, could also confuse the determination. Burial, leading to the deposition of calcite or gypsum on to the surface of the object could also elevate the lime content analysis.

Matin [66] noted that there is little known about the accidental discovery of a method for the production of faience. She observed that Flinders Petrie in 1909 had proposed that the method was invented from finding quartz pebbles fluxed by wood ashes in a hot fire. However, the present speculative theory proposed by Moorey [69] and by Tite and Shortland [106] is that glazed stones occurred when quartz pebbles, alkali-rich ashes and some copper were subjected to a hot fire. To assist in developing a theory, Matin prepared a number of glazes in an effort to determine if electron microscopy could elucidate more evidence. She determined that the addition of calcium carbonate (calcite) in the form of lime to sodium chloride derived from cattle dung materially assisted in the vaporising process when heated to 950 °C and this produced a shiny glaze. This researcher stated that cattle dung was a readily and universally available fuel in ancient times. Matin and Matin [67] reported that cattle dung has been used as a fuel from some time during the 8th Millennium BCE, and particularly in arid regions. Cattle dung contains significant concentrations of alkalis, urea and chlorine. The addition of cattle dung to clay improved its plasticity [97]. The presence of alkali carbonates was found to considerably reduce the vaporisation of copper. These researchers also commented on the preferential vaporisation of potash over soda.

Bouquillon et al. [9] suggested that presently available evidence does not enable a decision to be reached with confidence as to whether faience production first occurred in the Near East or in Egypt. They observed that in the Near East, faience first appeared during the final phases of the Ubaid period (5500–4000 BCE) and in Egypt during the Naqada I phase of the Middle Predynastic period (4000–3500 BCE). Peltenburg [83] favoured the almost simultaneous discovery during the fifth millennium BCE, namely during the Badarian period in Egypt and in Mesopotamia to the Late Halaf or Early Ubaid periods. McCarthy et al. [65] proposed that there is strong evidence that the glazing of stones and steatite originated in the Near East in about 4000 BCE and with time developed into the glazing being applied to clay and silica-based objects, now described as Egyptian faience. Peltenburg [83] proposed that in Egypt during the reign of Amenhotep III, or earlier, polychrome faience appears with their decoration inlaid in different glaze colours into a background glaze.

18th Dynasty faience from Amarna has been investigated by Shortland [93] and by Kappen et al. [47] using the synchrotron XAS beamline and at Memphis during the Roman period [71]. According to Nicholson lead-glazed pottery was fired in saggars and mostly produced in Asia Minor.

As stated by Lucas and Harris [57], the body or core material is always granular, generally friable, although it can be hard and very finely divided; it is usually white, or practically white in colour, but can be tinted brown, grey or yellowish. The core material is normally crushed quartz, or rock crystal as shown by its sharp, angular appearance. Brown, grey or yellowish body material was suggested to comprise powdered sand, sandstone, flint or chert, the colour being derived from impurities within the raw material. According to Clark and Gibbs [15] the crushed silica was mixed with water, together with additives such as lime and trona (natron). A thin outer layer of fine silica sand was placed over the body fabric and to this, the surface glaze was applied (in Lucas and Harris, Variant A); this fine quartz layer was used to enhance or modify the colour of the glaze. It has been postulated that the fine, white surface silica slip may have been obtained as a by-product from the drilling of stone vessels [[51])].

According to Caiger-Smith [12], Fustat, the old city of Cairo founded by the Kairouanese Fatimids (Kairouan, Ifriqiya is now in Tunisia) has been credited with providing the foundation for the lustre pottery of the Fatimid dynasty. This author provided a discussion based upon the religious wars and trading endeavours of the peoples at that time and its combined effect upon artistic and designs appearing on pottery until the final destruction of this city in 1171 CE by Saladin as he consolidated Islam against the Christian crusaders. The pottery site then became a rubbish dump for the city and for Cairo itself. Excavations of this site have located thirteenth and fourteenth century kilns and traces of Fatimid kilns. Pottery wasters dating from the twelfth to the fourteenth century CE include rare examples of the highly valued lustre-painted Fatimid tin-glazed pottery.

6.5 Examples of Phase Identification

Representative samples have been selected in order to present the Raman spectral fingerprint of the different phases and to compare with the ‘mean’ elemental compositions.

A sample from Amarna marked as “Egyptian blue” (inventory number A22/6/1966(2) held by the Bolton Museum, Manchester), but which is turquoise in colour, together with two shards of Egyptian green from Malkata (nomenclature M1 and M7), and wall samples from the Main Temple, Kellis, Western Desert, Egypt (W4) together with two cartonnage samples from Tomb 20, Kellis cemetery (K1), are shown in Fig. 6.1. Other samples are shown in Figs. 6.2 and 6.3.

Fig. 6.1 Egyptian green (“Green frit”): (**a**) is Optical microscope image of Egyptian green pigment A22/6/1966 (2), Bolton Museum. (**b**) FIB BSE image of the crystal growth in sample A22/6/1966 (E2), Bolton Museum. (**c**) Sherd M1 coated with a white slip and then coated with Egyptian green. Beeswax has been used to bond the Egyptian green to the slip substrate (**d**) Optical microscope image of the surface of sherd M1. Note the occasional blue crystal. The white slip is clearly visible, the red scale bar is 500 µm, (**e**) Sherd M7, Malkata sherd marked M73/K/5. (**f**) is optical microscope image. (The optical microscope image is of an area forming part of the shard surface. Residual blue, red and green pigments are applied to a white huntite slip surface. The very small size of the residual pigment crystals render their analysis to be highly influenced by the white, primarily huntite, substrate slip coating. (**g**) Sample W4, wall pigment W4, Kellis, detached from the plaster substrate (Main Temple, Kellis. D/2 Shrine I Vault). (**h**) and (**i**) Sample C4 Kellis, cemetery K1, Tomb No. 20. The green pigment as applied was most likely a mixture of malachite [$Cu_2CO_3(OH)_2$] and carbonatecyanotrichite [$Cu_4Al_2CO_3(OH)_2.2H_2O$]. Deterioration of the pigment has formed anatacamite [$Cu_2Cl(OH)_3$] antlerite $Cu^{2+}{}_3SO_4(OH)_4$] and a carbonaceous matter. The pigment has been applied on to a gypsum plaster surface, and treated with an organic complex of unknown composition, the organic material is most likely a mixture of Gum Arabic and a diterpenoid resin probably from the resin collected from Pinus pinea or P. haleopennsis)

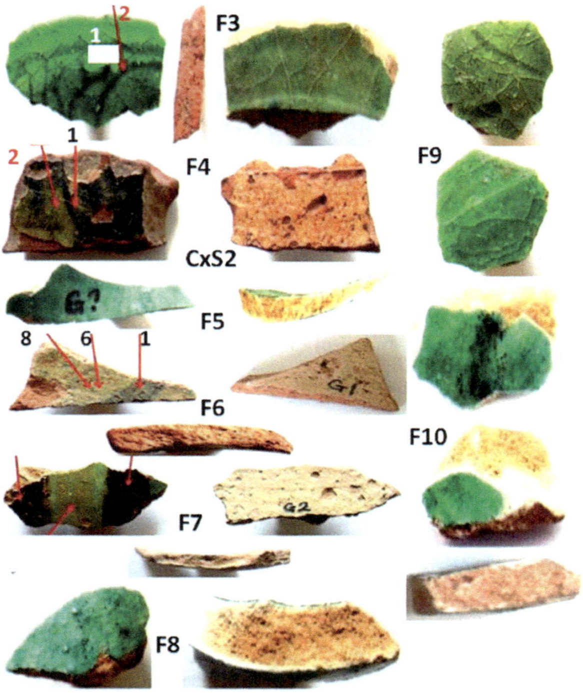

Fig. 6.2 Images of the low lead-alkali (F6) and lead-alkali glazed sherds. Red band = 1 cm

Fig. 6.3 Second century Roman period (or slightly later) alkali-lime lustre faience glazed wares (lustre induced by silver nanoparticles-rich layer in the glaze). Red band = 1 cm

Their mineralogy is deduced from the Raman spectra provided in Figs. 6.4, 6.5, 6.6, 6.7, 6.8, 6.9, 6.10, and 6.11 and in Tables 6.1 and 6.2. Elemental compositions are provided in Table 6.3 and SEM-EDS oxide ratios in Table 6.4. Eight glazed pottery samples are provided in Fig. 6.2; their SEM analyses are provided in Table 6.5. Their statistical Raman results are provided in Table 6.6.

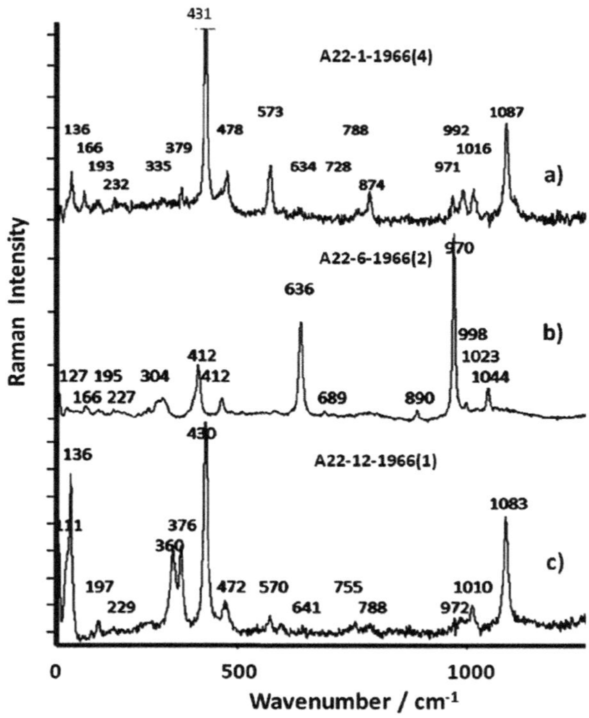

Fig. 6.4 Raman spectra of the Egyptian blue and Egyptian green samples held by the Bolton Museum, Greater Manchester. Sample (b) (A22/6/1966(2)) is turquoise in colour. All samples were obtained from Amarna ($\lambda = 514.5$ nm)

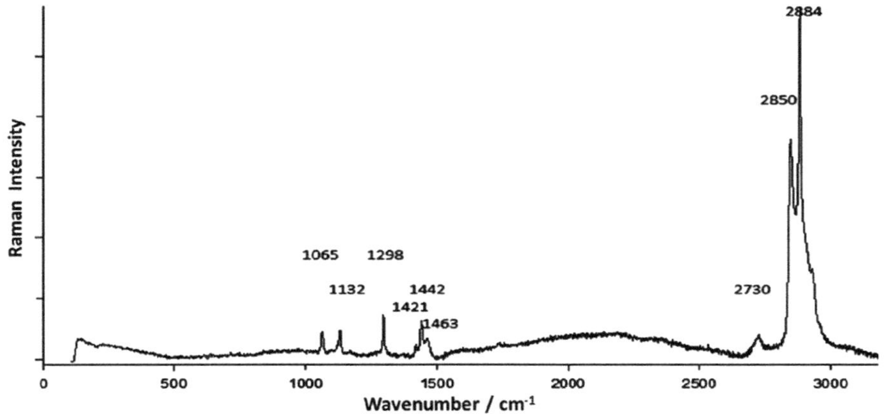

Fig. 6.5 Raman spectrum for beeswax used to bind the Egyptian green to the Malkata M1 pottery surface. ($\lambda = 514.5$ nm)

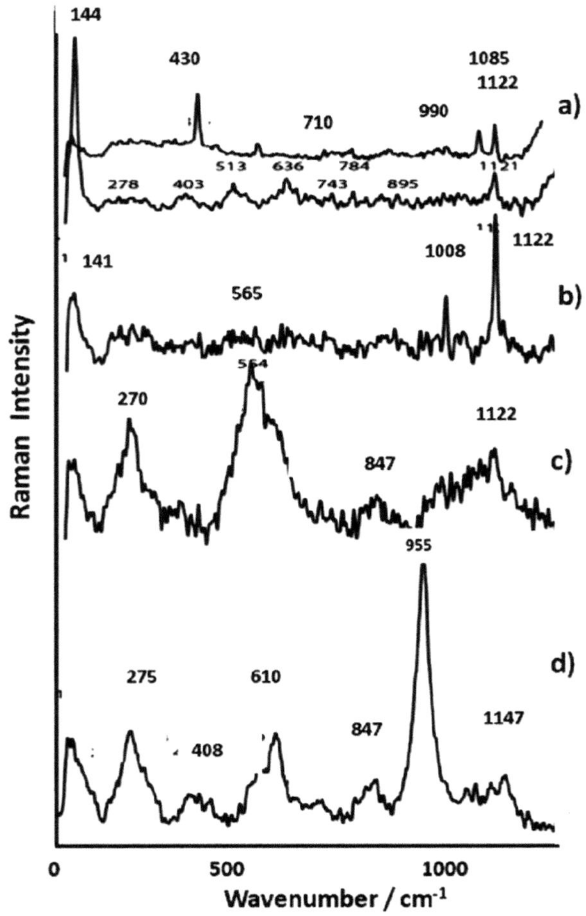

Fig. 6.6 Baseline subtracted Raman spectra, residual Egyptian green pigment on the surface of Malkata shard M7. ($\lambda = 514.5$ nm). (a) indicates the presence of traces of Egyptian blue within the green pigment. The band at 1122 cm^{-1} shows the presence of huntite. Gypsum, wollastonite, amorphous carbon and feldspars are detected. The strong band at 955 cm^{-1} is either an apatite or more likely, vivianite [$Fe^{2+}{}_3(PO_4)_2.8H_2O$] as indicated by their additional Raman bands

Two Roman period lustre sherds, shown as F1 and F2 in Fig. 6.3 with their SEM results being provided in Table 6.7.

6.5.1 Egyptian Green

As shown in Fig. 6.4a, c and in Table 6.1, all of the bands for the Raman signature for the mineral cuprorivaite (Egyptian blue [37]) are present in two of the samples from the Bolton Museum, Manchester. A comparison with the analyses of other Egyptian blue pigments by other workers is also provided in Table 6.1. However, the turquoise Egyptian green sample (A22/6/1966(2)) which is provided in Fig. 6.4 shows a strong Raman signature for wollastonite as indicated by the Raman bands

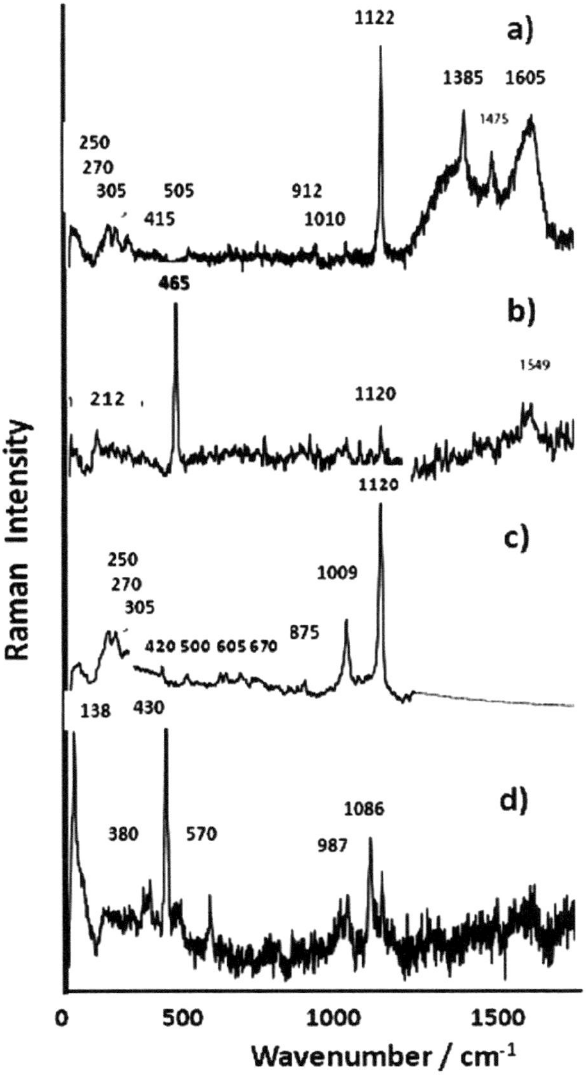

Fig. 6.7 Baseline subtracted Raman spectra, white areas on surface of shard M7. (λ = 514.5 nm). The strong band at 1122 cm^{-1} is huntite together with traces of gypsum, indicated by the band at 1009 cm^{-1}. The peaks at 430 and 1086 cm^{-1} indicates the presence of minor Egyptian blue. Quartz, tridymite and traces of feldspar are also present

at 636 and 970 cm^{-1}. As provided in Table 6.1, these Raman bands are assigned to α-wollastonite (para-wollastonite) and the bands at 984, 578 and 371 cm^{-1} to pseudo-wollastonite (β-wollastonite). The Raman spectral information for the three Bolton samples is given in Table 6.1.

The high MgO concentration indicated in Table 6.3 can be directly attributed to the huntite substrate onto which the Egyptian green has been applied (refer to the Raman spectroscopy results provided in Fig. 6.4b). The high sulphate concentration would have been derived from the gypsum forming part of the slip coating.

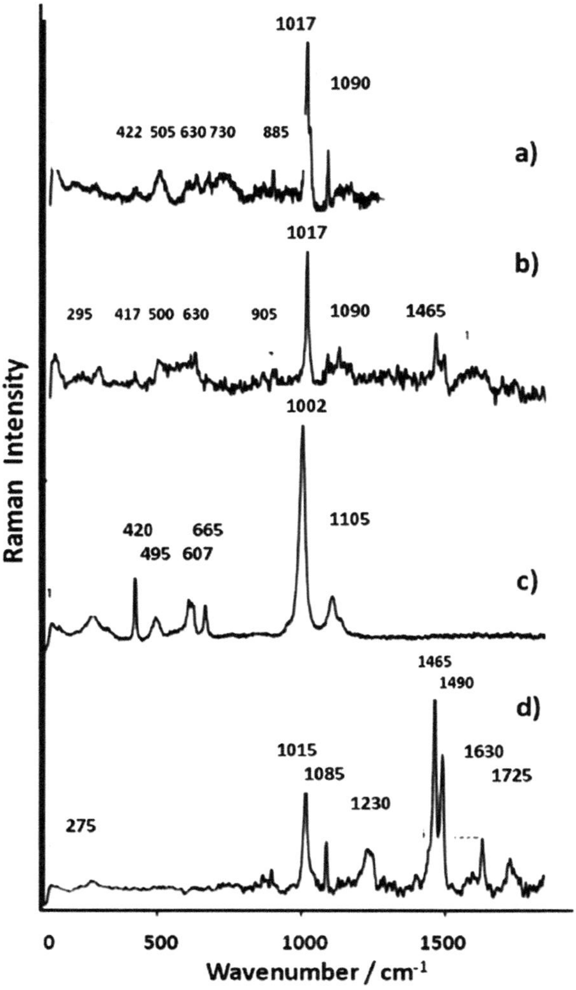

Fig. 6.8 Raman spectra, green pigment on wall decoration, sample W4 (λ = 514.5 nm). The Raman bands at 1463 and 1489 cm^{-1} arise from an organic phase

Other than for the wall sample W4, from the Main Temple, Kellis, the bulk copper oxide concentrations as provided in Table 6.4 are lower than the associated calcium oxide. This is reflected in the CaO/CuO ratios being >1.6 (Table 6.5). It is understood that the wall decoration from which sample W4 was obtained is from the Roman period. However, the presence of lead was not detected in this sample, as might have been expected; tin bronze scrap having been employed. W4 does contain high tin, copper and sulphate concentrations. The high copper analysis combined with the high sulphate could suggest the presence of a copper sulphate mineral such as langite. A Raman signature for this mineral was not detected. Furthermore, the high chlorine concentration, coupled with the detection of traces of atacamite would

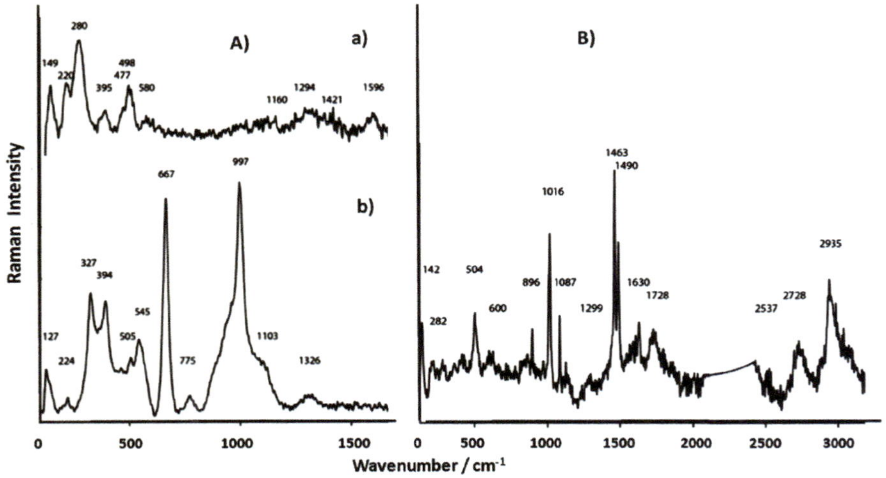

Fig. 6.9 (**A**) Baseline subtracted Raman spectrum, green pigment, wall decoration decomposition of sample W4 (λ = 514.5 nm); (**B**) Baseline subtracted Raman spectra of (**a**) red area, (**b**) brown area within the green glaze of Sample F6. (λ = 514.5 nm)

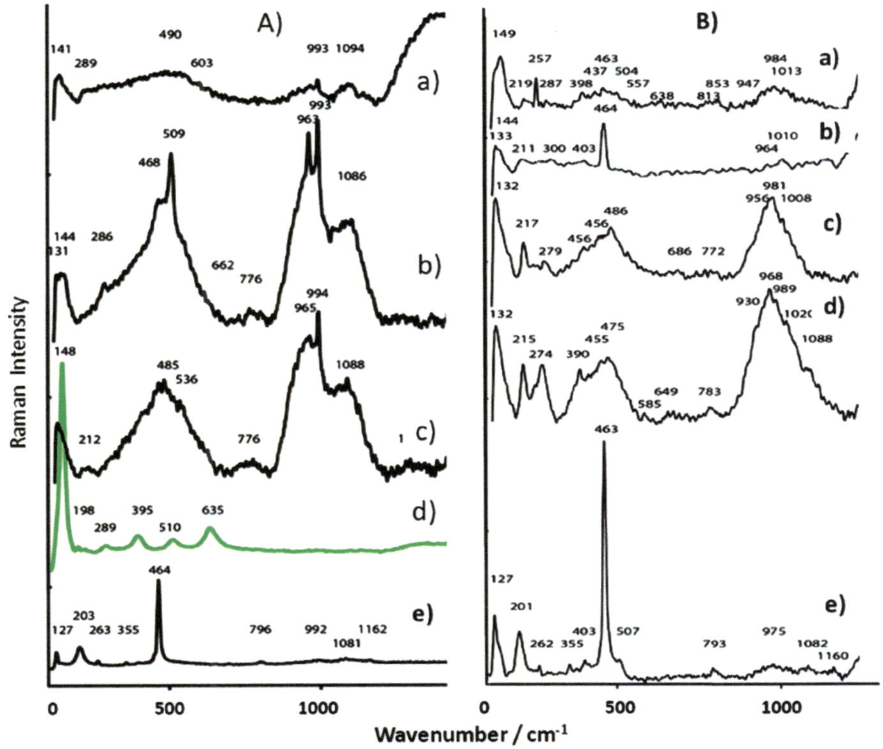

Fig. 6.10 (**A**) Baseline subtracted Raman spectra green glaze on sample F6 (λ = 514.5 nm); (**B**) Baseline subtracted Raman spectra sample F3 spots within the green glaze (λ = 514.5 nm)

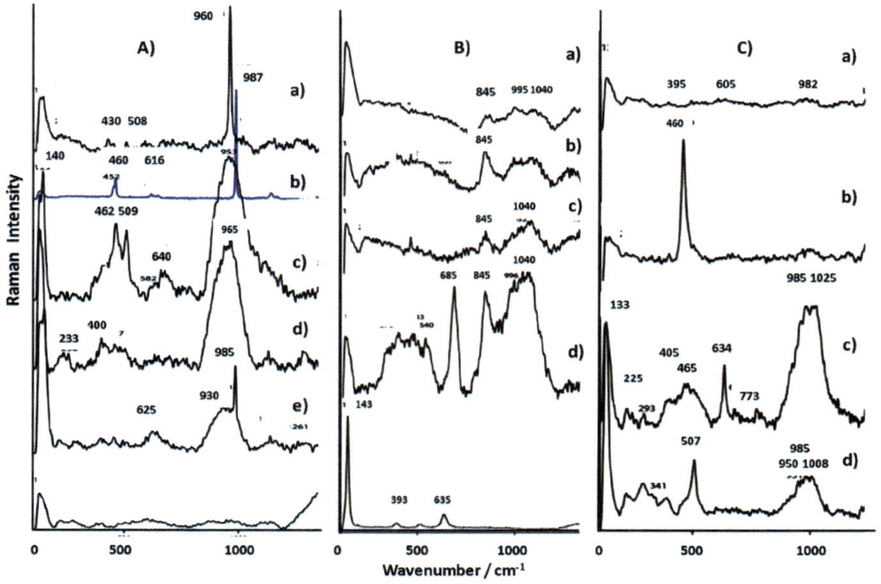

Fig. 6.11 (**A**) Baseline subtracted Raman spectra, (**a**–**e**), green areas in Fig. 2 for sample F4. (**f**) is black area (λ = 514.5 nm); (**B**) Baseline subtracted Raman spectra, green glaze applied to Sample F5 (λ = 514.5 nm). (**C**) Baseline subtracted Raman spectra Sample F7. (**a**, **b**) black glaze, (**c**, **d**) green glaze. (λ = 514.5 nm)

indicate that the reaction of the chlorine and copper and its subsequent precipitation on the wall surface would have increased the apparent copper concentration.

As shown in Figs. 6.10 and 6.11, weddellite (calcium oxalate dihydrate) and metastable calcium oxalate are present in wall sample W4. As indicated by Edwards and Farwell [28] and Edwards et al. [29], the presence of calcium oxalate monohydrate and dihydrate are identified by their strong Raman bands at 1463 and 1492 cm^{-1} respectively, indicative of equilibrium between the stable and the metastable forms of calcium oxalate at this site. Furthermore, the presence of the hydrated calcium oxalate in the sample is attributable to pollution (from animal waste) and/or lichen invasion, whose hyphae penetrate the calcareous substrates to form complexes with the oxalic acid secretions from the mycobiont. CH-stretching modes characterised by Raman bands in the region of 2935 cm^{-1} (and strongly visible in Fig. 6.11) support the assignment of these features to a biological source. Egyptian green, anhydrite, traces of feldspar, wollastonite [$CaSiO_3$], kinoite [$Ca_2Cu_2Si_3O_{10}.2H_2O$] and amorphous carbon are present in the pigment. The presence of tin in the elemental analyses indicates likely the use of bronze material as the copper source.

Table 6.1 Raman spectra comparison for the major Egyptian blue and Egyptian green bands for the samples from Bolton Museum, together with other published Raman spectral results for Egyptian blue

Samples	Laser, nm	Wavenumber assignment, cm^{-1}
Egyptian blue samples:	514.5	
A22/1/1966 (4)		1086, 1015, 991, 970, 790, 763, 573, 475, 466, 431, 379, 360, 232, 196, 165, 138, 114
A22/12/1966 (1)		1083, 1010, 972, 788, 755, 641, 570, 472, 430, 376, 360, 229, 197, 136, 111
Egyptian green sample		
A22/6/1966(2)		1044, 998, 970, 890, 636, 472, 412, 304, 127
Bell et al. [6]	514.5	1086, 1012, 992, 789, 762, 571, 475, 430, 377, 358, 230, 200, 137, 114
Bordignon et al. [8]	632.8	1085, 430, 376, 358, 137
Bruni et al. [10]	488	1085, 793 572, 478, 432
Burgio and Clark [11]	1064	1086, 465, 431
Rosalie David et al. [87]	514.5	1086, 1012, 992, 789 572, 470 430
Edwards et al. [30]	Unstated	1086, 571 430 360 162, 139
Mahmoud [59]	632.8	1086, 985, 574, 470, 432, 382, 229
Jorge-Villar and Edwards [45]	532	1087 430 114

Table 6.2 Minerals located within Malkata samples M1 and M7

Mineral	Formula	Wavenumber (cm^{-1})	Sample M1 Raman signature (cm^{-1})	Sample M7 Raman signature (cm^{-1})
Body				
α-quartz	SiO_2	128, 357, 402, **464**, 1084, 1160	129, 359, 403, **464**, 1084	
Tridymite	SiO_2	**209, 301, 351, 435**, 575, 790, 1071	356, **431, 572,** 790	
Parawollastonite	$CaSiO_3$	322, 337**, 637, 971,** 1045	322, 337, **638, 971, 1046**	**631, 977**
Anhydrite	$CaSO_4$	**1018,** 1163	1017	**1017**
Gypsum	$CaSO_4.2H_2O$	**1007**	**1006**	
Aragonite	$CaCO_3$	**1084**	**1085**	**1087**
Ferricopiapite	$Fe^{2+}Fe^{3+}{}_4(SO_4)_6(OH)_2.20H_2O$	268,301, 451, **479**, **993, 1019**, 1147	480, 993, 1017	
Langite	$Cu_4SO_4(OH)_6.2H_2O$	235**, 431**, **968/971**	234, **430**, 969/**971**	
Jarosite	$KFe^{3+}{}_3(SO_4)_2(OH)_6$	**223**, **299, 434, 624, 1007,**1102	224, 298, **626**, 1103	420, 621`, 1002, 1106
Cassiterite	SnO_2	**634, 776**	631	631, 776
Malayaite	$CaSnO(SiO_4)$	**323, 465, 573**	322, **465**, **572**	
Graphitic carbon	C	**~1360, ~1600**	**1360, 1600**	**~1350, ~1590**
Huntite	$CaMg_3(CO_3)_4$	1125	**1122**	**1122**
Burkeite	$Na_4(SO_4)(CO_3)$	995, 1063	**993, 1063**	
M1 Binder				
Beeswax	Mainly myricyl palmitate ester	2931, 2882, 2847, 2726, 1737, 1464, 1441, 1420, 1296, 1131, 1063, 891		
Beeswax[a]	Mainly myricyl palmitate ester	2881, 2849, 2723, 1460, 1440, 1418, 1369, 1295, 1129, 1062, 891		

[a]Burgio and Clark [11], see Fig. 6.5

Table 6.3 Raman spectroscopy assignment for Malkata sample M1

Mineral	Formula	Wavenumber (cm^{-1}) assignment.
α-quartz	SiO_2	1086, 465, 360
Tridymite	SiO_2	460, 430, 360
Wollastonite	$CaSiO_3$	969, 636
Vishnevite	$Na_8(AlSiO_4)_6O_{24}(SO_4).2H_2O$	993, 460, 256, 232, 199
kröhnkite	$Na_2Cu(SO_4)_2.2H_2O$	993, 460
Langite	$Cu_4(OH)_6SO_4.2H_2O$	969, 430, 232
Malachite	$Cu_2CO_3(OH)_2$	531, 430, 256
Chabazite-ca	$Ca(Si_4Al_2)O_{12}.6H_2O$	480, 460
Cassiterite	SnO_2	636
Apatite/calcium phosphate	$Ca_5(PO_4)_3(OH)/Ca_3(PO_4)_2$	969, 430
Marialite/scapolite	$(Na,ca)_4(Si,Al)_{12}O_{24}.(cl,CO_3,SO_4)$	764, 531, 460, 256, 166
Aragonite/calcite	$CaCO_3$	1086
Sanidine	$K(AlSi_3)O_8$	511, 475
M1—Slip		
Ankerite	$CaFe^{II}(CO_3)_2$	1100, 290
Aragonite/calcite	$CaCO_3$	1086
Dolomite	$CaMg(CO_3)_2$	1100, 290, 166
Phillipsite-Na	$Na_6(Si_{10}Al_6)O_{32}.12H_2O$	482, 329
α-quartz	SiO_2	465
Albite	$NaAlSi_3O_8$	1100, 763, 509, 482, 417, 290, 166
Oligoclase	$Na_{0.9-0.7}Ca_{0.1-0.3}Al_{1.1-1.3}Si_{2.9-2.7}O_8$	509, 482, 417
Labradorite	$Na_{0.5-0.3}Ca_{0.5-0.7}Al_{1.5-1.7}Si_{2.5-2.3}O_8$	509, 482, 417
Bytownite	$Na_{0.3-0.1}Ca_{0.7-0.9}Al_{1.7-1.9}Si_{2.3-2.1}O_8$	509, 482
Anorthoclase	$(Na,K)AlSi_3O_8$	509, 482
M1—Binder		
Beeswax	Mainly myricyl palmitate ester	2931, 2882, 2847, 2726, 1737, 1464, 1441, 1420, 1296, 1131, 1063, 891
Beeswax[a]	Mainly myricyl palmitate ester	2881, 2849, 2723, 1460, 1440, 1418, 1369, 1295, 1129, 1062, 891

[a]see Table 6.2

6.5.2 *Glaze Classification*

A number of lead-based shards have been examined and are illustrated in Fig. 6.2, and lustre faience surface samples from ancient Kellis, Dakhleh Oasis, with an alkali-lime glaze and dated to the second century Roman period or probably later, are illustrated in Fig. 6.3.

The faience glaze itself was recognised by Lucas and Harris [57] to comprise at least 6 variants in which an alkali flux with lime and colourants was applied over a highly siliceous body. These alternative formulae are discussed below and they

Table 6.4 EDS analyses of the principal elements present (wt. %) in the various Egyptian green samples analysed

Oxide	A22/6/1966	M1	M7	W4	Hatton [36]
CaO	18.3	17.5	26.7	22.1	13.5
CuO	11.3	8.0	10.5	32.3	5.2
SiO_2	63.8	49.2	11.8	11.0	73.1
SO_3	0.6	3.0	9.9	13.6	
MgO	0.1	2.3	26.4	1.8	
SnO_2	–	tr	–	3.6	
As_2O_3	–	tr	Variable 0–12	–	

Table 6.5 Element oxide ratios applicable to the samples analysed

Ratio	A22/6/1966	M1	M7	W4	Hatton [36]
CaO/CuO	1.6	2.2	2.5	0.7	2.6
CaO/SiO_2	0.3	0.4	2.3	2.0	0.2
SiO2/CuO	5.6	2.8	0.4	0.5	5.4
$Na_2O + K_2O$	3.3%	6.8%	2.4%	2.2%	1–11%
MgO	0.1%	2.3%	26.4%	1.8%	bd-6.1%

Table 6.6 Statistical analysis of Raman spectra: I_P = Polymerisation Index = A_{500}/A_{1000}; Al/Si = Ionicity Ratio = ½(Al_2O_3)/SiO_2; CR = Covalence Ratio = SiO_2/½(Al_2O_3; IOC = Ionic Oxide Content = Σ(½Na_2O + ½K_2O + CaO + PbO)

Sample	I_P	Al/Si	CR	IOC
Low lead alkali glaze				
F6	0.58	0.06	16.9	11.6
Lead alkali glaze				
F3(dark green)	0.38	0.07	13.7	26.1
F3 (green)	0.66	0.1	10.4	20.3
F4	0.17	0.19	5.3	28.7
F5	0.17	0.06	15.9	19.5
F7	0.31	0.16	6.2	30.1
F8	0.55	0.1	10.2	18.7
F9	0.41	0.05	18.6	30.9
Lustre glaze				
F1	1.22	0.066	15.2	6.0
F2	1.19	0.03	32.7	11.95

generally relate to the method of application, for example, the introduction of sodium chloride for efflorescent glazing, or later, the introduction of lead to modify the glaze.

As As proposed by Hodges [42] and by Rice [86], glazes are able to be clasclassified by their modifiers into three distinct groups, namely:

Table 6.7 Average elemental analysis (EDS) of the eight undated surface finds (Faience shards) analysed

Green, Sample F3

Na_2O	MgO	Al_2O_3	SiO_2	P_2O_5	SO_3	Cl	K_2O	CaO	TiO_2	MnO_2	FeO	CoO	CuO	SnO_2	PbO
3.9	2.4	10.7	55.9	0.9	0.0	0.6	1.8	2.6	0.7	0.0	3.6	0.0	2.4	0.0	14.8

Black, Area 1, Sample F4

Na_2O	MgO	Al_2O_3	SiO_2	P_2O_5	SO_3	Cl	K_2O	CaO	TiO_2	MnO_2	FeO	CoO	CuO	SnO_2	PbO
0.9	3.7	21.9	51.5	0.6	1.1	0.5	2.1	4.6	1.1	0.1	8.9	0.0	0.3	0.0	2.6

Green, area 2, sample F4*

Na_2O	MgO	Al_2O_3	SiO_2	P_2O_5	SO_3	Cl	K_2O	CaO	TiO_2	MnO_2	FeO	CoO	CuO	SnO_2	PbO
1.5	2.1	14.6	39.0	0.5	3.1	0.6	0.8	1.7	0.7	0.3	3.3	0.4	4.5	0.0	25.8

* + ZnO = 1.1%

Green, Sample F5#

Na_2O	MgO	Al_2O_3	SiO_2	P_2O_5	SO_3	Cl	K_2O	CaO	TiO_2	MnO_2	FeO	CoO	CuO	SnO_2	PbO
4.7	1.8	6.0	47.6	0.5	1.9	0.2	1.9	3.8	0.7	0.0	5.9	0.0	11.3	0.0	12.5

+ Cr_2O_3 = trace

Very pale green, Sample F6 (very thin coating)

Na_2O	MgO	Al_2O_3	SiO_2	P_2O_5	SO_3	Cl	K_2O	CaO	TiO_2	MnO_2	FeO	CoO	CuO	SnO_2	PbO
4.2	2.7	8.5	62.3	0.3	0.9	0.5	4.9	3.5	0.8	0.0	5.0	Tr	0.6	0.0	2.9

Black area, Sample F7

Na_2O	MgO	Al_2O_3	SiO_2	P_2O_5	SO_3	Cl	K_2O	CaO	TiO_2	MnO_2	FeO	CoO	CuO	SnO_2	PbO
1.3	2.3	15.1	42.3	0.8	3.5	0.4	1.6	3.9	0.5	1.0	5.5	0.0	0.5	0.0	21.3

Darker green area, Sample F7

Na_2O	MgO	Al_2O_3	SiO_2	P_2O_5	SO_3	Cl	K_2O	CaO	TiO_2	MnO_2	FeO	CoO	CuO	SnO_2	PbO
1.5	2.2	13.7	42.2	0.6	3.3	0.4	1.7	3.6	0.5	0.8	4.6	0.0	0.6	0.0	24.8

(continued)

Table 6.7 (continued)

Green, sample F8**

Na_2O	MgO	Al_2O_3	SiO_2	P_2O_5	SO_3	Cl	K_2O	CaO	TiO_2	MnO_2	FeO	CoO	CuO	SnO_2	PbO
3.1	2.7	10.6	55.3	0.6	1.7	0.2	2.4	3.6	0.4	0.0	3.7	0.0	3.2	0.0	12.2

** + Cr_2O_3 = 0.4%

Green, Sample F9

Na_2O	MgO	Al_2O_3	SiO_2	P_2O_5	SO_3	Cl	K_2O	CaO	TiO_2	MnO_2	FeO	CoO	CuO	SnO_2	PbO
5.2	1.5	9.7	53.2	0.1	1.8	0.8	1.2	2.5	0.3	0.0	3.2	0.0	1.7	0.0	18.7

Darker green area, Sample F10

Na_2O	MgO	Al_2O_3	SiO_2	P_2O_5	SO_3	Cl	K_2O	CaO	TiO_2	MnO_2	FeO	CoO	CuO	SnO_2	PbO
3.8	2.1	9.5	57.2	0.2	0.0	Tr	2.1	3.9	0.3	0.0	2.7	0.0	3.2	0.0	14.7

- alkaline glazes, (Na_2O ~ 16%);
- high lead glazes (up to 60% PbO);
- lead-alkali glazes.

These various glaze formulae would have had a maturing temperature in the range of 800—1000 °C [26] or 850—980 °C [100].

Tite [99] proposed a more complex classification of Islamic glazes, namely:

- alkali-lime glazes (<2 wt.% PbO);
- low lead-alkali glazes (2.0–9.9 wt.% PbO);
- lead-alkali glazes (10–35 wt.% PbO);
- high-lead glazes (>35 wt% PbO);

with each of these glaze types being further subdivided into transparent and opacified. This concept has been adopted here. Their statistics, namely, Polymerisation Index, Ionicity Ratio and Ionic Oxide Content are provided in Table 6.6. They will be commented upon later.

As discussed by Mason and Tite [62], during the ninth century most of the Islamic world was ruled by the Abbasid caliphs from Iraq with most of these wares being glazed with turquoise alkaline glaze or splashed high-lead glazes and only rarely the prestigious tin opacified lead glaze.

Colomban and co-workers have developed what they term a Polymerisation Index (I_P) which is described as being the ratio of the Si-O bending envelope of the area divided by the Si-O stretching envelope ($I_P = A_{500}/A_{1000}$) [17, 19, 23]. These authors proposed that the relationship between the glass composition and its melting temperature is well-established and can be related as follows:

- Family#1 = I_P (<0.3–0.5) generally corresponds to Islamic lead-bearing or lustre pottery and some Punic/Roman glasses. Processing temperature ~600 to 900 °C;
- Family#2 = ($0.5 < I_P < 0.3$) are lead-based soft paste porcelain enamels, some Punic/Roman glasses and Iznik glazes;
- Family#3 = ($0.8 < I_P < 1.1$) corresponds to most ancient glasses and eighteenth century soft paste porcelain enamels. Processing temperature ~1000 °C;
- Family#4 = ($1.1 < I_P < 1.3$) is indicative of celadon Ca-based enamels;
- Family#5 = (($1.3 < I_P < 2.5$) corresponds to Ca-based porcelain enamels;
- Family#6 = ($2.5 < I_P < 7$) corresponds to K-based hard paste porcelain glazes. Processing temperature ~1400 °C.

6.5.3 *Islamic Lead Glazes*

The lead-containing sherds are illustrated in Fig. 2. All analytical results are summarised in Table 5 and the Raman spectroscopy mineralogical results in Figs. 12 and 14 and Figs. 15 to 21.

Low Lead-Alkali Glaze, F6 Sample

The elemental analyis of a low lead-based green glaze is provided in Table 6.7 and the Raman spectra in Figs. 6.12 and 6.11B. As shown by the broad Raman bands observed at ~1300 and ~1600 cm^{-1} wavenumbers the glaze contains traces of amorphous carbon. As such, reducing conditions would have prevailed during the firing of the glaze. Wollastonite and probably pseudo-wollastonite, anatase and quartz, together with lead mineralisation, as plattnerite (PbO_2), minium ($Pb^{2+}{}_2Pb^{4+}O_4$), leadhillite [$Pb_4(SO_4)(CO_3)_2(OH)_3$], a lead-based mineral as discussed by Batonneau et al. [4] being nPbO.$PbSO_4$ mineralisation is present. The presence of the arsenical mineral beudantite [$PbFe^{3+}{}_2(AsO_4)(SO_4)(OH)_6$] is indicated by the strong Raman

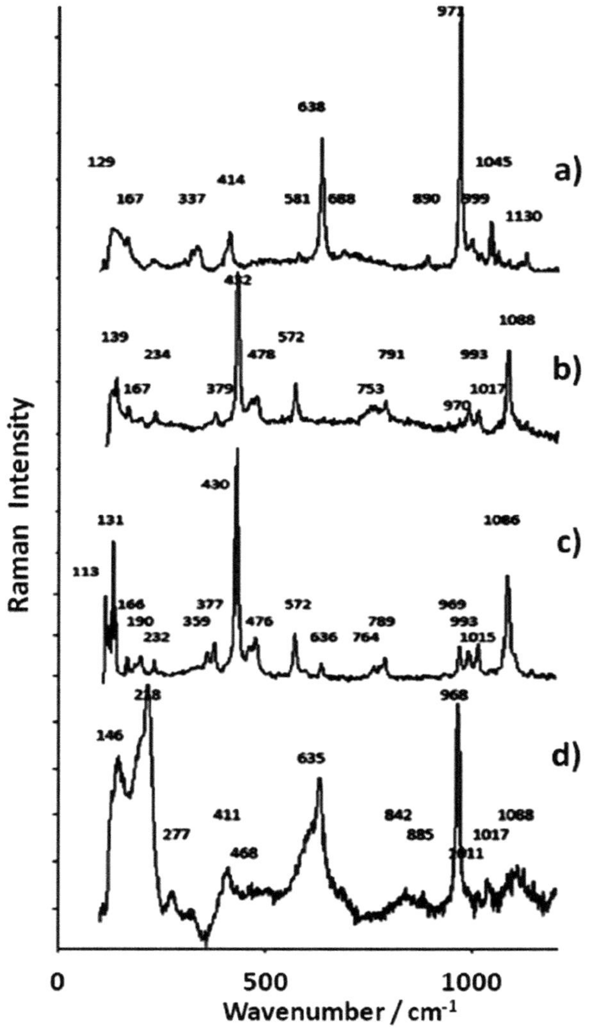

Fig. 6.12 Raman spectra for Egyptian green pigment in Malkata sample M1 (identified as M14) (λ = 514.5 nm). The number of crystals all exhibiting various pigment shades are clearly evident (d spectrum: baseline is subtracted)

peak at 997 cm^{-1}. Whilst As was not reported in the SEM-EDS analyses it could be present but not detected due to the ovelap of the main As Kα line at 10.532 keV and the Pb Lα line at 10.551 keV. The copper oxide tenorite (CuO) may have contributed to the green colour of the glaze due to the divalent copper ion being present in a disordered octahedral structure. Ferrous iron formed under reducing conditions would contribute to the green pigmentation.

The signature for a decomposed lead-tin Type I mineralisation [$PbSnO_4$] is indicated in Fig. 6.12d, although tin was not detected in the glaze analyses. The peak at wavenumber 148 cm^{-1} is indicative of the lead (II) oxide mineralisation and the peak at 635 cm^{-1} indicates cassiterite (Sn(IV) oxide). Clark et al. [14] indicated that decomposition of the lead-tin pigment will occur at temperatures >950 °C to form their respective elemental oxides in a glassy matrix. As such, the presence of these individual elements would suggest a relatively high fired glaze composition. The presence of tridymite would support a firing temperature in the order of 1000 °C.

Brown chamosite [Fe^{2+},Mg,Al,Fe^{3})$_6$(Si,Al)$_4O_{10}$(OH,O)$_8$ is indicated from its Raman signature provided in Fig. 6.11B by the strong bands at wavenumbers 545 and 667 cm^{-1}. Labradorite, anorthite and aegirine [$NaFe^{3+}Si_2O_6$] are present together with possibly diopside ($MgCaSi_2O_6$), natrochalcite [$NaCu_2(SO_4)_2(OH).H_2O$] and devilline [$CaCu_4(SO_4)_2(OH)_6.3H_2O$].

The Si-O bending maximum for the green glaze is at ~485 cm^{-1} and the Si-O stretching maximum is at ~1010 cm^{-1}. The Raman Polymerisation Index (I_p) as calculated from the A_{500}/A_{1000} areas under the respective peaks and is reported in Table 6 together with additional calculated statistics.

Lead-Alkali Glazes, Samples F3 to F5 and F7 to F10

Elemental analyses in Table 6.5 are similar to those found for the alkali-lead glazes from North Jordan as provided by al-Saad [2] and other researchers in Table 6.8; although, samples F3, F4 and F7 are slightly lower in soda and F4 and F7 are higher in alumina but similar in lime, potash and lead and samples F3, F6 and F8 are lower in lead. As stated by al-Saad [2] and by Tite et al. [105] for faience samples and as reported by Jay et al. [44] for similar lead-tin glaze compositions applied to delftware, part of the lead in the glaze diffuses into the body and part of the silica from the pottery body is enhanced within the glaze.

The cementation formula supplied by Tite et al. [105] in Table 6.9 contained ~6% K_2O, and no lead. Whereas the potash concentration in Sample F6 was similar, being only 1% lower than the Tite glaze, however Sample F6 also contained a low lead content.

Sample F6 is a "thin" application of faience glaze on to the body. According to the faience reproduced by Tite et al. [105] and provided in Tables 6.8 and 6.9, the soda content in both formulae is significantly higher than found in sherd F6. This may be related to the loss of sodium during burial. Both sherds had lead in their composition as a modifier.

The alumina concentration in the sherds at ~5–15% is higher than in the North Jordan glazes. The presence of the clay would have raised the viscosity of the glaze melt, controlling its ability to drain off the object during firing. Because of the

Table 6.8 Average analysis of faience alkali-based glaze from Northern Jordan, 18th, 22nd and late period Egypt, and the average of several samples of 'Ishkor' samples from Akhsiket, Uzbekistan (*na, not analysed, #, efflorescences)

Blue and purple [2]															
Na_2O	**MgO**	**Al_2O_3**	**SiO_2**	**P_2O_5**	**SO_3**	**Cl**	**K_2O**	**CaO**	**TiO_2**	**MnO**	**FeO**	**CoO**	**CuO**	**SnO_2**	**PbO**
16.0	2.1	2.0	66.8	0.0	Na	Na	2.2	6.5	Na	0.0	1.2	0.0	2.4	Na	0.0
Amarna sample Variant D 469, 18th Dynasty [93]															
Na_2O	**MgO**	**Al_2O_3**	**SiO_2**	**P_2O_5**	**SO_3**	**Cl**	**K_2O**	**CaO**	**TiO_2**	**MnO**	**FeO**	**CoO**	**CuO**	**SnO_2**	**PbO**
8.9	1.5	5.2	79.0	n.a	n.a	n.a	1.8	1.7	n.a	0.44	0.4	0.41	0.0	n.a	n.a
Amarna sample E5045, 18th Dynasty [105]															
Na_2O	**MgO**	**Al_2O_3**	**SiO_2**	**P_2O_5**	**SO_3**	**Cl**	**K_2O**	**CaO**	**TiO_2**	**MnO**	**FeO**	**CoO**	**CuO**	**SnO_2**	**PbO**
14.6	0.3	0.14	72.2	n.a.	n.a.	n.a.	2.53	1.8	n.a.	0.03	0.16	0.01	8.23	0.02	n.a.
Amarna sample BMRL 16321, 18th Dynasty (surface) [103]															
Na_2O	**MgO**	**Al_2O_3**	**SiO_2**	**P_2O_5**	**SO_3**	**Cl**	**K_2O**	**CaO**	**TiO_2**	**MnO**	**FeO**	**CoO**	**CuO**	**SnO_2**	**PbO**
3.3#	–	–	70.0	0.4	Na	1.2	1.5	0.4	Na	Na	–	Na	9.5	Na	Na
Amarna sample BMRL 16321, 18th Dynasty (glaze inner) [103]															
Na_2O	**MgO**	**Al_2O_3**	**SiO_2**	**P_2O_5**	**SO_3**	**Cl**	**K_2O**	**CaO**	**TiO_2**	**MnO**	**FeO**	**CoO**	**CuO**	**SnO_2**	**PbO**
16.0	0.2	–	69.3	–	Na	1.0	5.7	0.2	Na	Na	0.3	Na	7.4	Na	Na
Amarna sample BMRL 16319, New Kingdom [103]															
Na_2O	**MgO**	**Al_2O_3**	**SiO_2**	**P_2O_5**	**SO_3**	**Cl**	**K_2O**	**CaO**	**TiO_2**	**MnO**	**FeO**	**CoO**	**CuO**	**SnO_2**	**PbO**
-#	0.2	–	73.1	0.5	Na	1.0	–	1.4	Na	Na	0.3	Na	9.7	Na	Na
Amarna sample BMRL 16319, New Kingdom (glaze) [103]															
Na_2O	**MgO**	**Al_2O_3**	**SiO_2**	**P_2O_5**	**SO_3**	**Cl**	**K_2O**	**CaO**	**TiO_2**	**MnO**	**FeO**	**CoO**	**CuO**	**SnO_2**	**PbO**
>11.6	0.5	1.5	70.5	0.5	Na	0.5	3.2	2.8	Na	Na	0.9	Na	6.5	Na	Na
Amarna sample BMRL 16323, 21st Dynasty [103]															
Na_2O	**MgO**	**Al_2O_3**	**SiO_2**	**P_2O_5**	**SO_3**	**Cl**	**K_2O**	**CaO**	**TiO_2**	**MnO**	**FeO**	**CoO**	**CuO**	**SnO_2**	**PbO**
17.5	0.3	0.8	70.4	0.7	Na	1.0	1.4	4.7	Na	Na	0.5	Na	3.2	Na	Na

Abydos sample 2221/1, 22nd Dynasty [105]															
Na_2O	**MgO**	**Al_2O_3**	**SiO_2**	**P_2O_5**	**SO_3**	**Cl**	**K_2O**	**CaO**	**TiO_2**	**MnO**	**FeO**	**CoO**	**CuO**	**SnO_2**	**PbO**
12.63	0.12	0.84	72.8	Na	Na	Na	2.18	1.0	Na	0.19	0.5	0.04	9.66	0.05	Na
Amarna sample BMRL 16322, Late period (glaze surface) [103]															
Na_2O	**MgO**	**Al_2O_3**	**SiO_2**	**P_2O_5**	**SO_3**	**Cl**	**K_2O**	**CaO**	**TiO_2**	**MnO**	**FeO**	**CoO**	**CuO**	**SnO_2**	**PbO**
-#	1.2	3.2	63.7	0.4	Na	1.6	–	4.9			1.0		3.9		
Amarna sample BMRL 16322, Late period (glaze inner) [103]															
Na_2O	**MgO**	**Al_2O_3**	**SiO_2**	**P_2O_5**	**SO_3**	**Cl**	**K_2O**	**CaO**	**TiO_2**	**MnO**	**FeO**	**CoO**	**CuO**	**SnO_2**	**PbO**
18.6	1.1	4.8	61.0	–	Na	1.8	0.6	4.3	Na	Na	1.7	Na	2.8	Na	Na
'Ishkar' lilac glaze, Uzbekistan [41]															
Na_2O	**MgO**	**Al_2O_3**	**SiO_2**	**P_2O_5**	**SO_3**	**Cl**	**K_2O**	**CaO**	**TiO_2**	**MnO**	**FeO**	**CoO**	**CuO**	**SnO_2**	**PbO**
8.3	2.8	4.9	69.4	0	0.1	0.5	6.4	6.0	0.1	1.5	1.5	Na	0.3	Na	Na

Table 6.9 Efflorescence and cementation formulae used in trial production of faience by these methods [105]

Efflorescence formula, E1 glaze [105]															
Na_2O	**MgO**	**Al_2O_3**	**SiO_2**	**P_2O_5**	**SO_3**	**Cl**	**K_2O**	**CaO**	**TiO_2**	**MnO_2**	**Fe_2O_3**	**CoO**	**CuO**	**SnO_2**	**PbO**
14.46	0.02	–	80.66	–	–	–	0.06	1.92	–	–	–	–	–	–	–
Cementation formula, C1 glaze [105]															
Na_2O	**MgO**	**Al_2O_3**	**SiO_2**	**P_2O_5**	**SO_3**	**Cl**	**K_2O**	**CaO**	**TiO_2**	**MnO_2**	**Fe_2O_3**	**CoO**	**CuO**	**SnO_2**	**PbO**
13.06	1.14	0.81	69.5	–	–	–	5.95	8.58	–	–	–	–	–	–	–

complexity of the mineralisation revealed by Raman spectroscopy, the mineralogy applicable to samples 3–5 and 7–10 is described separately.

A plot of Polymerisation Index (I_p) versus Si-O bending moment is provided in Fig. 6.22.

Sample F3 Mineralogy, Raman Spectra, Fig. 6.12B

The lead-based green glaze on sample F3 contains carbon as shown in Fig. 6.12B by the broad Raman bands at ~1300 and ~1600 cm^{-1} wavenumbers. As such, reducing conditions would have prevailed during the firing of the glaze. Under reducing conditions the iron present in the glaze would be Fe^{2+} and as such, would contribute to a green colour. Raman signatures provided in Fig. 6.12B(a) and 6.12B(c) would indicate that some of the alumina has formed anorthite within the glaze. Excess silica (Raman band at 463 cm^{-1}) is observed in Fig. 6.12B(e) together with a small peak at 507 cm^{-1} which could indicate that crushed chert containing moganite, may have formed part of the silica used in the glaze preparation. The peaks at ~430 and at 1088 cm^{-1} would suggest that Egyptian green (green frit) formed part of the glaze mineralisation, indicative of a firing temperature approaching 1050 °C. Wollastonite and lead mineralisation, either as plattnerite (PbO_2) or minium ($Pb^{2+}{}_2Pb^{4+}O_4$) are present. Copper minerals indicated as likely to be present are the copper oxides tenorite (CuO) and/or cuprite ($Cu^{1+}{}_2O$) (shown by peaks in Fig. 6.12B(a) at wavenumbers 219, 287, 638 cm^{-1}). The lack of sulphur in the analyses would suggest that copper sulphate-based mineralisation was not present.

As shown by Pagès-Camagna et al. [81], Lau et al. [52] and Crupi et al. [27] tenorite is the more likely copper oxide present and when the copper is in an ionic form as Cu^{2+} this chromophore when located in a disordered octahedral structure in an amorphous glassy phase is green in colour.

The Si-O bending maximum for the dark green glaze is at 467 cm^{-1} and the Si-O stretching maximum is at 990 cm^{-1} and the Si-O bending maximum for the lighter green glaze is at 462 cm^{-1} and the Si-O stretching maximum is at 980 cm^{-1}. The Raman Polymerisation Index (Ip) is provided in Fig. 6.12 and Table 6.6 together with additional statistics. Raman signatures provided in Figs. 6.15B(a) and 6.15B (c) would indicate that some of the alumina has formed anorthite within the glaze.

Sample F4, Mineralogy, Raman Spectra, Fig. 6.5

The lead-based green glaze on sample F4 contains traces of carbon as is shown in Fig. 6.15 by the broad bands at ~1300 and ~1600 cm^{-1} wavenumbers. As such, reducing conditions would have prevailed during the firing of the glaze. Wollastonite, or more likely pseudo-wollastonite as indicated by the strong peak at wavenumber 986 cm^{-1} and either calcium phosphate/apatite or phosphoferrite [$Fe^{3+}{}_3(PO_4)_2.2H_2O$] as indicated by the strong peak at a wavenumber of 961 cm^{-1}. The Raman band at 143 cm^{-1} is indicative of lead mineralisation, either as litharge (PbO) (plattnerite (PbO_2) or minium ($Pb^{2+}{}_2Pb^{4+}O_4$) is present or more likely due to anatase (TiO_2, Raman band signature 143, 400, 517 cm^{-1}), quartz (462 cm^{-1}), and the feldspars labradorite and/or oligoclase. Copper minerals indicated as likely to be present are the copper oxides tenorite (CuO) and/or cuprite ($Cu^{1+}{}_2O$) together with probably zincolibethenite [$CuZnPO_4OH$] (zinc was detected in all elemental analyses)

(Raman signature, 256, 517, 964, 1012 cm^{-1}). Raman signatures for phosphatic minerals are surprising as no strong indication for the presence of phosphorus was detected in the SEM-EDS analyses, although no attempt was made to analyse individual spots throughout the glaze by electron microscopy. Anatase is a very good Raman scatterer and is detected at levels not detected by others techniques (Fig. 6.13).

The Si-O bending maximum for the green glaze is at 457 cm^{-1} and the Si-O stretching maximum is at 960 cm^{-1}. The Raman Polymerisation Index (Ip) is reported in Fig. 6.24 and Table 6.6 together with additional statistics.

The signature for mineralisation similar to that of a lead-tin Type I [$PbSnO_4$] is indicated in Fig. 6.15(e), although tin was not detected in the glaze analyses. Iron-containing minerals staurolite [$Fe^{2+}Al_9Si_4O_{23}(OH)$] and possibly andalusite [Al_2OSiO_4]containing traces of iron are indicated by their respective Raman signatures.

The Si-O bending maximum for the green glaze is at 457 cm^{-1} and the Si-O stretching maximum is at 960 cm^{-1}. The Raman Polymerisation Index (I_p), together with additional statistics is provided in Table 6.6 and plotted in Fig. 6.12.

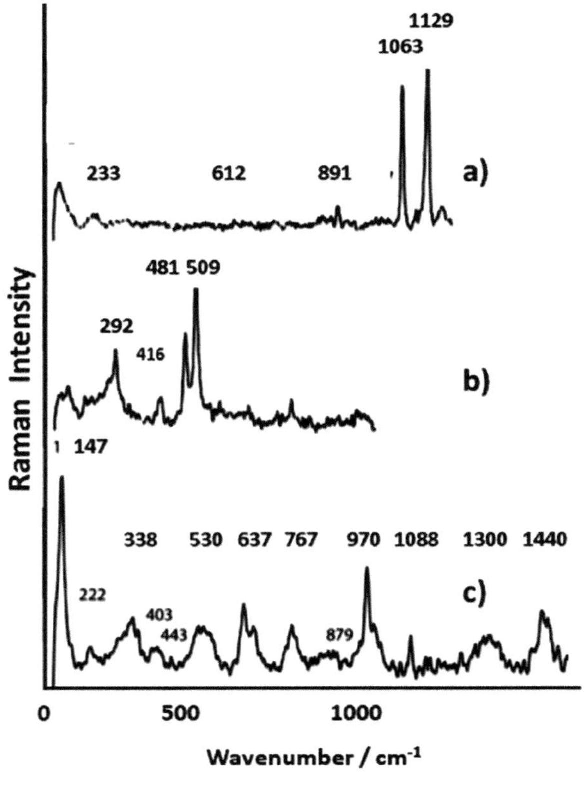

Fig. 6.13 Baseline subtracted Raman spectra of white slip coating on the surface of sherd M1. The coating is huntite together with calcite, wollastonite and possible gypsum. The presence of feldspar, as shown in (b), the more likely member being oligoclase rather than labradorite

Sample F5, Mineralogy: Raman Spectra, Fig. 6.15B

The SEM-EDS analysis of this lead-based green glaze is provided in Table 5. Its Raman spectra provided in Fig. 6.15B indicates the presence of carbon as shown by the broad bands at ~1300 and ~1600 cm^{-1} wavenumbers. As such, reducing conditions would have prevailed during the firing of the glaze. Wollastonite, anatase and quartz, together with lead mineralisation, either as plattnerite (PbO_2) or minium ($Pb^{2+}{}_2Pb^{4+}O_4$) is present. Magnetite [$Fe^{2+}Fe^{3+}{}_2O_4$], copiapite [$Fe^{2+}Fe^{3+}{}_4(SO_4)_6(OH)_2.20H_2O$] and possibly eastonite [$KAlMg_2(Si_2Al_2)O_{10}(OH)_2$] and monticellite [$CaMgSiO_4$] are indicated. The presence of arsenical minerals traces are shown by the strong Raman peak at 847 cm^{-1}. However, whilst arsenic was not reported in the SEM-EDS analyses it could be present but not detected due to the ovelap of the As Kα line at 10.532 keV and the Pb Lα line at 10.551 keV. The copper oxide tenorite (CuO) may have contributed to the green colour of the glaze due to the divalent copper ion being present in a disordered octahedral structure [24].

The signature for a decomposed lead-tin Type I mineralisation [$PbSnO_4$] is indicated in Fig. 6.15B, although tin was not detected in the glaze analyses. The peak at wavenumber 143 cm^{-1} in Fig. 16(e) is indicative of anatase and the peak at 636 cm^{-1} indicates cassiterite (Sn(IV) oxide). Clark et al. [14] indicated that decomposition of the lead-tin pigment will occur at temperatures >950 °C to form their respective elemental oxides in a glassy matrix.

The Si-O bending maximum for the green glaze is at ~440 cm^{-1} and the Si-O stretching maximum is at ~1000 cm^{-1}. The Raman Polymerisation Index (I_p) is provided in Fig. 6.22 and Table 6.6 together with additional statistics.

Sample F7, Mineralogy: Raman Spectra, Fig. 6.15C

The SEM-EDS analysis of this lead-based aluminosilicate green glaze is given in Table 6.5. Its Raman spectra provided in Fig. 6.15C indicates the presence of traces of amorphous carbon as shown by the broad bands at ~1300 and ~1600 cm^{-1} wavenumbers. The glaze is low in copper and high in iron, the ferrous ion contributing to much of the green pigmentation; the darker green region indicating a higher copper concentration. Wollastonite, anatase, quartz under tension ("compressive stress induced by the glaze matrix") is indicated by the lower wavenumber peak at 459 cm^{-1}, cassiterite and probably calcium phosphate. Labradorite and anorthite are expected.

The signature for a decomposed lead-tin Type I mineralisation [$PbSnO_4$] is indicated in Fig. 6.15B(e), although tin was not detected in the glaze analyses. The peak at wavenumber 128 cm^{-1} is indicative of the lead-tin Type I mineralisation and the peak at 634 cm^{-1} indicates the presence of cassiterite (Sn(IV) oxide).

The Si-O bending maximum for the green glaze is at ~479 cm^{-1} and the Si-O stretching maximum is at ~1005 cm^{-1}. The Raman Polymerisation Index (I_p) is plotted in Fig. 22 together with additional statistics is provided in Table 6.

Sample F8 Mineralogy

This lead-based aluminosilicate green glaze contains amorphous carbon as shown by the broad bands at ~1300 and ~1600 cm^{-1} wavenumbers. As such, reducing conditions would have prevailed during the firing of the glaze. The glaze contains both copper and iron, both of which are able to contribute to the green glaze colouration. Anatase, calcite and probably chalcocite, together with lead mineralisation, as minium ($Pb^{2+}{}_2Pb^{4+}O_4$) or plattnerite (PbO_2), possibly augite [$(Ca,Mg,Fe)_2(Si,Al)_2O_6$] and diopside are present. The elemental analysis for chromium provided in Table 5 is considered to be genuine, indicating the liklihood of traces of a chromite ore being present.

The Si-O bending maximum for the green glaze is at ~409 cm^{-1} and the Si-O stretching maximum is at ~1090 cm^{-1}. The Raman Polymerisation Index (I_p) is plotted in Fig. 6.22 and together with additional statistics is provided in Table 6.6.

Sample F9, Mineralogy: Raman Spectra Fig. 6.14

This lead-based aluminosilicate green glaze contains traces of amorphous carbon as shown by the broad bands at ~1300 and ~1600 cm^{-1} wavenumbers. The glaze contains both copper and iron, both of which are able to contribute to the green glaze colouration. Anatase, feldspar and wollastonite are present.

The Si-O bending maximum for the green glaze is at ~468 cm^{-1} and the Si-O stretching maximum is at ~989 cm^{-1}. The Raman Polymerisation Index (I_p) is plotted in Fig. 6.22 and together with additional statistics is provided in Table 6.6.

Sample F10 Mineralogy: Raman Spectra, Fig. 6.15

This lead-based aluminosilicate green glaze contains traces of amorphous carbon as shown by the broad bands at ~1300 and ~1600 cm^{-1} wavenumbers. The glaze contains both copper and iron, both of which are able to contribute to the green glaze colouration. Anatase, together with plattnerite, magnesioferrite [$MgFe^{3+}{}_2O_4$], labradorite, hematite and wollastonite are present. Quartz, together with the peak at 503 cm^{-1} could be indicative of the presence of moganite; it is therefore likely that ground chert (flint) formed part of the silica. The strong bands in the region of

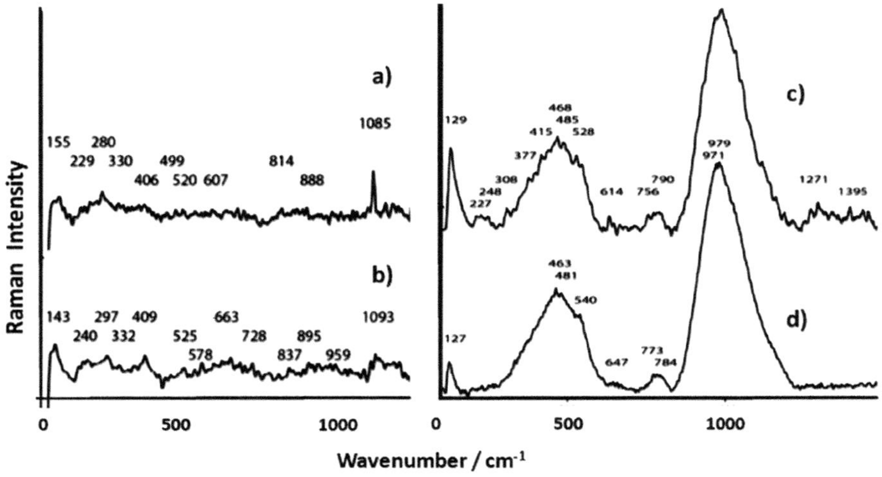

Fig. 6.14 Baseline subtracted Raman spectra, green glaze on sample F9 (λ = 514.5 nm

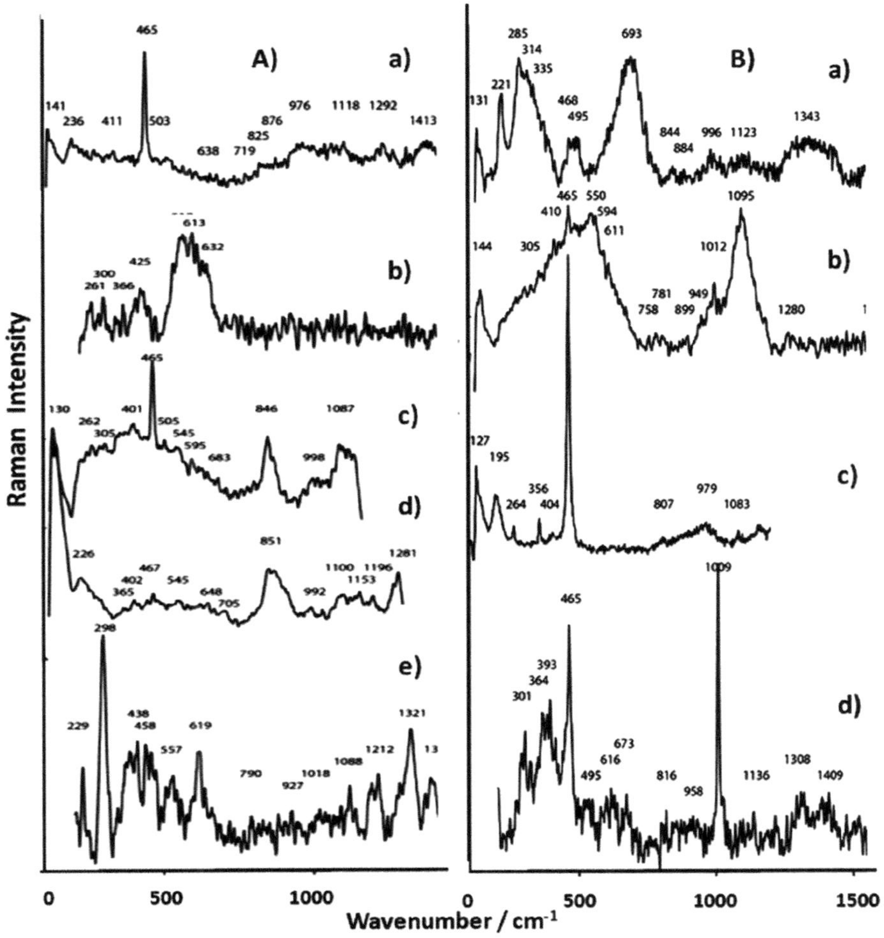

Fig. 6.15 (**A**) Baseline subtracted Raman spectra, green glaze on sample F10: (**a**) and (**d**) λ = 782 nm, (**b**) and (**c**) λ = 514.5 nm); (**B**) Baseline subtracted Raman spectra, faience glaze surface of sample F1(λ = 514.5 nm)

wavenumbers 846–851 cm^{-1} are indicative of the presence of an arsenical mineralisation whilst arsenic was not reported in the SEM-EDS analyses (Fig. 6.16).

Chromium in the analyses is considered to indicate the presence of a chromium ore rather than sample contamination. The statistics obtained from the Raman spectra were too poor to enable determination of the Polymerisation Index.

High Lead Glaze

Table 6.10 provides the average glazes for 6 samples analysed either by al-Saad [2] or by Henshaw [41]. Lead-glazed pottery has been reviewed by Tite [97, 98], Tite and Bimson [100, 101] and Walton and Tite [112]. No similar high lead-containing glazes were analysed from the Dakhleh Oasis samples.

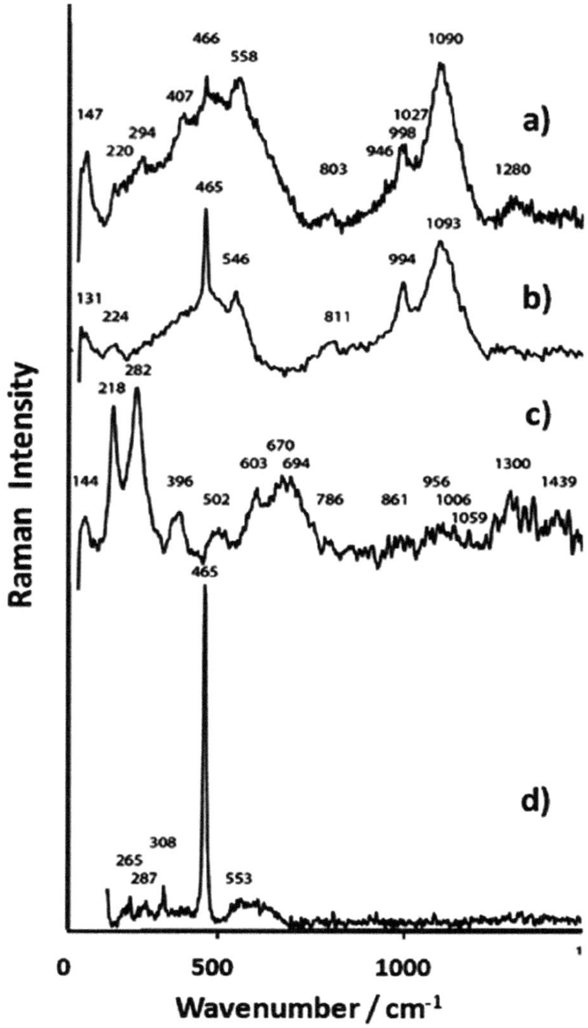

Fig. 6.16 Baseline subtracted Raman spectra, faience glaze surface of lustre sample F2 (λ = 514.5 nm)

Faience Lustre Glaze, Fig. 3

Tite et al. [103] produced faience under laboratory conditions as provided in Table 6.9, to obtain a bright turquoise complex following the technique conducted at Qom, Iran. According to Wulff et al. [116], and contrary to the proposal of Noble [72] a mixture of finely ground quartzite and gum tragacanth dissolved in water was fired at approximately 1000 °C whilst buried in a glazing mixture consisting of powdered quartz, hydrated lime, plant ash, copper oxide and charcoal to produce a turquoise glaze.

Table 6.10 Average analysis of six faience sherds with a high lead-based glaze from North Jordan [al-Saad [2]] and from Uzbekistan [41]

Average elemental analysis, Faience brown and green samples, high lead glaze type [al-Saad [2]].

Na_2O	MgO	Al_2O_3	SiO_2	P_2O_5	SO_3	Cl	K_2O	CaO	TiO_2	MnO_2	Fe_2O_3	CoO	CuO	SnO_2	PbO
0.5	0.38	6.74	30.21	n.a.	n.a.	n.a.	0.33	1.53	n.a.	0.0	0.8	0.0	0.6	0.0	57.86

Average elemental analysis, Faience samples, high lead glaze type, Uzbekistan, [41] [a].

Na_2O	MgO	Al_2O_3	SiO_2	P_2O_5	SO_3	Cl	K_2O	CaO	TiO_2	MnO_2	Fe_2O_3	CoO	CuO	SnO_2	PbO
0.3	0.1	3.0	40.3	Na	Na	Na	1.3	0.8	Na	Na	0.4	Na	Na	Na	53.8

[a] na = not analysed

The two lustre faience surface samples from Dakhleh Oasis identified as F1 and F2 are illustrated in Fig. 6.3. Both of these samples have an alkaline sky-blue azure glaze, which has been applied to the surface of a coarse, off-white siliceous material.

The EDS analyses of these sherds are summarised in Table 6.7 and their mineralogy is provided by Raman spectroscopy. As shown in Fig. 6.3, a finely divided silica slip interlayer was evident between the body fabric and the faience glaze in both samples F1 and F2. The copper pigments are incorporated into a lead-free, aluminosilicate glaze in both cases. The concentration of fluxing minerals (predominately Na_2O, plus some K_2O) is considered to be low in F1. The presence of about 1.5 wt. % of each of MgO and K_2O is indicative that some plant (or kiln) ash was included in the glaze formulation [refer Colomban [[17, 19, 22, 25]]. In sample F2 the concentration of fluxing minerals, particularly the soda flux is significantly higher as shown in Table 7 and confirmed by the much higher Ionic Oxide Content (IOC) calculated from the Raman spectroscopy data provided in Tables 6.6 and 6.11.

Shard F1 also had calcite crystals located on its surface; again, this would elevate the reported lime concentration.

In sample F2, broad Raman bands at wavenumbers ~1300 and ~1600 cm^{-1} indicate amorphous carbon and therefore reducing conditions during firing. As such the iron present in the glaze would be in the ferrous state (Fe^{2+}) and this would contribute to the greenish appearance of the glaze. Similarly, whilst the copper would have been present as either Cu^{+1} (colourless) or Cu^{2+}. Cu^{2+} in a square planar arrangement is blue and when in a disordered octahedral site in a glassy phase is green [52, 81].

The glaze on samples F1 and F2 has a matt or semi-gloss appearance. This may be either due to the formulation, or more likely to weathering.

Other than the lower soda and potash content, in Sample F1, probably due to weathering, both sherds are similar to that of the alkaline glazes discussed by al-Saad [2], Tite [1998, 2007], Shortland [93] and Henshaw [41] as provided in Table 6.8. Sample F2 is very similar in elemental composition to the Late period shabti analysis provided by Tite et al. ([103] sample BMRL 16322).

Whilst al-Saad [2] suggested that the Jordanian alkali glazes he examined were subject to cracking, this was not observed in either sample F1 or sample F2. He suggested that this cracking resulted from the glaze makers using a highly siliceous slip on the surface of the body to separate it from the glaze. Such a white slip is visible in both sample F1 and F2. Samples F1 and F2 are free from crazing. As such, these two samples are different to the al-Saad observation. Samples F1 and F2 were low or free from the presence of manganese and this ensured that these samples remained blue, rather than converting to a rich purple colour.

Coptic and Islamic lustre chemical compositions have been sometimes assigned to Egyptian and to Syrian glassworkers in the sixth or seventh century CE although, according to Kingery and Vandiver [49] most of preserved lustreware samples date

Table 6.11 Representative average elemental analysis (EDS) of the two Islamic lustre sherds analysed

Blue, Sample F1

Na_2O	MgO	Al_2O_3	SiO_2	P_2O_5	SO_3	Cl	K_2O	CaO	TiO_2	MnO_2	FeO	CoO	CuO	SnO_2	PbO
3.2	2.1	9.4	71.4	Tr	1.0	1.6	1.4	3.7	0.5	0.0	3.4	0.0	2.5	0.0	0.0

Blue, Sample F2. FIB EDS analysis

Na_2O	MgO	Al_2O_3	SiO_2	P_2O_5	SO_3	Cl	K_2O	CaO	TiO_2	MnO_2	FeO	CoO	CuO	SnO_2	PbO
14.2	1.2	4.0	65.3	Tr	0.9	1.4	0.9	4.4	0.3	0.2	2.4	0.0	4.0	0.0	0.3

from the ninth century CE. This lustre effect is obtained by the firing of a metal such as copper or silver under reducing conditions on the glaze surface to form the metallic lustre [16, 18, 21, 89]. This was later particularly conducted onto the surface of tin-opacified white glaze grounds. Abūl Qāsim's Persian recipe specified the addition of vinegar to the metal to provide a paintable suspension. The vinegar would react with copper and silver on heating to form their respective acetates. This reaction aids in more uniformly distributing the metal before it decomposes and deposits the metal as an oxide at about 240 °C [1]. Incorporation of elemental silver on a glass surface is known to produce a yellow-stained glass when fired [18, 22, 25].

The turquoise pigmentation has possibly been derived from an analogue similar to lazurite [$Na_8(Al_6Si_6O_{24})S_n$] or lapis lazuli as shown by the Raman bands at 550 (the S^{3-} symmetric stretching mode), 806 and 1095 cm^{-1} (see Chap. 10). The mineral lapis lazuli, consists of grains of the cubic mineral, lazurite; and is generally associated with one or more of the minerals including pyroxene (diopside), haüyne, sodalite, wollastonite, forsterite, calcite, phlogopite and pyrite [20]. In the present samples, lazurite, quartz, hematite, diopside, titanium dioxide as anatase and probably wollastonite and sodalite are considered to be present. Particles of silver, tin, ilmenite, bismuth and zinc were determined by SEM-EDS analysis. The presence of tin may indicate that bronze scrap was employed as a copper source. Bronze scrap was certainly the copper source used at that time for Egyptian blue production. Calcite crystals are located on the surface of sample F1 (refer to Fig. 6.3) indicated by areas found on the shard surface. These crystals contain high CaO (up to 78.5%), high P_2O_5 (up to 4.1%) and high K_2O (up to 4.2%) indicative that the vessel surface has also come into contact with material probably from a lime kiln and containing bone ash. These calcite crystals have also elevated the reported lime concentration. Both the silver and the calcium surface crystallisation are shown in Fig. 6.3.

Separate surface areas containing silver and tin were determined in the analysis of sample F1. In Fig. 6.3, sample F1 showed areas of insoluble silver on the surface. It most likely indicates an attempt to produce a lustre effect. The area covered by the silver, does not indicate contamination. This sherd may have originally emanated from a vessel which had been intended to have been heated under reducing conditions in an attempt to obtain a lustre effect.

In sample F1 the Si-O bending maximum is at 500 cm^{-1} and the Si-O stretching maximum is at 1095 cm^{-1}. The Raman Polymerisation Index (I_p) plotted in Fig. 6.17 and given in Table 6.6 corresponds to Family #4, namely Ca-based glassy silicates as described by Colomban and co-workers ([22, 25], 2008). Although, in the present samples, the CaO concentration is in the order of 4% and as such, it is not significantly higher than the sole-alkali glaze samples. In sample F2 the Si-O bending maximum is at 502 cm^{-1} and the Si-O stretching maximum is at 1090 cm^{-1} (Table 6.12).

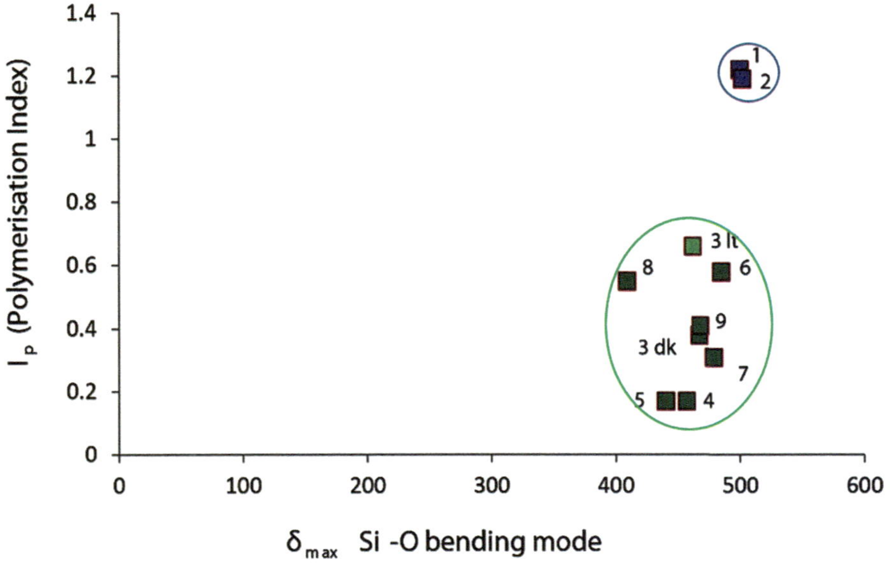

Fig. 6.17 Plot of wavenumber bending multiplets Vs Polymerisation Index (I_p) for glaze samples recovered at Dakhleh Oasis, Western Desert, Egypt. The blue envelope data are the two blue alkaline lustre glaze samples (F1 and F2) and the green envelope are the lead-alkali glazes. The green numbers 3, 6 and 8 (corresponding to Samples F3, F6 and F8) are all lower in soda and lead by comparison with the other samples analysed

6.6 Conclusions

The samples of lime-alkali glazed faience are from Dakhleh Oasis and the Islamic glazed pottery are undated surface finds from Qasr al-Dakhleh. The Dakhleh Oasis shards are lead-free glazes whereas the lead-alkali glazed pottery were lower in total alkali and often with a significant lead concentration. The two lime-alkali lustre glazed turquoise sherds are dated to the second century Roman period.

In the formation of Egyptian green, the presence of tridymite rather than cristobalite together with parawollastonite would indicate a firing temperature in the region of 1000 °C. In the complex copper minerals observed, the formation of Egyptian blue (cuprorivaite) was not detected.

A plot of the Polymerisation Index (I_P) versus the Si-O bending mode provided in Fig. 6.24 indicates a substantial difference between the alkali-lead glazes and the Islamic lustre lead-free aluminosilicate glazes.

Table 6.12 Average analysis of six faience sherds with a high lead-based glaze from North Jordan [al-Saad [2]] and from Uzbekistan [41]

Average elemental analysis, Faience brown and green samples, high lead glaze type [2].															
Na_2O	**MgO**	**Al_2O_3**	**SiO_2**	**P_2O_5**	**SO_3**	**Cl**	**K_2O**	**CaO**	**TiO_2**	**MnO_2**	**Fe_2O_3**	**CoO**	**CuO**	**SnO_2**	**PbO**
0.5	0.38	6.74	30.21	n.a.	n.a.	n.a.	0.33	1.53	n.a.	0.0	0.8	0.0	0.6	0.0	57.86
Average elemental analysis, Faience samples, high lead glaze type, Uzbekistan, [41][a].															
Na_2O	**MgO**	**Al_2O_3**	**SiO_2**	**P_2O_5**	**SO_3**	**Cl**	**K_2O**	**CaO**	**TiO_2**	**MnO_2**	**Fe_2O_3**	**CoO**	**CuO**	**SnO_2**	**PbO**
0.3	0.1	3.0	40.3	Na	Na	Na	1.3	0.8	Na	Na	0.4	Na	Na	Na	53.8

[a]na = not analysed

References

1. Allan, J.W.: Abū'l Qāsim's treatise on ceramics. Iran. **11**(1), 111–120 (1973). https://doi.org/10.1080/05786967.1973.11834171
2. al-Saad, Z.: Chemical composition and manufacturing technology of a collection of various types of Islamic glazes excavated from Jordan. J. Archaeol. Sci. **29**(8), 803–810 (2002). https://doi.org/10.1006/jasc.2000.0576
3. Baraldi, P., Bondioli, F., Fagnano, C., Ferrari, A.M., Tinti, A., Vinella, M.: Study of the vibrational spectrum of cuprorivaite. Ann. Chim. **91**(11–12), 679–692 (2001). https://www.researchgate.net/profile/Pietro-Baraldi/publication/11523845_Study_of_the_vibrational_spectrum_of_cuprorivaite/links/5f218f8b299bf134048fc51b/Study-of-the-vibrational-spectrum-of-cuprorivaite
4. Batonneau, Y., Brémard, C., Laureyns, J., Merlin, J.C.: Microscopic and imaging Raman scattering study of PbS and its photo-oxidation products. J. Raman Spectrosc. **31**(12), 1113–1119 (2000). https://doi.org/10.1002/1097-4555(200012)31:12<1113::AID-JRS653>3.0.CO;2-E
5. Beck, H.C.: Glass before 1500 BC. Ancient Egypt. **XIX**, 7–21 (1934)
6. Bell, I.M., Clark, R.J., Gibbs, P.J.: Raman spectroscopic library of natural and synthetic pigments (pre-≈ 1850 AD). Spectrochim. Acta A Mol. Biomol. Spectrosc. **53**(12), 2159–2179 (1997)
7. Berke, H.: The invention of blue and purple pigments in ancient times. Chem. Soc. Rev. **36**(1), 15–30 (2007). https://doi.org/10.1039/B606268G
8. Bordignon, F., Postorino, P., Dore, P., Trojsi, G.: Raman identification of green and blue pigments in Etruscan polychromes on architectural terracotta panels. J. Raman Spectrosc. **38**(3), 255–259 (2007)
9. Bouquillon, A., Kaczmarczyk, A., Vandiver, P.B.: Faience production in the Near East and the Indus Valley. In: Tite, M.S., Shortland, A.J. (eds.) Production Technology of Faience and Related Early Vitreous Materials Oxford University School of Archaeology, Monograph 72. Oxford University, Oxford (2008)
10. Bruni, S., Cariati, F., Casadio, F., Toniolo, L.: Spectrochemical characterization by micro-FTIR spectroscopy of blue pigments in different polychrome works of art. Vib. Spectrosc. **20**(1), 15–25 (1999)
11. Burgio, L., Clark, R.J.H.: Library of FT-Raman spectra of pigments, minerals, pigment media and varnishes, and supplement to existing library of Raman spectra of pigments with visible excitation. Spectrochim. Acta. **A57**(7), 1491–1521 (2001). https://doi.org/10.1016/S1386-1425(00)00495-9
12. Caiger-Smith, A.: Tin-Glaze Pottery. Faber and Faber, London (1973)
13. Caley, E.R.: Analyses of Ancient Glasses 1790–1957. Corning Museum of Glass, Corning (1962)
14. Clark, R.J.H., Cridland, L., Kariuki, B.M., Harris, K.D.M., Withnall, R.: Synthesis, structural characterisation and Raman spectroscopy of the inorganic pigments Lead tin yellow types I and II and Lead Antimonate yellow: their identification on medieval paintings and manuscripts. J. Chem. Soc. Dalton Trans., 2577–2582 (1995)
15. Clark, J.R.H., Gibbs, P.J.: Non-destructive in situ study of ancient Egyptian faience by Raman microscopy. J. Raman Spectrosc. **28**(2–3), 99–103 (1997). https://doi.org/10.1002/(SICI)1097-4555(199702)28:2/3<99::AID-JRS75>3.0.CO;2-U
16. Colomban, P.: Secrets retrouvés du Lustre Abbasside. Revue de la Céramique et du Verre. **139**, 13–20 (2004)
17. Colomban, P.: On-site Raman identification and dating of ancient glasses: a review of procedures and tools. J. Cultural Herit. **9**, e55–e60 (2008). https://doi.org/10.1016/j.culher.2008.06.005

18. Colomban, P.: The use of metal nanoparticles to produce yellow, red and iridescent colour, from bronze age to present times in lustre pottery and glass: solid state chemistry, spectroscopy and nanostructure. J. Nano Res. **8**, 109–132 (2009). https://doi.org/10.4028/www.scientific.net/JNanoR.8.109
19. Colomban, P.: Non-destructive Raman analysis of ancient glasses and glazes. In: Janssens, K. (ed.) Modern Methods for Analysing Archaeological and Historical Glasses. Wiley (2013a)
20. Colomban, P.: Rocks as blue, green and black pigments/dyes of glazed pottery and enamelled glass artefacts – a review. Eur. J. Mineral. **25**(5), 863–879 (2013b). https://www.researchgate.net/profile/Philippe-Colomban/publication/272205215_Rocks_as_blue_green_and_black_pigmentsdyes_of_glazed_pottery_and_enamelled_glass_artefacts_-_A_review/links/54f5b0550cf2eed5d739a58e/Rocks-as-blue-green-and-black-pigments-dyes-of-glazed-pottery-and-enamelled-glass-artefacts-A-review.pdf
21. Colomban, P., Truong, C.: Non-destructive Raman study of the glazing technique in lustre potteries and faience (9–14th centuries): silver ions, nanoclusters, microstructure and processing. J. Raman Spectroscopy. **35**, 195–207 (2004). https://doi.org/10.1002/jrs.1128
22. Colomban, P., Etcheverry, M.-P., Asquier, M., Bounichou, M., Tournié, A.: Raman identification of ancient stained glasses and their degree of deterioration. J. Raman Spectrosc. **37**(5), 614–626 (2006a). https://doi.org/10.1002/jrs.1495
23. Colomban, P., Milande, V.: "On-Site Raman Analysis if Rare Ancient Ceramics: Medici Porcelain and Iznik Pottery" 1st International Workshop on: Science. Technology and Cultural Heritage, Cinema Festival Palace, Venice (2004)
24. Colomban, P., Schreiber, H.D.: Raman signature modification induced by copper nanoparticles in silicate glass. J. Raman Spectrosc. **36**(9), 884–890 (2005). https://doi.org/10.1002/jrs.1379
25. Colomban, P., Tournié, A., Bellot-Gurlet, L.: Raman identification of glassy silicates used in ceramics, glass and jewellery: a tentative differentiation guide. J. Raman Spectrosc. **37**(8), 841–852 (2006b). https://doi.org/10.1002/jrs.1515
26. Cooper, E., Lewenstein, E.: Clays and Glazes – The Ceramic Review Book of Clay Bodies and Glaze Recipes. Ceramic Review, London (1988)
27. Crupi, V., Majolino, D., Venuti, V., Barone, G., Mazzoleni, P., Pezzino, A., La Russa, M.F., Ruffolo, S.A., Bardelli, F.: Non-destructive identification of green and yellow pigments: the case of some Silician renaissance glazed pottery. Appl. Phys. A. **100**, 845–853 (2010). https://doi.org/10.1007/s00339-010-5660-x
28. Edwards, H.G.M., Farwell, D.W.: 'Dragon's blood' I—characterization of an ancient resin using Fourier transform Raman spectroscopy. J. Raman Spectrosc. **28**(4), 243–249 (2008). https://doi.org/10.1002/(SICI)1097-4555(199704)28:4<243::AID-JRS91>3.0.CO;2-W
29. Edwards, H.G.M., Vandenabeele, P., Colomban, P.: Raman Spectroscopy in Cultural Heritage Preservation. Springer Nature, Cham (2023)
30. Edwards, H.G., Jorge Villar, S.E., Eremin, K.A.: Raman spectroscopic analysis of pigments from dynastic Egyptian funerary artefacts. J. Raman Spectrosc. **35**(8–9), 86–795 (2004)
31. El Goresy, A.: Polychromatic wall painting decoration in monuments of Pharaonic Egypt: Composition, chronology and painting technology. In: Sherratt, S. (ed.) Proceedings of the First International Symposium, The Wall paintings of Thera, pp. 49–70. Thera Foundation, Piraeus (2000)
32. El Goresy, A., Jaksch, H., Abdel Razek, M., Weiner, K.L.: Ancient pigments in wall paintings of Egyptian tombs and temples: an archaeometric project. In: Trenkwalder, H., Purtscheller, F., Lukas, W., Seidl, H. (eds.) Interdisziplinäres Gespräch: Geistesweissenschaft-Naturwissenschaft-Technik anhand konkreter Projekte, pp. 1–57. Sonderheft des Bunddesministerium für Weissenschaft und Forschung, Wien (1986)
33. Freestone, I.C.: Glass production in late antiquity and the early Islamic period: a geochemical perspective. In: Maggetti, M., Messiga, B. (eds.) Geomaterials in Cultural Heritage Special Publications 257, pp. 201–216. Geological Society, London (2006)

34. Greene, K.: Late Hellenistic and early Roman invention and innovation: the case of lead-glazed pottery. Am. J. Archaeol. **111**(4), 653–671 (2007). https://www.journals.uchicago.edu/doi/abs/10.3764/aja.111.4.653
35. Groat, L.A., Kek, S., Bismayer, U., Schmidt, C., et al.: A synchrotron radiation, HRTEM, -ray powder diffraction, and Raman spectroscopic study of malayaite, CaSnSiO5. Am. Mineral. **81**(5–6), 595–602 (1996). https://doi.org/10.2138/am-1996-5-606
36. Hatton, G.D.: Production of Egyptian blue and green frits. In: Tite, M.S., Shortland, A.J. (eds.) Monograph 72, pp. 147–185. Oxford University School of Archaeology (2008)
37. Hatton, G.D., Shortland, A.J., Tite, M.S.: Egyptian blue: where, when and how? In: Ives, R., Lines, D., Naunton, C., Wahlberg, N. (eds.) Current Research in Egyptology III BAR International Series 1192. BAR (2001)
38. Hatton, G.D., Shortland, A.J., Tite, M.S.: The production technology of Egyptian blue and green frits from second millennium BC Egypt and Mesopotamia. J. Archaeol. Sci. **35**(6), 1591–1604 (2008). https://doi.org/10.1016/j.jas.2007.11.008
39. Hedges, R.E.M.: Pre-Islamic Glazes in Mesopotamia—Nippur. Archaeometry. **18**(2), 209–238 (1976). https://doi.org/10.1111/j.1475-4754.1976.tb00162.x
40. Hedges, R.E.M., Moorey, P.R.S.: Pre-Islamic ceramic glazes at Kish and Nineveh in Iraq. Archaeometry. **17**(1), 25–43 (1975). https://doi.org/10.1111/j.1475-4754.1975.tb00113.x
41. Henshaw, C.M.: Early Islamic Ceramics and Glazes of Akhsiket, Uzbekistan. PhD Thesis, UCL, London (2010)
42. Hodges, H.: Artifacts: An Introduction to Early Materials and Technology. John Baker, London (1976)
43. Jaksch, H., Scipel, W., Weiner, K.L., El-Goresy, A.: Egyptian blue—cuprorivaite a window to ancient Egyptian technology. Naturwissenschaften. **70**, 525–535 (1983). https://link.springer.com/article/10.1007/BF00376668
44. Jay, W.H., Cashion, J.D., Blenkinship, B.: Lancaster delftware: a Raman spectroscopy, electron microscopy and Mössbauer spectroscopy compositional study. J. Raman Spectrosc. **46**(12), 1265–1282 (2015). https://doi.org/10.1002/jrs.4750
45. Jorge-Villar, S.E., Edwards, H.G.: Green and blue pigments in Roman wall paintings: a challenge for Raman spectroscopy. J. Raman Spectrosc. **52**(12), 2190–2203 (2021)
46. Kaczmarczyk, A., Hedges, R.E.M.: Ancient Egyptian Faience: an Analytical Survey of Egyptian Faience from Predynastic to Roman Times. Aris and Phillips, Warminster (1983)
47. Kappen, P. Eccleston, M., Wallwork, P.: XAC investigation of ancient Egyptian faience from the site of Amarna (2010). www.photon-science.desy.de/annual_report/files/2010/20101327
48. Kerr, R., Wood, N.: "Science and Civilisation in China", Vol. 5, Chemistry and Chemical Technology, Part XII: Ceramic Technology. Cambridge University Press, Cambridge (2004)
49. Kingery, W.D., Vandiver, P.B.: Ceramic Masterpieces. Collier Macmillan Publishers, London (1986)
50. Kühne, K.: Äegyptische Fayencen, Part I. In: S. Wenig (ed.) Grabungen der Deutschen Orient-Gesellschaft (1969)
51. Landgráfová, R.: "Faience Inlays from the Funerary Temple of King Raneferef", ABUSIR XIV. Czech Institute of Egyptology, Prague (2006)
52. Lau, D., Kappen, P., Strohschneidr, M., Brack, N., Pigram, P.J.: Characterisation of green copper phase pigments in Egyptian artifacts with X-ray absorption spectroscopy and principal component analysis. Spectrochim. Acta B. **63**(11), 1283–1289 (2008). https://doi.org/10.1016/j.sab.2008.09.013
53. Le Fur, D.: La Conversation des Peintures Murales des Temples de Karnak. Recherches sur les Civilisations. CNRS Erec, Paris (1994)
54. Lewis, G.: A Collector's History of English Pottery, 5th edn. Antique Collector's Club, Suffolk (1999)
55. Liang, H., Sax, M., Saunders, D., Tite, M.: Optical coherence tomography for the non-invasive investigation of the microstructure of ancient Egyptian faience. J. Archaeol. Sci. **39**(12), 3683–3690 (2012). https://doi.org/10.1016/j.jas.2012.06.007

56. Lucas, A.: Ancient Egyptian Materials and Industries. Edward Arnold & Co, London (1962)
57. Lucas, A., Harris, J.R.: Ancient Egyptian Materials and Industries, 4th edn. Histories & Mysteries of Man Ltd, London (1989)
58. MacKenzie, D.A.: Colour symbolism. Folklore. **33**(2), 136–169 (1922). https://doi.org/10.1080/0015587X.1922.9719237
59. Mahmoud, H.M.: A preliminary investigation of ancient pigments from the mortuary Temple of Seti I, El-Qurna, Luxor, Egypt. Mediterr Archaeol Archaeo. **11**(1), 99–106 (2011)
60. Maltoni, S., Silvestri, A., Marcante, A., Molin, G.: The transition from Roman to late antique glass: new insights from the *Domus of Tito macro* in Aquileia (Italy). J. Archaeol. Sci. **73**, 1–16 (2016). https://doi.org/10.1016/j.jas.2016.07.002
61. Mason, R.B.: Shine like the Sun: Lustre-Painted and Associated Pottery from the Medieval Middle East. Mazda Publishers, Costa Mesa (2004)
62. Mason, R.B., Tite, M.S.: The beginnings of Islamic stonepaste technology. Archaeometry. **36**(1), 77–91 (1994). https://doi.org/10.1111/j.1475-4754.1994.tb01066.x
63. Mason, R.B., Tite, M.S.: The beginnings of the tin-opacification of pottery glazes. Archaeometry. **39**(1), 41–58 (1997). https://doi.org/10.1111/j.1475-4754.1997.tb00789.x
64. McCarthy, B., Paynter, S.: Production of glazed pottery and brickwork in the Near East. In: Tite, M.S., Shortland, A.J. (eds.) Production Technology of Faience and Related Early Vitreous Materials Oxford University School of Archaeology, Monograph 72, pp. 187–198. Oxford University, Oxford (2008)
65. McCarthy, B., Vandiver, P.B., Gibson, M.: Innovation and continuity of Southwest Asian monochrome blue glazes. In: Vincenzini, P. (ed.) The Ceramics Cultural Heritage Academy of Ceramics, Proc. Intl. Symp.: The Ceramic Heritage of the 8th CIMTEC-World Ceramic Congress and Forum on New Materials, pp. 207–218, Florence (1995)
66. Matin, M.: An experimental investigation into the accidental invention of ceramic glazes. Archaeometry. **56**(4), 591–600 (2013). https://doi.org/10.1111/arcm.12039
67. Matin, M., Matin, M.: Egyptian faience glazing by the cementation method part I: an investigation of the glazing powder composition and glazing mechanism. J. Archaeol. Sci. **39**(3), 763–776 (2012). https://doi.org/10.1016/j.jas.2011.11.013
68. Moorey, P.R.S.: Materials and Manufacture in Ancient Mesopotamia: The Evidence of Archaeology and Art Metals and Metalwork, Glazed Materials and Glass BAR International Series 237. BAR, Oxford (1985)
69. Moorey, P.R.S.: Ancient Mesopotamian Materials and Industries: The Archaeological Evidence. Clarendon Press, Oxford (1994)
70. Nicholson, P.T.: Brilliant Things for Akhenaten. In: The Production of Glass, Vitreous materials and Pottery at Amarna Site O45.1 Excavation Memoir 80. Egypt Exploration Society, London (2007)
71. Nicholson, P.: Working in Memphis: The Production of Faience at Roman Period Kom Helul Excavation Memoir 105. Egypt Exploration Society, London (2013)
72. Noble, J.V.: The technique of Egyptian faience. Am. J. Archaeol. **73**(4), 435–439 (1969). https://www.journals.uchicago.edu/doi/abs/10.2307/503999?journalCode=aja
73. Noll, W.: Mineralogy and Technology of the Painted Ceramics of Ancient Egypt. In: Hughes, M.J. (ed.) Scientific Studies in Ancient Ceramics Occasional Paper No. 19, pp. 143–154. British Museum (1981a)
74. Noll, W.: Bemalte Keramiken Altägyptens: Material, Rohstoffe und Herstellungstechnik. In: Arnold, D. (ed.) Studien zur altägyptischen Keramik, pp. 103–138. Philipp von Zabern, Mainz am Rhein (1981b)
75. Noll, W., Hangst, K.: Grün- und Blaupigmente der Antike. N. Jb. Miner. Mh. H. **12**, 529–540 (1975b)
76. Noll, W., Hangst, K.: Zuz Kenntnis der altägyptischen Blaupigmente. N. Jb. Miner. Mh. H. **12**, 309–314 (1975a)
77. Oppenheim, A.L., Brill, R.H., Barag, D., von Saldern, A.: Glass and Glassmaking in Mesopotamia. Corning, New York (1988)

78. Pagès-Camagna, S., Colinart, S.: The Egyptian green pigment: its manufacturing process and links to Egyptian blue. Archaeometry. **45**(4), 637–658 (2003). https://doi.org/10.1046/j.1475-4754.2003.00134.x
79. Pagès-Camagna, S., Colinart, S.: Authors' reply. Archaeometry. **48**(4), 710–713 (2006). https://doi-org.inc.bib.cnrs.fr/10.1111/j.1475-4754.2006.282_2.x
80. Pagès-Camagna, S., Colinart, S., Coupry, C.: Fabrication processes of archaeological Egyptian blue and green pigments enlightened by Raman microscopy and scanning electron microscopy. J. Raman Spectrosc. **30**(4), 313–317 (1999). https://doi.org/10.1002/(SICI)1097-4555(199904)30:4<313::AID-JRS381>3.0.CO;2-B
81. Pagès-Camagna, S., Reiche, C., Brouder, C., Cabaret, D., Rossana, S., Kanngieβer, B., Erke, A.: New insights into the colour origin of archaeological Egyptian blue and green by XAFS at the Cu K-edge. X-Ray Spectrosc. **35**(2), 141–145 (2006). https://doi.org/10.1002/xrs.885
82. Paynter, S., Tite, M.: The evolution of glazing technologies in the ancient near East and Egypt. In: Shortland, A. (ed.) The Social Context of Technological Change, Egypt and the Near East, 1650–1550BC. Oxbow Books, Oxford (2001)
83. Peltenburg, E.: Some early developments of vitreous materials. World Archaeol. **3**(1), 6–12 (1971). https://doi.org/10.1080/00438243.1971.9979489
84. Peltenburg, E.J.: Early faience: recent studies, origins and relations with glass. In: Bimson, M., Freestone, I.C. (eds.) Early Vitreous Materials British Museum Occasional Paper 56, pp. 5–30. British Museum (1992)
85. Rehren, T.: A review of the factors affecting the composition of early Egyptian glasses and faience: alkali and alkaline earth oxides. J. Archaeol. Sci. **35**(5), 1345–1354 (2008). https://doi.org/10.1016/j.jas.2007.09.005
86. Rice, P.M.: Pottery Analysis: A Source Book. The University of Chicago Press (1987)
87. Rosalie David, A., Edwards, H.G.M., Farwell, D.W., De Faria, D.L.A.: Raman spectroscopic analysis of ancient Egyptian pigments. Archaeometry. **43**(4), 461–473 (2001)
88. Sapsford, M.: The use of sodium salt deposits in medical and medically associated industries in ancient Egypt. PhD Thesis, Department of Applied Science, Security and Resilience, Cranfield University (2009)
89. Sciau, P., Noé, L., Colomban, P.: Metal nanoparticles in contemporary potters' master pieces: lustre and red "pigeon blood" potteries as models to understand the ancient pottery. Ceram. Int. **42**, 15349–15357 (2016). https://doi.org/10.1016/J.ceramint.2026.06.179
90. Schiegl, S., El Goresy, A.: Comments on S. Pagès-Camagna and S. Colinart, 'The Egyptian green pigment: its manufacturing process and links to Egyptian blue', Archaeometry, 45(4) (2003), 637–58*. Archaeometry. **48**(4), 707–713 (2006). https://doi.org/10.1111/j.1475-4754.2006.282_1.x
91. Schiegl, S., Weiner, K.L., El-Goresy, A.: Discovery of Copper Chloride Cancer in Ancient Egyptian Polychromic Wall Paintings and Faience: a Developing Archaeological Disaster, p. 15. Max-Planck-Institut Für Kernphysik, Heidelberg (1989) plus appendices
92. Schibille, N., Gratuze, B., Ollivier, E., Blondeau, E.: Chronology pf early Islamic glass compositions from Egypt. J. Archaeol. Sci. **104**, 10–18 (2019). https://doi.org/10.1016/j.jas.2019.02.001
93. Shortland, A.: Vitreous Materials at Amarna BAR Intl. Series 827. BAR, London (2000)
94. Shortland, A.J., Tite, M.S.: Raw materials of glass from Amarna and implications for the origins of Egyptian glass. Archaeometry. **42**(1), 141–151 (2000). https://doi.org/10.1111/j.1475-4754.2000.tb00872.x
95. Smith, G.D., Clark, R.J.H.: Raman microscopy in archaeological science. J. Archaeol. Sci. **31**(8), 1137–1160 (2004). https://doi.org/10.1016/j.jas.2004.02.008
96. Sparavigna, A.C.: Faience: the ceramic technology of ancient Egypt, *ARCHAEOGATE*, **1973–2953** (2012). https://core.ac.uk/download/pdf/11424403.pdf
97. Tite, M.S.: Ceramic production, provenance and use—a review. Archaeometry. **50**(2), 216–231 (2008). https://doi.org/10.1111/j.1475-4754.2008.00391.x

98. Tite, M.S.: The production technology of Italian maiolica: a reassessment. J. Archaeol. Sci. **36**(10), 2065–2080 (2009). https://doi.org/10.1016/j.jas.2009.07.006
99. Tite, M.S.: The technology of glazed Islamic ceramics using data collected by the late Alexander Kaczmarczyk. Archaeometry. **53**(2), 329–339 (2011). https://doi.org/10.1111/j.1475-4754.2010.00546.x
100. Tite, M.S., Bimson, M.: Faience: an investigation of the microstructures associated with the different methods of glazing. Archaeometry. **28**(1), 69–78 (1986). https://doi.org/10.1111/j.1475-4754.1986.tb00375.x
101. Tite, M.S., Bimson, M.: Identification of early vitreous materials. In: Black, J. (ed.) Recent Advances in the Conservation and Analysis of Artifacts, pp. 81–85. Institute of Archaeology, London (1987)
102. Tite, M.S., Bimson, M.: Glazed steatite: an investigation of the methods of glazing used in ancient Egypt. World Archaeol. **21**(1), 87–100 (1989). https://doi.org/10.1080/00438243.1989.9980092
103. Tite, M.S., Freestone, I.C., Bimson, M.: Egyptian faience: an investigation of the methods of production. Archaeometry. **25**(1), 17–27 (1983). https://doi.org/10.1111/j.1475-4754.1983.tb00658.x
104. Tite, M.S., Freestone, I., Mason, R., Molera, J., Vendrell-Saz, M., Wood, N.: Lead glazes in antiquity – methods of production and reasons for use. Archaeometry. **40**(2), 241–260 (1998). https://doi.org/10.1111/j.1475-4754.1998.tb00836.x
105. Tite, M.S., Manti, P., Shortland, A.J.: A technological study of ancient faience from Egypt. J. Archaeol. Sci. **34**(10), 1568–1583 (2007). https://doi.org/10.1016/j.jas.2006.11.010
106. Tite, M.S., Shortland, A.J.: Production Technology of Faience and Related Early Vitreous Materials Oxford University School of Archaeology> Monograph 72. Oxford University, Oxford (2008)
107. Toffolo, M.B., Klein, E., Elbaum, R., Aja, A.J., Master, D.M., Boaretto, E.: An early iron age assemblage of faience beads from Ashkelon, Israel: chemical composition and manufacturing process. J. Archaeol. Sci. **40**(10), 3626–3635 (2013). https://doi.org/10.1016/j.jas.2013.05.010
108. Ullrich, D.: Egyptian blue and green frit. Pact 17 II.3.1, 323–332 (1987)
109. Vandiver, P.: Technological change in Egyptian faience. In: Olin, J.S., Franklin, A.D. (eds.) Archaeological Ceramics, pp. 167–179. Smithsonian Institution Press, Washington, DC (1982)
110. Vandiver, P.: "The manufacture of faience", Appendix A. In: Kaczmarczyk, A., Hedges, R.E.M. (eds.) Ancient Egyptian faience, pp. A1–A144. Aris and Phillips, Warminster (1983)
111. Vandiver, P.B., Kingery, W.D.: Egyptian faience, the first high-tech ceramic. In: High-Technology Ceramics: Past, Present, and Future-the Nature of Innovation and Change in Ceramic Technology, vol. 3, pp. 19–34. The American Ceramic Society, Westerville (1986)
112. Walton, M.S., Tite, M.S.: Production technology of Roman lead-glazed pottery and its continuance into late antiquity. Archaeometry. **52**(5), 733–759 (2010). https://doi.org/10.1111/j.1475-4754.2009.00506.x
113. Weatherhead, F., Buckley, A.: Artists pigments from Amarna. In: Kemp, B. (ed.) Amarna Reports V, pp. 202–238. Egypt Exploration Society, London (1989)
114. Wedepohl, K.H., Simon, K., Kronz, A.: Data on 61 chemical elements for the characterisation of three major glass compositions in late antiquity and the middle ages. Archaeometry. **53**(1), 81–102 (2011). https://doi.org/10.1111/j.1475-4754.2010.00536.x
115. Wilkinson, G.: "Some Accounts of the Natron Lakes of Egypt"; in a letter to W.R. Hamilton, Esq. 1842. J. R. Geogr. Soc. Lond. **13**, 113–118 (1843). https://www.jstor.org/stable/1798137
116. Wulff, H.E., Wulff, H.S., Koch, L.: Egyptian faience, a possible survival in Iran. Archaeology. **21**(2), 98–107 (1968)

Chapter 7
Egyptian Glass Fourth to Sixth Centuries

Abstract Glass production in the Dakhleh Oasis, Egypt, during the fourth to sixth century CE is reviewed. The compositions of cobalt- and copper-based blue glass sherds, together with other glass sherds and two black glass bracelets have been analysed using Raman spectroscopy and electron microscopy. In addition, the composition of blue and green bracelets dated to post tenth century CE and discovered at Qasr al-Dakhleh are also reported. Glass working is thus determined and which also tentatively indicates that glass production may also have occurred at this Oasis. The mineralogy of the orange and yellow decoration applied to the blue and green bracelets has been determined.

Keywords Dakleh oasis · Egypt · Cobalt glass · Copper glass · Blue glass · Bracelets

7.1 Glass Types and Previous Studies

Glass beads have been recovered from archaeological contexts dated to the early third millennium BCE onwards in Mesopotamia and the 5th Dynasty (2484–2345 BCE) in Egypt [58, 61, 79]. The study of Roman and early Mediaeval glass has been reviewed by Smith and Clark [85] and a number of authors have reviewed glass manufacture in the Near East [4, 58, 65, 66]. Glass production dated between the first and the seventh century CE from Roman and Byzantine glass industries has been investigated and discussed in detail by Maltoni et al. [53]. And among others, both Colomban [17, 18, 20] and Edwards et al. [29] have conducted comprehensive and detailed studies into the identification of glassy silicates using Raman spectroscopy.

Glass making in ancient Egypt has been reviewed by Rehren [70], who observed that glass production first commenced in Mesopotamia around 1600 BCE and in Egypt in about 1550 BCE. The origin of glass production was suggested to be in

H. G. M. Edwards, P. Colomban, *Blue by Fire: A Marker of the Technical History of Glass and Ceramics*, Cultural Heritage Science,
https://doi.org/10.1007/978-3-031-77122-4_7

accordance with the use of plant ash as flux [7]. Glass and faience compositions are more likely to be due to the different methods of production rather than the use of different plant ashes. Brill proposed that glass was manufactured using only two raw materials, namely crushed quartz pebbles and plant ash, such as the unrefined ash of halophytic plants. Furthermore, he rejected the suggestion that calcareous sand or lime were used as raw materials. Rehren suggested that calcium entered the glass from the calcium lining of the vessels used in melting the charge. Thus, whilst only two raw materials were employed, he suggested that another 12 elements could enter the glass via the plant ash and at concentrations >0.05%. He indicated that prior to 1000 BCE, the glass produced typically contained 18 ± 2 wt % Na_2O, ~2 wt % K_2O, 5–10 wt % CaO and 3.5 wt % MgO. The presence of Al_2O_3 was not reported.

Freestone [32] generally agreed with this proposal by Rehren but noted that if the flux was plant ash, then a lime-free silica was used: however, if a soda-based flux was used, then a calcareous quartz sand was used. Wedepohl et al. [100] suggested that the term plant ash should not be used and should be replaced with the term soda-ash glass as wood ash can also be used (wood ash glass); and that trona [$Na_3H(CO_3)_2.2H_2O$] rather than natron is the major sodium mineral in low-Mg glass (soda-lime glass). Furthermore, Wedepohl and co-workers suggested that the starting materials for soda-lime glass are quartz, trona and lime. Caley [13] suggested that perhaps the ash of the fuel employed to melt the glass was often incorporated into successive batches as one of the glass raw materials.

Rehren [69] divided glass into two general compositional groups, the high-magnesia (HMG) glasses of the Late Bronze Age in the Early Islamic world and the low-magnesia (LMG) glasses of the Roman period. The HMG are typically high in potash whereas, the LMG are often low in potash. Shortland and Tite [79] reported that various authors have cited the soda, potash and magnesia contents as being indicators for the source of the alkali flux.

Shugar and Rehren [81] reported that for more than 2000 years, two compositional groups of soda-lime-silica glass dominated glass composition, namely the high magnesium and high potassium, plant ash based, Late Bronze Age (LBA) glass of Egypt and Mesopotamia and the low magnesium and low potassium, mineral natron based Hellenistic and Roman glass. Minor compositional changes were considered to reflect the slight differences in local raw materials. Most LBA Egyptian glass tends towards a higher calcium rich composition, whereas Roman glass tends towards a more silica rich composition. They proposed a smelting model for glass making that includes the formation of a cotectic glass melt in the presence of a significant quantity of crystalline material which acts as a buffer of reservoir material from which the melt draws upon as it forms. Such a process results in an upper layer of crystalline material floating on top of the glass. On cooling, simple crushing and hand selection would enable quality glass to be obtained. This proposed model would produce a uniform quality glass and thus the importance of minor component heterogeneity in ultimately provenancing glass becomes more obvious.

Sayre and Smith [74] on the basis of the analysis of some 400 glass samples from a wide range of sources from the second millennium BCE to the twelfth century CE determined that the correlation between magnesia and alkali fluxes indicated high magnesia and high potash (HMG), with no correlation between magnesia and lime. They stated that from 1500 to 800 BCE glass produced in the Eastern Mediterranean, Egypt and Mesopotamia was of the high magnesia (3–7%), high potash (1–4%) type whereas in the following period up to the latter part of the first millennium CE, the glasses were low magnesia (0.5–1.5%), low potash (0.1–1.0%) (LMG). Nicholson and Henderson [62] stated that in typical LMG glass the magnesia and potash concentrations are <1%, indicating the use of natron. This variation has led various authors to link this magnesia/soda change to a change in the source of the alkali, the earlier HMG glass being plant ash-based and the later LMG to be natron-based glass [7, 40, 51].

The composition of glass objects found in the European region and the Near East remained similar for the period of the Roman Empire and beyond into the eighth and ninth centuries, a period of about 800 years. Alumina content is generally ~2.5 wt. % and CaO between 6–8% and the glasses are characterised by low magnesia and low potash (LMG), each being below <1.5%. Soda in the form of natron was considered to have replaced plant ash as shown by the low magnesia and low potash. The constant alumina content and also calcium oxide could have been related to the silica source. Thus, Aerts et al. [2] conducted analyses of silica from the eastern Mediterranean coast and in southern Syria with silica from further afield in Israel, Germany and Belgium to determine if changes occurred in the silica sources used for the production of glass in the Roman and post-Roman periods. They observed that changes in the titanium and zircon content could have reflected political instability in the fourth to fifth century CE exemplified by the fragmentation of the Roman Empire.

As reported by Shortland et al. [80] an evident break in the supply of natron (trona) from Wadi Natrun occurred towards the end of the eighth century CE; this led to the substitution of wood ash for the alkali source in Europe. This change in alkali flux may have been caused by weather events, or by political events. The change in flux also changed the glass composition with potash, magnesia, phosphate and calcium increasing by a factor of three in soda ash glass than in soda lime glass [100].

Freestone [32] reported that raw glass was widely traded as lumps of colourless or naturally coloured glass in the Roman and early medieval periods and as ingots in the Late Bronze Age. Thus, primary glass production may have been conducted separately from secondary glass-working sites where workers used the glass to produce various shaped artefacts. Freestone suggested that in conformity with the work of Sayre and Smith [73] and of Nicholson and Henderson [62] ancient soda-lime-silica glasses can be placed into two general categories, high potash plus high magnesia in which each material is present at >1.5% and the low potash, low magnesia glasses in which both raw materials are present at <1.5%.

Wedepohl et al. [100] observed that glass can also be subdivided into classes based upon its iron content, where 1.6% Fe_2O_3 glasses were classified as high-iron and 0.7% Fe_2O_3 as low iron glasses. Furthermore, he observed that a relatively homogeneous profile of alkali metal and alkaline earth usage in these glasses over about a 500 year period in all of Egypt, is compositionally indicative for the use of a relatively constant plant ash.

Rehren [70] disagreed with the proposal by Shortland and Eremin [77] that all of the elements, other than silica, are derived from plant ash; he noted the poor correlation between the alkali metals and the alkaline earth elements offered by this concept with their concentrations in the finished glass. In a number of studies, the melting of glass in small batches was investigated [56, 68, 71, 75, 81, 91, 92]. At 800–900 °C these various workers postulated that an initial melt was necessary to produce a semi-finished glass in which the various raw materials would be incorporated, followed by a high temperature firing at 1000–1100 °C to produce the final glass. They proposed that in the initial lower temperature melt, the concentration of CaO is ~6%, increasing to ~10% at the maximum temperature. They attributed this increase in calcium concentration to the chemical attack of the glass on the high calcium internal layer lining of the volumetrically small reaction vessel, limiting the efficiency of the parting layer as a refractory material. Conservatively, they estimated that this contributed 25–30% of the CaO in the final glass. If this observation is correct, then raw glass produced in large batches should either be lower in calcium or, if at the same concentration as ingot-produced glass, then lime must have been added to the batch in large raw glass batch production. This is demonstrated by an early ninth century 8 tonne slab of glass found at Beth She'arim, Israel, and which was rendered unusable because of excess lime in the batch.

The presence of magnesia has been shown to form diopside-like crystals at the interface between the glass batch and the lime-dominated parting layer. Such crystal formation contained approximately equal proportions of MgO and CaO and limits the MgO concentration in the final glass to about 4–5% MgO after which diopside crystallisation occurs. Thus, the parting layer firstly provides added lime into the glass melt and secondly to act as a magnesia sink limiting the MgO concentration in the final glass batch.

The difference in the alkali metal concentrations can probably assist in provenancing early Mesopotamian and Egyptian glasses. It has been postulated by Brill [8] and by Shortland and Eremin [77] that Mesopotamian glass is richer in potash and lower in soda concentrations than Egyptian glass from the same period. However, it must be borne in mind that these alkali metals can leach from glass and therefore the SEM-EDS elemental results directly obtained from the shard surface may underestimate the original soda and potash concentrations.

Shortland and Tite [79] examined 19 blue glass sherds from Amarna in Egypt and reported that two different types of glass were identified. Namely, a cobalt glass with a natron-based alkali flux and a copper-based blue glass with a plant ash alkali flux made in accordance with the Mesopotamian tradition. Lilyquist and Brill [51] had previously indicated that analyses of the earliest glass recovered from the tomb of three foreign wives of Thutmose III (1479–1426 BCE) had indicated the presence of the same two glass types.

Shortland and Eremin [77] and later, Shortland et al. [78] considered that electron beam analysis (SEM-EDS) of glass has proven equivocal because the glasses all exploit the same range of raw materials and therefore natural variation in raw materials can alter the analyses. For this reason, they therefore used LA-ICP-MS for their analyses. They concluded that Amarna and Malkata had glass of identical composition and therefore Egypt was producing colourless, copper and cobalt blue glass of a characteristic type from one or more glassmaking workshops. This analytical technique however is moderately destructive insofar that it does form a visible crater at the point where the glass is volatilised. Walton et al. [99] suggested that a better analytical approach would be to use LA-ICP-TOFMS as this will only produce a discrete 60 μm wide × 50–100 μm deep crater. LIBS may also represent an alternative ablation technology. Each of these methods would likely underestimate the sodium and potassium present at the surface of the glass due to the leaching of these ions during burial of the artefact.

Farnsworth and Ritchie [30] were the first to detect cobalt in Egyptian-manufactured blue glass. It has been proposed that the darker shade of blue obtained from the use of cobalt, was used in imitation of lapis lazuli and that the lighter copper-based blue glasses imitated turquoise. Importantly, the Dakhleh (Dakhla) Oasis is one of only two known sites at which cobalt mineralisation in alum deposits has been discovered in Egypt. Its use has been proposed as the raw material source from which cobalt blue pigments and their subsequent use as a blue glass pigment were produced. Mössbauer spectroscopy has been applied to kiln slag recovered at Dakhleh Oasis. Thus, the local knowledge of high temperature processes is established from this discovery of kiln slags and could indicate the existence of as yet undiscovered glass kilns.

An additional problem investigated by Tanimoto and Rehren [92] was the effect of sodium chloride on the melt. They reported that the partitioning of the soda and of the potash obtained from the plant ash between the glass and a separate salt melt or scum (known as "gall(e)") with the potash increasingly partitioning in the gall as the sodium chloride percentage increased. Thus, higher potash concentrations are favoured by low soda concentrations. Evidence in support of the salt-melt formation was supplied by Stapleton and Swanson [89] who studied the effect of sulphates in glass production and observed droplets of an alkali sulphate surrounded by silicates. These sulphate inclusions were considered to have formed during the melting or incorporation of poorly prepared plant ash. Rehren [70] suggested that this observation requires further investigation.

The adsorption of phosphates on to the surface of porcelain under burial conditions has been discussed [33]. These researchers reported that as opposed to fired clay, fired porcelain shows a much greater capacity to adsorb phosphate from typical soil clays. The levels of phosphate reported in this study are consistent with this observation. Whilst the environmental conditions in Egypt are semi-arid, rain events do occur, therefore these buried glass shards have potentially had a long exposure to moisture, so providing the conditions to support surface leaching and/or surface deposition.

Shortland and Eremin [77] reported that the drier conditions prevailing in Egypt have allowed the glass fragments recovered there to be exceptionally well preserved, without any indication of the leaching of mobile elements as has occurred in Mesopotamia and in Greece. This model is considered simplistic as over the millennia the rainfall pattern has changed and burial itself will assist in maintaining moisture around the artefact as demonstrated by the significant deposition of calcite onto the surface of the cobalt blue Sample 4 from Qasr Abu Matta. The subject of leaching of glass and of glazes has been discussed [1, 29, 34, 44–46, 93] and as will be discussed, the loss of Na and particularly K is complex [94].

In a study by Melcher and Schreiner [55] they reported that glass when exposed to environmental conditions undergoes surface leaching. Specifically, the network modifier ions, K, and Ca leached, with K^+ leaching slightly, but to significantly higher leaching depths than the divalent Ca^{2+} ion. K^+ and H_3O^+ ions have a similar size that promotes exchange and leaching by acidic water [94]. Smets and Lommen [82] reported that the leaching of sodium silicate glasses is governed by the diffusion of water into the glass and that this water diffusion is more complex than an ion exchange process. They suggested that diffusing molecules are immobilised at the sodium (and potassium) NBO's (Non Bridging Oxygens as shown in Fig. 7.1) by the formation of silanol groups.

The sodium ions (and the potassium ions) freed by this reaction are no longer retained by the glass network. However, these researchers also reported that in aluminosilicate glasses some of the sodium ions are bonded to AlO_4^- tetrahedra groups and cannot be mobilised by reaction with water, only being mobilised by a much slower ion exchange process with hydronium ions (H_3O^+). The incorporation of relatively small amounts of Al_2O_3 into the glass may result in water conducting paths becoming closed resulting in a drastic decrease in the water diffusion coefficient [82]. As discussed by Abd-Allah [1], the depth of penetration into the glass is a function of glass polymerisation and is particularly effective in glasses that are rich in potassium. As a function of glass composition the in-depth corrosion rate typically ranges between 1 and 10 μm/century that can be used for a dating comparison [50, 94]. It should be noted that as explained by Jay et al. [46] the ionic radius of the sodium ion is smaller than that of the potassium ion and therefore the sodium ion is more tightly bound to the silica network (through the NBO's) and therefore less able to leach. Surprisingly, Jackson et al. [44] reported that in glasses in which

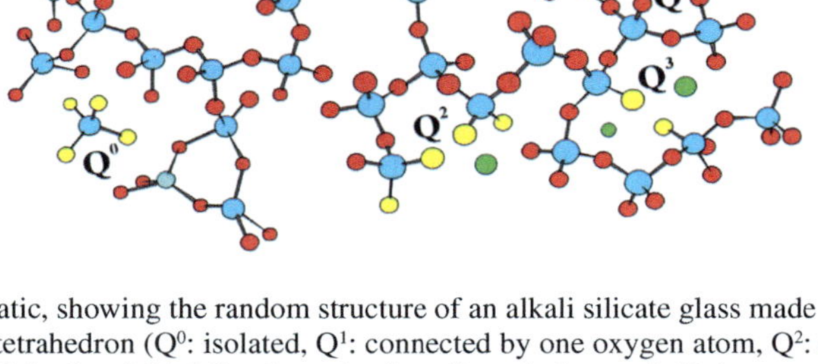

Fig. 7.1 Schematic, showing the random structure of an alkali silicate glass made of more or less connected SiO_4 tetrahedron (Q^0: isolated, Q^1: connected by one oxygen atom, Q^2: connected by 2 oxygen atoms, Q^3: by 3 and Q^4: full polymerized); yellow circle: non-bridging oxygen atom; green circle: alkali or earth alkali or lead atoms

the Na concentration was about 12% and K at about 3% that a reduction in corrosion is apparent.

The above reactions are further complicated by the soil pH profile and the changes brought about at the glass surface with these leached metal ions forming salts at the surface, creating a crust and modifying the degree of deterioration [35, 44, 88, 96]. Garcia-Valles and co-workers stated that the two main environmental controls on stained glass decay are water availability and warm to hot environmental conditions. Phosphorus, a common component of medieval glasses, can also act as a network former and according to Garcia-Vallès et al. [35] can be attributed to the ashes used as a source of the alkali raw material. As discussed by Verney-Carron et al. [96] and Tournié et al. [94], whilst the inter-diffusion during which the glass modifier cations such as alkali ions are replaced by protons, dissolution by the hydrolysis of ionic-covalent bonds (Si-O-Si, Si-O-Al) occurs and this process leads to the formation of altered layers within the glass, which can then become amorphous after hydrolysis and condensation processes.

Whilst opaque glasses result from the presence of excess phases, it is normally considered to be undesirable in the production of uncoloured glass. Crystallisation occurs when diffusion rates are sufficient to enable nucleation and crystal growth based upon temperature-dependent limits of solubility. The consequence of this temperature-dependent solubility is the ability of molten glass to absorb most oxidic materials it contacts providing their wetting properties allow sufficient interaction between glass melt and oxide material and that the solubility of the oxide material is not exceeded.

The mineralogy and the application of Raman spectroscopy and electron microscopy to fourth century CE glass shards and post tenth century opaque Islamic glass bracelet sherds which were recovered within the Oasis were investigated in an attempt to determine if these glass sherds provide support for glass production or for glass working.

Visually, some of the glass sherds examined appear to have possibly undergone degradation during burial. Analyses of the sand/soil chemistry at the excavation site are not available. Furthermore, the published literature contains references to variable and conflicting leaching of elements from buried glasses [6, 22, 25, 85, 94]. Calcium phosphate species have been previously identified on the surface of potash glasses, but no similar deposits were detected on the surface of soda-lime glasses of comparable age [26] and high lead-containing glazes were not discussed.

7.2 Representative Analytical Fingerprints

Fourth to sixth century CE glass shards located in excavations conducted at Kellis by the Dakhleh Oasis Project, Egypt, are listed in Table 7.1 and are illustrated in Figs. 7.2 and 7.3. Their elemental analyses are provided in Table 7.2. Later dated Islamic glass bracelets discovered at Qasr al-Dakhleh are provided in Table 7.2 and in Fig. 7.4.

Table 7.1 Glass shard recovered within Dakhleh Oasis Project (Monash University concessions)

Sample	Figure	Location	Identifier
Glass			
1 blue glass	7.3b	House 4 (A/6) Room 4, Kellis	A/6/94.514
3 blue glass	7.3a	House 4 (A/6) Room 4, Kellis	A/6/94.494
4 blue glass	7.3c	House 4, (A/6) Room 3A, Kellis	A/6/94.693
4 blue glass	7.3d	Trench 14 JM—8/1/11, Dayr Abu Matta	S11.0008
5 clear glass	7.3f	House 4 (A/6) Room 4, Kellis	A/6/94.514
6 clear glass	7.3g	House 4 (A/5) context (4), Kellis	A/5/1.00044
7 turquoise run on 8	7.3h	Main Temple (D/8) n w corner, Kellis	D/8/93.47
8 clear glass +7	7.3h	Main Temple (D/8) n w corner, Kellis	D/8/93.47
9 manganese glass	7.3i	House 4 (A.6) Room 1B, Kellis	A/6/94.141
10 manganese glass	7.3j	House 4 (A/6) Room 6, Kellis	A/6/94.440
11 manganese	7.3k	House 4 (A.5) Room 4, Kellis	A/5/91.311
13 manganese	7.3l	House 4 (A/6) Room 3, Kellis	A/6/94.348
14 amber	7.3n	Main Temple (D/1) Zone XV, Kellis	D/1/94.306
14 dark amber	7.3o	Main Temple (D/1) Zone XV, Kellis	D/1/93.18
15 green	7.3m	House 4 (A/6) Room 1B, Kellis	A/6/94.864
15 red	7.3m	House 4 (A/6) Room 1B, Kellis	A/6/94.864
15 brown	7.3m	House 4 (A/6) Room 1B, Kellis	A/6/94.864
15 white	7.3m	House 4 (A/6) Room 1B, Kellis	A/6/94.864
Bracelets			
16 black	7.4a	Roman, surface find, Kellis	LAL
17 black	7.4b	Roman, surface find, Kellis	LAS
18 blue	7.4c	Islamic, Qasr al-Dakhleh	33 Blue
19 blue	7.4d	Islamic, Qasr al-Dakhleh	43 Blue
30 blue	7.4e	Islamic, Qasr al-Dakhleh	49 Blue
31a green	7.4f	Islamic, Qasr al-Dakhleh	38 Green
31b orange	7.4f	Islamic, Qasr al-Dakhleh	38 Green
31c yellow	7.4f	Islamic, Qasr al-Dakhleh	38 Green
33a green	7.4g	Islamic, Qasr al-Dakhleh	44 Green
33b orange	7.4g	Islamic, Qasr al-Dakhleh	44 Green
33c yellow	7.4g	Islamic, Qasr al-Dakhleh	44 Green

The combination of SEM-EDS elemental analyses with Raman spectroscopy has been shown to be an important method for analysing glass. Research has identified the presence at Kellis and elsewhere, of glasses varying in their mineral composition. The data has strongly indicated that Raman spectroscopy is able to non-destructively analyse polymorphous glasses, faience and glazed pottery and provide information in respect of its melting profile and its relationship to other glass types. The soda content of Dakhleh glass shards [28] is significantly lower than for other LBA glass produced elsewhere. Presently, insufficient samples have been analysed to allow any definitive statement to be made in respect of glass production within

Fig. 7.2 Gladiator jug, Ismant el-Kharab, ancient Kellis excavation in Enclosure 4, Area D. (Reproduced with permission from Ref. [43])

the Dakhleh Oasis. However, tentatively, there does seem to be the possibility that the glass used at ancient Kellis is different to that published in respect of glass production from other sites within Egypt and therefore, may have been locally manufactured.

The application of Raman spectroscopy to glasses has been investigated in detail [14, 17–21, 29, 52]. As shown in the Raman spectra of each glass shard, two significant broad bands are evident, one centred around 500 cm^{-1} associated with the bending vibrations of Si-O-Si bonds (δ Si-O) and the other centred around 1000 cm^{-1} associated with the stretching vibrations of more or less depolymerised silicate species (ν Si-O). Thus, as discussed by Colomban [17, 18] the relative Raman intensity of the Si-O bending and stretching envelopes varies with composition. A polymerisation index (Ip) can be developed in which the area under the 500 cm^{-1} bending massif (A_{500}) and the area under the 1000 cm^{-1} stretching massif (A_{1000}) provides information in respect of both the glass composition and the processing temperature as discussed in detail by Colomban [17, 18], Colomban et al. [23], and Edwards et al. [29]. Thus, the connectivity of the SiO_4 polymeric units can be investigated by the relative intensities of the bending and stretching modes. In highly connected structures, the larger the amplitude of motion of oxygen perpendicular the to

Fig. 7.3 (**a**) and (**d**) are cobalt-based glasses, (**b**) and (**c**) are copper-based glass, (**e**) is a vessel indicating the presence of the blue boss as provided in (**d**), (**f**) and (**g**) are clear glass shards, (**h**) is a run of turquoise glass on clear glass, (**i**) to (**l**) are manganese coloured glass shards, (**m**) is a multi-coloured glass shard, (**n**) and (**o**) are amber glass shards

the Si-O-Si link, the higher is the polarisation change in this vibration and therefore the Raman intensity around 500 cm^{-1}. In less connected structures the larger the amplitude of oxygen atoms in the stretching mode, the larger is the Raman intensity around 1000 cm^{-1}.

As shown by their respective SiO_4 bending and stretching vibrations of glassy or crystalline silicates and of aluminates, the Raman cross-section of the Si-O bending and stretching modes is orders of magnitude stronger that of the Al-O bond [17]. This arises from the more covalent character of the Si-O bond. Thus, the mode at 500 cm^{-1} of amorphous structures composed of connected SiO_4 tetrahedra or in aluminosilicate networks promoted by the use of feldspars as fluxing units is strong. In contrast, the highest intensity of the stretching mode envelope (around 1000 cm^{-1}) is observed in glassy networks containing significant concentrations of elements such as alkali, alkaline earths, lead or zinc ions which break many of the Si-O bonds. The different components of the SiO_4 stretching band (Q^0 to Q^4) [17, 52] are a fingerprint of the glassy network. And, as reported by Colomban et al. [22, 25], the peak at about 950 cm^{-1} can be related to the calcium content of the glass.

Table 7.2 Elemental analysis of cobalt- and copper-based blue glass shards from Ancient Kellis and from Dayr Abu Matta, Dakhleh Oasis, Egypt, together with additional coloured glass shards (Fig. 7.4)

		Na3O	MgO	Al3O4	SiO3	P3O5	SO4	Cl	K3O	CaO	TiO3	MnO	FeO	CoO	NiO	CuO	ZnO	PbO	SnO3	As3O4
Sample 1	Fig. 7.4a	3.9	1.4	4.8	79.3	0.4	0.4	0.5	1.7	6.1	0.3	0.0	3.1	0.4	0.0	0.4	0.0	0.0		
Sample 3	Fig. 7.4b	3.6	1.3	4.6	76.7	0.0	0.8	1.4	1.9	5.8	0.0	0.1	0.9	0.3	0.3	4.6	0.4	0.0		
Sample 4	Fig. 7.4c	5.8	1.5	4.3	74.4	0.5	0.9	3.7	1.5	7.1	0.3	0.8	1.3	0.0	0.0	0.4	0.0	0.0		
Sample 4	Fig. 7.4d	4.4	0.9	4.4	78.7	tr	1.4	1.8	1.5	8.0	0.1	tr	0.8	tr	tr	0.3	0.0	0.0	0.0	tr
Additional glass samples																				
Sample 5	Clear A/6/94/514	5.8	1.0	4.8	79.8	0.0	1.0	1.7	0.9	5.6	0.0	0.0	0.6	0.0	0.0	0.0	0.0	0.0		
Sample 6	Clear A/5/1/00044	11.4	0.4	1.9	71.3	0.0	0.7	0.0	0.6	14.6	0.0	tr	0.5	0.0	0.0	0.0	0.0	0.0		
Sample 7	Turquoise 93.47	6.0	1.0	4.8	74.4	0.4	0.9	0.8	1.0	5.7	0.0	0.0	3.4	0.0	0.0	4.5	0.4	0.0		
Sample 8	Clear 93.47	3.8	3.5	9.4	70.7	0.7	1.1	1.0	3.0	6.5	0.4	tr	3.6	0.0	0.0	0.0	0.0	0.0		
Sample 9	Manganese 94.141	7.4	1.3	4.9	76.5	0.0	0.4	1.5	0.8	6.4	0.0	1.6	0.8	0.0	0.0	0.0	0.0	0.0		
Sample 10	Manganese 94.440	4.4	1.1	4.3	80.4	0.1	0.9	1.5	1.0	6.4	0.3	0.3	0.8	0.0	0.0	0.0	0.0	0.0		
Sample 11	Manganese A/5/91.31	4.0	0.9	5.1	84.4	0.0	0.6	1.7	0.9	1.8	0.4	0.5	0.9	0.0	0.0	0.0	0.0	0.0		
Sample 13	Manganese A/6/94.348	3.9	0.8	4.4	78.3	0.0	0.7	1.4	1.5	7.6	0.4	1.4	1.1	0.0	0.0	0.0	0.0	0.0		
Sample 14	Amber D/1/94.306	10.4	1.7	4.8	68.3	1.3	0.3	0.6	3.6	7.3	0.4	0.1	3.9	0.0	0.0	0.0	0.0	0.0		
Sample 14	Dark Amber	5.0	1.5	4.9	74.7	0.0	0.7	1.3	1.6	9.4	0.4	1.7	1.4	0.0	0.0	0.0	0.0	0.0		
Sample 15	green	9.4	1.1	4.0	69.6	0.0	0.8	0.9	1.4	5.4	0.0	0.7	3.4	0.0	0.0	0.9	0.0	4.7		
	red	9.6	4.5	4.1	64.6	1.4	0.9	1.9	3.3	7.5	0.3	0.4	1,3	0.0	0.0	1.7	0.0	1.3		
	brown	8.1	1.4	4.6	71.4	0.0	1.1	1.3	1.4	6.0	0.0	1.3	3.4	0.0	0.0	0.0	0.0	1.4		
	white*	6.9	1.9	4.5	73.8	0.0	0.9	1.5	1.4	6.7	0.0	1.4	1.8	0.0	0.0	0.0	0.0	0.0		

[a]Cassiterite particle, SnO_3 = 19.1%, PbO = 43.4%

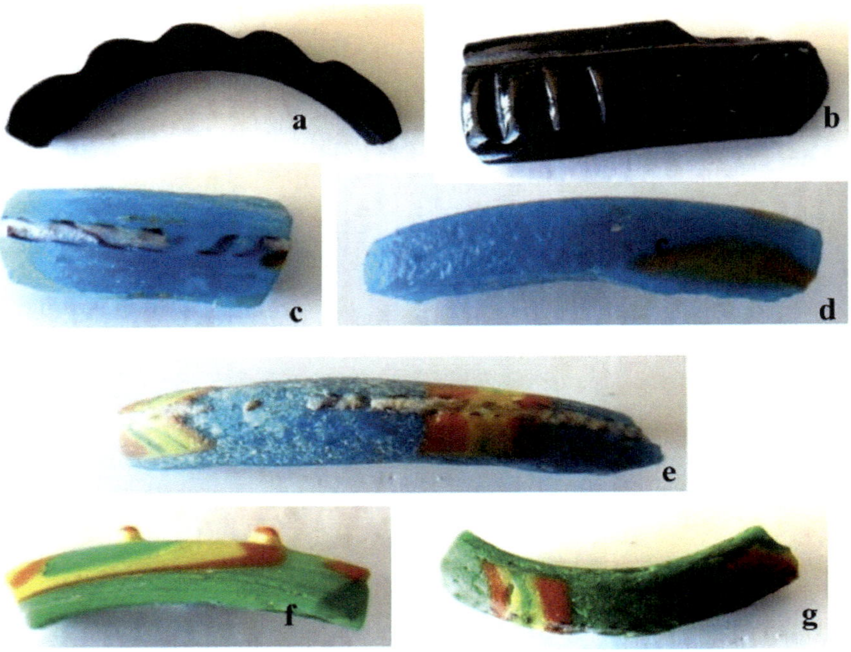

Fig. 7.4 (**a**) and (**b**) are fourth century CE bracelets. (**c–g**) are post Roman glass bracelets

The position of this peak is close to that of wollastonite, a calcium mesosilicate that precipitates very easily in calcium-saturated glasses and glazes. The effect of breaking of the Si-O bonds is clearly demonstrated by Colomban [17, 18] for a range of "Family" types. He reported that from the polymerisation index the following conclusions could be drawn:

Ip = ~7, processing temperature is ~1400 °C
Ip = ~1.4, processing temperature is ~1000 °C
Ip = ~0.4, processing temperature is ~600 °C

Thus, high Ip values correspond to silica-rich and alkali-poor compositions, whilst low Ip values correspond to modifier-rich and silica-poor compositions. Colomban stated that more research is required to fully understand the relationship of the Ip to glass structure, noting that there is evidence from Raman spectroscopy, radial distribution function and NMR data that divalent cations perturb the network more than monovalent ones and that ring size distribution depends upon the [Al]/[Al + Si] ratio. Colomban [17] also reported that Ip values <0.4 mostly correspond to Islamic lead-containing or lustre potteries and some Punic/Roman glasses; those between 0.4 < Ip <0.8 are nineteenth century lead-based porcelain enamels and some blue, green and colourless Punic/Roman glasses; those between $0.8 < Ip < 1.1$ correspond to most ancient glasses and eighteenth century porcelain enamels.

It should be noted that depending upon the environmental conditions to which the glass shards have been exposed, post-production disposal can result in significant surface modification. For example, the weathering crusts on potash glasses are said to contain calcium phosphate species, whereas soda-lime glasses of comparable age did not contain such inclusions [26]. According to Bertoncello et al. [6] heavily weathered surfaces were observed on soda-lime Mediaeval glass, but not on potash glass. However, the above comments in respect of potassium leaching would question this finding. Smith and Clark [85] reported that natural ground-waters penetrate the outer surface of northern European glasses and leach alkaline and alkaline earth oxides, whereas, Middle Eastern glasses have been reported to be much more robust.

7.3 Characterization

7.3.1 Fourth to Sixth Century Glass

Other than in one specific instance, the clear glass samples were recovered from excavations conducted within the Dakhleh Oasis at Kellis. The production of glass and/or glass working or faience production at Kellis has been discussed in detail by Eccleston [28] and he concluded that in the examination of a selection of possible production waste he identified shards which may be considered to be from glass blowing [28] i.e., glass working.

The elemental analytical results for glass samples are provided in Table 7.2. It should be noted that the analyses were directly obtained from the surface of the glass without any attempt being made to remove any surface leaching or surface deposition. As discussed by Freestone [32], it may be necessary to remove all weathered material. The decision was made not to damage the surface of the shards, recognising that if required, future quasi non-destructive microscopic sampling method as proposed by Bronk and Freestone [10] should be considered. Alternatively, LIBS or other laser ablation methods might be investigated.

Calcium and phosphorus can be deposited on to the surface of buried glass as identified in Fig. 7.5. Smith and Clark [85] concluded that glass weathering is extremely complex and may well be highly specific to the burial environment. As such, analysis of the soil chemistry should form part of the excavation site recovery procedure. Thus, depending upon the treatment of the shard after recovery, such as washing, can result in additional surface modification. As such, the direct elemental analysis of the glass shard surface using SEM or XRF techniques can result in errors and so the results provided herein are considered to be semi-quantitative.

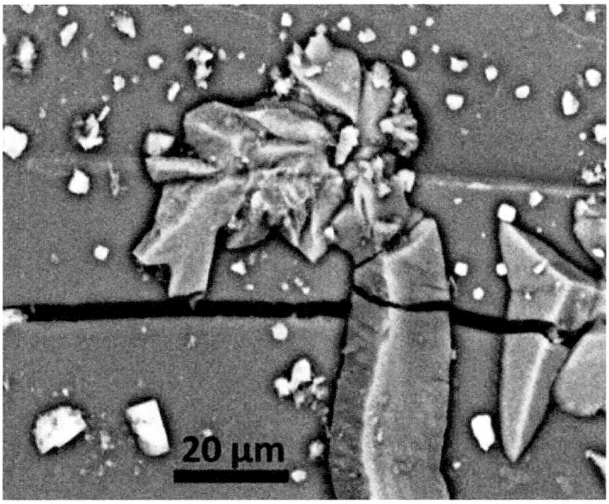

Fig. 7.5 Surface growth on surface of shard from Dayr Abu Matta in Fig. 7.4d. This surface deposit is high in MgO (4.4%) and CaO (34.4%)

Crystal growth, possibly vaterite and certainly calcite, were evident on the surface of some of the glass shards, particularly the blue glass shard from Dayr Abu Matta, indicating that groundwater had most likely penetrated the external surfaces of the shards to a greater or lesser degree. Also, the cleaning of the individual glass shards after recovery may have removed some of the surface growths. Freestone [31] conjectured that in the weathering of pottery shards, particularly of the porous biscuit, that calcium and potassium would have leached preferentially relative to aluminium. Thus, the change in chemistry and its effect on the shards under examination cannot be quantified.

7.3.2 Post Tenth Century Islamic Bracelets

Opaque blue sample images of three bracelets recovered at Qasr al-Dakhleh are provided in Fig. 7.4c–e and their SEM-EDS elemental analyses are provided in Table 7.3. Two green pigmented opaque glass bracelets were also found at this site and their images are provided as Fig. 7.4f, g and their EDS analyses are provided in Table 7.3. All samples were surface finds.

The three blue samples are low soda glasses (3.5–3.8% Na_2O) with high iron, high titanium but low manganese and therefore considered to be probably post Roman production. Calcium oxide content varies between 4.8 and 5.1% and magnesium oxide is in the order of 3%. The incorporation of very high alumina content (7.5–13.9%) would account for the opacity of the resultant glass. Feldspars would most likely be the source of the alumina incorporated into the glass.

Table 7.3 Bracelets, Fig. 7.4a, b are Kellis surface finds, fourth century CE, Fig. 7.4c–g are post Roman samples from Qasr al-Dakhleh

		Na3O	**MgO**	**Al3O4**	**SiO3**	**P3O5**	**SO4**	**Cl**	**K3O**	**CaO**	**TiO3**	**MnO**	**FeO**	**CoO**	**NiO**	**CuO**	**ZnO**	**PbO**	**SnO3**	**As3O4**
Sample 16	Fig. 7.4a	3.4	3.4	14.9	63.7	0.4	0.8	0.6	4.1	4.3	0.5	0.3	9.1	0.4	0.0	0.0	0.0			
Sample 17	Fig. 7.4b	11.8	1.0	4.1	75.6	tr	0.8	0.7	0.8	7.3	0.4	1.8	14.3	0.4	0.0	0.0	0.0	1.4		
Sample 18	Fig. 7.4c	3.5	1.9	7.5	74.5	0.4	1.0	1.1	1.3	5.1	0.5	0.3	3.8	0.4	0.0	1.4	0.0	0.3		
Sample 19	Fig. 7.4d	3.8	3.1	9.3	73.4	0.3	1.0	1.4	3.4	4.8	0.5	0.0	4.0	tr	0.0	1.8	0.0	0.0		tr
Sample 30	Fig. 7.4e	3.6	3.4	13.9	66.6	0.3	1.3	1.4	1.9	4.0	0.5	0.0	4.5	0.0	0.0	0.9	0.1	1.9	tr	
Sample 31A	Fig. 7.4f	4.8	3.1	13.3	65.9	0.5	1.7	0.7	1.5	4.4	0.4	tr	4.4	0.0	0.0	0.6	0.0	0.7	tr	
Sample 31B	Fig. 7.4f	3.8	1.1	4.4	68.4	0.5	3.5	1.6	0.8	4.4	0.4	0.0	1.3	tr	0.0	0.3	0.9	10.43	1.7	
Sample 31C	Fig. 7.4f	5.3	1.9	10.6	64.7	0.6	3.4	1.1	1.5	4.4	0.4	0.0	3.7	tr	0.0	0.0	0.4	5.9	0.0	
Sample 33	Fig. 7.4g	3.4	3.0	10.5	68.3	tr	3.4	1.0	1.1	4.8	0.4	0.4	3.6	0.0	0.0	0.5	0.4	4.3	0.6	

7.4 Colouring Agents

7.4.1 Roman Glass

Glasses and glazes consist of a controlled modification of a three dimensional network as shown in Fig. 7.1 and which involve the replacement of Si^{4+} covalent bonded atoms by other ions, thereby decreasing the number of Si-O bridges and the connectivity of the network. Soda glasses are water soluble and require the presence of calcium or another alkaline earth metal or of aluminium to render them insoluble. In this structure, the silica atoms act as network formers and the alkalis act as network modifiers. Lead, depending on its concentration in the glass structure can act as either network formers or network modifiers. The network modifiers, particularly Na^{+} and K^{+}, cause depolymerisation of the silica network, rupture bonds and the metal cations can then react with the silica to create non-bridging oxygen ions. Such reactions lower the melting point of the glass and modify the viscosity of the molten polymer. The addition of alkaline metals, particularly Ca^{2+} and Mg^{2+} decrease the aqueous solubility by stabilising the glass structure and Al^{3+} and Fe^{2+}, Fe^{3+} can substitute for Si as network formers.

It should be noted that the colouring of glasses will introduce additional features. As stated by Colomban [17, 18], the final hue for a given colouring ion also depends upon the glaze composition, the firing temperature, the kiln atmosphere, and the grinding of the frit. Thus, the technology adopted by the glass worker will determine the visual appearance of the glass object. The various methods adopted will influence the Raman spectrum as follows:

- dispersion of transition metal ions such as Co^{3+}, Co^{2+}, Cu^{2+}, Mn^{3+}, Mn^{4+}, Fe^{2+}, Fe^{3+}, Ni^{2+}, etc. in the glass will generally not provide a sensitive Raman signature;
- precipitation of crystals such as cassiterite will opacify the glass matrix and will modify the Raman spectrum;
- dispersion of insoluble pigments into the glaze will provide characteristic Raman signatures;
- dispersion of metal nanoparticles such as gold (purple of Cassius, Perrot ruby), silver (yellow, lustreware) or copper (red, lustreware) provides very low wavenumber features and broad luminescence (background). Raman signatures at such low wavenumbers are difficult to obtain without sample preparation (polishing) and use of a high resolution instrument.

Diffusion coefficients of various polyvalent elements in glass melts over a temperature range from 850 to 1550 °C have been studied [16, 97]. They have demonstrated that at the same temperature, diffusion coefficients varied by a factor of as much as 3800. Polyvalent elements showing two redox steps (As, V, Cr) always possess higher diffusion coefficients in the lower valence state. Such observations have been confirmed in Na_2O-$3SiO_2$ or Na_2O-CaO-SiO_2 melts. As reported by Gerlach et al. [36, 37], an increase in the alkali cation radius also leads to an increase in viscosity, and viscosity also increases with increasing alumina content and a decrease in the self-diffusion coefficient.

Sample 8 in Fig. 7.3h, a turquoise run of glass on the surface of a clear glass, is indicative of various batches of glass being used together in the one workshop and at the same time. The soda content of the turquoise glass is significantly higher than the clear glass, lowering its melting point and assisting in its fusion to the clear glass. This analysis is confirmed from the higher Ip of the clear glass. A careful study of Fig. 7.3h will indicate the coloured glass run to have softened the clear glass but it still remains sitting on the surface of the clear glass without significantly wetting the clear glass surface (a hydrophobic appearance); this is indicative of minimal intermixing of the two glasses.

One cobalt blue coloured shard from Kellis, (Sample 4, House 4 (A/6), Room 4, Context 5) has indicated the inclusion of the locally prepared cobalt blue glass pigment into a glass. As such, this fourth century CE shard has been produced significantly later than the change in cobalt source claimed by Kaczmarczyk [47]. Although from the elevated copper content, when related to the low cobalt concentration, would create some uncertainty in respect of a manufacturing date, and particularly in respect of possible trading in cobalt-containing and copper- containing glasses. The elemental analysis of the cobalt mineralisation in another shard, being a manganese-free, high-iron glass, has clearly evidenced the importation of a cobalt mineral probably from Iran. This finding would be in accordance with the published work of Kaczmarczyk [47]. An opaque shard from Qasr al-Dakhleh has indicated that a cobalt mineral from Iran was also employed as a pigment. The presence of the locally produced cobalt pigment in Kellis Sample 1 (House 4 (A/6), Room 4, Context 5) could suggest an earlier production date than Sample 3 (House 4 (A/6) Room 4, Context 4) or possibly the use of recycled glass from earlier production, or the unlikely use of residual 18-19th Dynasty cobalt pigment. As such, this would suggest an alternative site within the oasis for glass working or the later known shortage of cobalt from the Oasis deposit, which had led to the pigment importation from Persia. Insufficient data is presently available to resolve this issue.

Raman spectroscopy has enabled the melting point of the various glasses to be determined from their respective Polymerisation Indexes. This is a new technique in respect of Egyptian glasses and the development of a significant database may well enable various glasses to be separated by date and by place of manufacture.

The pigmented shards examined may be grouped as follows:

- clear (colourless);
- blue;
- green;
- red;
- amber;
- black.

7.4.1.1 Colourless Glass

Clear glass Samples 5 and 6 illustrated in Figs. 7.3f, g, with their analyses provided in Table 7.2, are both low in MgO and low in K, indicative of Roman glass production. The Raman spectrum for Sample 5 is provided in Fig. 7.6b. Sample 6 (Fig. 7.6a) has slightly lower alumina and elevated calcium (CaO = ~14.5%) as against the expected concentration of 6–8%. This has resulted in a significant shift in their respective Polymerisation Indexes. Their bending and stretching multiplets differ significantly and this is identified with the higher soda and calcium concentrations in Sample 6 lowering the Ip and this is reflected in a lower melting point of the resultant glass. Its Raman spectrum is provided in Fig. 7.6a. The clear glass forming

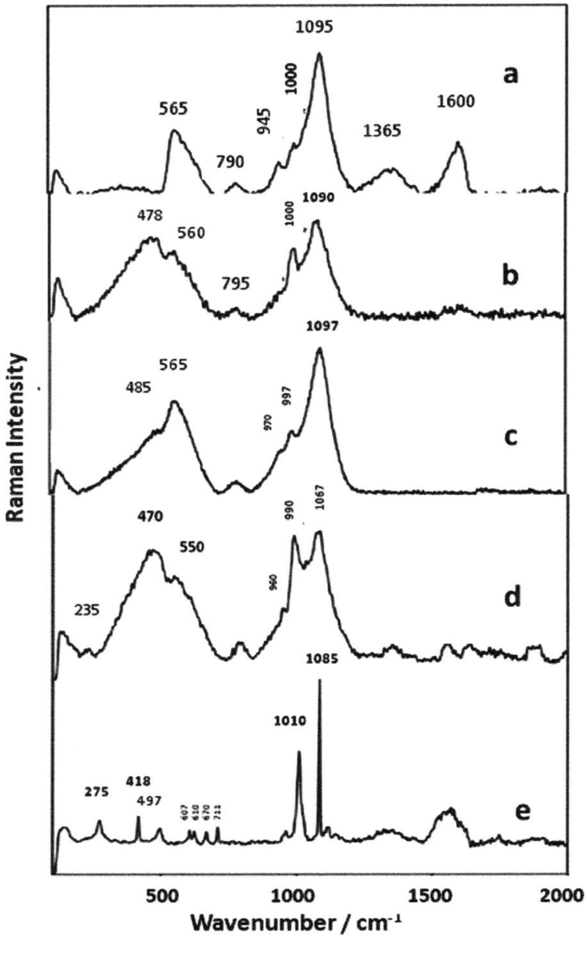

Fig. 7.6 Representative Raman (baseline subtracted) spectra. Location: Kellis (**a**) Clear glass sample 6, (**b**) clear glass sample 5, (**c**) blue glass sample 3. Location: Dayr Abu Matta (**d**) blue glass sample 4, (**e**) is calcite deposit on surface of sample 4. (λ = 514.5 nm)

part of Sample 8 (Fig. 7.3h) and its analysis is intermediate in calcium concentration between Samples 5 and 6 and this is reflected in an Ip that lies between the other two clear glass samples.

7.4.1.2 Blue Glass

The Dakhleh Oasis is one of only two known sites at which cobalt has been discovered in Egypt and it has been proposed as the raw material source from which cobalt blue glasses and/or some faience glazes were produced. As previously mentioned, a slag sample from a kiln there has been analysed and this could well relate to faience production.

All analyses are provided in Tables 7.2 and 7.3 and the Raman spectra are provided in Fig. 7.7. The opaque glass bracelet samples are either cobalt or copper-pigmented glasses and are surface finds at Qasr al-Dakhleh and probably date from the post tenth century CE.

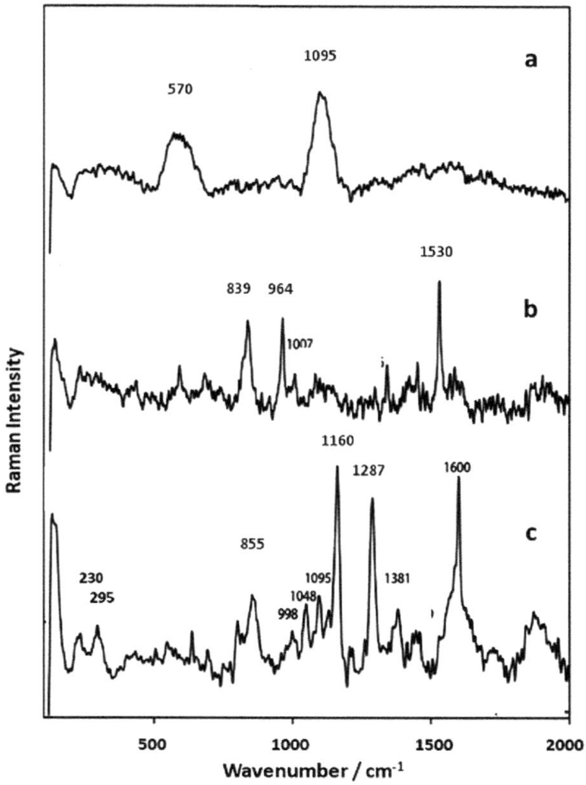

Fig. 7.7 Raman (baseline subtracted) spectra, blue striped glass Sample 4 ($\lambda = 514.5$ nm)

Samples 3 and 4 provided in Fig. 7.3a, d are cobalt blue coloured glasses in which the cobalt in Sample 3 has most likely been obtained from the Dakhleh Oasis. They contain manganese, nickel and zinc mineralisation in accordance with the alums located within this oasis and the glass can be classified as being low in iron. Whereas the blue glass, Sample 1, is different; it does not indicate the presence of manganese, nickel or zinc and is high in iron, which indicates the use of an alternative source of the cobalt. Cobalt ore-bodies are generally complex, thus to attribute the cobalt source to a specific site is not straightforward. However, as this sample is manganese-free and high in iron, this is probably indicative of a Persian (Iranian) deposit. An Islamic opaque glass sample from Qasr al-Dakhleh, (Sample 18) contains erythrite, suggesting that the cobalt in this sample was also probably sourced from Persia (Iran). Kaczmarczyk [47] stated that Egyptian sources of cobalt were only in use from the sixteenth to seventeenth century BCE and after that date, all cobalt was sourced elsewhere, probably Iran. Sample 3, being dated to the fourth century CE and containing Egyptian-sourced cobalt, would, as discussed below, question this Kaczmarczyk assertion. Alternatively, it is possible that the glass could have been recycled.

However, according to Hope et al. [42], the mining of alum in Egypt may have continued into the early Late Period or the Roman period and this could account for a continuing supply of cobalt. Alternatively, recycling of glass could possibly provide the cobalt glass with the unique Dakhleh/Kharga Oases cobalt mineralisation. The presence of significantly elevated copper does suggest that the shard could be related to the CoCu or $Co^{+}Cu$ glasses as discussed by Shortland and Eremin [77], Walton et al. [99], Smirniou et al. [84], Smith and Clark [85], Walton et al. [98] and Smirniou and Rehren [83], although again, the dating of these Co-Cu glasses still requires further investigation. The Raman signature for copper-blue glass in sample 4 is provided in Fig. 7.8.

Sample 4, a fifth or sixth century glass shard from Dayr Abu Matta, and which is probably a boss from another vessel, as provided in Fig. 7.3e, shows considerable deposition of calcite onto its surface as provided in Fig. 7.5. This clearly evidences the presence of calcium in the sand and also moisture over an extended period in the soil profile from which the shard was recovered. Its Raman spectrum is provided in Fig. 7.7d and in Fig. 7.7d, c and the calcite deposition on its surface is clearly indicated by the band at 1085 cm^{-1} and in Fig. 7.5.

The respective Ip results for Samples 1 and 3 would indicate a melting point of ~1000 °C. Sample 4 is a low soda, low magnesia, low potash glass, in which the Ip is suggestive of a melting point in the region of 800–900 °C, although this does not accord with the SEM-EDS analyses which would suggest a higher melting point. The low alkali concentration has modified the bending and stretching modes of the glass. The cobalt in this sample is probably present as Co^{2+}. However, the growth of calcite and gypsum crystals on the surface of the shard Sample 4 as shown in Fig. 7.5 has possibly modified the apparent Si-O stretching mode indicated in Fig. 7.7d, increasing it to higher wavenumbers. Such a change would be reflected in a higher Polymerisation Index.

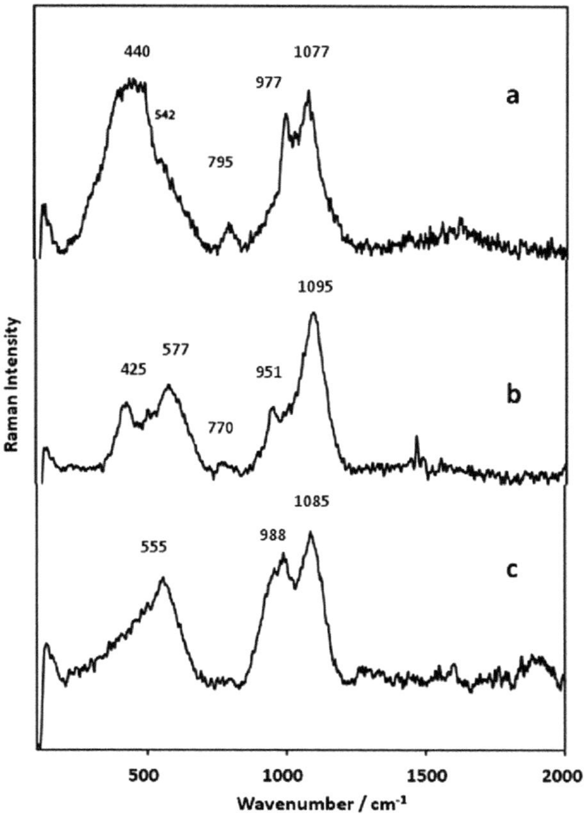

Fig. 7.8 Representative Raman (baseline subtracted) spectra, (**a**) manganese-containing glass Sample 10; (**b**) amber glass Sample 14 being D/1/94.306; (**c**) Sample 15, black bracelet, Kellis. ($\lambda = 514.5$ nm)

The blue striped Sample 4 in Table 7.2 and Fig. 7.7 is of a copper-based blue glass. As indicated by Colomban and Schreiber [24], copper in glass can be present in three forms, namely Cu^{2+} which is blue, Cu^{+} which is colourless, and the reduced copper species (Cu^{o}) which is red. The divalent blue copper ion does not provide a strong Raman signature. Colloidal dispersions of metallic nanocrystals of copper in glass do however provide a narrow characteristic Raman signature at approximately 570 cm^{-1}. It was indicated that in a glass containing 0.25 wt. % Cu that a Raman band at 490 cm^{-1} was obtained using the 514.5 nm laser beam. As shown in Fig. 7.7c, a similar band was located within the blue region in Sample 4. The Polymerisation Index would indicate a slightly lower melting point than that of the cobalt blue glasses and this is reflected in its higher soda concentration lowering its melting point.

It has been reported that the incorporation of copper ions into the glass melts is aided by increasing concentrations of Na_2O with the oxidised species being incorporated into the glass as a tetragonally distorted octahedron [48]. In order to achieve electron neutrality, Cu^{+} is coordinated with one non-bridging oxygen and Cu^{2+} by two non-bridging oxygens. Thus, both Cu^{+} and Cu^{2+} are octahedrally coordinated but with different lattice distortion. Both copper species act as network modifiers in silicate glasses and as shown by these researchers, both copper and calcium ions (which also act as network modifiers) provide similar diffusion coefficients.

Importantly, the average elemental analysis of blue glass samples from Amarna [79] reported that all glasses in the 18th Dynasty are soda-lime type, typically containing 15–31% soda and 4–11% lime together with added colourants. The presence of antimony would suggest contamination from traces of a yellow pigment.

7.4.1.3 Manganese Glass

Sample 9 is a low potash, low magnesia glass in which the soda and calcium concentrations are also lower than that expected for a Roman glass of the period. This is reflected in its Raman spectrum which, from its Ip is indicative of a glass with a higher melting point and with bending and stretching mode wavenumbers outside of those indicated for most of the other glass shards. Its Raman signature also indicates undissolved silica in the glass (band at 465 cm^{-1}). Sample 10 (provided in Figs. 7.3 and 7.8a) is much lower in MnO_2 (0.3%) than sample 9 (1.6% MnO_2). Sample 10 is low in potash, low magnesia and low soda and calcium. Its Raman spectrum is provided in Fig. 7.8a. The Ip of Samples 9 and 10 reflect this difference in composition to Sample 14 (provided in Fig. 7.8b). Both samples have similar iron concentrations and this has probably resulted in the presence of the Fe^{3+} ions providing the more apparent yellow appearance of the shard Sample 9.

7.4.1.4 Amber (Iron-Sulphur Chromophore) Glass

Sample 14 (whose Raman spectrum is given in Fig. 7.8b) is a low potash, low magnesia glass. As provided in Table 7.2, the only significant concentration difference is the doubling of the soda and iron concentrations in Sample 14 and a significantly higher MnO_2 at 1.7% in sample 14. Both would have melting points in the region of 700–800 °C.

Sample 14 contains 3.9% Fe, probably as Fe^{3+}; it is low in magnesia, potash (3.6%), and CaO 7.4%. Its Raman spectrum provided in Fig. 7.8b would indicate the presence of feldspar and this is possibly indicated by the alumina concentration of 4.8%. The higher soda concentration (10.4%) has lowered the Ip and indicates a glass melting point in the region of 600 °C. Probably, wollastonite has formed within the glass and this would suggest a higher melting point. Another glass shard from the same find spot as Sample 14 is darker in visual appearance. It is high in MnO_2 at 1.7% and is higher than the previous sample in both magnesia and potash. Its Ip is indicative of a Roman glass of the period with a melting point probably in the region of 700–800 °C.

7.4.1.5 Mixed-pigment Glass

Shard sample 15 shown in Fig. 7.3m, contains a mixture of colours including green, red, brown and white. Its Raman signature is provided in Fig. 7.8c. The glass contains undissolved lead and cassiterite seen as high SnO_2 and high PbO spots within the matrix. Whilst the tin and lead could come from bronze scrap it is also possible that tin was dissolved in lead to form a crude low melting point solder as described by Ogden [63]. Certainly, the addition of solder to porcelain glazes was adopted in England before the 1760s [45] and this would have had a genesis from the production of much earlier tin-glazed delftware and probably Islamic glasses or glazes. The alumina concentration was in excess of 4% and there is clear evidence from the Raman spectra of feldspar being present.

The red glass is generally obtained by the presence of Cu^o but Cu_2O (cuprite) and α Fe_2O_3 (hematite) also gives red colour. There is a peak situated at 564 cm^{-1} and Colomban and Schreiber [24] suggest that a bending peak at about 570 cm^{-1} is characteristic of a glassy matrix around nanocrystals in glass. The results obtained from the examination of shard Sample 15 as provided in Table 7.2 and Fig. 7.9, indicate either wüstite or magnetite are present together with an unidentified iron complex. The copper has been reduced to either cuprite or nanoparticles of copper to produce the red colouration.

Both the iron spinel magnetite (Fe_3O_4) and wüstite have a signature band at 660 cm^{-1} which is downshifted slightly for wüstite (FeO) [27, 54, 64, 90] and the bands at 313 and 374 cm^{-1} are downshifted for hematite (signature bands at 337 and 394 cm^{-1}) and as suggested by de Faria et al. [27] these two bands represent artefacts derived from the laser breakdown of the wüstite. Probably, this hematitic iron complex has contributed to the red colour.

The Raman spectrum for wüstite is well separated from hematite and they cannot be in equilibrium. Thus, any formation of hematite-like iron phases has emanated from thermal heating by the laser beam. It is uncertain as to the composition of this iron phase. However, at the temperatures within the molten glass, as shown by the presence of tridymite, it is presumed that the iron complex is metastable.

According to Guido et al. [39] the introduction of 1.4–3.4% iron oxide (of which 50% was as iron metal) into the melt will facilitate the dissolution of the copper and provides the appropriate reducing conditions for the precipitation of either cuprite (Cu_2O) crystals or nanoparticles of Cu^o (metallic copper). These comments in respect of the appropriate reducing conditions resulting from the addition of iron are repeated by Nicholson and Henderson [62]. It must be noted that Guido et al. [39] conducted their glass experiments in a gas fired furnace at 1410–1413 °C, well above the melting point of the Egyptian glass under examination. As shown by phase diagrams, at such high temperatures the reactivity of the metals present is significantly different to their reactions expected at 1000 °C or lower [59]. Furthermore, 50% of the iron metal is already in the reduced state and at this temperature is molten, and is more likely to directly enter into reactions with the silica

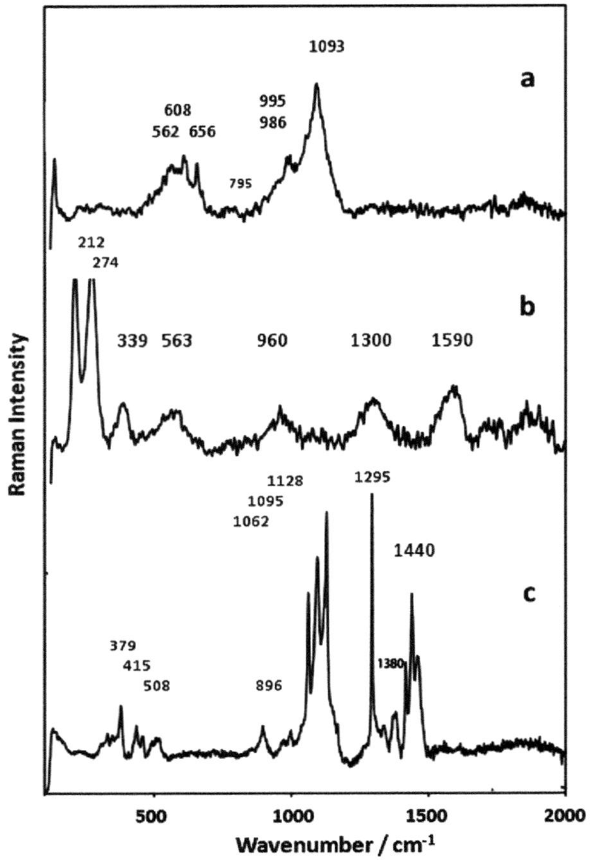

Fig. 7.9 Raman spectra (baseline subtracted) of multi-coloured glass shard sample 15: (**a**) is a green area; (**b**) red area; (**c**) another green area ($\lambda = 514.5$ nm)

at these very high temperatures thereby acting as a bond breaker, as would all other metals present in the melt. Guido has investigated what might be described as "blast furnace" slag. As such, the work of Murad and Cashion [60] in respect of Mössbauer studies of blast furnace slags is instructive, where these authors have shown that depending upon the iron content, there is a tendency for increasing iron content in the slag to act as a network former (tetrahedral Fe^{3+}) rather than as a network modifier (octahedral Fe^{3+}). Depending upon whether the slags are lime rich, or sodium rich, so this can influence the coordination of the Fe^{3+}. The oxygen potential was found to have little effect on the ratio of activity coefficients. The reduction of copper would require an alternative reductant to be present at the temperature likely to have been used for the production of the present shard.

The brown glass a$_2$reas are dominated by high manganese (MnO_2, 1–1 .5%) and high iron (FeO up to 4.0%) and PbO of around 4.5%, in a low magnesia, low potash, moderate soda (8–10% Na_3O) glass.

A number of the large area analyses of green glass did report higher MgO (4.5%), higher potash (K_2O = 3.3%), and increased phosphate (P_2O_5 = 1.4%) accompanied by CuO analyses of up to 1.7%. Soda was normally in the range of 6–10% Na_2O. It has been reported [39, 41] that a bright 'sealing wax' red or a dull brown red colour (both of which are evident in this shard) is obtained when the cuprous ion is present and Guido noted that the 'green' colour is only produced in the presence of lead oxide (as observed in this shard) in the glass, although another researcher has noted that the copper-green colour is also present in glasses in which no detectable PbO was found [41]. Furthermore, copper and zinc are often associated with copper in ancient glasses resulting from the use of recycled scrap glass, or bronze or brass scrap. Henderson [41] reported that 0.1% PbO in the glass greatly facilitates the incorporation of copper into the glass and in its precipitation of crystalline copper out of the glass melt. As reported by Henderson, all Egyptian opaque red glasses which predate the ninth century BCE contain PbO at <4% and after this date, the PbO levels generally rise producing a brighter 'sealing wax' red.

The white glass is a high lead glass containing significant cassiterite. Individual spot analyses were >10% SnO_2 and >10% PbO (precipitation of cassiterite requires at least ~8 wt % SnO_2). This is different to the opaque white glass discussed by Nicholson and Henderson [62] who reported on the use of calcium antimonate and that XRD was the only method available to distinguish between the two forms, namely $Ca_2Sb_2O_7$ and $Ca_2Sb_2O_6$.

Nicholson and Henderson [62] stated that the opaque white, turquoise and yellow Egyptian glasses all contain opacifiers based upon antimony. In the present work, antimony was not detected in any opaque sample analysed. As shown by Raman spectroscopy, opacification was principally obtained from the addition of feldspar.

7.4.1.6 Black Glass

Two black fourth century CE bracelet shards, Samples 16 and 17, both exhibiting a slight brown translucency, have been analysed. Their images are provided in Fig. 7.4a, b. Sample 16 does not contain lead, but is high in iron (FeO = 9%) accompanied by 0.3% MnO_2, 0.5% TiO_2, and 0.4% CoO. The chemical composition of Sample 17 is indicative of a typical of LBA Roman glass [67]. The colour of this type of glass is derived from a high iron concentration (up to 14.3% FeO) and is supplemented by traces of manganese (1.8% MnO_2), titanium (0.4% TiO_2), and particularly cobalt (0.4% CoO) in a low lead-containing (1.4% PbO) glass. This combination provides a slight brown translucency.

The design of these bracelets has been discussed by Spaer [86, 87] who provided illustrations of similar vertically-ribbed designs. She proposed that the original design is pre-Islamic, which would confirm that its antecedents were to be found in Roman glass production. The fact that these shards were find spots at Kellis indicates that the overall bracelet design extended over several centuries. The chemistry of these deep black glasses is discussed in detail by Van der Linden et al. [95] and by Cagno et al. [11, 12]. These workers proposed that two types of deeply coloured glass produced, black and the more abundant deep green colour. Two glass compositions were adopted, a low iron group (1–3% Fe_3O_4) which is associated with the Early Roman Imperial period up to about 150 CE, and a high iron group (4–10% Fe_3O_4) after 150 CE leading to the High Iron, Manganese, Titanium (HIMT) glass of the Late Roman Period. The present two glasses fit the composition proposed for this period. Importantly, all of these black glasses contain a minor addition of cobalt.

Furthermore, self-diffusion coefficients for iron are larger in glasses containing larger alkali content. Thus, as demonstrated by Claußen et al. [16] in a soda-lime-silica melt, the Fe^{3+} self-diffusion coefficient increases with an increase in soda content. These authors related this observation to the ionic charge, noting the Cr^{6+} has the smallest self-diffusion coefficient, and copper and iron the largest self-diffusion coefficients, the self-diffusion coefficients varying by more than six orders of magnitude at 800 °C. In silicate glasses Fe^{3+} mainly occurs in a distorted fourfold coordination tetrahedron, formally possessing a negative charge. This negative charge is balanced by two oxygens coordinated with an alkali ion and this complex is more effectively stabilised by smaller alkali cations. Thus the iron ion has larger self-diffusion coefficients if larger cations are present in the melt at the same viscosity, that is, in the same mobility of the network. This is caused by the improved stabilisation of the Fe^{3+} tetrahedra in glass networks containing smaller cations. If larger cations are present the network complex is less stabilised, otherwise, an increase in the alkali content causes a better stabilisation. It is this difference which explains the decrease in the self-diffusion coefficients in melts of the same viscosity containing larger concentrations of alkali cations.

As reported by Gerlach et al. [36, 37] an increase in the alumina content leads to larger iron self-diffusion coefficients in melts with the same viscosity. Thus, in glasses in which the Al_2O_3-content is smaller than the sum of the alkali metal and alkaline earth content, aluminium is mainly incorporated in fourfold coordination. Therefore, larger Al_2O_3 contents have the same effect as smaller alkali contents. In both cases, the incorporation of Fe^{3+} into the glass network decreases, and thus, the self-diffusion coefficients of iron in melts with the same viscosity increases. The substitution of magnesium oxide for alumina provides similar self-diffusivity outcomes [37]. As explained by Weidenroth and Rüssel [101] self-diffusivities show a small decrease with increasing MgO concentration, its effect being much less pronounced than those of Na_2O and Al_2O_3 and which is in contrast to the effect of CaO stabilisation of the FeO_4 tetrahedra. Increasing the MgO concentration leads to a shift of the redox equilibrium to the oxidised state. This may be explained by the similarities in ionic radii of Mg^{2+} and Fe^{3+} and also metal-oxygen bond lengths. Thus, Mg^{2+} and Fe^{3+} should occupy similar sites in the melt structure. Increasing MgO concentration should therefore lead to the incorporation of Fe^{3+} into energetically less advantageous sites and hence to a shift in the redox ratio to the oxidised state.

7.4.2 Post-Tenth Century Islamic Bracelets

Importantly, cobalt ores, including those containing arsenic, sulphur, iron, nickel, copper and manganese when fired under reducing conditions in high temperature glasses and glazes, will enable fine shades of blue to develop. Whereas, if fired under oxidising conditions and particularly in high lead glazes, these same ores will produce browns, greys and blacks rather than strong blues [49].

Iran is famous for its manganese-free, low iron arseno-sulfide ores [49]. Kerr and Wood indicated that the ores from around Qamsar in Central Iran vary widely in their cobalt mineralisation and in their accompanying elements with various ores being high in either nickel, iron, copper or manganese, thus, attribution of cobalt to this location is not trivial. The analysis of the cobalt detected in Sample 19 (Fig. 7.4d), provided in Table 7.3, contains arsenic and is free from manganese, nickel and zinc, thus it is likely to have been obtained from this general Iranian source.

Lead is present in Samples 30–33 (Figs. 7.4e–g) and their analyses are provided in Table 7.3. Sample 30 also contains significant tin; as shown in Fig. 7.4e this particular shard contains splashes of orange and yellow and Fig. 7.4e indicates an area of insoluble lead-tin at the glass surface.

Bracelet Samples 31 and 33 contain a mixture of colours including green, orange, brown and yellow. The glasses contain undissolved lead and cassiterite which occur as spots or areas which are both high in SnO_2 and PbO. Whilst, as previously stated, the tin and lead could come from bronze scrap, the very high concentrations of PbO reported in some spot analyses would suggest that the tin was dissolved in lead to form a crude form of low melting point solder as described by Ogden [63] and Jay and Cashion [45]. Calcined tin and lead were used in Islamic glasses or glazes.

The lead and tin were blended together in a ratio of about 3:1. At this ratio, tin is not soluble in lead, but will form a solid metal mixture. As discussed by Jay and Cashion [45] tin is only soluble to about 4% in lead metal at room temperature. Allan [3] reported that in a fourteenth century treatise on ceramics, Abu'l Qasim described a method for the preparation of a glaze pigment by which "lead and tin were melted together until they form a white substance to which quartz and plant ashes were added to make a glaze". Tite et al. [93] reported that during the Roman period a similar Pb:Sn ratio of 3:1 was used in the production of soft solders. It would appear that both the blue and the green pigmented glass bracelet sherds examined here employed a method related to this process.

The alumina concentration in these glass bracelets was in excess of 10% and there is clear evidence from the Raman spectra of feldspar being present. The presence of soda, potash and phosphate would suggest the use of plant ash as the flux.

A number of the large area analyses of green glass did report MgO (3.0%), higher potash (K_2O = 1.1–1.5%), and increased phosphate (P_2O_5 = 0.5%). Soda was normally in the range of 3–5% Na_2O. It has been reported [39, 41] that a dull brown colour which is evident in shard Sample 31 occurs when the cuprous ion (Cu^+) ion is present, and Guido noted that the 'green' colour is only produced in the presence of lead oxide (as observed in these shards) in glass, although other researchers have noted the copper-green colour to be reported in glasses in which no detectable PbO was found [41]. Furthermore, copper and zinc are often associated with copper in ancient glasses resulting from the use of scrap glass or scrap bronze (Cu/Sn) or brass (Cu/Zn). Henderson [41] reported that 0.1% PbO in the glass greatly facilitates the incorporation of copper into the glass melt.

In Sample 31, different pigmented glasses were combined to produce the green glass with the yellow/orange blended coloured glass trails on its surface. Other than the PbO content (6% in yellow, 4% in green glass) the soda, potash, magnesia lime and alumina concentrations are similar to each other. However, the orange glass, which may have been used to highlight the yellow, is significantly different in chemical composition. It has only 40% of the alumina (4.5%) and significantly higher lead (10% PbO). This would have lowered the viscosity and the melting point of this glass when compared with the yellow and green glass. The lowering of the Raman stretching multiplets (ν_{max}) for the high PbO glass when compared with the green and yellow glasses confirm such viscosity and melting point changes. The green colouration would have also resulted from the high iron concentration present, which under reducing conditions is green.

Nicholson and Henderson [62] reported that yellow glass is produced when antimony is mixed with iron in concentrations above impurity levels (these impurity levels were not quoted). They also noted that manganese and antimony can produce yellows, amber, brown and black glass. Brill et al. [9] reported that Egyptian workers employed a lead source different to that used in Mesopotamia.

Lead antimonate yellow ($Pb_2Sb_2O_7$) and lead-tin yellow are amongst the oldest synthetic pigments with their synthesis going back at least 4500 years. Shortland [76] reported that at Malkata and Amarna in Egypt, the lead antimonate yellow and calcium antimonate white appear concurrently with the introduction of glass. Molina et al. [57] noted that a lead-antimony-silica mixture comparable to that used in Venice in the eighteenth and nineteenth centuries CE was the desired form for Roman yellow glasses. They noted that towards the end of the Roman period from about the fourth century CE onwards, lead stannate replaced lead antimonate in the production of opaque yellow glasses. Tite et al. [93] also indicated that for a short period during the second to the first centuries BCE and then again throughout the Roman and Byzantine Empires from the fourth century onwards, lead stannate yellow was used in place of lead antimonate yellow. Tite et al. [93] proposed that the lead-tin yellow was the cubic form, $PbSnO_4$ and not the orthorhombic form, Pb_2SnO_4; whereas Bell et al. [5] proposed that there are two yellows, namely lead-tin yellow Type I, Pb_2SnO_4 and Type II, a silicon-substituted lead(II) stannate $PbSn_{1-x}Si_xO_4$. Clark et al. [15] provided a number of references to support their contention that lead-tin yellow Type I is Pb_2SnO_4 and not $PbSnO_4$. Clark et al. [15] provided the chemical and thermal basis for the formation of each form of lead tin yellow and for their respective decomposition.

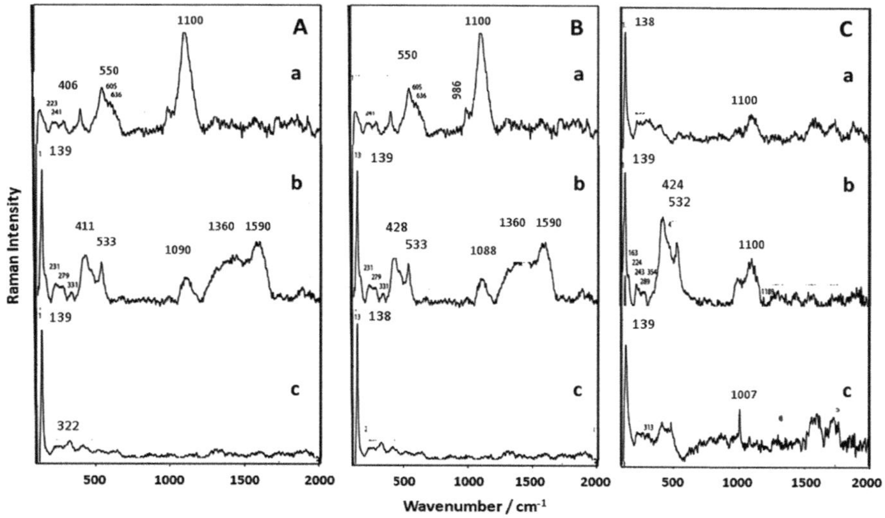

Fig. 7.10 Representative Raman spectra (baseline subtracted) of (**A**) Sample 19: (**a**) blue bracelet area (**b**) orange decoration (**c**) yellow decoration, (λ = 514.5 nm); (**B**) Sample 31 green bracelet: (**a**) is green area, (**b**) is orange area, (**c**) is yellow area. (λ = 514.5 nm); (**C**) green bracelet Sample 33: (**a**) is green area (**b**) is orange decoration (**c**) is yellow decoration. (λ = 514.5 nm)

As reported in Table 7.3, both lead and tin are present in Samples 31 and 33 from their EDS analyses and specifically within the yellow and orange regions as shown for Sample 31. As indicated by their respective Raman signatures provided in Fig. 7.10, a very strong band is present at ~140 cm^{-1} and a broad band at 410–420 cm^{-1}. Bell et al. [5] assign this Raman signature to lead-tin yellow Type II being a silicon-substituted lead(II) stannate of the general formula $PbSn_{1-x}Si_xO_4$. Gill and Rehren [38] support the mineralogical structure for both Type I and Type II lead tin yellow as provided by Bell et al. [5]. Thus, as shown in Fig. 7.10 the strong Raman band at ~140 cm^{-1} accompanied by a broad weak band at 433–437 cm^{-1} confirms lead tin yellow Type II mineralisation.

All of the bracelet shards contain zinc. The role for this zinc in the chemistry of the various pigments is not fully understood. It is, however, known that the introduction of zinc or tin into the lead antimonate structure modifies the colour of the glass. Tin-containing antimonate specimens provide a purer yellow; zinc and orange hues and a darker yellow hue occur in the case of lead antimonates with a Pb:Sb ratio exceeding 1 [72]. These changes are attributed to the changes in ionic radii of the various elements contributing to a change in the lattice size; the octahedral coordination cation radius increases in the order Sb^{5+}, Sn^{4+}, Zn^{2+}. Gill and Rehren [38] suggest that an alternative to the lead tin yellow Type II structure is the addition of zinc, which they described as a ternary oxide of lead, tin and zinc which they named as lead-tin orange. They were also unable to determine where the Zn^{2+} ions fit into the structure. However, all of the glass bracelets contained zinc at various concentrations, namely blue (spots at 0.4% ZnO), yellow (<0.5% ZnO), orange (0.6–3.9% ZnO plus ~4% FeO) and green (0.5% max.). These concentrations are lower than the 6–7% ZnO observed in the seventeenth century Indian glaze examined by Gill and Rehren [38].

7.5 Conclusions

The samples of Roman glass are from Dakhleh Oasis and dated to the fourth to sixth century CE as are two glass bracelet shards from Qasr al-Dakhleh. The opaque blue and green Islamic glass bracelets are undated surface finds from Qasr al-Dakhleh and are considered to be post-tenth century production.

As shown from their SEM-EDS analyses, the cobalt containing glass shards found at Kellis in ancient Egypt appear to have their cobalt sourced from different deposits, probably, one from Dakhleh Oasis, the others from Iran.

The bracelets are coloured with splashes of orange and yellow glass. Yellow glass has been stated by Nicholson and Henderson [62] to form from the combination of antimony and iron. However, as discussed in some detail, in the present research, the bracelet samples were found to contain the yellow lead-tin-Type II pigment together with some zinc oxide. This lead-tin-silica-zinc combination has not been previously reported to have been produced or found within the Western Desert region of Egypt. The earliest reporting of such a pigment was by Gill and Rehren [38] for a seventeenth century Indian glaze. Based upon the Raman signature for a yellow zinc-modified, silicon-substituted, Pb-Sn Type II glass, a production date of after fourth century CE is proposed.

As provided in Table 7.3, lead and tin were blended together in a ratio of about 3:1. At this ratio, tin is not soluble in lead, but will form a solid metal mixture. As discussed by Jay and Cashion [45] tin is only soluble to 4% in lead metal at room temperature. Allan [3] reported that in the fourteenth century treatise on ceramics, Abu'l Qasim described a method for the preparation of a glaze pigment by "lead and tin were melted together until they form a white substance to which quartz and plant ashes were added to make a glaze". It would appear that both the blue and the green pigmented glass bracelet sherds examined employed a method related to the above process.

The light Rare Earth element cerium has been detected in several of the analyses. The presence of the Rare Earths has been considered to be a potential marker for the source of the silica. However, without further analyses, this theory cannot be advanced.

Both lead and tin are reported to be present in Samples 31 and 33 as shown from their SEM-EDS analyses specifically within yellow and orange regions as shown for Sample 31. And, as indicated by their respective Raman signatures provided in Fig. 7.10 a very strong band is present at 148 cm^{-1} and a broad band at 434 cm^{-1}. Bell et al. [5] assign this Raman signature to lead-tin yellow Type II being a silicon-substituted lead(II) stannate of the general formula $PbSn_{1-x}Si_xO_4$. Gill and Rehren [38] support the mineralogical structure for both Type I and Type II as provided by Bell et al. [5].

All of the bracelet shards contain zinc. The role for this zinc in the chemistry of the various pigments is not fully understood. It is however known that the introduction of zinc or tin into the lead antimonate structure modifies the colour of the glass. Tin-containing antimonate specimens provide a purer yellow, and zinc, an orange hue and a darker yellow hue in the case of lead antimonates with a Pb:Sb ratio exceeding 1 [72].

References

1. Abd-Allah, R.: Damage assessment and chemical characterisation of glass objects excavated from Gadara, Northern Jordan. In: Ioannides, D., Fritsch, J., Leissner, R., Davies, F., Remondino, R.C. (eds.) Progress in Cultural Heritage Preservation, EUROMEXD 2012, pp. 377–383. Springer, Limassol (2012)
2. Aerts, A., Velde, B., Janssens, K., Dijkman, W.: Change in silica sources in Roman and post-Roman glass. Spectrochim. Acta B. **58**(4), 659–667 (2003). https://doi.org/10.1016/S0584-8547(02)00287-2
3. Allan, J.W.: Abu'l –Qasim's Treatise on Ceramics. Iran. **11**(1), 111–130 (1974). https://doi.org/10.1080/05786967.1973.11834171
4. Beck, H.C.: Glass before 1500 BC. Ancient Egypt. **XIX**, 7–31 (1944)
5. Bell, I.M., Clark, R.J.H., Gibbs, P.J.: Raman spectroscopic library of natural and synthetic pigments (pre- ~1850 AD). Spectrochim. Acta A. **53**(12), 2159–2179 (1997). https://doi.org/10.1016/S1386-1425(97)00140-6
6. Bertoncello, R., Milanese, L., Russo, U., Pedron, D., Guerriero, P., Barison, S.: Chemistry of cultural glasses: the early Medieval glasses of Monselice's Hill (Padova, Italy). J. Non- Cryst. Solids. **306**(3), 249–262 (2003). https://doi.org/10.1016/S0022-3093(02)01181-X
7. Brill, R.H.: The chemical interpretation of the texts. In: Oppenheim, A.L., Brill, R.H., Barag, D., von Saldern, A. (eds.) Glass and Glassmaking in Ancient Mesopotamia, pp. 105–140. Corning Museum of Glass, New York (1970)
8. Brill, R.H.: Chemical Analyses of Early Glasses. Corning Museum of Glass, New York (1999)
9. Brill, R.H., Barnes, I.L., Adams, B.: Lead isotopes in some ancient Egyptian objects. In: Bishay, A. (ed.) Recent Advances in Science and Technology of materials, vol. 4, pp. 9–37. Plennum, London (1974)
10. Bronk, H., Freestone, I.C.: A quasi non-destructive microsampling technique for the analysis of intact glass objects by SEM/EDXA. Archaeometry. **43**(4), 517–527 (2001). https://doi.org/10.1111/1475-4754.00034
11. Cagno, S., Cosyns, P., Van der Linden, V., Schalm, O., Izmer, A., Deconinck, I., Vanhaecke, F., Nowak, A., Wagner, B., Bulska, E., Nys, K., Janssens, K.: Composition data of a large collection of black-appearing Roman glass. Open J. Archaeom. **1**(1), e22 (2014a). https://doi.org/10.4081/arc.2013.e22

12. Cagno, S., Cosyns, P., Izmer, A., Vanhaeke, F., Nys, K., Janssens, K.: Deeply coloured and black-appearing Roman glass: a continued research. J. Archaeol. Sci. **42**, 128–139 (2014b). https://doi.org/10.1016/j.jas.2013.11.003
13. Caley, E.R.: Analyses of Ancient Glasses 1790–1957. Corning Museum of Glass, Corning (1963)
14. Clark, R.J.H.: Raman microscopy as a structural and analytical tool in the fields of art and archaeology. J. Mol. Struct. **834–836**, 74–80 (2007). https://doi.org/10.1016/j.molstruc.2007.01.031
15. Clark, R.J.H., Cridland, L., Kariuki, B.M., Harris, K.D.M., Withnall, R.: Synthesis, Structural Characterisation and Raman Spectroscopy of the Inorganic Pigments Lead Tin Yellow Types I and II and Lead Antimonate Yellow: Their identification on Medieval Paintings and manuscripts. J. Chem. Soc. Dalton Trans. **16**, 2577–2582 (1995). https://doi.org/10.1039/DT9950002577
16. Clauβen, O., Gerlach, S., Rüssel, C.: Self-diffusivity of polyvalent ions in silicate liquids. J. Non-Cryst. Solids. **253**(1–3), 76–83 (1999). https://doi.org/10.1016/S0022-3093(99)00345-2
17. Colomban, P.: Polymerisation degree and Raman identification of ancient glasses used for jewellery, ceramic enamels and mosaics. J. Non-Cryst. Solids. **323**(1–3), 180–187 (2004a). https://doi.org/10.1016/S0022-3093(03)00303-X
18. Colomban, P.: Raman spectrometry, a unique tool to analyse and classify ancient ceramics. Appl. Phys. A Mater. Sci. Process. **79**, 167–170 (2004b). https://doi.org/10.1007/s00339-004-2512-6
19. Colomban, P.: On-site Raman identification and dating of ancient glasses: a review of procedures and tools. J. Cultural Herit. **9**(Suppl), e55–e60 (2008). https://doi.org/10.1016/j.culher.2008.06.005
20. Colomban, P.: Non-destructive Raman analysis of ancient glasses and glazes. In: Janssens, K. (ed.) Modern Methods for Analysing Archaeological and Historical Glasses. Wiley (2014a)
21. Colomban, P.: Rocks as blue, green and black pigments/dyes of glazed pottery and enamelled glass artefacts—A review. Eur. J. Mineral. **25**(5), 863–879 (2014b). https://doi.org/10.1127/0935-1221/2013/0025-2305
22. Colomban, P., Etcheverry, M.-P., Asquier, M., Bounichou, M., Tournié, A.: Raman identification of ancient stained glasses and their degree of deterioration. J. Raman Spectrosc. **37**(5), 614–626 (2006a). https://doi.org/10.1002/jrs.1495
23. Colomban, P., Kirmizi, B., Franci, G.S.: Cobalt and associated impurities in blue (and green) glass, glazes and enamel: relationships between raw materials, processing, composition, phases and international trade. Fortschr. Mineral. **11**, 644 (2021) https://soi.org/104490/min1060644
24. Colomban, P., Schreiber, H.D.: Raman signature modification induced by copper nanoparticles in silicate glass. J. Raman Spectrosc. **36**(9), 884–890 (2005). https://doi.org/10.1002/jrs.1379
25. Colomban, P., Tournié, A., Bellot-Gurlet, L.: Raman identification of glassy silicates used in ceramics, glass and jewellery: a tentative differentiation guide. J. Raman Spectrosc. **37**, 841–852 (2006b). https://doi.org/10.1002/jrs.1515
26. Cox, G.A., Ford, B.A.: The long-term corrosion of glass by groundwater. J. Mater. Sci. **28**, 5637–5647 (1993). https://doi.org/10.1007/BF00367840
27. de Faria, D.L.A., Venâncio Silva, S., de Oliveira, M.T.: Raman microspectroscopy of some iron oxides and oxyhydroxides. J. Raman Spectrosc. **28**(11), 873–878 (1997). https://doi.org/10.1002/(SICI)1097-4555(199711)28:11<873::AID-JRS177>3.0.CO;2-B
28. Eccleston, M.A.J.: Metalworking at Kellis: a preliminary report. In: Hope, C.A., Bowen, G.E. (eds.) Dakhleh Oasis Project: Preliminary Reports on the 1994–1995 to 1998–1999 Field Seasons, pp. 144–149. Oxbow, Oxford (2003)

29. Edwards, H.G.M., Vandenabeele, P., Colomban, P.: Raman Spectroscopy in Cultural Heritage Preservation. Springer Nature, Cham (2023)
30. Farnsworth, M., Ritchie, P.D.: Spectrographic studies on ancient glass mainly of the eighteenth dynasty, with special reference to its cobalt content. Tech. Stud. Field Fine Arts. **6**(3), 155–173 (1938)
31. Freestone, I.C.: A technical study of limehouse ware. In: Drakard, D. (ed.) Limehouse Ware Revealed, pp. 68–74. English Ceramic Circle (1994)
32. Freestone, I.C.: Glass production in Late Antiquity and the Early Islamic Period: a geochemical perspective. In: Maggetti, M., Messiga, B. (eds.) Geomaterials in Cultural Heritage Special Publications 357, pp. 301–316. Geological Society, London (2006)
33. Freestone, I., Middleton, A., Meeks, N.: Retention of phosphate in buried ceramics: an electron microbeam approach. Archaeometry. **27**(2), 161–177 (1985). https://doi.org/10.1111/j.1475-4754.1985.tb00359.x
34. Freeth, S.J.: A chemical study of some bronze age pottery sherds. Archaeometry. **10**(1), 104–119 (1967). https://doi.org/10.1111/j.1475-4754.1967.tb00621.x
35. Garcia-Vallès, M., Gimeno-Torrente, D., Martínez-Manent, S., Fernández-Turiel, J.L.: Medieval stained glass in a Mediterranean climate: Typology, weathering and glass decay, and associated biomineralisation processes and products. Am. Mineral. **88**(11–12), 1996–2006 (2003). https://doi.org/10.2138/am.2008.2583
36. Gerlach, S., Clauβen, O., Rüssel, C.: Self-diffusivity of iron in alkali-magnesia-silica glass melts. J. Non-Crystalline Solids. **226**(1–2), 11–18 (1998a). https://doi.org/10.1016/S0022-3093(98)00368-8
37. Gerlach, S., Clauβen, O., Rüssel, C.: Self-diffusivity of iron in alkali-lime-aluminosilicate glass melts. J. Non-Cryst. Solids. **240**(1–3), 110–117 (1998b). https://doi.org/10.1016/S0022-3093(98)00704-2
38. Gill, M.S., Rehren, T.: The intentional use of lead-tin orange in Indian Islamic glazes and its preliminary characterisation. Archaeometry. **56**(6), 1009–1023 (2014). https://doi.org/10.1111/arcm.12062
39. Guido, M., Henderson, J., Cable, M., Bayley, J., Biek, L.: A Bronze Age glass bead from Wilsford, Wiltshire: Barrow G.43 in the Lake group. Proc. Prehistoric. Soc. **50**(1), 245–254 (1984). https://doi.org/10.1017/S0079497X00007544
40. Henderson, J.: The scientific analysis of ancient glass and its archaeological interpretation. In: Henderson, J. (ed.) Scientific Analysis in Archaeology Oxford University Committee for Archaeology, Monograph 19, pp. 40–63. Oxford University, Oxford (1989)
41. Henderson, J.: Glass. In: The Science and Archaeology of Materials: An Investigation of Inorganic Materials. Taylor and Francis, Florence (2004)
42. Hope, C.A., Bowen, G.E., Cox, J., Dolling, W., Milner, J., Pettman, A.: Report on the 2009 season of Excavations at Mut el-Kharab, Dakhleh Oasis. Bull. Aust. Centre Egyptol. **20**, 47–86 (2009) https://figshare.mq.edu.au/articles/journal_contribution/Report_on_the_2009_season_of_excavations_at_Mut_el-Kharab_Dakhleh_Oasis/26876713?file=48891373. Accessed 6 Sept. 2024
43. Hope, C.A., and Whitehouse, H., (2004), “The Gladiator Jug from Ismant el-Kharab”, in: The Oasis Papers, 4, Proceedings of Third International Conference of the Dakhleh Oasis Project (Eds. C.A. Hope and G.E. Bowen), Oxbow Books, Oxford, 391-410
44. Jackson, C.M., Greenfield, D., Howie, L.A.: An assessment of compositional and morphological changes in model archaeological glasses in an acid environment. Archaeometry. **54**(3), 489–507 (2012). https://doi.org/10.1111/j.1475-4754.2011.00632.x
45. Jay, W.H., Cashion, J.D.: Raman spectroscopy of Limehouse porcelain sherds supported by Mössbauer spectroscopy and electron microscopy. J. Raman Spectrosc. **44**(12), 1718–1732 (2013). https://doi.org/10.1002/jrs.4404
46. Jay, W.H., Cashion, J.D., Blenkinship, B.: Lancaster delftware: a Raman spectroscopy, electron microscopy and Mössbauer spectroscopy compositional study. J. Raman Spectrosc. **46**(12), 1265–1282 (2015). https://doi.org/10.1002/jrs.4404

47. Kaczmarczyk, A.: The Source of Cobalt in Ancient Egyptian Pigments. In: Olin, J.S., Blackman, M.J. (eds.) Proceeding of the 34th International Archaeometry Symposium, pp. 469–476. Smithsonian Institution Press, Washington, DC (1986)
48. Kaufmann, J., Rüssel, C.: Diffusion of copper in soda-silicate and soda-lime-silicate melts. J. Non-Cryst. Solids. **356**(23–24), 1158–1162 (2010). https://doi.org/10.1016/j.jnoncrysol.2010.03.015
49. Kerr, R., Wood, N.: Science and Civilisation in China. In: Chemistry and Chemical Technology, Part XII: Ceramic Technology, vol. 5. Cambridge University Press, Cambridge (2004)
50. Kırmızı, B., Chen, S., Colomban, P.: The Raman signature of protonic species as a potential tool for dating or authentication of glazed pottery. J. Raman Spectroscopy. **50**(5), 696–710 (2019). https://doi.org/10.1002/jrs.5558
51. Lilyquist, C., Brill, R.H.: Studies in ancient Egyptian glass. Metropolitan Museum of Art, New York (1994)
52. McMillan, P.: Structural studies of silicate glasses and melts—applications and limitations of Raman spectroscopy. Am. Mineral. **69**(7–8), 622–644 (1984) https://pubs.geoscienceworld.org/msa/ammin/article-abstract/69/7-8/622/104856/Structural-studies-of-silicate-glasses-and-melts
53. Maltoni, S., Silvestri, A., Marcante, A., Molin, G.: The transition from Roman to Late Antique glass: new insights from the *Domus of Tito Macro* in Aquileia (Italy). J. Archaeol. Sci. **73**, 1–16 (2016). https://doi.org/10.1016/j.jas.2016.07.002
54. Marciuš, M., Ristić, M., Ivanda, M., Musić, S.: Formation of Iron Oxides by Surface Oxidation of Iron Plate. Croat. Chem. Acta. **85**(1), 117–134 (2012). https://doi.org/10.5562/cca1943
55. Melcher, M., Schreiner, M.: Evaluation procedure for leaching studies on naturally weathered potash-lime-silica glasses with medieval composition by scanning electron microscopy. J. Non-Cryst. Solids. **351**, 1210–1225 (2005). https://doi.org/10.1016/j.jnoncrysol.2005.02.020
56. Merkel, S., Rehren, T.: Parting layers, ash trays, and Rameside4 glassmaking: an experimental study. In: Pusch, E.B., Rehren, T. (eds.) Hochtemperatur-Technologie in der Ramses-Stadt. Verlag Gebrüder Gerstenberg, Hildesheim (2007)
57. Molina, G., Odin, G.P., Pradell, T., Shortland, A.J., Tite, M.S.: Production technology and replication of lead-antimonate yellow glass from New Kingdom Egypt and the Roman Empire. J. Archaeol. Sci. **41**, 171–184 (2014). https://doi.org/10.1016/j.jas.2013.07.030
58. Moorey, P.R.S.: Materials and Manufacture in Ancient Mesopotamia: The Evidence of Archaeology and Art Metals and Metalwork, Glazed Materials and Glass BAR international Series, 237. BAR, Oxford (1985)
59. Morton, G.R., Wingrove, J.: Constitution of Bloomery Slags: Part I: Roman. J. Iron Steel Inst. **207**, 1556–1564 (1969)
60. Murad, E., Cashion, J.: Mössbauer Spectroscopy of Environmental Materials and their Industrial Utilisation. Kluwer Academic publishers, Boston (2004)
61. Nicholson, P.T.: Brilliant Things for Akhenaten, "The Production of Glass, Vitreous materials and Pottery at Amarna Site O45.1", Excavation Memoir 80, Egypt Exploration Society, London (2007). https://www.academia.edu/download/83587462/pk-2009-4w-nicholson-akhenaten-2007.pdf. Accessed 8 Sept. 2024
62. Nicholson, P.T., Henderson, J.: Glass. In: Ancient Egyptian Materials and Technology, pp. 195–336. Cambridge University Press, Cambridge (2000)
63. Ogden, J.: Metals. In: Nicholson, P.T., Shaw, I. (eds.) Ancient Egyptian Materials and Technology, pp. 148–176. Cambridge University Press, Cambridge (2000)
64. Oh, S.J., Cook, D.C., Townsend, H.E.: Characterisation of iron oxides commonly formed as corrosion products on steel. Hyperfine Interact. **112**, 59–66 (1998). https://doi.org/10.1023/A:1011076308501

65. Oppenheim, A.L., Brill, R.H., Barag, D., von Saldern, A.: Glass and Glassmaking in Mesopotamia. Corning, New York (1988)
66. Peltenburg, E.J.: Early faience: recent studies, origins and relations with glass. In: Bimson, M., Freestone, I.C. (eds.) Early Vitreous Materials British Museum Occasional Paper 56, pp. 5–30. British Museum, London (1987)
67. Polikreti, K., Murphy, J.M.A., Kantarelou, V., Karydas, A.G.: XRF analysis of glass beads from Mycenaean palace of Nestor at Pylos, Peloponnesus, Greece: new insight into the LBA glass trade. J. Archaeol. Sci. **38**(11), 2889–2896 (2011). https://doi.org/10.1016/j.jas.2011.05.003
68. Pusch, E.B., Rehren, T.: Hochtemperatur-Technologie in der Ramses-Stadt. Rubinglas für den Pharao, Forschung in der Ramses-Stadt, 6, Katalog. Verlag Gebrüder Gerstenberg, Hildesheim (2007). https://www.academia.edu/download/31734749/Pusch_Rehren_2007_Rubinglas.pdf. Accessed 9 Sept. 2024
69. Rehren, T.: Rationales in Old World base glass compositions. J. Archaeol. Sci. **27**(12), 1225–1234 (2000). https://doi.org/10.1006/jasc.1999.0620
70. Rehren, T.: A review of the factors affecting the composition of early Egyptian glasses and faience: alkali and alkaline earth oxides. J. Archaeol. Sci. **35**(5), 1345–1354 (2008). https://doi.org/10.1016/j.jas.2007.09.005
71. Rehren, T., Pusch, E.B.: Late Bronze Age Glass Production at Qantir-Piramesses, Egypt. Science. **308**(5729), 1756–1758 (2005). https://doi.org/10.1126/science.1110466
72. Rosi, F., Manuali, V., Miliani, C., Brunetti, B.G., Sgameliotti, A.: Raman scattering features of lead pyroantimonate compounds. Part I: XRD and Raman characterisation of $Pb_2Sb_2O_7$ doped with tin and zinc. J. Raman Spectrosc. **40**(1), 107–111 (2009). https://doi.org/10.1002/jrs.2092
73. Sayre, E.V., Smith, R.W.: Compositional categories of ancient glass. Science. **133**(3467), 1824–1826 (1961). https://doi.org/10.1126/science.133.3467.1824
74. Sayre, E.V., Smith, R.W.: Analytical studies in ancient Egyptian glass. In: Bishay, A. (ed.) Recent Advances in the Science and Technology of Materials, pp. 47–70. Plenum Press, New York (1974)
75. Schoer, B., Rehren, T.: The Composition of Glass and Associated Ceramics from Qantir. In: Pusch, E.B., Rehren, T. (eds.) Hochtemperatur-Technologie in der Ramses- Stadt. Verlag Gebrüder Gerstenberg, Hildesheim (2007)
76. Shortland, A.J.: The use and origin of antimonate colourants in early Egyptian glass. Archaeometry. **44**(4), 517–530 (2002). https://doi.org/10.1111/1475-4754.t01-1-00083
77. Shortland, A.J., Eremin, K.: An analysis of second millennium glass from Egypt and Mesopotamia, Part 1: New WDS analyses. Archaeometry. **48**(4), 581–603 (2006). https://doi.org/10.1111/j.1475-4754.2006.00274.x
78. Shortland, A.J., Rogers, N., Eremin, K.: Trace element discrimination between Egyptian and Mesopotamian Late Bronze Age glasses. J. Archaeol. Sci. **34**(5), 781–789 (2007). https://doi.org/10.1016/j.jas.2006.08.004
79. Shortland, A.J., Tite, M.S.: Raw materials of glass from Amarna and implications for the origins of Egyptian glass. Archaeometry. **42**(1), 141–151 (2000). https://doi.org/10.1111/j.1475-4754.2000.tb00872.x
80. Shortland, A.J., Schachner, L., Freestone, I., Tite, M.: Natron as a flux in the early vitreous materials industry: sources, beginnings and reasons for decline. J. Archaeol. Sci. **33**(4), 521–530 (2006). https://doi.org/10.1016/j.jas.2005.09.011
81. Shugar, A., Rehren, Th.: Formation and composition of glass as a function of firing temperature. Glass Technol. **43C**, Proc. XIX Int. Congr. Glass, Edinburgh, 1–6 July 2001, pp. 145–150 (2003). https://discovery.ucl.ac.uk/id/eprint/93659/1/Rehren_ShugarRehren%202002%20Glass%20firing%20temperature%20Glass%20Technol%2043C.pdf. Accessed 9 Sept. 2024

82. Smets, B.M.J., Lommen, T.P.A.: The leaching of sodium containing glasses: ion exchange or diffusion of molecular water? J. Phys. Colloques. **43**, C9-649–C9-652 (1982). https://doi.org/10.1051/jphyscol:19829129
83. Smirniou, M., Rehren, T.: Shades of blue—cobalt-copper coloured blue glass from New Kingdom Egypt and the Mycenaean world: a matter of production or colourant source? J. Archaeol. Sci. **40**(12), 4731–4743 (2013). https://doi.org/10.1016/j.jas.2013.06.029
84. Smirniou, M., Rehren, T., Sismani, A., Asderaki, E., Gratuze, B.: Mycenaean beads from KAZANAKI, Volos: a further node in the LBA glass network. In: Ignatiadou, D., Antonaras, A. (eds.) Annales du 18e Congrès de l'Association Internationale pour L'Histoire du Verre. AIHV, Thesaloniki (2009)
85. Smith, G.D., Clark, R.J.H.: Raman microscopy in archaeological science. J. Archaeol. Sci. **31**(8), 1137–1160 (2004). https://doi.org/10.1016/j.jas.2004.02.008
86. Spaer, M.: The pre-Islamic glass bracelets of Palestine. J. Glass Stud. **30**, 51–61 (1988) https://www.jstor.org/stable/24190807
87. Spaer, M.: The Islamic glass bracelets of Palestine: Preliminary findings. J. Glass Stud. **34**, 44–62 (1992) https://www.jstor.org/stable/24190979
88. Spring, M., Higgitt, C., Saunders, D.: Investigation of pigment-medium interaction processes in oil paint containing degraded Smalt. Natl. Gallery Tech. Bull. **26**, 56–70 (2005) https://www.jstor.org/stable/42616313
89. Stapleton, C.P., Swanson, S.S.: In: Vandiver, P.B., Goodway, M., Druzik, J.R., Mass, J.L. (eds.) Batch material processing and glassmaking technology of the 9th century B.C. artifacts excavated from the site of Hasanlu, Northwest Iran MRS Online Proceedings (OPL) 712, pp. 117–114. MRS (2002) https://www.cambridge.org/core/journals/mrs-online-proceedings-library-archive/article/abs/batch-material-processing-and-glassmaking-technology-of-9th-century-bc-artifacts-excavated-from-the-site-of-hasanlu-northwest-iran/F3696BA1CC4ECCE533655C4259FD2655
90. Strobel, R., Pratsinis, S.E.: Direct synthesis of maghemite, magnetite and wustite nanoparticles. Adv. Powder Technol. **20**(2), 190–194 (2009). https://doi.org/10.1016/j.apt.2008.08.002
91. Tanimoto, S.: Experimental study of Late Bronze Age glass-making practice. PhD thesis, UCL Institute of Archaeology, University of London (2007)
92. Tanimoto, S., Rehren, T.: Interactions between silicate and salt melts in LBA glassmaking. J. Archaeol. Sci. **35**(9), 2566–2573 (2008). https://doi.org/10.1016/j.jas.2008.04.016
93. Tite, M.S., Pradell, T., Shortland, A.: Discovery, production and use of tin-based opacifiers in glasses, enamels and glazes from Late Iron Age onwards: a reassessment. Archaeometry. **50**(1), 67–84 (2008). https://doi.org/10.1111/j.1475-4754.2007.00339.x
94. Tournié, A., Ricciardi, P., Colomban, P.: Glass corrosion mechanisms: a multiscale analysis. Solid State Ionics. **179**(38), 2142–2154 (2008). https://doi.org/10.1016/j.ssi.2008.07.019
95. Van der Linden, V., Cosyns, P., Schalm, O., Cagno, S., Nys, K., Janssens, K., Nowak, A., Wagner, B., Bulkska, E.: Deeply coloured and black glass in the northern provinces of the Roman Empire: Differences and similarities in chemical composition before and after AD 150. Archaeometry. **51**(5), 822–844 (2009). https://doi.org/10.1111/j.1475-4754.2008.00434.x
96. Verney-Carron, A., Gin, S., Libourel, G.: A fractured Roman glass block altered for 1800 years in seawater: Analogy with nuclear waste glass in a deep geological repository. Geochim. et. Cosmochim. Acta. **72**(22), 5372–5385 (2008). https://doi.org/10.1016/j.gca.2008.08.018
97. von der Gönna, G., Rüssel, C.: Diffusivity of various polyvalent elements in a $Na_3O.3SiO_3$ glass melt. J. Non-Cryst. Solids. **261**(1–3), 204–210 (2000). https://doi.org/10.1016/S0022-3093(99)00598-0
98. Walton, M., Eremin, K., Shortland, A., Degryse, P., Kirk, S.: Analysis of Late Bronze Age glass axes from Nippur – a new cobalt colourant. Archaeometry. **54**(5), 835–852 (2012). https://doi.org/10.1111/j.1475-4754.2012.00664.x

99. Walton, M., Shortland, A., Kirk, S., Degryse, P.: Evidence for the trade of Mesopotamian and Egyptian glass to Mycenaean Greece. J. Archaeol. Sci. **36**(7), 1496–1503 (2009). https://doi.org/10.1016/j.jas.2009.02.012
100. Wedepohl, K.H., Simon, K., Kronz, A.: Data on 61 chemical elements for the characterisation of three major glass compositions in late antiquity and the Middle Ages. Archaeometry. **53**(1), 81–102 (2011). https://doi.org/10.1111/j.1475-4754.2010.00536.x
101. Weidenroth, A., Rüssel, C.: The effect of MgO on the diffusivity of iron and incorporation of iron in soda-magnesia-aluminosilicate melts. J. Non-Cryst. Solids. **347**(1–3), 180–186 (2004). https://doi.org/10.1016/j.jnoncrysol.2004.06.018

Chapter 8
Renaissance and Afterwards

Abstract The use of blue pigments in the Renaissance from the fourteenth and fifteenth Centuries and afterwards: cobalt blues, zaffre, smalts, and their mineralogical differentiation. Blue glasses and glazes. Analytical distinction which identifies source areas for pigment raw materials and ores.

Keywords Cobalt blue · Zaffre · Smalt · Blue glass · Blue glaze · Renaissance blue pigments

8.1 Context

The periods of the Quattrocento and Cinquecento in Italy and the Renaissance in France, traditionally beginning after the great caesura of the Black Death (ca. 1350 which saw the reduction of more than 30% of the European population) and ending at the beginning of the seventeenth century, corresponds to a second technical revolution in Europe, the first having been that initiated by the flowering of numerous monasteries in the twelfth century. This second period is characterized by a re-appropriation of the Graeco-Roman heritage combined with a new tool for disseminating information (Gutenberg's printing press), contributing to the development of technical, even scientific practices and to a certain globalization with the great discoveries which mark the turning point from the fifteenth to the sixteenth century. It is significant that the great artists of the time were also technicians, engineers and heads of teams with many companions. If the first revolution impact on metallurgy is rather well documented [39], its impact on the ceramic and various Arts-of-Fire is quite weak if we exclude the development of stained glass production compared to the innovations of the fifteenth century and sixteenth centuries, attested in particular by the famous books of Picolpasso for pottery [22, 86] and Neri [80] for glass, and the generalization of production sites for earthenware and glassware for

H. G. M. Edwards, P. Colomban, *Blue by Fire: A Marker of the Technical History of Glass and Ceramics*, Cultural Heritage Science,
https://doi.org/10.1007/978-3-031-77122-4_8

Table 8.1 First significant European works concerning the creation of coloured decorations published from the end of the Middle-Ages to the Renaissance

Authors [Ref.]	Title	Date
Theophilus [56, 109]	Schedula diversarum artius	11th c.
Anonymous [22, 29]	El Lapidario del Rey Alphonso X	1279
J. d'Outremeuse [22, 29]	Le quatrième livre du 'Trésorier de philosophie naturelle des pierres précieuses'	ca. 1370
A. Al-Qâsem Kâshâni	Arâyes al-javâher	14th c.
A. de Pise [51]	L'art du vitrail	1400
G. Rosetti [22]	Plictho de l'arte de tentori	1548
G. Bauer (G. Agricola)[22, 29]	De re metallica	1556
C. Picolpasso [22]	Li tre libri dell'arte del vasaio	1557
C.A. Cennini [22]	Il libro dell'arte	16th C.
B. Perez de Vargas [22]	De re metallica	1568
A. Neri [22, 80]	De arte vitraria	1612

a mainly urban clientele. Table 8.1 lists the main works concerning pigments usable in painting and Arts-of-Fire, [22, 23, 29]. A more complete list was recently published [42]. Indeed the development of printing multiplied the sources, which were much rarer when manuscripts had to be copied manually to be distributed. These sources must, however, be considered with caution, the terms used to designate the ingredients being often difficult to associate with modern names and concepts and these documents being written by scholars and not true practitioners.

As indicated in Chap. 2, blue minerals are rare and only two minerals can really be used as a pigment, namely azurite, a hydroxylated copper carbonate and lazurite (a feldspathoid, i.e. a sodium and calcium aluminosilicate, the main constituent of the rock called lapis lazuli). Only the last one is stable enough to be used in fire art. We will focus on glasses that are coloured blue as well as blue pigments used in painting (in the broad sense, from illumination, through frescoes and wall paintings, easel painting and fire arts (enamels on different substrates)) from the end of the Middle Ages to the present period. We will see that two other technical revolutions will concern coloring agents, that of the second half of the nineteenth century with the development of innovative commercial and industrial production of inorganic and organic colouring agents and the second half of the twentieth century with a new generation of inorganic pigments using rare earths [24, 43] and organic paints with a polymeric matrix, which lie outside the scope of this work [19, 58, 64, 65, 96]. This will be the subject of Chap. 8. The presentation is made in chronological order of use in general (and not only of use in the Arts-of-Fire), namely, lapis lazuli, glasses coloured by cobalt Co^{2+} ions, i.e. *saffre* and smalt [41, 74, 106, 109], synthetic lapis lazuli (ultramarine) and pigments from the nineteenth and twentieth centuries ("modern pigments", cobalt-based, or cobalt-free). More exhaustive lists of pigments, whatever their uses and compositions, have been published [4, 11, 22, 23, 29, 42].

8.2 Lapis Lazuli

8.2.1 Uses

The use of lapis lazuli as a sculpted or polished stone dates back to at least 6000 BCE as evidenced by objects from excavations in the Indus Valley and Sumerian cities, the most beautiful objects coming from Ur (2500 BCE); Egyptian objects only date back to the 18th Dynasty (~1500 BCE) [22, 23, 29].

As a pigment for wall paintings, lapis lazuli is observed throughout the Silk Road in numerous Buddhist paintings and frescoes, the most famous being those in the Mogao caves (near the Dunhuang oasis, Taklamakan desert, Gansu Province, China) dating from the four to the fourteenth century [83, 98, 117], those from the nineth century presenting for the first time representations of landscapes in perspective, that is to say several centuries before its 'so-called' invention of the Quattrocento in Italy. Manuscripts found in these caves also use lapis lazuli in their illustrations [104].

The identification of lapis lazuli on painted pottery from the Northern Wei Dynasty (fourth century) [69] shows its wide use on other supports throughout the Silk Road on the Asian side. On the western side, hundreds of m^2 of wall paintings are found from Armenian and Georgian churches (tenth century, Tokali, Georgia) and (Agtamar and Tatev, Armenia) [110] as well as later in Moldavia, Italy and then France [22, 23]. In France we can cite, for example, the frescoes of the abbey of Saint-Savin (Vienne, France, twelfth century) or the church of Berzé-la-Ville (thirteenth century, Saône-et-Loire, France) which use the lapis lazuli [95], but to a lesser extent than on the Silk Road. We can think that the availability and the cost were a direct function of the distance from the *Sar-e-Sang* mine, a mining place already described by Marco Polo in Badakhshan Mountains of the Pandjchir valley (Hindou Kouch, Afghanistan, see ch 2) [22, 23].

The study of microsamples of pigments from illuminated manuscripts by Coupry was one of the first applications of Raman microspectroscopy which demonstrated the use of this pigment in competition with other blue pigments [30–33]. This determination of the pigment palette in association with stylistic criteria contributes to the identification of illuminated manuscript production workshops. The exceptional sensitivity of resonance Raman from S_2^- and S_3^- polysulphide ions, the chromophores of lazurite [2, 17, 18, 21], explains the fact that many works have since used Raman analysis to detect the presence of lapis lazuli (we will discuss later the possible confusion with synthetic ultramarine and techniques to differentiate natural and synthetic varieties) in manuscript illustrations and in painting. After the pioneering work of Coupry [30–33] and Guineau [53–55], then Aceto et al. [1], the book by Panayoyova et al. [85] gives a recent state of the art concerning the study of pigments concerning European illuminated manuscripts. Other types of painting include walls [8], paintings on wood or canvas [34, 40, 105], tempera [79] and paper [113].

Episodes complicating or blocking trade (end of the Sassanids (seventh century), Crusades (eleventh-thirteenth centuries), etc.) and those favouring trade (Pax Mongolica (thirteenth-fourteenth centuries; Part of Persia belonged to the Yuan Empire), and the Ottoman Empire (>thirteenth century)) modulate access to lapis lazuli and contribute towards its cost. However, it must be considered that the intended uses require very different qualities of pigment, beautiful pieces without cracks for jewellery, and powder for frescoes and illumination. For certain applications such as enamelling, the presence of 'white' phases contributes to the quality of the colour. The use of lapis lazuli in painting has been the subject of numerous studies [23].

For a very long time, the use of lapis lazuli as a glass colouring agent was not even considered, probably due to the fact that archaeometry was—and sometimes still is—dominated by elemental analysis. Lazurite being an aluminosilicate like the vitreous matrix and the chromophore a quasi-impurity (polysulphide ions) the elemental analysis techniques were blind to its presence. Some authors had noted their impossibility of detecting cobalt in Islamic blue glasses [50] but implicitly attributed this to the difficulty in detecting cobalt. The coloring power of cobalt ions is very strong: 0.5 wt% of Co is enough to strongly colour a glass [29]. Furthermore, in XRF the main peak of cobalt $K\alpha$ is almost identical to the secondary peak Kß of iron; iron is always present and its peaks are more intrinsically intense.

In 1997 Clark, Curli and Laganara [17] analyzed by Raman microspectroscopy fragments of medieval glass and pottery (thirteenth century) in Frederickian sites in southern Italy (Lucera, Melfi, Castel del Monte, the residences of Emperor Frederick II, King of Sicily) and observed the intense signal (because of resonance Raman enhancement) of the S_n chromophores characteristic of lapis lazuli [13, 71] and have been widely studied by the main author. Then, in 2003 Colomban studied contemporary Lajvardina pottery (Iran) but of common manufacture while the Frederickian fragments corresponded to refined objects [21]. More surprisingly, the enamel of Lajvardina is colored by both cobalt ions and lapis lazuli, which means that attributions that were made based on simple elemental analysis are certainly erroneous. It took a few more years for the importance of lapis lazuli as a colouring agent for glasses and enamels to be recognized by archaeometricians who had read and practiced Raman spectroscopy little before the 2010s. Since then it has been recognised that the use of lapis lazuli grains, alone or in association with cobalt as a blue or green colouring agent (in association with a yellow pigment), is well characterized for glasses (Lagides, Begram treasure, Afghanistan) [12] and Ptolemaic pottery [15] and Mamluke enameled glasses [28]. Note that Egypt was the place of production for the last two periods.

8.2.2 Characterization

Raman spectroscopy is the only effective method for detecting lapis lazuli S_n chromophores, the spectrum being in resonance for a wide range of laser excitation wavelengths, from green to red [14, 21]. With these excitations the Raman signature of the host network is not visible. To observe it you need UV or deep blue or NIR excitation. This makes it possible to unambiguously differentiate lapis lazuli from synthetic ultramarine. Other methods have been proposed such as infrared detection of the CO_2 band, trapped in the sites or pores of the feldspathoid [103] host or the relative presence of other minerals characteristic of the lapis lazuli rock of the site considered (calcite, dolomite, scapolite, pyrite, diopside, sodalite, wollastonite, etc.) as well as other spectroscopic criteria [3, 21, 45, 70]. The identification of traces of lapis lazuli on the teeth of deceased victims made it possible to prove the presence of women as manuscript illustrators in scriptoria [92].

8.3 Blue Glasses

8.3.1 Coloration Due to Electronic Transition

As indicated in the previous chapter, cobalt ions dissolved in silicate glass were the first colouring agents to produce a synthetic material imitating precious stones, first an opaque glass paste (3000 BCE) then translucent glass (1500 BCE –2000 BCE) and finally transparent (200–300 BCE). These dates of course vary greatly depending on the location and depending on new discoveries and precision made in the dating. The characterizations of Co^{2+} and Co^{3+} ions in vitreous silicate matrices and in aluminosilicate crystalline phases show absorptions at respectively 524, 596 and 650 nm (Co^{2+}, tetrahedral site) and 270, 344, 470, 506, 544, 600, 636, 644 and 1196 nm (Co^{3+}, octahedral site) [62, 100, 115, 116].

The first glass objects were moulded or worked in a viscous state [94]. The Romans set up quasi-industrial production with the primary production of large ingots (several meters) in the Levant and shaping by blowing and blow-moulding in a multitude of workshops. Studies of Roman wrecks have shown the evidence for the widespread recycling of broken glass, particularly blue [49]. In fact, glass recycling is much older (1500 BCE) as proven by other shipwrecks [89, 90] Blue, turquoise, jade green, red, amber and white are the main colours. The oldest enamelled glass objects are Roman.

In the state of current knowledge the use of cobalt in bulk glass material goes back to the oldest Egyptian Dynasties and is well established with the early work of the French chemist Alexandre Brongniart, the former director of the Sèvres Factory (from 1800 to 1847) and the founder of the first museum dedicated to pottery and glass—the National Museum of Ceramics—at the very beginning of the nineteenth-Century [9, 10]. The evidence of cobalt use in blue Mycenaean (Greece) [35, 121] and Mesopotamian Late Bronze Age glasses was also asserted [118, 119] and the long-distance trade activity during this period was in evidenced [91]. Large glass production stopped for a few hundred years around 1200 B.C., a period of several wars accompanied by a trade disruption [61]. Similar interruptions occurred with the decline of the Roman Empire.

In Asia, glass seems to be a little less old in the state of knowledge [16, 52, 60]. In China, South India the history of vitreous deposits on substrate, glass, metal or ceramic is different depending on the support used. In the Western world, the first enamels on metal date back to at least 1500 BCE in Cyprus whereas an older written mention of an enamelled shield is found in the Iliad (eighteenth century BCE). From various areas, famous artefacts include an eleventh century BCE enamelled bronze wine ladle from the early Western Zhou dynasty (British Museum) or numerous Celtic enamels on brass, bronze, gold, and iron made from at least the fourth Century BCE with blue, red, green, yellow, and black hues.

Then, cobalt was identified in various blue coloured artefacts such as glazed pottery, glass objects, stained and lustre glass produced from Europe to Asia by many scholars [73, 87, 88, 72, 35, 46, 57, 59, 78, 82, 93, 101, 102, 111, 112].

More recently, we can also cite the bricks of the famous Ishtar gate (Mesopotamia, Babylon, ca. 575 BCE) glazed in blue colour and similar blue glazed bricks at Susa in Iran during the first millennium BCE [46]. Indeed, it seems that the oldest glass coloured in blue by cobalt dates to the Late Bronze Age Egypt (~sixteenth- to the eleventh-Century BCE) [32, 48] but the number of analysed artefacts remains limited, especially regarding the masterpieces coloured in blue Prehistoric (according to [36]) Egyptian cobalt by product of Erzgebirge silver, false Oasis Dakhla alums rich in Co ($\sim > 0.4$ wt%) [59].

8.3.2 Coloration Due to Light Scattering

The pale blue colour can be obtained by creating heterogeneity on the micron or submicron scale. When the heterogeneity size (pore, bubble, particle, phase) is smaller than the wavelength of the incident visible light (380–780 nm), the opalescence phenomenon occurs. When the particle size is less than 100 nm, the Rayleigh scattering phenomenon occurs, forming a blue colour. The intensity of scattering light is related to the intensity of incident light and the volume of particles [99, 122]. There are many examples of liquid-liquid phase-separated glazes in the high-temperature glazes of ancient Chinese stoneware and porcelain. Visualization of heterogeneity by SEM or TEM is often difficult without chemical attack to reveal the different phases. For example, fancy glazes and Tenmoku glazes with obvious opacifying and opalescence effects produced by Huangdao kiln, Wuzhou kiln and Changsha kiln in the Tang Dynasty, Jun kiln in the Song Dynasty, Jian kiln and Jizhou kiln in the Song and Yuan Dynasties. Especially, the Jun kiln of the Song Dynasty is famous for its wonderful and colourful flow patterns of purple, red, blue, and white produced during high-temperature kiln-change. The earliest high-temperature kiln-changed glazed porcelain in China could be traced back to Lushan speckle porcelain known as the "Jun porcelain of the Tang Dynasty" [44, 66, 67, 68, 108].

8.4 *Saffre* and *Smalt*

Saffre or *zaffre* and *smalt*, two stages of the cobalt-rich colouring material used in Europe initially as a pigment, are sometimes confused. In fact, the metal cobalt was only identified by G. Brandt in Sweden in 1742 or 1745 and even today the extraction and refining of cobalt remains complex and expensive. G. Brandt was the first chemist to show that cobalt is a "semi-metal". He proved that the blue colour communicated to the glass is due neither to bismuth nor to iron, but to cobalt itself [6, 7]. In 1957, Morral gave a definition of his two products, *saffre* and *smalt* [76]. According to Morral, in 1470, in Saxony (the Saxon site of Erzgebirge, literally metal-rich mountains, now the Southeast of Germany), it was discovered that the brittle lumps obtained during the extraction of silver could be ground to "*saffre*," which was sold to the pottery and glass industries of Europe [41, 51, 74, 106, 109]. In 1540, C. Schlirer, in Bohemia (the other side of Erzgebirge, now the Czech Republic), produced the glassy silicate of cobalt and potassium by mixing *saffre* with potash glass (i.e. using plant ash as a flux) which, after melting and granulation in water (technically fritting), resulted in "smelt" (or "*smalt*"). The term "*smalt*" derives from the proto-German word *smaltian* which means to smelt. After Valmont de Bomare [114] the cobalt-rich matter is mixed with variable quantities of quartz, sand and/or flint and potash and then melted again to obtain different grades of *smalt* which is a potash glass coloured in blue. It would seem according to Delamarre [38] that the first identified mention of the product providing what we call cobalt is found in the memoir of Antoine de Pisa (ca. 1395) under the name of *chafarone* and a German origin is mentioned as well as its high cost. Under the name *caffaranum* it is mentioned in a Venetian text from the fourteenth century as coming from German merchants [38]. Note that in the text of Theophilus from the eleventh century [56] the blue glass is only obtained by the re-use of Roman blue glass and that on the other hand Abbot Suger, project manager of the construction of the Saint-Denis royal burial basilica close to Paris, France, [75] mentions in 1142 the use of Saxon *saffre* for stained glass windows [38]. Indeed, the production of medieval stained glass using few colours, mainly blue, red and green, required large quantities of cobalt for the first time [25, 26, 27]. Then Morral cites several works on the processing of cobalt ores: Pirotechnia by Biringuccio (1540), the famous De Re Metallica by G. Bauer alias Agricola (1556). He notes that in an alchemical text by C.E. Gellcrest, the symbol "K" is used for cobalt. In 1792 F. Kapffauth produced a work entitled "*Beitrage zur Geschichte des Kobolts, Kobolt bergbanes in der Blaufarben werke*" i.e., "Contributions to the History of Cobalt, Cobalt Mining in Blue-Dye Manufacturing". This is confirmed in the study of the history of the production and marketing of *smalt* and in particular on the almost monopolistic control of this trade by the Dutch Merchants [77]. According to Morral, *smalt* contains around 6% wt cobalt and Mimoso indicates that the *zaffre* contains ~70% wt of CoO [cited by [29]. To date, microanalyses carried out on *smalt* grains identified in layers of paint showed that the content was/could be much higher than 6%wt, with the following compositions: (wt.%): 66–72 SiO_2, 10–21 K_2O, 2–18 CoO and 0–8 As_2O_3 [84].

These differences in cobalt content are in agreement with the work of Valmont de Bomare [114] which indicates that a large number of qualities of *smalt* (grade) of different costs are available.

The first stage of processing European ores is roasting which leads to the sublimation of part of the arsenic which re-condenses in the chimney and is recovered, which makes this stage dangerous. Let us recall that European deposits, of the vein type [21, 29, 63] are based on arsenide and cobalt sulphides which gives an As/Co ratio which can reach 4. According to some authors the addition of sand is made to obtain zaffre [38, 80]. The texts cited by Delamare indicate the production of bismuth around 1500 in Schneeberg (Saxony). Around 1580 (from a treatise by Lazarus Ercker, the superintendent of mines of the Holy Germanic Empire and the Kingdom of Bohemia, residing in Annaberg) a description of the recovery by fusion of bismuth contained in the ores was published, and we can read: "*The nodular residue obtained by both smelting processes [in the open air or in a furnace] is used for the production of large quantities of blue pigment; it is used to colour glass blue. It is sold and shipped to glassworks everywhere.* ».

If it is *saffre*, as its use by glassmakers suggests, this product is not blue, but gray but Ercker speaks specifically of "blue pigment". It is therefore likely that he confused the two products under the same name of *saffre*, which would tend to suggest the existence of *smalt* at this time. We can conclude that the use of *smalt* by painters was common from 1550. This was the time when the Turks invaded Hungary, which stopped exports of azurite of which this country was then a major producer. Becoming rare and expensive, azurite is then replaced by *smalt*. This prosperous period for *smalt* production continued during the seventeenth century. *Smalt* was used much earlier by Flemish and German painters [120]. Among their works, the first recognized occurrence of *smalt* was for a long time the Entombment by the Flemish painter Dieric Bouts the Elder, dated around 1450, and kept at the National Gallery in London. But this work was recently dethroned in this pioneering role by a Carrying the Cross painted around 1400 by an unidentified German painter, a painting which is kept in the Old Pinakothek in Munich. Another painting shares this privilege with him, but this time, it is a wall-painting mural, also painted around 1400, located in the church of St Johann and St Salvator in Donaustauf (Bavaria).

Given the significant number of paintings and decorations studied so far, this date of 1400 seems credible for the appearance of *smalt*. If we compare it to that determined above for the appearance of *saffre,* Delamarre notes that, far from being concomitant with it, *smalt* would have appeared in Western Europe at least two centuries after it. From the sixteenth century, this temporal gap completely disappeared from collective memory, as we can see in the various technical treatises in which it is difficult to separate *saffre* and *smalt*. In summary It therefore seems that the appearance of *saffre*, then of *smalt*, took place in Saxony or Bohemia, in any case in the immediate vicinity of silver and bismuth mines. And certainly, there is no shortage of texts evoking the creation of *saffre* and *smalt* factories in this region of the Erzgebirge.

Historical data has been collected by F. Delamare concerning several apparently Saxon artisans [37, 38]. Christian Lehmann, who was a priest in Scheibenberg, where he died in 1688, had carried out research into the history of the Misnie mines. He had written notes at the start of the 30 Years' War. According to these, the colour mills were, at his time, around a hundred years old, which would place their establishment in the years 1540–1560. Prepared for potters and glass painters, this material is *saffre*. Then mills appear for the grinding of blue glass. These colours are now *smalt*. Eight colour grinding mills of the same type were soon built in Holland. They received the roasted cobalt ore from Schneeberg in barrels. And the Dutch were soon more skilled than the Saxons in the art of preparing, and especially grinding, this colour. The Elector Johann Georg had two colour manufacturers sought out in Holland, and offered them a thousand florins so that they could improve the state of this art in his States. He made this decision following a remark from the Schneeberg workers: during roasting, most of the cobalt contained in the ore was lost. Before long, other colour mills were built around Schneeberg. Hans Burghard, merchant and chamberlain in Schneeberg, built a mill which replaced the eleven mills of Platten. Paul Nordhoff, a Frisian, who lived in the Zwitter-mill, made numerous experiments to improve the colour. He became ruined there and had to leave this place where he had been employed for 10 years in the colour factory. He settled in Annaberg, where he created in 1649 with funds from a Leipzig merchant, a colour factory which he directed. Thus, Annaberg cobalt came to be used. Consumption of this colour decreased over time. In 1659, while there were colour mills in all the towns surrounding the mines, he had more than 8000 quintals left in stock.

The determining role of the Dutch is well explained by Mimoso [77] and Delamare [37, 38]. They let the Saxons process cobalt ore at a great expense of wood and with the emission of toxic arsenical fumes (*saffre*), and concentrated their efforts on the manufacture of materials with high added value, in this case, the different grades of *smalt*. The cobalt veins of Freiberg began to be exploited in the twelfth century, and the *saffre* produced was sold and exported. Did the appearance of *saffre* and then *smalt* in Saxony result from a transfer of technique from Byzantium, or perhaps the Normans of Sicily or Venice which recovered Byzantine glassmakers via the Crusades?

Indeed, established in Torcello from the first half of the seventh century, organized into specialized corporations in 1271 and settled in Murano in 1291, Venetian artisans produced blue glass [107]). Mimoso notes that Italian workers work in the Saxon mines. For dark blues, the Venetians probably imported their cobalt from a non-European source located, for example in the Byzantine Empire, or in Kashan, in Persia, as long as that of Saxony was not available and competitive. At the same time, Norman glassmakers produced remarkable enamelled glass where the blue and green were obtained either with cobalt, or with lapis lazuli, or both [13]. The Venetian colour merchants, the *vendecolori*, gave moreover a wide variety of names for the product: *caffaranum, chafarone, cafarone, gafaro, saferina, saffra, zaffera, zaffra, zafro,* and *zaffaro*.

Did Byzantine glassmakers make cobalt-tinted glass (soda-lime glass), or did they just recover it? A first answer can be provided by observing the Byzantine mosaics with the naked eye, where the cobalt blues, very present, are very well distinguished from the turquoise blues which owe their colour to copper. Among the oldest (they date from the middle of the fifth century), and the most spectacular, we can cite those adorning the vaults and walls of the "mausoleum of *Galla Placidia*" in Ravenna. The cobalt blue tesserae occupy a total surface area of nearly two hundred square meters. Such abundance testifies, not to the recovery of antique glass, but to the production of cobalt-coloured glass intended for the manufacture of tesserae and, ipso facto, to access to one or more sources of this metal. According to Delamare [37] the study of various pictorial layers showed that Byzantine painters also used blue glass tinted with cobalt, ground into powder, therefore a *smalt* or *zaffre*, as pigment. The analysis of blue tesserae from Hierapolis, Turkiye, 6th–ninth century [81] soda-lime with natron or mixed natron-potash uses cobalt without arsenic, therefore not Saxon. The clearer, mixed may have been made by recycling. The darker ones don't. Here we have another proof of the use by the Byzantines of an ore other than that obtained from the Saxon mines. In different cases the thread passes through Byzantium which must have inherited Roman know-how and perhaps also Ptolemaic know-how.

8.5 Synthetic Ultramarine

Jean Baptiste Guimet is considered by most historians as the inventor of the ultramarine blue pigment synthesis process in 1828. For that, a mixture of metakaolin, sodium carbonate, sulfur and a reducer (carbon?) is fired in several steps up to 800 ° C, first in the absence of air, followed by an oxidation step at a lower temperature. Pink and green pigments were synthesized. In fact it seems that the invention is actually older and dates from the 1770s with the work of Joseph de Montgolfier, one of the inventors of hot air balloons, the Montgolfières, the first aircraft [20]. Now calcination in the absence of air of a mixture comprised of zeolite, sodium sulphide and sulphur, the ratios produced S/zeolite >5.6%lead blue and < 2% lead green. There are cheaper inorganic alternatives such as Prussian Blue ($Fe_7(CN)_{18}(H_2O)_x$).

8.6 Nineteenth Century Innovation: Synthetic Blue Spinels

The second half of the nineteenth century saw the rise of the chemical industry in Europe and associated teaching and numerous publications, books and patents concerning in particular new colored compositions [47, 48, 97]. Concerning thermally stable blue pigments, the period sees the exploitation of the possibilities offered by the spinel structure. The structural reference is $MgAl_2O_4$; it is characterized by a compact stack of layers of oxygen ions defining tetrahedral and octahedral sites for the 2^+ and 3^+ ions. The distribution can be as in the structural reference (called direct spinel) but the same site can be occupied by 2^+ and 3^+ ions which can lead to oxygen vacancies and therefore non-stoichiometry, for example for Fe_3O_4. Oxygen vacancies can induce protonation which confers high reactivity but a risk of thermal decomposition. The main bivalent ions are Mg, Fe, Zn, Ni, Mn, Co and the trivalent ions Fe, Cr, Mn and Co; cobalt being essential to obtain blue color. This type of pigment has already been widely used from natural products (ferrites, chromites) by Ottoman potters, the dominant presence of one of the ions leading to the black colour to separate the coloured areas. These pigments have been studied in detail [5].

8.7 Twentieth Century Blue Pigments

From the 1970s a technical revolution developed in the field of pigments for glass and ceramics. For reasons of simplification of the manufacturing process and regulatory constraints concerning the chemical stability of glasses and enamels, the possibilities of coloring thermochemically stable matrices, namely zircon (ZrO_2), zircon ($ZrSiO_4$), cassiterite (SnO_2), rutile (TiO_2), garnets and spinels by minor substitution with transition elements (V, Mn, Nb mainly) or rare earths (Pr mainly) [24, 43].

References

1. Aceto, M., Agostino, A., Fenoglio, G., Baraldi, P., Zannini, P., Hofmann, C., Gamillscheg, E.: First analytical evidences of precious colourants on Mediterranean illuminated manuscripts, *Spectrochim.* Acta Part A-Mol. & Biomol. Spectrosc. **95**, 235–245 (2012). https://doi.org/10.1016/j.saa.2012.04.103
2. Ali, E.M.A., Edwards, H.G.M.: Analytical Raman spectroscopy in a forensic art context: the non-destructive discrimination of genuine and fake lapis lazuli, *Spectrochim.* Acta Part A-Mol. & Biomol. Spectrosc. **121**, 415–419 (2014). https://doi.org/10.1016/j.saa.2013.11.005
3. Bacci, M., Cucci, C., Del Federico, E., Ienco, A., Jerschow, A., Newman, J.M., Picollo, M.: An integrated spectroscopic approach for the identification of what distinguishes Afghan lapis lazuli from others. Vib. Spectrosc. **49**(1), 80–83 (2009)
4. Bell, I.M., Clark, R.J.H., Gibbs, P.J.: Raman spectroscopic library of natural an synthetic pigments (pre-~1850 AD). Spectrochim, Acta Part A. **53**, 2159–2179 (1997). https://doi.org/10.1016/S1386-1425(97)00140-6
5. Bouchard, M., Gambardella, A.: Raman microscopy study of synthetic cobalt blue spinels used in the field of art. J. Raman Spectrosc. **41**(11), 1477–1485 (2010). https://doi.org/10.1002/jrs.2645
6. Brandt, G.: Cobalti nova species examinata et descripta. Acta Societatis regiæ scientiarum upsaliensis Stockholm. **3**, 33–41 (1742a)
7. Brandt, G.: Dissertatio de Semi-Metallis, Acta literaria et scientiarum Sueciæ, vol. IV., 1735, pp. 1–12, Uppsala (1742b)
8. Brysbaert, A.: Lapis lazuli in an enigmatic 'Purple' pigment from a thirteenth-century BC Greek wall painting. Stud. Conserv. **51**(4), 252–266 (2006). https://www.academia.edu/49469935/Lapis_lazuli_in_an_enigmatic_purple_pigment_from_a_thirteenth_century_BC_greek_wall_painting?uc-g-sw=14823591. Accessed 4 Mar 2024
9. Brongniart, A.: Mémoire sur la Peinture sur Verre. Imprimerie Sellingue, Paris (1829)
10. Brongniart, A.: Traité des Arts Céramiques ou des Poteries Considérées Dans Leur Histoire, leur Pratique et leur Théorie, 3rd edn. Libraire Faculté de Médecine, Paris (1887) Two volumes

11. Burgio, L., Clark, R.J.H.: Library of FT-Raman spectra of pigments, minerals, pigment media and varnishes, and supplement to existing library of Raman spectra of pigments with visible excitation. Spectrochim. Acta Part A. **57**(7), 1491–1521 (2001). https://doi.org/10.1016/S1386-1425(00)00495-9
12. Caggiani, M.C., Colomban, P., Valotteau, C., Mangone, A., Cambon, P.: Mobile Raman spectroscopy analysis of ancient enamelled glass masterpieces. Anal. Meth. **5**(17), 4345–4354 (2013). https://doi.org/10.1039/C3AY40648B
13. Caggiani, M.C., Acquafredda, P., Colomban, P., Mangone, A.: The source of blue colour of archaeological glass and glazes: the Raman spectroscopy/SEM-EDS answers. J. Raman Spectrosc. **45**(11–12), 1251–1259 (2014). https://doi.org/10.1002/jrs.4492
14. Caggiani, M.C., Colomban, P.: Raman microspectroscopy for cultural heritage studies. In: Sabbatini, L., van der Werf, I.D. (eds.) Chemical Analysis in Cultural Heritage. De Gruyter, Berlin (2018). https://doi.org/10.1515/9783110457537
15. Catalano, I.M., Genga, A., Laganara, C., Laviano, R., Mangone, A., Marano, D., Traini, A.: Lapis lazuli usage for blue decoration of polychrome painted glazed pottery: a recurrent technology during the middle ages in Apulia (Southern Italy). J. Archaeol. Sci. **34**, 503–511 (2007). https://doi.org/10.1016/j.jas.2006.05.010
16. Carter, A.K.: The production and exchange of glass and stone beads in Southeast Asia from 500 BCE to the early second millennium CE: an assessment of the work of Peter Francis in light of recent research. Archaeol. Res. Asia. June 1. **6**, 16–29 (2016a)
17. Clark, R.J.H., Dines, T.J., Kurmoo, M.: On the nature of the sulfur chromophores in ultramarine blue, green, violet, and pink and of selenium chromophore in ultramarine selenium—characterization of radical-anions by electronic and resonance Raman-spectroscopy and the determination of their excited state geometries. Inorg. Chem. **22**(19), 2766–2772 (1983). https://doi.org/10.1021/ic00161a024
18. Clark, R.J.H.: Raman microscopy – application to the identification of pigments on medieval manuscripts. Chem. Soc. Rev. **24**(3), 187–193 (1995). https://doi.org/10.1039/CS9952400187
19. Claybourn, M., Agbenyega, J.K., Hendra, P.J., Hellis, G.: Fourier-Raman spectroscopy in the study of paints. In: Urban, M.W., Craver, C.D. (eds.) Structure-Property Relations in Polymers: Spectroscopy and Performance., Adv. Chem. Series, vol. 236, pp. 443–482 (1993)
20. Clement, P.-L.: Les Montgolfières, 2nd edn, p. 20. Tardy, Paris (1982)
21. Colomban, P.: Lapis lazuli as unexpected blue pigment in Iranian Lâjvardina ceramics. J. Raman Spectrosc. **34**(6), 420–423 (2003). https://doi.org/10.1002/jrs.1014
22. Colomban, P.: The destructive/non-destructive identification of enameled pottery, glass artifacts and associated pigments—a brief overview. Arts. **2**, 77–110 (2013a). https://doi.org/10.3390/arts2030077
23. Colomban, P.: Rocks as blue, green and black pigments/dyes of glazed pottery and enamelled glass artefacts–a review. Eur. J. Mineral. **25**(5), 863–879 (2013b). https://doi.org/10.1127/0935-1221/2013/0025-2305
24. Colomban, P.: Full spectral range Raman signatures related to changes in enameling technologies from the 18th to the 20th century: guidelines, effectiveness and limitations of the Raman analysis. Materials. **15**, 3158 (2022). https://doi.org/10.3390/ma15093158
25. Colomban, P., Etcheverry, M.-P., Asquier, M., Bounichou, M., Tournié, A.: Raman identification of ancient stained glasses and their degree of deterioration. J. Raman Spectrosc. **37**, 614–626 (2006). https://doi.org/10.1002/jrs.1495
26. Colomban, P., Tournié, A.: On-site Raman identification and dating of ancient/modern stained glasses at the Sainte-Chapelle, Paris. J. Cult. Herit. **8**, 242–256 (2007). https://doi.org/10.1016/j.culher.2007.04.002
27. Colomban, P., Tournié, A., Ricciardi, P.: Raman spectroscopy of copper nanoparticle-containing glass matrices: ancient red stained-glass windows. J. Raman Spectrosc. **40**, 1949–1955 (2009). https://doi.org/10.1002/jrs.2345
28. Colomban, P., Tournié, A., Caggiani, M.C., Paris, C.: Pigments and enamelling/gilding technology of Mamluk mosque lamps and bottle. J. Raman Spectrosc. **43**(12), 975–984 (2012)

29. Colomban, P., Kırmızı, B., Simsek Franci, G.: Cobalt and associated impurities in blue (and green) glass, glaze and enamel: relationships between raw materials, processing, composition, phases and international trade. Fortschr. Mineral. **11**, 633 (2021). https://doi.org/10.3390/min11060633
30. Coupry, C.: Les oeuvres d'art et les artistes à l'épreuve du laser. Revue du Palais de La Déccouverte. **196**, 15–36 (1992)
31. Coupr, C.: Les pigments utilisés pour l'enluminure à Fécamp, XI-XIIe siècle. In: Bouet, P., Dosdat, M. (eds.) Manuscrits et enluminures dans le monde normand (X-XVe siècles), pp. 69–79. Presses Universitaires de Caen, Caen (1999)
32. Coupry, C., Brissaud, D.: Applications in art, jewelry and forensic science. In: Turrell, G., Corset, J. (eds.) Raman Microscopy, pp. 421–453. Academic (1996) ISBN 0-12-189690-0
33. Coupry, C., Gousset, M.T.: Les pigments des enluminures. Pour la Science. **293**, 12–18 (2002). https://www.pourlascience.fr/sr/article-fond/les-pigments-des-enluminures-4657.php. Accessed 4 Mar 2024
34. Daniilia, S., Andrikopoulos, K.S., Sotiropoulou, S., Karapanagiotis, I.: Analytical study into El Greco's baptism of Christ: clues to the genius of his palette. Appl. Phys. A – Mater. Sci. Process. **90**(3), 565–575 (2008). https://doi.org/10.1007/s00339-007-4328-7
35. Dayton, J.E.: Geological evidence for the discovery of cobalt blue glass in Mycenaean times as a by-product of silver smelting in the Schneeberg area of the bohemian Erzgebirge. In: Proceedings of the Actes du XXe Symposium International d'Archéométrie, Paris (1981) 26–29 March 1980; Volume III
36. Dayton, J., Bowles, J., Shepperd, C.: Egyptian blue or kyanos and the problem of cobalt. In: Minerals Metals Glazing & Man, pp. 451–462, Londres (1978)
37. Delamare, F.: Saffre, smalt, bleu d'esmail et azur. In: *Bleus en poudres, 5 000 ans d'innovations techniques*. Ecole des Mines, Paris (2008) ISBN-10. 2911762827
38. Delamare, F.: Aux origines des bleus de cobalt: les débuts de la fabrication du saffre et du smalt en Europe occidentale. Compt. rendus séances Acad. Inscriptions et Belles-Lettres. **153**(1), 297–315 (2009). https://doi.org/10.3406/crai.2009.92472
39. Dillmann, P., Pérez, L., Verna, C.: Les aciers avant Bessemer. Presses universitaires du Midi, Toulouse (2011). https://doi.org/10.4000/books.pumi.37523
40. Duran, A., Siguenza, M.B., Franquelo, M.L., Imenez de Haro, M.C., Justo, A., Perez-Rodriguez, J.L.: Murillo's paintings revealed by spectroscopic techniques and dedicated laboratory-made micro X-ray diffraction. Anal. Chim. Acta. **671**(1–2), 1–8 (2010). https://doi.org/10.1016/j.aca.2010.05.004
41. Ercker, L.: Beschreibung allerfürnemisten mineralischen Ertzt und Bergk wercksarten. In: Sisco, A.G., Smith, C.S. (eds.) Sigmundt Feyrabend, Frankfurt-am-Main, 1580. Traduction Anglaise Treatise on Ores and Assaying, 2nd edn, p. 278. University of Chicago Press., 1951 (1580)
42. Edwards, H.G.M., Vandeneebele, P., Colomban, P.: Pigments, Dyes and Colouring Agents, Ch. 5, in Raman Spectroscopy in Cultural Heritage Preservation, pp. 53–106. Springer, Cham (2023). https://doi.org/10.1007/978-3-031-14379-3_5
43. Eppler, R.A., Eppler, D.R.: Glazes and Glass Coatings. The American Ceramic Society, Westerville (2000) ISBN 9781574980547
44. Fang, Z.J., Chen, Z., Peng, W.C., Shao, C.Y., Zheng, S.P., Hu, L.L., Qiu, J.R., Guan, B.O.: Phase-separation engineering of glass for drastic enhancement of upconversion luminescence. Adv. Opt. Mater. **7**, 1801572 (2019). https://doi.org/10.1002/adom.201801572
45. Favaro, M., Guastoni, A., Maroni, F., Bianchin, S., Gambirasi, A.: Characterization of lapis lazuli and corresponding purified pigments for a provenance study of ultramarine pigments used in works of art. Anal. Bioanal. Chem. **402**(6), 2195–2208 (2012). https://doi.org/10.1007/s00216-011-5645-4
46. Feller, R.L.: Artist's Pigments: a Handbook of their History and Characterization, Volume 1; Oxford University Press: New York. (1986). https://www.nga.gov/research/publications/pdf-library/artists-pigments-vol-1.html. Accessed 4 Mar 2024

47. Fiedler, I., Bayard, M.A.: Cadmium yellows, oranges, and reds. In: Feller, R.L. (ed.) Artists' Pigments A Handbook of their History and Characteristics, vol. 1, pp. 65–108. National Gallery of Art, Washington (1986)
48. Fornacelli, C., Colomban, P., Turbanti Memmi, I.: Toward a Raman/FORS discrimination between art nouveau and contemporary stained glasses from CdSxSe1-x nanoparticles signatures. J. Raman Spectrosc. **46**, 1129–1139 (2015). https://doi.org/10.1002/jrs.4758
49. Foy, D., Jézégou, M.-P.: L'épave antique Ouest-Embiez 1. Archéologia. **407**, 22–33 (2004)
50. Freestone, I.C., Stapleton, C.P.: Composition and Technology of Islamic enamelled glass of the 13–14th centuries (ch 24). In: Ward, R. (ed.) Gilded and Enamelled Glass from the Middle East, pp. 122–128. British Museum Press, London (1998) ISBN-10.0714114782
51. Gallo, D., Sandron, D.: Le Traité d'Antoine de Pise, fac-similé et transcription du manuscrit Le vitrail vers 1400. In: Lautier, C., Sandron, D. (eds.) Corpus Vitrearum-France, Études VIII, pp. 67–78. CTHS, Paris (2008) EAN 9782735506590
52. Gan, F., Li, Q., Henderson, J.: Recent advances in the scientific research on ancient glass and glaze, world century—world scientific, Singapore (2016). https://nottingham-repository.worktribe.com/output/785545
53. Guineau, B.: Microanalysis of painted manuscripts and of colored archeological materials by Raman laser microprobe. J. Forensic Sci. **29**(2), 471–485 (1984). https://doi.org/10.1520/JFS11695J
54. Guineau, B.: L'étude des pigments par le moyen de la microspectrométrie Raman. In: Delamare, F., Hackens, T., Helly, B. (eds.) Datation-caractérisation des peintures parietals et murals PACT-17, pp. 259–294. European University Centre, Ravello (1987)
55. Guineau, B., Coupry, C., Gousset, M.T., Forgerit, J.-P., Vezin, J.: Identification de bleu de lapis-lazuli dans six manuscrits à peintures du XIIe siècle provenant de l'abbaye de Corbie. Scr. Theol. **40**(2), 157–171 (1986). https://www.persee.fr/doc/scrip_0036-9772_1986_num_40_2_1443. Accessed 4 Mar 2024
56. Hendrie, R.: Theophili qui et Rugerus… libri III. de diversis Artibus, seu diversarum artium schedula. An Essay upon various Arts, by Theophilus… forming an encyclopaedia of Christian Art of the eleventh century. Translated, with notes, by R. Hendrie. Lat. & Eng. John Murray. (1847). Available on line https://books.google.fr/books?hl=en&lr=&id=fdIV_jAultwC&oi=fnd&pg=PP8&dq=Schedula+diversarum+artius+theophilus&ots=zp9jYT0NTw&sig=zk4eTZR_YMRdO86nfBGcE04SMYE&redir_esc=y#v=onepage&q=Schedula%20diversarum%20artius%20theophilus&f=false. Accessed 4 Oct 2023
57. Hodgkinson, A.K., Rohrs, S., Muller, K., Reiche, I.: The use of cobalt in 18th dynasty blue glass from Amarna: the results from an on-site analysis using portable XRF technology. Sci. Tech. Archaeol. Res. **5**, 1–17 (2019). https://www.google.fr/url?sa=t&rct=j&q=&esrc=s&source=web&cd=&ved=2ahUKEwjShuXBtdqEAxUQTaQEHfiLBocQFnoECBUQAQ&url=https%3A%2F%2Frefubium.fu-berlin.de%2Fbitstream%2Fhandle%2Ffub188%2F29021%2FThe%2520use%2520of%2520Cobalt%2520in%252018th%2520Dynasty%2520Blue%2520Glass%2520from%2520Amarna%2520the%2520results%2520from%2520an%2520on%2520site%2520analysis%2520using%2520portable%2520XRF%2520technology.pdf%3Fsequence%3D1%26isAllowed%3Dy&usg=AOvVaw1vs3ZG8AaZoD3GrLMutO8t&opi=89978449. Accessed 19 Apr 2024
58. Izzo, F.C., van den Berg, K.J., van Keulen, H., Ferriani, B., Zendri, E.: Modern oil paints–formulations, organic additives and degradation: some case studies. In: van den Berg, K., et al. (eds.) Issues in Contemporary Oil Paint, pp. 75–104. Springer, Cham (2014). https://doi.org/10.1007/978-3-319-10100-2_5
59. Kaczmarczyk, A.: The source of cobalt in ancient Egyptian pigments. In: Olin, J.S., Blackman, M.J. (eds.) Proceedings of the 24th International Archaeometry Symposium, pp. 369–376. Smithsonian Institution Press, Washington, DC (1986)
60. Carter, A.K.: The production and exchange of glass and stone beads in Southeast Asia from 500 BCE to the early second millennium CE: an assessment of the work of Peter Francis in light of recent research. Archaeological Research in Asia. **6**, 16–29 (2016b)

61. Kessler, A.T.: Song Blue and White Porcelain on the Silk Road, vol. 27. BRILL (2012)
62. Kime, M.B., Makgoale, D.: Characterization of copper-cobalt ores and quantification of Cu2+, Co^{2+}, Co^{3+}, and Fe^{3+} in aqueous leachates using UV/visible spectrophotometry. Chem. Engn Comm. **203**(12), 1648–1655 (2016). https://doi.org/10.1080/00986445.2016.1230102
63. Kissin, S.A.: Five-element (Ni-Co-As-Ag-Bi) veins. Geosci. Can.. 19, 113–124 (1992). Available online: https://journals.lib.unb.ca/index.php/gc/article/view/3768/4282/. Accessed 15 Dec 2019
64. Kuptsov, A.H.: Applications of fourier-transform Raman-spectroscopy in forinsic-sciences. J. Forinsic Sci. **39**(2), 305–318 (1994). https://doi.org/10.1520/JFS13604J
65. Learner, T.: Modern paints. In: Sackler NAS Colloquium: Scientific Examination of Art: Modern Techniques in Conservation and Analysis, pp. 137–151 (2005). https://nap.nationalacademies.org/read/11413/chapter/11
66. Li, W., Li, J., Wu, J., Guo, J.: Study on the phase-separated opaque glaze in ancient China from Qionglai kiln. Ceram. Int. **29**, 933–937 (2003). https://doi.org/10.1016/S0272-8842(03)00048-8
67. Li, W.D., Luo, H.J., Li, J.N., Li, J.Z., Guo, J.K.: Studies on the microstructure of the black-glazed bowl sherds excavated from the Jian kiln site of ancient China. Ceram. Int. **34**, 1473–1480 (2008). https://doi.org/10.1016/j.ceramint.2007.04.004
68. Li, Y.Q., Yang, Y.M., Zhu, J., Zhang, X.G., Jiang, S., Zhang, Z.X., Yao, Z.Q., Solbrekken, G.: Colour-generating mechanism of copper-red porcelain from Changsha kiln (AD 7th–10th century), China. Ceram. Int. **42**, 8495–8500 (2016). https://doi.org/10.1016/j.ceramint.2016.02.072
69. Liu, Z., Han, Y., Han, L., Cheng, Y., Ma, Y., Fang, L.: Micro-Raman analysis of the pigments on painted pottery figurines from two tombs of the northern Wei dynasty in Luoyang. Spectrochim. Acta A Mol. Biomol. Spectrosc. **109**, 42–46 (2013). https://doi.org/10.1016/j.saa.2013.02.015
70. Lo Giudice, A., Re, A., Calusi, S., Giuntini, L., Massi, M., Olivero, P., Pratesi, G., Albonico, M., Conz, E.: Multitechnique characterization of lapis lazuli for provenance study. Anal. Bioanal. Chem. **395**, 2211–2217 (2009). https://doi.org/10.1007/s00216-009-3039-7
71. Mangone, A., Caggiani, M.C., Giannossa, L.C., Laviano, R., Mutino, S., Sabbatini, L., Traini, A.: Islamic gilded and Enamelled glasses from Melfi (Southern Italy): an Archaeometric study. In: Proceedings of the 5th International Congress "Science and Technology for the Safeguard of Cultural Heritage in the Mediterranean Basin", Istanbul (2011) 22–25 November 2011
72. Matin, M., Pollard, A.M.: From ore to pigment: a description of the minerals and experimental study of cobalt ore processing from the Kâshân mine, Iran. Archaeometry. **59**(4), 731–746 (2017). https://doi.org/10.1111/arcm.12272
73. Matin, M., Pollard, A.M.: Historical accounts of cobalt ore processing from the Kashan mine, Iran. Iran. **53**, 171–183 (2015). https://doi.org/10.1080/05786967.2015.11834755
74. Melzer: Berglauftige Beschreibung der Stadt Schneeberg, pp. 405 (1684)
75. Merson, O.: *Les vitraux* (Vol. 29). Librairies-impr. réunies. Paris (1895)
76. Morral, F.R.: From Cabul to cobalt: a historical view of the "mischievous metal". J. Chem. Educ. **34**(4), 185 (1957). https://doi.org/10.1021/ed034p185
77. Mimoso, J.M.: Origin, early history and technology of the blue pigment in azulejos. In Proceedings of the GlazeArch2015, Lisbon., 2–3 July 2015; pp. 357–376 (2015). Available online: http://azulejos.lnec.pt/AzuRe/links/07%20Origin_of_blue_pigment.pdf. Accessed 20 Jan 2021
78. Naef-Galuba, I.: Les bleus d'Egypte. Rev. Céramique & Verre. **69**, 41–45 (1993)
79. Navas, N., Romero-Pastor, J., Manzano, E., Cardell, C.: Raman spectroscopic discrimination of pigments and tempera paint model samples by principal component analysis on first-derivative spectra. J. Raman Spectrosc. **41**(11), 1486–1493 (2010). https://doi.org/10.1002/jrs.2646
80. Neri, A.: Arte vetraria distinta in libri sette, Florence (1612)

81. Neri, E., Morvan, C., Colomban, P., Guerra, M.F., Prigent, V.: Late Roman and byzantine mosaic opaque "glass-ceramics" tesserae (5th–9th century). Ceram. Int. **42**(16), 18859–18869 (2016)
82. Nicholson, P.T.: Glassmaking and glass working at Amarna: some new work. J. Glass Stud. **37**, 11–19 (1995). https://www.jstor.org/stable/24190772. Accessed 4 Mar 2024
83. Ogura, D., Nakata, Y., Hokoi, S., Takabayashi, H., Okada, K., Su, B., Xue, P.: Influence of light environment on deterioration of mural paintings in Mogao Cave 285, Dunhuang. In: AIP Conference Proceedings, vol. 2170, p. 020013 (2019). https://doi.org/10.1063/1.5132732
84. Panighello, S., Kavcic, A., Vogel-Mikus, K., Tennent, N.H., Wallert, A., Hocevar, S.B., van Elteren, J.T.: Investigation of smalt in cross-section of 17th century paintings using elemental mapping by laser ablation ICP-MS. Microchem. J. **125**, 105–115 (2016). https://doi.org/10.1016/j.microc.2015.11.015
85. Panayoyova, S., Jackson, D., Ricciardi, P.: Colour. The art & Science of Illuminated Manuscripts. Harvey Miller Publishers, Cambridge, UK (2016)
86. Picolpasso, C.: Les trois livres de l'art du potier (1557), Revue de la Céramique et du Verre. Vendin-le-Viel (2007)
87. Porter, Y.: Textes persans sur la céramique. In: Vesel, Z. (ed.) La Science dans le Monde Iranien, pp. 165–189. Institut Français de Recherche en Iran, Teheran (1998) ISBN 978-2-909961-10-1
88. Porter, Y.: Le cobalt dans le monde Iranien (IXe-XVIe siècles). Taoci. **1**, 5–14 (2000). https://www.academia.edu/12424228/Le_cobalt_dans_le_monde_iranien_IXe_XVIe_si%C3%A8cles_Notes_sur_son_utilisation_en_c%C3%A9ramique_et_son_commerce_TAOCI_1_2000_pp_5_14. Accessed 4 Mar 2024
89. Pulak, C.: The Uluburun shipwreck: an overview. Inter. J. Nautical Archaeol. **27**, 188–224 (1998). https://doi.org/10.1111/j.1095-9270.1998.tb00803.x
90. Pulak, C.: The cargo of the Uluburun ship and evidence for trade with the Aegean and beyond. In: Bonfante, L., Karageorghis, V. (eds.) Italy and Cyprus in Antiquity, 1500–450 BCE, pp. 13–60. The Costakis and Leto Severis Foundation, Nicosia (2001)
91. Purowski, T., Kępa, L., Wagner, B.: Glass on the Amber road: the chemical composition of glass beads from the bronze age in Poland. Archaeol. Anthropol. Sci. **10**, 1283–1302 (2018)
92. Radini, A., Tromp, M., Beach, A., Speller, C., McCormik, M., Dudgeon, J.F., Collins, M.J., Ruhli, F., Kroger, R., Warinner, C.: Medieval women's early involvement in manuscript production suggested by lapis lazuli identification in dental calculus. Sci. Adv. **5**(1), eaau7126 (2019). https://doi.org/10.1126/sciadv.aau7126
93. Rehren, T.: Aspects of the production of cobalt-blue glass in Egypt. Archaeometry. **43**, 483–489 (2001). https://doi.org/10.1111/1475-4754.00031
94. Richet, P. (ed.): Encyclopedia of Glass Science, Technology, History, and Culture, 2 volume set. Wiley (2021)
95. Rollier-Hanselmann, J.: Berzé-la-Ville, la Chapelle-des-Moines: découverte d'un Christ caché sous les repeints. Bull. Monum. **163**(3), 243–249 (2005). https://doi.org/10.3406/bulmo.2005.1285
96. Scherrer, N.C., Stefan, Z., Francoise, D., Annette, F., Renate, K.: Synthetic organic pigments of the 20th and 21st century relevant to artist's paints: Raman spectra reference collection. Spectrochimica Acta Part A: Mol. Biomol. Spectrosc. **73**(3), 505–524 (2009). https://doi.org/10.1016/j.saa.2008.11.029
97. Schulte-Schrepping, K.-H., Piscator, M.: Ullmann's Encyclopedia of Industrial Chemistry. Wiley-VCH, Weinheim (2000). https://doi.org/10.1002/14356007
98. Seccaroni, C., Michael, J.U.N.G.: The provenance of Lapis Lazuli, the written sources, and its distribution as pigment in the wall paintings of Central Asia and the Middle East. Acta Via Serica. **7**(2), 89–108 (2022). https://doi.org/10.22679/avs.2022.7.2.004
99. Shi, P., Wang, F., Zhang, B., Luo, H., Zhu, J., Fei, G.: Study on the coloring mechanism of the Ru celadon glaze in the northern song dynasty. Ceram. Int. **46**, 23662–23668 (2020). https://doi.org/10.1016/j.ceramint.2020.06.139

100. Shim, J., Venkata Reddy, C., Sarma, G.V.S.S., Narayana Murthy, P., Ravikumar, R.V.S.S.N.: Effect of Co^{2+} and Ni^{2+}-doped zinc borate nano crystalline powders by co-precipitation method. Spectrochim. Acta Part A-Mol. Biomol. Spectrosc. **142**, 279–285 (2015). https://doi.org/10.1016/j.saa.2015.02.008
101. Shortland, A.J., Tite, M.S.: The interdependence of glass and vitreous faience production at Amarna. In: McCray, P. (ed.) Prehistory and History of Glassmaking Technology, pp. 251–265. The American Ceramic Society, Westerville (1998)
102. Shortland, A.J., Tite, M.S., Ewart, I.: Ancient exploitation and use of cobalt alums from the western oases of Egypt. Archaeometry. **48**, 153–168 (2006). https://doi.org/10.1111/j.1475-4754.2006.00248.x
103. Smith, G.D., Klinshaw II, R.J.: The presence of trapped carbon dioxide in lapis lazuli and its potential use in geo-sourcing natural ultramarine pigment. J. Cult. Herit. **10**(3), 415–421 (2009). https://doi.org/10.1016/j.culher.2008.12.001
104. Stanley, T.: The examination and analysis of Dunhuang and Turfan manuscript materials at Princeton University Library's east Asian library. J. Am. Inst. Conserv. **56**(3–4), 194–210 (2017). https://doi.org/10.1080/01971360.2017.1392105
105. Stanzani, E., Bersani, D., Lottici, P.P., Colomban, P.: Analysis of artist's palette on a 16th century wood panel painting by portable and laboratory Raman instruments. Vib. Sprectr. **85**, 62–70 (2016). https://doi.org/10.1016/j.vibspec.2016.03.027
106. Suger, ca. 1200. De administratione, XXXIII, trad. Lecoy de La Marche, Paris. M. Bur, Paris (1867)
107. Tait, H.: The Golden Age of Venetian Glass, Londres (1979)
108. Tang, L., Krishnan, N.A., Berjikian, J., Rivera, J., Smedskjaer, M.M., Mauro, J.C., Zhou, W., Bauchy, M.: Effect of nanoscale phase separation on the fracture behavior of glasses: toward tough, yet transparent glasses. Phys. Rev. Mater. **2**, 113602 (2018). https://doi.org/10.1103/PhysRevMaterials.2.113602
109. Theophilus: Essai sur divers arts, livre II, chap. VII et VIII, édition et traduction C. de l'Escalopier, Paris (1863)
110. Thierry, N., Thierry, M.: Peintures murales de caractère occidental en Arménie: Eglise Saint-Pierre et Saint-Paul de Tatev, Byzantion, **38**, 180–240 (1968). https://www.jstor.org/stable/44169294. Accessed 4 Mar 2024
111. Tite, M.S., Shortland, A.J.: Production technology for copper- and cobalt-blue vitreous materials from the New Kingdom site of Amarna—a reappraisal. Archaeometry. **45**, 285–312 (2003). https://doi.org/10.1111/1475-4754.00109
112. Tite, M.S.: The technology of glazed Islamic ceramics using data collected by the late Alexander Kaczmarczyk. Archaeometry. **53**, 329–339 (2011). https://doi.org/10.1111/j.1475-4754.2010.00546.x
113. Tomasini, E.P., Marte, F., Careaga, V.P., Landa, C.R., Siracusano, G., Maier, M.S.: Virtuous colours for Mary. Identification of lapis lazuli, smalt and cochineal in the Andean colonial image of our lady of Copacabana (Bolivia). Philos. Trans. Roy. Soc.-A Math. Phys. Engn Sci. **374**(2082), 20160047 (2016). https://doi.org/10.1098/rsta.2016.0047
114. Valmont de Bomare, M.: Minéralogie, ou Nouvelle Exposition du Règne Minéral, 2 Vincent Imprimeur-Libraire: Paris, pp. 76–96 (1774). https://gallica.bnf.fr/ark:/12148/bpt6k486073x. Accessed 4 Mar 2024
115. Vieira Ferreira, L.F.; Ferreira Machado, I.; Ferraria, A.M.; T.M. Casimiro, ; Colomban, P. 2013 Portuguese tin-glazed earthenware from the 16th century: a spectroscopic characterization of pigments, glazes and pastes. Appl. Surf. Sci. 285 Part B, 144–152. https://doi.org/10.1016/j.apsusc.2013.08.016
116. Visinescu, D., Paraschiv, C., Ianculescu, A., Jurca, B., Vasile, B., Carp, O.: The environmentally benign synthesis of nanosized $Co_xZn_{1-x}Al_2O_4$ blue pigments. Dyes Pigm. **87**, 125–131 (2010). https://doi.org/10.1016/j.dyepig.2010.03.006

117. Wainwright, I.N., Moffatt, E.A., Sirois, P.J., Young, G.S.: Analysis of wall paintings fragments from the Mogao and the Bingling Temple Grottoes. In: Proceedings of the International Conference on the Conservation of Grotto Sites, Mogao Grottoes, Dunhuang, *October 1993*, pp. 334–340 (2004, June)
118. Walton, M.S., Shortland, A., Kirk, S., Degryse, P.: Evidence for the trade of Mesopotamian and Egyptian glass to Mycenaean Greece. J. Archaeol. Sci. **36**(7), 1496–1503 (2009)
119. Walton, M., Eremin, K., Shortland, A., Degryse, P., Kirk, S.: Analysis of late bronze age glass axes from Nippur—a new cobalt colourant. Archaeometry. **54**(5), 835–852 (2012)
120. Weber, F.W.: Artist's Pigments, their Chemical and Physical Properties. Van Nostrand, New York (1923)
121. Zacharias, N., Kaparou, M., Oikonomou, A., Kasztovszky, Z.: Mycenaean glass from the Argolid, Peloponnese, Greece: a technological and provenance study. Microchem. J. **141**, 404–417 (2018)
122. Zhou, X., Lv, H., Cui, J., Dong, X., Wang, Y.: Fluorite used in ancient Chinese glassmaking during the 10th to 12th centuries: evidence from glass products excavated in the capital city site of the Liao dynasty. Archaeometry. **64**(5), 1138–1147 (2022). https://doi.org/10.1111/arcm.12768

Chapter 9
The Role of Analysis in the Holistic Provenancing of Ceramics

Abstract The role of analysis in the attribution and provenancing of porcelains and shards. The selection of appropriate analytical techniques for the characterisation of the elemental and molecular species present in high-temperature fired ceramics. Selection of case studies to illustrate the importance of chemical analysis in porcelain studies, including English delftwares, shards from marine archaeological excavations (Belitung, Tang, Tek Sing, Ming), identification of Chinese blue and white porcelains, the San José galleon, the Holy Grail of shipwrecks.

Keywords Marine archaeology shards · Tek Sing treasure · Belitung Island · San José · English delftwares · South African Portuguese shipwrecked carracks

9.1 Holistic Analysis and the Provenancing of Art Works

In the attribution of artworks to an artistic creator the primary centre of gravity hitherto has always been the judgemental opinion of a connoisseur who brings to bear many years of experience to formulate a decision upon whether a work of art is genuinely of a particular period, originating from a particular studio or atelier and is the work of a particular artist. Over the past few decades, the application of analytical techniques to the elucidation of the materials comprising a work of art and especially the pigments used in its composition has been fundamental is assisting this process and contributing significantly to the generation of spectral data and other information relevant to the establishment of the origins of artworks. The holistic process that is now advocated for the provenancing of an art work that is so critical for its acceptance into the genre therefore involves three components, which should all be considered as equally vital to this process: namely, analytical science, historical documentation and connoisseurship—the exclusion of one or more of these in a provenancing exercise for an unknown piece when available for whatever reason is no longer acceptable [14]. Certainly, the hard scientific data arising from analyses and confirmed documentation should never be ruled as subordinate to the

H. G. M. Edwards, P. Colomban, *Blue by Fire: A Marker of the Technical History of Glass and Ceramics*, Cultural Heritage Science,
https://doi.org/10.1007/978-3-031-77122-4_9

soft science based solely on a connoisseur's opinion, which may be based upon evidentially insupportable suppositions that have been held for many years that do not really stand up to the proper forensic treatment and methodology that is so necessary for establishing an irrefutable provenance for an art work. Basically, the questions that arise in a forensic chemical analytical investigation can be summarised as follows: what is the qualitative chemical composition of the specimen, how much of each component is present (the quantitative analytical component), what has happened to the specimen since it was made (is there deterioration or degradation), where did the materials comprising the specimen originate and finally, perhaps the most difficult component of all to assess, why are the analytical signatures there? Clearly, the analytical data require interpretation and this means that the conclusions must be properly derived so that the due logical processes are demonstrated to the readers. The input of analytical science to the holistic attribution and determination of the origin of ceramic artefacts might be a recent development but it has already shown its worth, as will be demonstrated through the several case studies cited and following later in this Chapter.

9.2 The Selection of an Appropriate Technique for Ceramics Analysis

Earlier, a comprehensive coverage has been undertaken of the analytical techniques that are used for ceramics and their pigment decoration: it should be appreciated that an analytical technique must be appropriate for its destined purpose—for example, it would not be useful to embark upon accelerator mass spectrometric radiocarbon dating for ceramic bodies, their glazes and pigments because few specimens, if any, would yield any carbon necessary for the C^{12}/C^{14} isotope measurement required to determine the age of the specimen. Also, the technique is destructive of specimen anyway, which negates its consideration for adoption here to determine its age. However, a technique that has been used to date ceramic bodies in several areas is thermoluminescence but care must be taken in its use as the following example illustrates.

9.2.1 Case Study I. An English King Charles II Pottery Mug

In December 2022, an apparently early tin-glazed pottery mug from an unknown factory, measuring some 3.75″ high and with a diameter of between 2.75 and 2.85″, was offered at auction (Fig. 9.1). The polychrome decoration, comprising essentially blue and white pigments with some ochre, purple and gilding, depicted a bust of King Charles II wearing armour and was accompanied by the cipher C II R and dated 1660. It was made to record the accession of King Charles II to the throne of

Fig. 9.1 English tin-glazed delftware armorial tankard, height 10 cm, diameter 7 cm, decorated in blue, ochre and purple pigments showing King Charles II in armour and bearing the monogram CDR 2, wearing the regal crown and dated 1660. Made in 1660 to celebrate the Restoration of the Monarchy. A rare example attributed to the Southwark earthenware manufactory, London. (Reproduced with permission from the Collection of E. and H. Manners Ltd., Church Street, Kensington, London)

England and the restoration of the Monarchy after the Commonwealth and Protectorate of Oliver Cromwell had been in control since the end of eth Civil War in 1646. The vendors had taken the mug to Oxford Authentication for specialist antique authentication using thermoluminescence analysis and the auctioneers cited the following report from the analysts:

> Using standard methods and techniques we are unable to draw any conclusions regarding the authenticity of this piece (sample number N122j75). The clay appeared to contain a spurious component which gave rise to the background signal, making any dating unreliable

This must have been a disappointing result for the vendors, who probably wished to remove any doubt about the authenticity of their piece, nevertheless, the mug eventually sold at auction in excess of its estimate for £12,000, plus fees. The spurious component alluded to was not specified in the analysis report but a photograph of the mug indicated the presence of two small holes that were drilled into the base to extract specimens for the analysis. The mug was coloured with a creamy-white slip and the sample extraction would have necessitated this being part of the sampling for analysis: the composition of the creamy-white pigment was not defined but could have contained calcite, dolomite or white clays. The phenomenon of spurious thermoluminescence has been known for some time and has been evaluated in more [29, 31] recent studies: rather than being attributable to one particular cause it seems that spurious thermoluminescence can arise from several things, including, the presence of calcite and carbonates in general, the presence of organic matter, the presence of plagioclase and adsorbed water, the buildup of surface charge on a specimen during fine grinding procedures and the introduction of chemical phase transformations in the specimen upon heating. The source of this spurious thermoluminescence background signal was identified as mineral carbonates, mica, quartz and feldspars and more particularly to minor component transition metals such as manganese therein. The pottery mug would certainly be a candidate for having one or more of these materials present which would be contributory to its observed spurious thermoluminescent signals during the analysis.

An ancillary comment relates to the process of authentication that was the prime purpose of this particular analysis: most analytical techniques used for ceramics give data and information about the raw materials used, the minerals formed at elevated temperatures in the firing kilns and the pigments used in their decoration and the thermoluminescence technique is the only one that can give direct information about the age of the specimen and thence its authenticity, as was required in the case of the King Charles II pottery mug. Connoisseurship is based upon expert knowledge gained from a mentally acquired database and the visual appreciation of the specimen's aesthetic standing in that scenario, whereas scientific analysis relies upon an instrumental database with its detection of the presence of out of context materials of a potential chronological significance: neither approach alone can verify the authenticity of a specimen infallibly or absolutely especially where minimal data are available from either opinion or fact to consider. A recent article has expounded upon this theme [38] in which a classic controversy has arisen regarding a papyrus fragment which supposedly purports to date from the fourth Century CE and which describes a saying of Jesus to His wife [23] and known as the Gospel of Jesus' Wife (with the acronym, GJW). The opinions of connoisseurs and analytical scientists are controversial and not definitive enough for its authenticity to be declared and has raised sequential questions in the art world of the role of scientific analysis as a potential provenancing for forged artworks. In this case, the science has determined that the ink used was carbon lampblack and the accelerator mass spectrometric dating gave a papyrus parchment date of the eighth Century CE, which themselves are deemed not sufficient to unequivocally assign the specimen to be authentic and several connoisseurs are hence of a mixed opinion as to the artefact.

9.2.2 Case Study II: Marine Archaeology of Chinese Porcelains

Much of the Chinese porcelain imported into Europe in the Seventeenth, Eighteenth and Nineteenth Centuries was carried by ships of the British Honourable East India Company (HEIC), the Dutch East India Company (VOC), and to a lesser extent by the Swedish East India Company (SOIC) and Portuguese carracks under charter to these companies (that were later called the "Bombay ships"). Hitherto, much trade between China and the Middle East and Europe, including the carriage of ceramics, was accomplished overland via the Silk Road and using dhows across the Indian Ocean into the Islamic Abbasid sultanates of East Africa. There was also a vigorous local trade in ceramics carried by large ocean-going junks of up to one thousand tons burthen: unlike the heavily armed and crewed European vessels of the HEIC and VOC, the Chinese local traffic was susceptible to piracy in the dangerous South China Sea and many such wrecks have now been located, with their cargoes of porcelain and ceramics comprising complete items and many shards. A list of 22 identified wrecks which carried cargoes of porcelain and other ceramics is provided in Table 9.1: this is by no means an exhaustive compilation of marine archaeological excavations of this type but it gives an idea of the size and diversity of wrecks, the date of their sinking and of the origins of their ceramic cargoes [21]. Many of the ceramic cargo artefacts comprised blue and white decorated earthenwares, stonewares and porcelain but little analytical work has been undertaken at the present time on the blue pigments involved, hence a resume now follows of the analytical chemical studies on blue and white decorated ceramics from these marine archaeological excavations. The oldest specimen extant of Chinese blue and white decorated ceramic is a Tang Dynasty glazed stoneware bowl from Hunan decorated with blue and white motifs, which was recovered from an Arab dhow which sank off Belitung Island in the Singapore Straits in 830 CE (Fig. 9.2). There now follows case studies of two selected accounts of the analyses of several marine archaeological excavations involving Chinese blue and white porcelain artefacts.

9.2.2.1 The *Tek Sing* Treasure

The *Tek Sing* (*True Star*) was a large, 3-masted Chinese ocean-going junk (50 m long, 10 m wide and 1000 tons burthen) of the Fukien type, which left Amoy (Xiamen, Fujian) in January 1822 heavily laden with a mixed cargo of porcelain, silks and spices, with a crew of 200 and 1600 passengers bound for Batavia in the Dutch East Indies. After 1 month at sea her captain, Io Tauko, exasperated at their very slow progress and concerned at the ongoing discomfort of his passengers, attempted to take a short cut through the uncharted Gaspar Strait, north of Java, between the islands of Bangka and Belitung, when the ship hit a reef in the Belvidere Shoals and sank in 30 m of water with a large loss of life. The accident has been called the “Titanic of the East”. The next morning, February seventh, 1822, the *Indiana* an HEIC East Indiaman sailing to Borneo (Master Captain James Pearl) encountered much debris and flotsam at the wreck site and rescued 190 survivors.

In May 1999, a marine salvage expert, Michael Hatcher, discovered the wreck of the *Tek Sing* which contained some 350,000 pieces of porcelain, representing the largest cargo of Chinese porcelain ever recovered from a marine excavation. Most of the cargo had been packed tightly in the hold surrounded by tea in chests and some had even been strapped in tea chests to the outside of the hull. The porcelain was auctioned in Dr. Fritz Nagel’s Kunstauktionhaus in Stuttgart in November 2000 over a period of some 8 days, the auction sales running for 24 h a day! The utilitarian cargo of Qing Dynasty porcelain made mainly during the reign of the seventh Qing Emperor, Daoguang (1820–1850), comprising bowls, dishes and plates which were decorated mostly in underglaze blue and white (Fig. 9.3.), came from three

Table 9.1 Selection of Shipwrecks from which Chinese and other Asian Porcelains and Ceramics have been recovered

Ship/Cargo[a]	Date of sinking /CE	Cargo[b]
Beliting/Tang/Batu Hutam	830	Tang Dynasty porcelain and stonewares
Tanjung Simpang	960–1127?	Sung porcelain and celadons
Turiang	1370	Yuan/Ming porcelains, celadons from Thailand, Vietnam and China
Nanjiang	1380	Thai ceramics, Sisantchantai[c], Sukhothai[c]
Longquan	1400	Celadons, Ming white wares, Sisantchantai, Sukhothai
Pandaman	ca.1450	Chinese porcelain (Ming: Yuan), Vietnamese (Chu Dau: Go Sanh), and Thai Sukhothai ceramics
Royal Nanhai	1460	Sisantchantai, Sukhothai celadons
Hoi An	1480	Vietnamese porcelains
Sing Tai	1550	Sisantchantai ceramics
Binh Tuan	1608	Chinese, Vietnamese porcelains
Wanli	1625	Blue/white porcelain
Hatcher	1643	Blue and white Chinese porcelain and Kraak
Vung Tau	1690	Blue/white porcelain and blanc de chine (Dehua)
San Jose[d]	1708	Blue /white porcelain
Ca Mau	1725	Blue/white porcelain and polychrome
Svecia	1740	Blue/white porcelain: Swedish vessel, Goteborg
Nanking/*Geldermalsen*[e]	1750	Blue/white porcelain: Dutch East Indiaman
Sandana Island	1765	Chinese porcelain
Brederode	1785	Blue/white porcelain: Dutch East Indiaman
Diana	1816	Blue/white porcelain: British HEIC East Indiaman
Tek Sing	1822	Blue/white porcelain
Desaru	1830	Blue/white porcelain, Yixing teapots

[a]Where known, the name of the ship describes the cargo, otherwise a town on the mainland is chosen to identify the wreck and its cargo

[b]The cargo is frequently a mixture of porcelain from several sources and other pottery ceramics, stonewares, and celadons

[c]Sisantchantai and Sukhothai are sites of porcelain and celadon production in Thailand

[d]This galleon was sunk by a British warship whilst heading home to Spain from Cartagena de India with a fabulous treasure and a Chinese porcelain blue and white dinner and tea/coffee service intended for King Philip V of Spain from present day Colombia: it is the only wreck cited here in the Atlantic Ocean and it is conjectural how the Chinese porcelain crossed the Pacific Ocean to Cartagena

[e]Italic script signifies that the actual ship wrecked has been positively identified. The *Geldermalsen* cargo of porcelain is alternatively and more commonly known as the *Nanking* treasure, from the port of embarkation

Fig. 9.2 Oldest example of a Chinese blue and white decorated ceramic in existence: a Tang Dynasty glazed stoneware dish from the Chaysha (Hunan) kilns, *ca.* 830 CE, with double lozenge and foliate motifs. Recovered from the wreck of an Arab dhow which sank off Belitung Island in the Singapore Straits carrying a cargo of assorted ceramic artefacts to Arabia. (Reproduced with permission of the Asian Civilisations Museum, Singapore, from their Tang Shipwreck Collection, accession number 2005.1-00473)

Fig. 9.3 The *Tek Sing* Treasure: showing here a collection of porcelain rice bowls from Dehua decorated with lotus flowers in blue and white. In total some 350,000 pieces of porcelain were recovered from the wreck site in 1999

sites in southern China, namely, Dehua (in Fujian), Yixing (in Jiangsi) and Swatow (in Guangdong), and was consistent chronologically with known recent ceramic productions at these sites. However, several other items were clearly of a more antique classification and could be identified as fifteenth Century Ming Dynasty porcelain from the Imperial porcelain manufactory at Jingdezhen which were ascribed as being the personal possessions of the passengers. Basically, the types of porcelain discovered comprised standard blue and white decorated bowls, plates and dishes, but also some creamy-white "blanc de Chine" items, celadons and olive-glazed porcelain vases. The majority were classed as wares from the Dehua site; the Yueji wood-burning kiln there had already then been operating at Dehua for 400 years and still produces porcelains for small, local artisans, being fired up now perhaps only four times a year.

The porcelains generally have a high standard of decoration for provincial wares and provide a good idea of the requirements for ceramics designed for the Chinese expatriates in the East Indies in the early nineteenth Century. Although mostly utilitarian items, several figures were found and identified as coming from Jingdezhen in the 1740s and some earlier Ming porcelains and celadons which were thought unlikely to have been kept in a warehouse for several centuries and therefore were probably more likely to have been the personal treasured possessions of the passengers. It was surprising that there were relatively few white wares in the consignment considering that this was the basis of the production that was famed at Dehua and perhaps reflects the disappearance of the popularity of the *blanc de Chine* in the eighteenth Century. Other more unusual items were sets of cosmetic boxes and celadon wine cups. For the blue and white wares from Dehua, a great variety of symbolic botanical motifs were representatively portrayed on the china including the peony (wealth, rank), magnolia (purity), bamboo (humility, fidelity), chrysanthemum (longevity), pomegranate (fecundity), lotus flower (purity, harmony) and the peach (immortality, long and healthy life) and follow the earlier seventeenth Century decorations emanating from Jingdezhen. The Swatow—derived wares generally comprised blue-glazed dishes and polychrome bowls in red, green and yellow pigments. In the study reported here, three porcelain rice bowls, of diameter 15.5, 16.0 and 15.0 cm and of height 6.5, 7.0 and 6.0 cm, from the *Tek Sing* cargo, decorated in cobalt blue and white enamel, were presented for nondestructive Raman spectroscopic analysis. The transmitted light has a distinctly green translucency which is reminiscent in colour of the finest contemporary Swansea and Coalport European porcelains of the second decade of the nineteenth Century, which is normally associated with the presence of iron oxide in the china clays used for the recipe for the porcelain body and can be scientifically related to the charge-transfer electronic absorption between iron (III) ions and the silicate matrices in the visible region (absorption of the red wavelengths) of the electromagnetic spectrum [13]. This can at first be considered quite surprising as Dehua porcelain was always famed for its white body and its production of *blanc de chine* in the eighteenth Century, which was attributed to the very low content of iron oxide in the component china clays, and could therefore potentially reflect the usage of poorer quality china clays in their porcelain production in the 1820s, to match with the cargo of the *Tek Sing* and its sinking in February 1822 [1, 11]. The use of these special white clays also meant that the Dehua kilns were fired in an oxidising atmosphere to give their typical ivory white body unlike the Jingdezhen porcelains which had to utilise a reducing atmosphere of carbon dioxide in the kiln. However, an estimate of the percentage transmission (translucency) of each of these *Tek Sing* porcelain specimens can be given as approximately 25, 10 and 5%, which is consistent with their thickly potted characteristics. Although the first two bowls clearly belong to a similar Dehua set, the third and slightly smaller bowl could be from a different site as it is of a much heavier potting and glazing. The glaze is very thickly applied and is rather crazed in parts, particularly on two of the bowls: on one bowl, the presence of deep brown tea stains can be seen—witness to the statement that the *Tek Sing* porcelain cargo had been packed into tea chests for safe transit, especially those which had been attached to the ship's freeboard to complement the storage capability in the main hold. It has been proposed by various authors previously that the survivability of the *Tek Sing* porcelain for two centuries on the sea-bed with the shifting sand and currents can be attributed to the way it was packed together so tightly in the hold of the ship, but clearly the porcelain items being packed in tea chests surrounded by tea would also have provided some additional significant insulation against mechanical damage.

Previous Analyses of Relevant Chinese Porcelains

Li et al. [26] have published two papers on their chemical analyses of Dehua porcelains from five kiln sites at Wanpinling, Quduogong, Zinlonggong, Jiabeisian, and Xingjiao which range from the Song Dynasty through the Yuan and Ming to the last Qing Dynasty in a very broadly dated period from 960 to 1911. The Qing Dynasty specimens from the later Xingjiao single-firing kiln site are broadly representative of the period of the *Tek Sing* treasure porcelains and their results are very interesting for the interpretation of the observed translucency of these porcelains. Li et al. find that although the Dehua porcelains were renowned for their pale white translucency and porcelain body as typified by their *blanc de Chine* output in the Kangxi period (1644–1722) of the early Qing Dynasty, it seems that the manufacture had to resort to a poorer quality clay later in the Dynasty which was characterised by a significantly higher content of iron (III) oxide and this resulted in a pale greenish translucency of the porcelains. Their analyses showed that the earliest Dehua porcelains during the Ming Dynasty contained only a fraction of a percent (0.1–0.2%) iron (III) oxide whereas later analyses showed an order of magnitude greater than this, i.e. approximately 2%. The porcelain stone that was used later at Dehua was predominantly a sericite, of formulation $KAl_3Si_3O_{10}(OH)_2$, containing kaolinite, quartz, sericite, potassium feldspar and mica, rather than the purer kaolinite and china stone used at Jingdezhen. Sericite derives its name from the Latin *sericus*, meaning silky, and describes the silky, soapy texture of this micaceous potassium aluminosilicate. The analytical composition of this Dehua sericite stone was 64% silica, 26% alumina, 4% potash, 2% iron (III) oxide, with less than 1% each of soda ash, magnesia, and lime and was used in their "Unitary Formula" single—stage site firing process. Because of the increased content of the alkaline flux used in later Dehua porcelains this meant that the kiln firing temperatures could be reduced to between 1280 and 1300 °C compared with the earlier Ming Dynasty firing temperatures of 1350–1390 °C. The lower alumina content compared with that of the earlier Ming porcelains also meant that the translucency was somewhat diminished. XRD determinations also demonstrated that the glaze on later Dehua porcelains primarily contained tridymite and mullite, as expected for a glaze on a hard paste body and the absence of anorthite and cristobalite consequent upon the lower firing temperatures adopted in the kiln.

Table 9.2. collects some of the key analytical elemental oxide studies made for Chinese porcelains in six literature investigations: although most of these refer to earlier Dynastic period specimens, an interesting conclusion is that there is actual little significant difference in the formal elemental oxide analytical composition between the earlier Yuan and the later Qing Dynasty porcelains, whatever the particular site investigated seems to indicate. For example the average silica content over the specimens for all six studies is 74 +/− 4%, alumina is 20 +/− 3%, soda 2 +/− 1%, potash 4 +/−2% and iron oxide 1+/−1%. When one considers that the single standard deviation on their silica determination quoted by Li et al. [26] is +/−2% and for alumina +/−3% then the conclusion must be that each single analytical percentage alone is not sufficient to unequivocally determine the site of origin or the period of manufacture without the consideration of other historical documentary factors—i.e. a holistic approach to the attribution of specimens is mandatory!

Although the recovery of Chinese porcelains and other early ceramics from shipwrecks is not unusual as the listing in Table 9.1. will confirm, there is little that one can compare with currently for the *Tek Sing* items. The Table 9.1. gives an indication of the salvage operations that have been carried out to recover quantities of Chinese porcelains from shipwrecks, from which it is seen that the *Tek Sing* is one of the most recent.

It was paramount that the first Raman spectroscopic analyses presented here were carried out non-destructively and to this extent the dictates of the spectrometer sample chamber were a controlling factor. Initial Raman spectroscopic analysis at 785 nm demonstrates that spectra can be collected from the samples. However, limitations with chamber size and fluorescence have made more a comprehensive analysis more difficult. This paper will demonstrate the first preliminary Raman spectroscopic analysis of surviving porcelain from the *Tek Sing* shipwreck.

A typical Raman spectrum of the porcelain bowls was obtained with 830 nm excitation: the spectrum is dominated by the broad diffuse scattering between 600 and 1100 cm^{-1} with maxima near 670 and 820 cm^{-1} that are characteristic of the silicate matrix and from which it is possible to determine the firing temperature of the kiln in which the porcelain was synthesised [4]. Distinct Raman bands at 525 and 990 cm^{-1} are assignable to the silicate mineral microcline, $KAlSi_3O_8$, and features at 457, 200 and 120 cm^{-1} are characteristic of quartz, SiO2. A band at 400 cm^{-1} is assignable to para-wollastonite, and those at 390 and 355 cm^{-1} are features of diopside, $CaMgSi_2O_6$. The bands at 320 and 260 cm^{-1} are assignable to the complex aluminosilicate, illite. No spectral evidence is presented for the presence of key minerals such as mullite, cristobalite or tridymite which are detectable in porcelain bodies that have been thermally treated to very high kiln temperatures so it can be concluded that the specimens from the *Tek Sing* cargo have indeed been subjected to only a medium-high kiln temperature at the Dehua and Swatow manufacturing sites. This is confirmed by the ratio of the relative intensities of the broad silicate features at 500/600 and 800/900 of 1.25, which expresses the DP, the degree of

Table 9.2 Elemental percentage compositions/% for selected Chinese Porcelains

Dynasty	Emperor	Period	Location[a]	SiO_2	Al_2O_3	CaO	MgO	Na_2O	K_2O	P_2O_5	Fe_2O_3	PbO	Reference
Ming	Kangxi	1662–1722	Jaibei	73	20	0	0	0	6	0	0	0	[36]
Ming	Kangxi	1662–1722	–	72	23	1	0	2	2	0	0	1	[12]
Yuan	–	1271–1368	–	79	16	0	0	1	3	0	1	0	[37]
Yuan	–	1271–1368	–	74	20	0	0	2	2	0	1	0	[37]
Ming	Kangxi	1368–1644	–	70	25	0	0	3	3	0	0	0	[24]
Qing	Daoguang	1820–1850	Xingjiao	76	17	0	0	0	5	0	1	0	[26, 27]
		Average	74+/−3	20+/−3	0	0	2+/−1	4+/−2	0	1+/−1	0		

[a]The kiln site location is often not specified, understandably, in the analytical studies: the Imperial manufactory at Jingdezhen, for example, at one time had more than a hundred kilns operational at several sites in the locality. Jaibei is a Jingdezhen porcelain kiln site, whereas Xingjiao is a Dehua porcelain kiln site. Generally, the site location is confirmed by local archaeological excavation

polymerisation of the silicate matrix, which is indicative of a kiln firing temperature of approximately 1200–1250 °C [4, 5, 6]. This is significantly lower than the kiln temperatures achieved at the Imperial Chinese porcelain manufactory at Jingdezhen, where a kiln temperature of upwards of 1380 °C was used. To give a comparison, the green and white vitreous aluminosilicate glazed Transylvanian tiled faience from Sibiu in Romania gave a DP of 0.2 from its analogous Raman spectral band intensity ratio, which translated into a kiln firing temperature of approximately 700 °C. [2, 14, 15].

9.2.2.2 Ming Porcelain Shards from South African Portuguese Carrack Shipwrecks

In 1498 the Portuguese explorer Vasco da Gama sailed around the Cape of Good Hope from the South Atlantic and into the Indian Ocean, thereby opening up the sea route to China and facilitating the importation of Chinese porcelains, spices, teas and silks from India into Europe by sea, a trade that generated multiple voyages from predominantly Portuguese, Swedish, British and Dutch traders in the 17th, 18th and 19th Centuries. The Ming porcelains (Ming Dynasty, 1368–1644 CE) were particularly in demand in Europe for their beauty, fine blue and white decoration and novelty. Portuguese carracks and navetas returning home from China were often overloaded and the ships that had suffered from the 6 months long and arduous outward journey were heavily laden and in poor condition to survive the storms, piracy and treacherous seas when homeward bound. The British Honourable East India Company (HEIC) and the Dutch Veerenigde Oostindische Compagnie (VOC) designed special East Indiamen ships for trade with China which were heavily armed, often with up to 42 cannons on gun decks, and very fast so they did not fall prey easily to pirates [14, 15]. The British HEIC East Indiamen were often coopted into the Royal Navy as frigates during need at times of war and served as outliers and spotters for the approach of enemy naval forces and blockade runners. The Portuguese Companhia da India Oriental (Casa da India) has recorded that of the 840 carracks that set out from Portugal for China between 1600 and 1650 only 420 actually made it home safely to Portugal [34]. The coast of South Africa where the Indian Ocean meets the Atlantic Ocean is the graveyard of many ships and their cargoes that were plying the trade route to India and China and rounding the Cape of Good Hope: the stormy seas where the Indian ocean met the South Atlantic Ocean generally meant that the porcelain cargoes did not survive the foundering of the ships there and most wrecks have been located with only shards of Chinese porcelain remaining. From the records, it appears that at least 74 Portuguese ships were lost around South Africa on the way home from the Far East between 1600 and 1650 [34]. The location of at least 15 Portuguese shipwrecks which sank prior to 1650 have been identified near the coast where survivors managed to reach the shore, and nine of these have now been subjected to archaeological investigation [8–10].

In 2004, de Waal [8, 9] reported the Raman spectroscopic analysis of blue and white decorated glazed porcelain shards from the J.A. van Tilburg Collection curated by Dr. Valerie Esterhuizen at the University of Pretoria, that had been recovered from the wrecks of 9 Portuguese carracks dating from between 1550 and 1650 which sank off the coast of South Africa and South-East Africa. The blue cobalt pigment has been characterised as cobalt aluminate [8, 9, 25], Co_2AlO_4, and they were found to have been fired at a high temperature between 1100 and 1300 °C, as expected for the Chinese hard paste Ming porcelains from Jingdezhen. The names of the vessels and dates of their sinking were all known from survivors and the porcelain artefacts all belong to the Ming Dynasty (1368–1644 CE). A special type of porcelain was often carried by these Portuguese carracks predominantly for the Dutch market, called *kraak* ware (Fig. 9.4) or *kraakporselein,* which bore a characteristic blue and white decoration of a geometric design in panels, and was made during the reigns of the Ming emperors Wanli (1573–1620 CE), Tianqi (1620–1636 CE) and Chongzheng (1627–1644 CE). The name for this porcelain is believed to have been derived from the Dutch word for "carrack".

A further combined SEM/EDAXS and Raman spectroscopic study was later carried out by Carter *et al.* [3] on five shards excavated from the *Santa Espiritu* and *Santa Maria Madre de Deus* shipwrecks. The *Santa Espiritu* foundered in Morgan's Bay, Eastern Cape, in 1608 and the *Santa Maria Madre de Deus* sank in Bonza Bay, East London, in 1643, both carrying cargoes of Ming porcelain from China homeward to Portugal. Four shards were provided from the *Santa Maria Madre de Deus* and one shard from the *Santa Espiritu:* the latter shard, #5, was from the Wanli (Shenzong) Ming period (1573–1620 CE) and the former shards, #1-#4, were from the Chongzheng (Ssuzong) late Ming period (1628–1644 CE) (Fig. 9.5).

Fig. 9.4 Dutch East India Company *kraak* style large porcelain plate, diameter 12.5″ (31.3 cm), Japanese Arita ware, bearing the logo VOC in underglaze blue with a six-panelled geometric border, Edo period, *ca.* 1660. Metropolitan Museum of Art, New York, the bequest of Dr. and Mrs. Roger G. Gerry, Roslyn, New York, 2000, Accession number 2002.447.40. *Public domain*

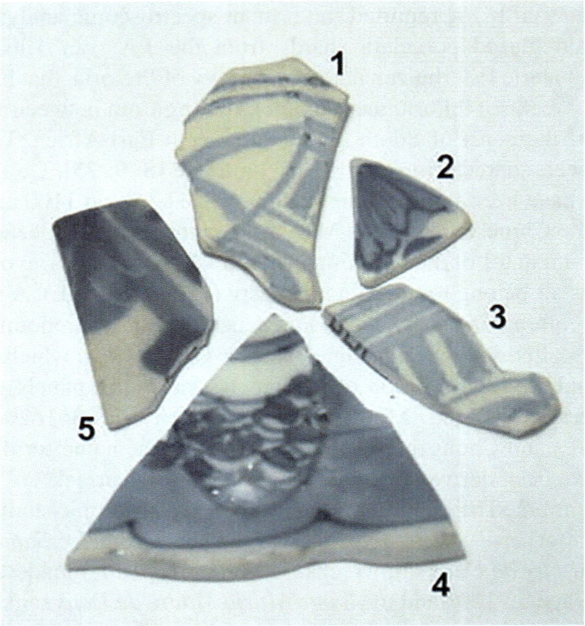

Fig. 9.5 Ming porcelain shards, *ca.* 1600, recovered from the wrecks of Portuguese carracks attempting to round the Cape of Good Hope from China with cargoes of Chinese hard paste porcelain for the European market. Raman spectroscopic analysis of the porcelain body reveals the presence of the mineral anatase, titanium (IV) oxide which occurs naturally along with kaolin, china clay. Shards #1- #4 were recovered from the wreck of the *Santa Maria Madre de Deus* which foundered in Bonza Bay, East London, in 1643 and Shard #5 was recovered from the wreck of the *Santa Espiritu* which foundered in Morgan's Bay off the Eastern Cape in 1608. (Reference: E.A. [3]). (Courtesy of Dr Valerie Esterhuizen, Curator of the J.A. van Tilburg Collection at the University of Pretoria, South Africa)

The combined Raman and SEM/EDAXS study by Carter et al. [3] of these five selected shards from the *Santo Espirito* and the *Santa Maria Madre de Deus* showed that there was a mixture of high-fired and low-fired porcelains from the presence of the mineral beta-wollastonite, the inosilicate $CaSiO_3$ in shards #1 and #3 from the *Santa Maria Madre de Deus*, indicted by Raman spectral signatures at 971, 638 and 340 cm $^{-1,}$ and also bands evident for anatase, namely, at 143, 398 and 638 cm $^{-1,}$ unlike the other shards which demonstrated the high temperature mineral mullite. An earlier shard excavated from another shipwreck, the *Sao Bento,* which sank off Msikalba, Eastern Cape, in 1554, again carrying a cargo of Ming porcelain (Jiajing period), gave the Raman spectral features [8, 9] of the higher temperature polymorph of titanium (IV) oxide, rutile, and not anatase which is unstable at higher temperatures in the kiln as would be expected for a high temperature hard paste firing process and converts kinetically to rutile [28, 16].

It was also found that different shades of blue decoration were achieved by the incorporation and mixing of china clay containing titanium dioxide, TiO_2, as the rutile polymorph and carbon as a darkening agent. In admixture with the blue pigment, spectral evidence for haematite in a solid solution with ilmenite, $FeTiO_3$, is also found and this would also serve to provide a different colour and tonal quality for the applied blue pigment.

A special feature of the Raman spectra excited with a laser wavelength of 785 nm was a broad band resonance fluorescence emission centred around 1320–1360 cm^{-1}, which was attributed to the lanthanide rare earth impurity elements found in china clays and which had been identified previously in Chinese porcelain shards of the Yuan, Ming and Qing Dynasties by Widjaja et al. [39]. Excitation at other laser wavelengths does not produce this band. The presence of the titanium dioxide polymorph, anatase, in the Raman spectra of the shards from the *Santa Maria Madre de Deus* was unusual in that it implies that a lower kiln firing temperature had been used as found in soft paste porcelains: this correlates with the discovery of anatase in ancient Chinese porcelains excavated in China [20, 40].

The two unusual shards exhibiting the characteristics of *soft paste* porcelain with the spectral presence of *beta*-wollastonite and the presence of anatase, the latter itself indicative of a lower kiln firing temperature for the porcelain body, raises an interesting question: did the Chinese in fact then manufacture soft paste porcelain in the Ming period in the seventeenth Century? Jorg and Wilson ([22]) write in their authoritative survey:

> Chinese soft paste porcelain, which is different from European soft paste porcelain, originated about 1700 and became popular in the second quarter of the 18th Century as part of the export assortment. Unlike ordinary porcelain it is not very translucent and often has a creamy-white appearance. The glaze is often finely cracked as a result of a difference in the cooling between the glaze and the body. The latter is made from a white-firing clay called "huashi" (transl. Slippery stone) the use of which is documented in the letters of 1717 and 1722 by the Jesuit priest Pere d'Entrecolles. As the clay was expensive, Chinese soft paste porcelains are usually small and thinly potted. They are also well-painted as the body is particularly suited for detailed drawing. Besides this, pieces are sometimes noted which have a hard paste body and a coating of huashi clay, which gives the same effect. Huashi clay is a plain type of kaolin which stiffens the paste and fires very white.

The additional point of interest with respect to the soft paste porcelain shards discovered in this particular case study cited here is their early date. Accounts of Chinese porcelain wherein the soft paste body variant is recognised, all refer to its appearance in Europe around the first or second quarter of the eighteenth Century, i.e. at the earliest, the later Kangxi period (1662–1722 CE) of the Qing (Manchu) Dynasty and not before—so its excavation from a shipwrecked cargo of Ming Dynasty porcelain which is documented faithfully and chronologically at 1643 is noteworthy and this soft paste variant does not seem to have been found associated with Ming porcelains hitherto. Secondly, the discovery of anatase in the soft paste porcelain body, although itself not remarkable as it is known to occur along with kaolin deposits in several geographical locations, does confirm the lower firing temperatures in the kiln required for soft paste porcelain manufacture compared with the more stringent and higher temperatures necessary for the hard paste body variant. Zuo et al. [40] have also found anatase in earlier period Chinese ceramics, but these comprised fired earthenwares and pottery from Xishan, where again the kiln firing temperatures would have been lower than those required for the firing of the hard paste porcelains. This would be a suitable area in which to undertake a comprehensive analytical search for soft paste Chinese porcelain analogues which may well exist as specimens in earlier seventeenth Century imported Chinese wares and perhaps as such until now have gone unrecognised.

9.2.3 The San Jose: *The Holy Grail of Marine Archaeological Excavations*

On June seventh, 1708, a Royal Naval fleet under the command of Commodore Sir Charles Wager, comprising the *Expedition* (70 guns), *Kingston* (60 guns), *Portland* (50 guns) and *Vulture* (28 guns) gave battle with a Spanish treasure fleet off Cartagena da India, in an engagement that would later be known as "Wager's Action" in the War of the Spanish Succession. Although the Royal Navy won the battle without the loss of a ship, and captured the galleons *San Joaquin* (64 guns), *Santa Cruz* (44 guns) and *Santa Cartagena* with their cargoes, the Spanish flagship *San Jose* (64 guns) sank when her magazine exploded (Fig. 9.6) and took her cargo of 200 tons gold, silver, emeralds and a very special Kangxi Chinese porcelain dinner service intended for the court of King Philip V of Spain with her. The Spanish

Fig. 9.6 Oil Painting by Samuel Scott (1702–1772) entitled "Wager's Action: The Taking of the Galeons by Sir Charles Wager", seventh June 1708, off Cartagena, showing the *San Jose* catching fire and exploding whist being engaged by the Royal Navy flagship *HMS Expedition*.. Now in the National Maritime Museum, Greenwich, London. *Public domain*

Admiral, Jose de Santillana went down with his ship. Only 11 of her crew of 600 survived the action: the wreck has been located in 2015 and she currently lies in 900 fathoms of water. Commodore Wagerand his fellow captains were court-martialled upon their return to England for failing to secure the treasure on board the *San Jose*, although they did retrieve a significant amount from the other ships that were boarded and captured. The treasure on board the *San Jose* is valued at $20 billion currently—and a complex international legal argument is raging as to who owns it: Colombia, whence it originated, although the silver and gold mines were located at Potosi, now in Bolivia, or Spain, who owned the *San Jose* and acquired its cargo. Underwater photography of the wreck site reveals some bronze cannon, impressed Seville 1655, and an immense quantity of gold bullion, coins and a blue and white porcelain dinner service, which indeed does appear to be in an excellent if not perfect condition.

Of particular relevance to the theme of the current text is the question: where did this Chinese blue and white dinner service come from — was it originally traded across the Pacific from China to Central South America? It would be expected that, since it was designed for a powerful European monarch, namely King Philip V of Spain, then it would be probably Kangxi Qing-era porcelain of the highest imperial quality and its future recovery would be eagerly awaited by museum conservators and collectors of Chinese porcelains.

9.3 Identification of the Earliest American Blue and White Porcelain of John Bartlam Cain Hoy China Manufactory, Charleston, South Carolina, 1765–1769, and the Bonnin and Morris Manufactory, Philadelphia, 1770–1773

The primary purpose of the American ceramics manufactories initially was to produce local porcelains of high quality to rival the porcelain imported into the American Colonies, especially before the Declaration of Independence from Great Britain was signed on July fourth, 1776, in Philadelphia, Pennsylvania, which declared that the 13 colonies were henceforth an independent country. The imported porcelains in this case were, however, mainly derived from China and England and only to a smaller extent from continental Europe, and these especially were mainly sourced from the two large output factories of Sevres and Meissen.

The claim for the first porcelain manufactory in America has generally been accorded hitherto historically to Bonnin & Morris in Philadelphia, although there was an account of an earlier John Bartlam foundation, which it was hitherto believed to have manufactured earthenwares only, in Cain Hoy, South Carolina, which operated between 1765–1769 and for which shards excavated from the site were analysed by Professor Victor Owen in 2007 [30]. Initially, it was believed that John Bartlam (1735–1781), a master potter who emigrated to America from Delph Lane,

Staffordshire, in 1763 established a factory in Cain Hoy, nine miles north of Charleston in South Carolina then moving a little later to Camden nearby, where he manufactured mainly creamwares and pearlwares [35]. This seemed to definitively cede the honour of the prime manufacturer of porcelain to the later Philadelphia manufactory of Bonnin & Morris which operated from 1770 onwards. However, Professor Owen's analyses of the shards excavated from the Cain Hoy site established conclusively that they comprised truly a soft paste phosphatic porcelain, which was actually of a very similar composition to that then being produced in Isleworth and later in Philadelphia by Gousse Bonnin, an Englishman of Huguenot extraction, and George Anthony Morris between 1770 and 1772 [30]. Recent comparative analyses of a shard from Bartlam's Cain Hoy factory and early Bow porcelain (1755–1770) as shown in Table 9.3. are interesting as they conclusively indicate that the two body recipes employed were very closely similar indeed, if not identical [32].

This area in South Carolina, near Charleston, had already been earmarked as a potential site for porcelain production much earlier in the 1730s when Andrew Duche apparently carried out experiments in manufacturing porcelain there using the local white Cherokee North Carolina clays, called *unaker* locally, which have been described as being "*A Clay as White as Lime*" [7, 33]. It has been suggested that Andrew Duche may have actually succeeded in making porcelain in Savannah in the late 1730s from the local Cherokee kaolin clay deposits (*unaker*) but there is no direct evidence of this in the form of surviving pieces or even shards or fragments. Duche left Savannah in February, 1743, and returned to England, where it is noted in their patent of December sixth, 1744, that Messrs. Edward Heylyn and Thomas Frye of Bow had produced porcelain "*from an earthy mixture – an earth the product of the Cherokee nation in America called by the natives unaker*", and it is postulated that Duche had arrived in England with supplies of *unaker* with which he liaised with Heylyn and Frye. In May, 1745, William Cookworthy, a geologist who is credited with the discovery of the Cornish source of high quality kaolin with which he eventually founded his own hard paste porcelain manufactory at Plymouth, recorded that he had been visited "*.....by the person who hath discovered the American china earth, bringing with him several examples of chinaware of their*

Table 9.3 Comparative analysis/% of Bow (1775–1770) and Bartlam (1765–1770) Porcelains

Elemental oxide	Bow	Bartlam Cain Hoy
SiO_2	54.4	45.3
TiO_2	0.2	0.2
Al_2O_3	4.8	4.5
Fe_2O_3	0.2	0.2
MgO	0.3	0.5
CaO	20.7	26.7
Na_2O	0.2	0.6
K_2O	0.5	0.6
SO_3	2.7	2.7
BaO	0	0.1

making with him which were I think equal to the Asiatic—he had found both the petuntse and the kaolin while searching for mines in the back of Virginia" . The restless Duche eventually returned to America and an entry in *The Pennsylvania Packet* of August third, 1772 refers to him clearly in assisting Bonnin and Morris with their fledgling manufactory in Philadelphia:

> *The proprietors of the China Manufactory in this City* (Author: Gousse Bonnin and George Anthony Morris) *have lately made experiments with some clay presented to them by a Gentleman* (Author: presumably, Andrew Duche?) *of Charlestown, South Carolina, which produces China superior to any brought from the East Indies, and will stand the heat beyond any kind of crucibles ever yet made.* (Hood, *Bonnin & Morris of Philadelphia: The First American Porcelain Factory, 1770–1772*, [17]).

John Bartlam obviously intended to compete with the underglaze decorated blue and white luxury porcelains then being imported from England using the local raw materials in South Carolina, which he achieved successfully between 1765 and 1769.

Only some seven examples of Bartlam's earliest American porcelain from the Cain Hoy manufactory are known to exist, comprising four teabowls and two saucers from the same set and an early teapot in blue and white, without its lid and somewhat damaged, depicting a local palmetto tree and a crane in a chinoiserie type man-on-the bridge decoration, two characteristic flora and fauna of the Carolinas but not shown on English porcelains (Fig. 9.7). This made the large sum of £575,000 ($806,000 including the buyer's premium) when sold recently at a Wooley & Wallis auction in England in 2018, the buyer being the Metropolitan Museum of Art, New York, where it now resides. The significance of this early teapot and its potential association with the so-called "Boston Tea-Party" of 1773, where colonists

Fig. 9.7 Teapot, *ca.* 1765–1769, John Bartlam's Cain Hoy Porcelain Manufactory, Charleston, South Carolina, USA. Soft paste porcelain, decorated in underglaze blue with local flora and fauna. Now in the Metropolitan Museum of Art, New York. Ronald S. Kane bequest, Louis V. Bell, Harris Brisbane, Dick Fletcher and Rogers Funds: Joseph Pulitzer bequest, Richard Chilton, Anthony Wu and Lulu C. Wang Gifts, 2018. Accession number 2018.156. *Public domain, Creative Commons CC0 1.0*

boarded and destroyed the tea cargo on British vessels in the harbour as a protest against taxation and ignited the American Revolution is immediately apparent.

The key to the identification of these early porcelains as being of John Bartlam's manufacture was the chemical elemental oxide analysis carried out by Professor Victor Owen and the identification of wasters from the site at Cain Hoy: it had been catalogued as Isleworth porcelain in a previous auction sale. A teabowl and saucer from the Bartlam china works of an unusual chinoiserie pattern was also sold at auction later, the auctioneer commenting that "*the science is there as well as the connoisseurship*" which certainly echoes the theme and advocacy of the current text precisely. Bartlam's manufacture of porcelain in America is hence seen to provide undoubtedly the first examples discovered and predates the Bonnin & Morris Philadelphia American China Manufactory by 5 years [18, 19]. This can be ascribed to the analytical confirmation that he was in fact making porcelain there as well as the creamware ceramics as thought previously historically which was a triumph for analytical science in correcting the well-established but apparently misinformed history. Further details and an account of the early Bonnin & Morris Philadelphia porcelains is given in a text by Edwards [14, 15].

References

1. Ayers, J., Kerr, R.: Blanc de Chine Porcelain from Dehua. Art Media Resources Publishing Ltd., London (2000)
2. Bratu, I., Marutoiu, C., Hernanz, A., Nemes, O.F., Tiplic, M., Edwards, H.G.M.: Firing temperature determination of some 18th Century Transylvanian stove tiles using spectroscopic techniques. Vib. Spectrosc. **116**, 1–22 (2021)
3. Carter, E.A., Wood, M.L., de Waal, D., Edwards, H.G.M.: Porcelain shards from Portuguese wrecks: Raman spectroscopic analysis of marine archaeological ceramics. Herit Sci. **5**, 1–8 (2017). https://doi.org/10.11886/s40494-017-0130-9
4. Colomban, P.: Differentiation of antique porcelains, celadons and faiences from the Raman spectra of their bodies, glazes and paintings. Asian Chem Lett. ***5***, 133 (2001)
5. Colomban, P.: A case study: glasses, glazes and ceramics. In: Edwards, H.G.M., Chalmers, J.M. (eds.) Raman Spectroscopy in Archaeology and Art History. Cambridge, Royal Society of Chemistry Publishing (2005)
6. Colomban, P.: Raman spectroscopy: a unique tool to analyze and classify ancient ceramics and glasses. Appl Phys A. **79**, 167–170 (2004)
7. Daniels, P.: The Origin and Development of Bow Porcelain, 1730–1747, Including the Participation of the Royal Society, Andrew Duche and the American Contribution. Resurgat Publishers, Oxford (2007)
8. De Waal, D.: Raman investigation of ceramics from 16th and 17th century Portuguese Shipwrecks. J. Raman Spectrosc. **35**, 646–649 (2004a)
9. De Waal, D.: Raman identification of pigment in blue and white Ming porcelain shards. Asian Chem Lett. **8**, 57–65 (2004b)
10. De Waal, D.: Micro-Raman and portable Raman spectroscopic investigation of blue pigments in selected Delft plates. J Raman Spectrosc. **40**, 162–2170 (2009)
11. Donnelly, P.J.: Blanc de Chine. Faber & Faber, London (1969)

12. Eccles, H., Rackham, B.: Analysed Specimens of English Porcelain in the Victoria and Albert Museum Collection. Victoria and Albert Museum Publications, South Kensington, London (1922)
13. Edwards, H.G.M.: Nantgarw and Swansea Porcelain: A Scientific Reappraisal. Springer Publishing, Dordrecht (2017)
14. Edwards, H.G.M.: Porcelain Analysis and its Role in the Forensic Attribution of Ceramic Specimens. Dordrecht, Springer-Nature (2022a)
15. Edwards, H.G.M.: The answer lies in the glaze. In: Porcelain Analysis and Its Role in the Forensic Attribution of Ceramic Specimens, pp. 381–398. Springer-Nature, Dordrecht (2022b)
16. Hanaor, D.A.H., Sorrell, C.C.: Review of the anatase to rutile phase transformation. J. Mater. Sci. **46**, 855–874 (2011)
17. Hood, G.: Bonnin & Morris of Philadelphia: The First American Porcelain Factory, 1770–1772. Williamsburg, Institute of Early American History and Culture at Williamsburg. Virginia/University of North Carolina Press, Chapel Hill (1972)
18. Hudgins, L.: John Bartlam's Porcelain at Cain Hoy: A Closer Look. In: Hunter, R. (ed.) *Ceramics in America*, pp. 203–208 (2007)
19. Hunter, R.: John Bartlam: America's first Porcelain manufacturer. In: *Ceramics in America*. Chipstone Foundation, Milwaukee (2007)
20. Jaen, Z., Chang'sui, W.B., Cunyi, X.: Non-destructive in-situ study of white and black coatings on painted pottery shards from the Bancun site (Henang, China) by Raman microscopy. Spectrosc. Lett. **31**, 1431–1440 (1998)
21. Jarvis, C.: Archaeological classification of age of Sail Shipwrecks based on genever's material culture. Heritage. **6**, 397–416 (2023)
22. Jorg, C., Wilson, P.: Chinese Porcelain in the collection of the Rijksmuseum. The Ming and Qing Dynasties, The Rijksmuseum, Amsterdam, Amsterdam (1997)
23. King, K.: "Jesus Said to His Wife …": a new coptic papyrus fragment. Harv. Theol. Rev. **107**, 131–159 (2014)
24. Kingery, W.D.: The development of European porcelain. In: Kingery, W.D. (ed.) Ceramics and Civilisation, High-Technology Ceramics, Past, Present and Future: The Nature of Innovation and Change in Ceramic Technology, Volume III, pp. 153–180. American Ceramic Society Press, Columbus, Ohio (1984)
25. Kock, L.D., de Waal, D.: Raman studies of the underglaze blue pigment on ceramic artefacts of the Ming Dynasty and of unknown origins. J. Raman Spectrosc. **38**, 1480–1487 (2007)
26. Li, W., Luo, H., Li, J., Lu, X., Guo, J.: The white porcelains from the Dehua kiln site of China, Part I. Chemical composition and the evolution regularity. Ceram. Int. **37**, 355–361 (2011a)
27. Li, W., Luo, H., Li, J., Lu, X., Guo, J.: The white porcelains from the Dehua kiln site of China, Part II. Microstructure and its physicochemical basis. Ceram. Int. **37**, 651–658 (2011b)
28. Lui, L.-G., Mernagh, T.P.: Phase transitions and Raman spectroscopy of anatase and rutile at high pressures and room temperature. Eur Mineral J. **4**, 45–52 (1992)
29. Madini, M., Sibilia, E., Calderon, T., Di Renzo, F.: Spurious thermoluminescence in archaeological ceramics: a study of affecting factors. Tracks Radiat Meas. **14**, 339–342 (1988)
30. Owen, J.V.: Geochemistry of High-Fired Bartlam Ceramics. In: Hunter, R. (ed.) *Ceramics in America*. Chipstone Foundation, Milwaukee (2007)
31. Pagonis, V., Maniatis, Y., Michael, C., Bassiakos, Y.: Spurious and Regenerated Thermoluminescence in Calcite Powder samples. Radiat. Meas. **27**, 37–42 (1997)
32. Ramsay, W. R. : Private communication. (2021)
33. Rauschenberg, B.L.: A Clay as White as Lime of which there is a Design formed by some Gentlemen to make China: The American and English Search for Cherokee Clay in South Carolina, 1745-1775. J Early South Decorative Arts. **17**, 67–80 (1991)
34. Rei, C.: Turning Points in Leadership, Shipping Technology in the Portuguese and Dutch Merchant Empires. Working Paper 11-W23, November. Vanderbilt University, Nashville (2011)

35. South, S.: The Search for John Bartlam at Cain Hoy: America's First Creamware Potter. In: South Carolina Institute of Archaeology and Anthropology, Research Manuscript Series Number 219. University of South Carolina Press, Columbia (1993)
36. Spataro, M., Bimson, M., Meeks, N.: The earliest English porcelains? , Part 3. A scientific investigation of three pieces of porcelain presumed 17th Century from Burghley House. Trans English Ceram Circle. **20**, 187–194 (2008)
37. Tite, M., Freestone, I.C., Bimson, M.: A technological study of Chinese porcelain of the Yuan Dynasty. Archaeometry. **26**, 139–154 (1984)
38. Walton, M.: The pitfalls of using science to authenticate archaeological artefacts. SAS Bull (Newsletter of the Society for Archaeological Sciences). **37**(4), 1–3 (2014)
39. Widjaja, E., Lim, G.H., Lim, Q., Mashadi, A.B., Garland, M.: Pure component Raman Spectral reconstruction from glazed and unglazed Yuan, Ming and Qing shards: A combined Raman and BTEM study. J. Raman Spectrosc. **42**, 377–382 (2011)
40. Zuo, J., Xu, C., Wang, C., Yushi, Z.: Identification of the pigments on painted pottery from the Xishan site by Raman microscopy. J. Raman Spectrosc. **30**, 1053–1055 (1999)

Chapter 10
Lapis Lazuli

Abstract The presence of lapis lazuli minerals in blue pigments, glasses and glazes, its principal sourcing in Badakhshan, Afghanistan, and the identification of its occurrence using the radical polysulfide anion chromophores detectable in low concentrations by resonance Raman spectroscopy. Modern alternatives.

Keywords Lapis lazuli · Ultramarine · Badakhshan original source mines · Resonance Raman spectroscopy · Polysulfide radical anion · Chromophore

10.1 Lapis Lazuli Rocks and Their Minerals

For a very long time and still today, many specialists in the history of art and technology are unaware of the widespread use in certain periods of lapis lazuli powder as a pigment in various matrices including glass. This is understandable because the composition of the blue minerals in lapis lazuli rock, comprising silicates, is little different from that of the glass in which they are dispersed to colour it. In addition, we will note that lapis lazuli can be associated with cobalt in certain cases [12, 13, 32]. This explains why for decades elemental analyzes carried out by 'traditional' methods (SEM-EDS, mass spectroscopy, X-ray fluorescence, PIXE-PIGE, etc.) did not identify (or even detect) the presence of lapis lazuli in glasses, glazes and enamels.

Lapis lazuli was used for millennia in Mesopotamia and Egypt both as a as gem [2, 11] and as a paint pigment [2]. Although the Badakhshan mines in Afghanistan (the most famous of these being Sar-e-Sang) [101] are widely considered as the only sources of the lapis lazuli that was used in ancient times [59, 63, 64, 97] other sources have been taken into consideration by certain authors: these are the Pamir mountains (Lyadzhuar Dara, Tajikistan) [17, 52, 63], Pakistan (Chagai Hills) [5, 17], Siberia (Irkutsk, near Lake Baikal) [63], Iran [63], Burma (Mogok), and the Sinai [82]. A Chilean source (Las Flores de los Andes, Ovalle) came into use during the seventeenth century and afterwards with the establishment of the South-America

H. G. M. Edwards, P. Colomban, *Blue by Fire: A Marker of the Technical History of Glass and Ceramics*, Cultural Heritage Science,
https://doi.org/10.1007/978-3-031-77122-4_10

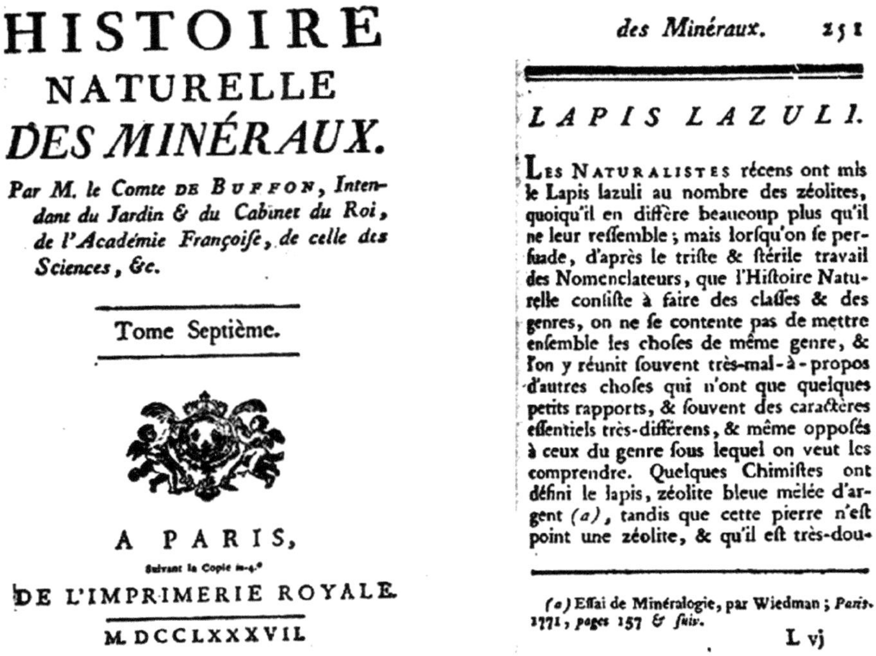

HISTOIRE
NATURELLE
DES MINÉRAUX.

Par M. le Comte DE BUFFON, Intendant du Jardin & du Cabinet du Roi, de l'Académie Françoiſe, de celle des Sciences, &c.

Tome Septième.

A PARIS,
Suivant la Copie in-4.°
DE L'IMPRIMERIE ROYALE.
M. DCCLXXXVII.

des Minéraux. 251

LAPIS LAZULI.

LES NATURALISTES récens ont mis le Lapis lazuli au nombre des zéolites, quoiqu'il en diffère beaucoup plus qu'il ne leur reſſemble ; mais lorſqu'on ſe perſuade, d'après le triſte & ſtérile travail des Nomenclateurs, que l'Hiſtoire Naturelle conſiſte à faire des claſſes & des genres, on ne ſe contente pas de mettre enſemble les choſes de même genre, & l'on y réunit ſouvent très-mal-à-propos d'autres choſes qui n'ont que quelques petits rapports, & ſouvent des caractères eſſentiels très-différens, & même oppoſés à ceux du genre ſous lequel on veut les comprendre. Quelques Chimiſtes ont défini le lapis, zéolite bleue mêlée d'argent *(a)*, tandis que cette pierre n'eſt point une zéolite, & qu'il eſt très-dou-

(a) Eſſai de Minéralogie, par Wiedman ; *Paris*, 1771, *pages* 157 & *ſuiv*.

L vj

Fig. 10.1 Notification of the use of lapis lazuli from the work of Georges Louis Leclerc, Comte de Buffon (1707–1788), published in 1787

to Europe trade route operated by the Spanish Kingdom [9, 29, 34, 35] (Fig. 10.1). More recently, new sites in Pakistan, Iran, Canada, California, Azerbaijan and Italy have been identified [65, 81]. Recent work shows that the heat treatment of a phase containing not S_n^- ions but sulphate ions, by heat treatment and decomposition of SO_4^{2-}ions gave a blue phase due to the formation of the S_n^- chromophores [78] that may explain the use of poor quality lapis lazuli for the coloration of glass.

If it is established that in Europe in the Middle Ages and the Renaissance, lapis lazuli powder was an expensive material for painters [54, 59, 85], whereas in other places and at other times lapis lazuli powder has been used to cover hundreds of m^2 of murals from Armenia and Georgia [10, 93, 96] to China along the Silk Road [8, 34, 35, 72, 91] with the remarkable example of the Mogao caves [71, 100] which are decorated with paintings with Buddhist themes from the fourth to the seventeenth centuries. The price of lapis lazuli in Europe [45–49, 58, 61] actually depends on the ease of establishing a trade route to Persia.

Lapis lazuli, a rock, is made up mainly of two parent minerals:

- Lazurite, a tectosilicate mineral with the addition of sulfate, sulfur and chloride with formula $Na_3Ca(Si_3Al_3)O_{12}(S, Cl, SO_4)$. It is a feldspathoid and is a member of the sodalite group with a cubic unit-cell (P-43n, a = 0.909 nm, Z = 2)

– Haüyne, a tectosilicate with a $Na_3Ca(Si_3Al_3)O_{12}(SO_4)$ composition with a slightly larger unit-cell parameter (0.89 to 0.915 nm), most likely arising from the larger size of the SO_4^{2-} ions.

Actually, both of the above phases have a more complex formula due to the partial substitution of some elements by others. The parent phases are sodalite ($Na_4(Al_3Si_3)O_{12}Cl$), nosean ($Na_8(Al_6Si_6)O_{24}(SO_4)\cdot H_2O$), tsaregorodtsevite ($N(CH_3)_4Si_4(SiAl)O_{12}$), tugtupite ($Na_4BeAlSi_4O_{12}Cl$) and vladimirivanovite ($Na\ Na_6Ca_2[Al_6Si_6O_{24}](SO_4,S_3,S_2,Cl)_2\cdot H_2O$) but all these phases form solid solutions and their characterization is ambiguous by many techniques.

The crystals are transparent to translucent, with a vitreous to greasy lustre. The colour of lazurite and haüyne is usually bright blue, but it can also be white, grey, yellow, green and pink, the other phase being mainly white and grey. Actually, the chromophores are S_n^- (S_3^- and S_2^-) polysulfur (radical) ions and the origin of the colour was only well understood after the studies of these ions in different phases [20, 21, 23, 24, 66], including synthethic ultramarine [22, 25, 65, 84] and lapis lazuli itself [22, 34, 35].

Although Buffon already mentioned that heat treatment improved the colour of lapis lazuli [9] and recent work shows that these natural phases—which even do not appear to be blue under visual examination—can be used as a glaze dye [12, 32, 78]. However, it is only a few decades ago with the work of Clark, Curli and Laganarra [26] concerning luxury medieval Frederician glass and pottery, and then Colomban with studies on common *Lajvardina* pottery (Iran) [32–34] that the importance of lapis lazuli as a colouring agent for glasses and enamels has been recognized.

10.2 Polysulfide Ions and Their Colour

Sulfur formed by polymerisation provides an extensive series of polysulfide dianions $[S_n]^{2-}$ ($n = 2$–9), where calculation demonstrated that their stability is possible up to $n = 18$, and related radical thiozone monoanions $[S_n]^{\cdot -}$ [95]. Series of alkali polysulfides are known and exhibit different colours from yellow to red. The anions can be isolated in aqueous solutions with appropriate cations or in crystalline (organic and inorganic) frameworks and structurally and have been spectroscopically characterized [68]. $S_3{}^{2-}$and $S_4{}^{2-}$ions have a C_{2v} symmetry but the $[S_3]^{\cdot -}$ radical anion has a D_{3h} symmetry [94]. The thiozone anion $S_3^{\cdot -}$ has attracted the most attention as this radical anion had been identified as the key chromophore of the brilliant blue pigment of natural lapis lazuli rock and of synthetic ultramarine (a zeolite with a structure different from that of sodalite). Note that $S_n^{\cdot +}$ cluster cations and protonated, nitrogenated (S_nN^-), etc. species also exist [94]. The addition of elemental sulfur to systems containing $S_2^{\cdot -}$ or $S_3^{\cdot -}$ results in the formation of more sulfur-rich species [94]:

$$S_2^{\cdot -} + 1/8S_8 \Leftrightarrow S_3^{\cdot -} \text{and} S_3^{\cdot} + 1/8S_8 \Leftrightarrow S_4^{\cdot -}$$

Table 10.1 Main absorption and colour of the different polysulfide ions

Ions	Absorption (nm)	Colour
$S_2^{\cdot-}$	430,610	Yellow
$S_3^{\cdot-}$	370,635	Blue
$S_4^{\cdot-}$ or S_4?	520–535,600	Pink
S_6^{2-}	430?	Yellow

The material hosting the S_n^- species are thus easily modified by thermal treatment. Although the smaller radical monoanions may be stabilized in zeolitic matrices, they are usually formed in solution via disproportionation or partial dissociation of the dianions as well as by electrochemical reduction of elemental sulfur. These anionic sulfur species play a crucial role, viz. in alkali-metal–sulfur batteries, organic syntheses, biological chemistry, geochemical processes including metal transport, coordination complexes, atmospheric chemistry and materials science [30, 86]. There are a variety of techniques used to characterize these polysulfide dianions and radical anions both in solution and in the solid state, namely Raman and UV-visible, EPR, NMR and X-ray absorption spectroscopy, X-ray crystallography, mass spectrometry, chromatography and high-level quantum-chemical calculations [95]. The most efficient techniques used are UV-vis-NIR absorption, Resonance Raman and EPR spectroscopy. Depending on the concentration ratio between the two radical anions the material will be either violet-blue (with a high ratio of $S_3^{\cdot-}$/ $S_2^{\cdot-}$), to a steel-blue or green (low ratio) [23, 88]. The exact colour of the materials hosting polysulfide (radical) ions depends upon the relative proportions of the species but the main species can be deduced roughly from the colour of the material (Table 10.1).

10.3 Raman Signatures of Polysulfide Ions and Their Host Framework

Raman spectroscopy is particularly effective for detecting polysulfide ions because the light-chromophore interaction is generally resonant, which leads to condensing the Raman intensity in certain modes and making harmonics and combinations intense enough that many orders are observable (Figs. 10.2 and 10.3). The spectrum therefore depends upon the wavelength of the laser used for the analysis (Table 10.2). This is classic for Raman resonance spectra: as usual the shift in band wavenumber is limited to a few cm^{-1} but the variation in intensities of the different peaks is significant [57]. Consequently, the contribution of the other phases present is thus too weak for their spectrum to be detectable. This is the case for the phases hosting the S_n^- chromophore ions (sodalite or zeolite) and for the associated phases. This is the reason that for most laser excitation conditions from blue to infrared, the contribution of the phases hosting the chromophores, sodalite for lazurite and haüyne in lapis lazuli rock, or the zeolite for synthetic ultramarine, is not detected. It is therefore generally not possible to distinguish from a single Raman spectrum whether a

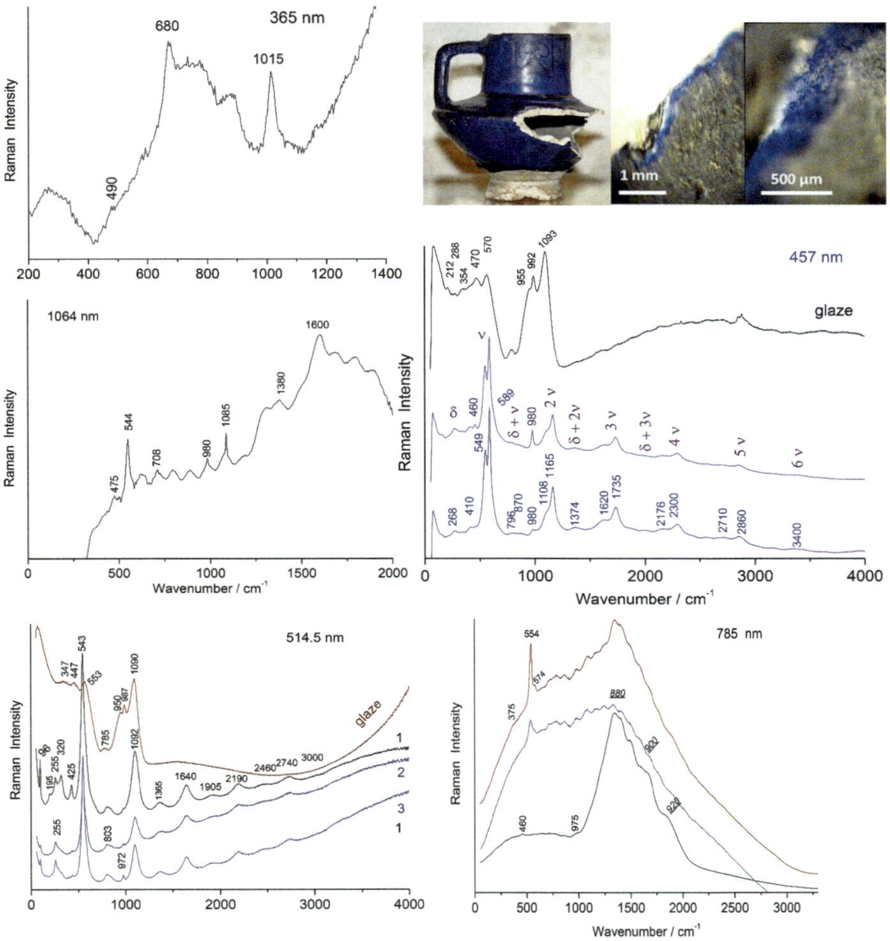

Fig. 10.2 Raman spectra obtained for different excitation lasers (UV: 365 nm; blue: 457 nm; green: 514.5 nm; near infrared: 785 nm; infrared: 1024 nm) on the ultramarine enamel (recorded on lapis lazuli –rich and –free spot) of a Lâjvardina type ewer (Iran, thirteenth century) [34, 35]

blue pigment is attributed to 'natural' lapis lazuli sodalite [4, 19, 79] or synthetic ultramarine zeolite [69]. This is only possible if the resonance effect is removed or sufficiently reduced to observe the Raman signatures of the other phases. Caggiani and Colomban [16] were the first to show that the resonance profile was maximum in the green to red range and therefore the resonance effect was minimized on the NIR side (excitation at 1064 nm) and beyond the blue (UV). A spectrum obtained with an excitation at 364 nm [12] then shows a significant contribution from the host network, in particular the vibrational modes of the SiO_4^{2-} tetrahedra around 980 and 1080 cm^{-1}. Strongest peaks of the framework are sometimes visible on spectra recorded under blue laser excitation (Fig. 10.2).

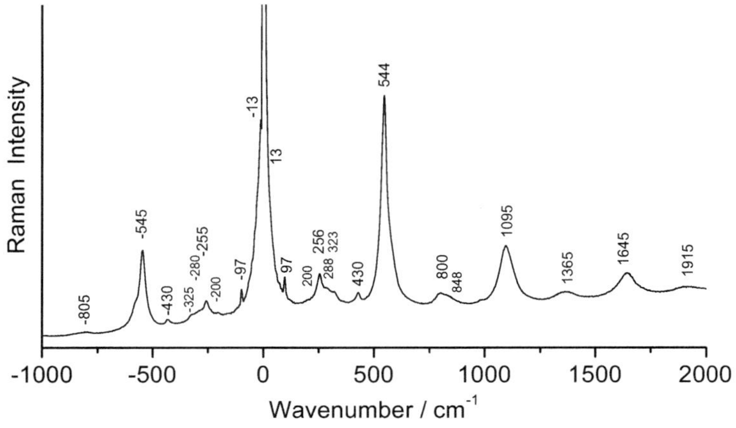

Fig. 10.3 Anti-Stokes and Stokes Raman spectrum excited by the green laser of the ultramarine enamel in Fig. 10.2

Table 10.2 Phases and elemental traces characteristic of lapis lazuli provenance

Traces	Phase	Formula	Reference	Provenance
	Diopside	$CaMgSi_2O_6$	[2, 3, 73, 90]	Afghanistan Siberia
	Pyrite	FeS_2	Idem	Afghanistan
	Wollastonite	$CaSiO_3$	Idem	Chili
	Ba-feldspar		[2, 87]	Siberia
	Sr-feldspar		[2, 87]	Siberia
	Cancrinite	$Na_6Ca_2CO_3{,}Al_6Si_6O_{24}{\cdot}2H_2O$	[87]	Pamir
Se in pyrite			[2]	Chile
Sr in diopside			[2]	Siberia
Ni in pyrite Ti, V and Cr in diopside			[2, 87]	Afghanistan
Cu in pyrite			[2]	Tajikistan
As			[87]	Pamir

Lazurite shows framework modes at about 330, 440, 615–660 (SiO_4^{4-} bending), and 990 and 1065 cm^{-1} (SiO_4^{4-} stretching) [19].

Zeolites present a main band (SiO_4^{4-} bending) around 475 to 485 (pollucite, analcime and gismondine), 485 to 510 (chabazite, mordenite) or 530–535 (natrolite, cowlesite) cm^{-1} [98], and some of these bands can be confused with the signature of the chromophore. However, due to the structural disorder (orientation disorder of the SiO_4^{4-} tetrahedron, due to partial Si/Al substitution), the band widths of the modes are wider than those of the S_n^- ions. In these structures the SiO_4^{4-} stretching modes are not very intense, a feature common for tectosilicates [33, 41].

The phases associated with lazurite (pyrite and diopside) present in the lapis lazuli rock can provide arguments for identifying the provenance. The level of certain trace elements (Ni, Cu, Se, Sr, Cr…) as well as luminescence signatures [83], in particular of wollastonite in relation with the Ti, Mn and Fe content are also used [2].

The $S_3^{\cdot -}$ radical anion is observed in several systems: non-aqueous polysulfide solutions, doped alkali halides, ultramarine pigments (UP) for which S_3^- is the blue chromophore and S_2^- is the yellow one and pigments of zeolite 4A structure. The S_3^- ion has C_{2V} symmetry, and therefore its three vibrational modes should be observed in the Raman and in IR spectra. In resonance Raman spectroscopy, only the symmetric stretching mode ν_1 and the bending mode ν_2 have been studied, whereas the anti-symmetric stretching mode ν_3 has never been observed whatever the system. However, our investigation of several blue samples, with various concentrations of S_2^-, shows that there is a superposition of two bands at *ca* 590 cm^{-1}: the first is assigned to ν (S_2^-) and the second to ν_3 (S_3^-). With the 457.9 nm excitation line, for which the resonance conditions are simultaneously fulfilled for S_2^- and S_3^-, the band at *ca* 590 cm^{-1} is the sum of the contributions of both ν (S_2^-) and ν_3 (S_3^-) vibrations, while, with the 647.1 nm line, which only satisfies the resonance conditions of S_3^-, the band at *ca* 584 cm^{-1} must be assigned only to ν_3 (S_3^-). Furthermore, ν_3 (S_3^-) is observed in green compound and in pigments of the zeolite structure. The ν_3 vibration of S_3^-, which is observed neither in polysulfide solutions nor in doped alkali halides in resonance Raman conditions, can therefore be observed when this species is inserted into the β-cages of the sodalite or of the zeolite 4A structures. So, the band at *ca* 590 cm^{-1} cannot always be assigned to S_2^- in these systems.

Figure 10.2 shows a jug enamelled in ultramarine blue which, upon elemental analysis, shows (Table 10.3) a glaze containing approximately 2% wt of cobalt oxide [32, 33], which is largely sufficient for the colour observed, 0.5% wt being necessary to achieve this. From such an analysis, without a Raman spectrum, cobalt staining would be the obvious conclusion. However, as shown in the spectra in Fig. 10.2 recorded for different laser excitation wavelengths, the resonant spectrum of the $S_n^{\cdot 2-}$ ions is obtained with variations in spectrum and intensity. At certain points with a high magnification objective (x50 or x100) the signature of the glaze,

Table 10.3 Composition of Lâjvardina glaze coloured by dispersion of lapis lazuli grains (see Fig. 10.2)

Oxide	wt%
SiO_2	56.7
Al_2O_3	3.3
MgO	1.8
CaO	15.9
Fe_2O_3	18.3
TiO_2	2
CoO	2

Table 10.4 Wavenumbers of fundamental vibration modes (δ,ν) and combinations and harmonics for Sn radical ions in different host networks and for different laser excitations

Assign.	Intensity	S_2	S_2	S_3	S_2	S_2	S_3	S_3	S_3	S_4
λ (nm)		325	325	325	413.1	457.8	457.8	514.5	785	488
		Lapis lazuli	Lapis lazuli	Lapis lazuli	Ultramarine	Lapis lazuli	Lapis lazuli	Lapis lazuli		
Ref		This work	[12]		[25]	This work	This work	This work		[70]
δ							268	256		303
1ν	1		590		591.2	589	549	544	554	575
2ν	0.52		1170		1179	1165	1108	1092		
3ν	0.29				1762	1735	1620	1645		
4ν	0.17				2340	2300	2176	2190		
5ν	0.10				2912	2860	2710	2740		
6ν					3475		3000	3000		
7ν					4035					
8ν					4590					
9ν					5150					

typical of a silicate glass rich in calcium [34] is obtained except for the excitation at 785 nm where the characteristic fluorescence of wollastonite and calcium glasses are dominant (bands from 880 to 920 nm) [35, 67]. The main peak of the signature of the $S_n^{\cdot 2-}$ radical ion is the stretching mode around 544 cm^{-1} in the green and its bending mode around 255 cm^{-1} [21, 24, 25]. Under green or blue laser excitation a large sequence of combination bands and overtones (δ + ν, 2ν, δ + 2ν, 3ν, δ + 3ν, 4ν, 5ν and 6ν) is observed [6, 21, 33]. An analysis with a very high resolution spectrometer highlights the external modes of the ions at 97 and 13 cm^{-1} (Fig. 10.3). Table 10.4 lists the main peaks of S_n^- ions for different excitations and frameworks. For certain experimental configurations the signatures of S_2 and S_3 $^-$ions are observed simultaneously. As indicated above, small shifts in cm^{-1} are observed but the variability mainly concerns the intensities of the fundamental modes and combinations and harmonics.

The narrow peak at ~973 cm^{-1} sometimes detected under blue and green irradiation is that of diopside, a phase present in lapis lazuli with an intense spectrum.

Fondamental (bending: δ; stretching: ν and overtones and combination modes are labelled for the spectrum recorded under blue excitation. The section view shows the concentration of lapis lazuli grains in an alkaline earth glaze. Italic-underlined numbers correspond to the wavelengths of the absolute scale of the fluorescence bands. The other numbers are the position of the peaks following the relative scale of the Raman vibrational modes.

10.4 Examples of the Use of Lapis Lazuli

The detection of the resonant Raman signature of S_n^- (radical) anions in Cultural Heritage masterpieces was first carried out by Guineau et al. [60, 61] and numerous manuscripts were analyzed by Coupry in the same laboratory [45–48]. This work was taken up on similar artefacts and published in English by Clark [7, 22, 27] which brought the interest in the Raman analysis of dyes to a wider audience. This work was carried out by microsampling with a pin, the microscopic fragment being sufficient for identification under the microscope of the MOLE, the first Raman microspectrometer [31, 51, 53]. This initial work used a green laser, a laser colour that gives strong resonance and therefore remarkable sensitivity for the S_n^-chromophores, even with the low levels of laser illumination allowed on a fragile substrate such as parchment or paper.

Figure 10.4 gives an example of the analysis of eighteenth century wallpaper where only the dark blue knot is made with lapis lazuli. This wallpaper having been installed before the death of Buffon the famous naturalist (Fig. 10.1) in 1788, before the availability of synthetic ultramarine, discovered independently by J. de Montgolfier (1770) [28] and then Guimet (1826) [62]. It is reported that W. Goethe reported having observed in 1787 blue deposits in the Italian lime kilns of Palermo (Sicili), which were recovered as pigments locally [56] and that P. Tassaert in 1814 observed the same in the kilns of the St Gobain Manufactory [Wikipedia] and gave a sample to Louis Nicolas Vauquelin for analysis [99]. At low power (0.1 mW at the sample) the spectrum is not very intense and the combinations and harmonics are not visible.

On the contrary, intense signatures are observed for enamels on glass or ceramic recorded with sample powers of several mW to 10 mW, these high powers being allowed by the high thermal resistance of these materials. Examples are presented in Fig. 10.5 for different periods, namely the Lagidis (Ptolemy, ~first century BCE, Egypt), Norman court (thirteenth century, Sicilia and South of Italy) and Saxon court (eighteenth century, Meissen, Saxony).

Fig. 10.4 Raman signature (excitation 532 nm) of lapis lazuli pigment in a eighteenth Century wallpaper (view and detail) found on a wall of Buffon's library during renovation works [36]

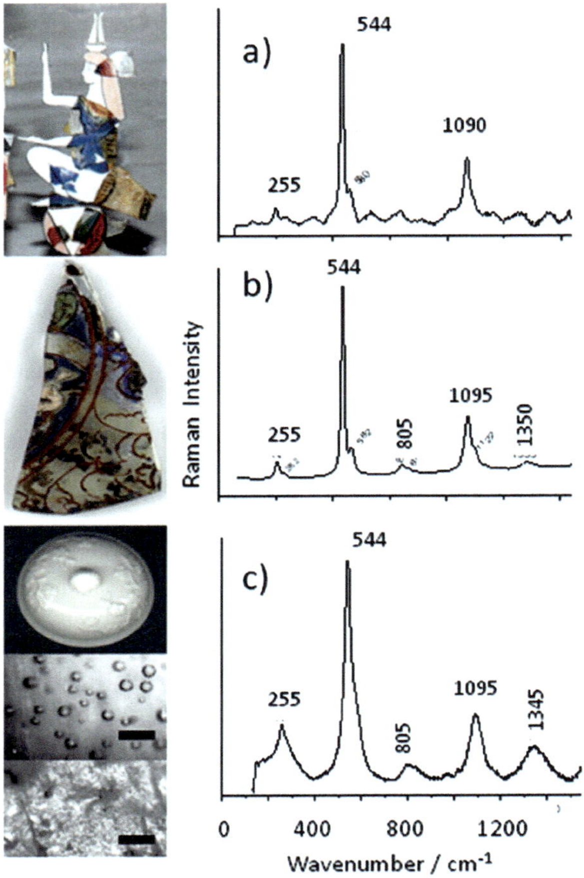

Fig. 10.5 Raman spectra (excitation 514.5 nm) of fragments of (**a**) enamelled glass from a goblet depicting Egyptian figures from the Begram treasure [50] (MG21180, first Century BCE to third Century CE, inv. MG21180, Coll. National Museum of Asian Arts-Guimet, Paris [12]), (**b**) a Frederician enamelled glass (thirteenth Century, inv. 345,874, Coll. Universit'a degli Studi di Bari-Aldo Moro [12, 15, 18]) and (**c**) of the enamel of a porcelain created by J.F. Böttger (Coll. National Museum of Ceramics, Sèvres, [38])

In the Begram hoard which formally dates from the Ptolemaic and successors era, namely from the second to third Century CE [80, 81], the blue enamelled fragments of two goblets [76] revealed the presence of lapis lazuli [12]. The support glass is soda-lime (Raman bands at ~570 and 1090 cm^{-1}) similar to Roman glass. White colouration is achieved using calcium antimonite. Elemental analyses of apparently comparable fragments by Dussubieux and Gratuze [55] indicate the use of mineral soda (natron), a classic 'Roman' procedure. [77] and showed that Ptolemaic ceramics from the Pompeii excavations, therefore, which can be reliably dated before the second century CE, used lapis lazuli as a pigment. The shape of the objects is consistent with Ptolemaic production from Memphis (Egypt) and the concomitant use of lapis lazuli and cobalt (and copper) is observed.

The original works of [26] were the first to show the use of lapis lazuli as a pigment to colour glass or glazes, in enamelled glass and ceramic production at the court of Fredric II Hohenstaufen (1194–1250). Fredric II heir to the Norman crown

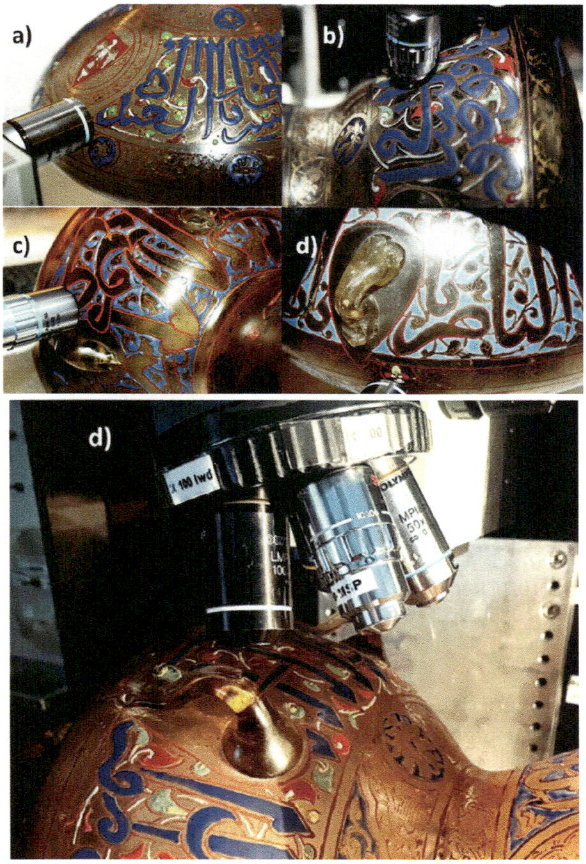

Fig. 10.6 View of original Mameluk period enamelled glass objects (**a**): OA3365; (**c**): OA7568; (**d**): 7880/66) and nineteenth century 'copies' (**b, d**). The objects presented in a, c and d are analyzed with the HE532 portable device [42] while the analyses in b and d are carried out with an HR800 laboratory spectrometer [42, 44]

through his mother Constance de Hauteville, daughter of Roger II king of Sicilia, and was elected Emperor of the Holy Roman Empire (Germanic) in 1196, succeeding his father in 1215. Subsequent work [12–14, 78] confirmed the wide use during this period of lapis lazuli as a heat-stable pigment for glasses and glazes (their thermal stability for several hours at 900 °C was demonstrated by Colomban [43]) and even the local presence of neighbouring non-coloured phases (nosean) but becoming blue by heating led to the hypothesis of using these phases in place of lazurite [78]. Any particles of lazurite being visible at the micron scale (optically) it was assumed that solubilized sulfur radical ions have been formed in the glass in some mosque lamps [89] and in Della Robbia blue glazes [92], the presence of the S_3- ion Raman signature having been identified despite the fact that no inclusions are visible.

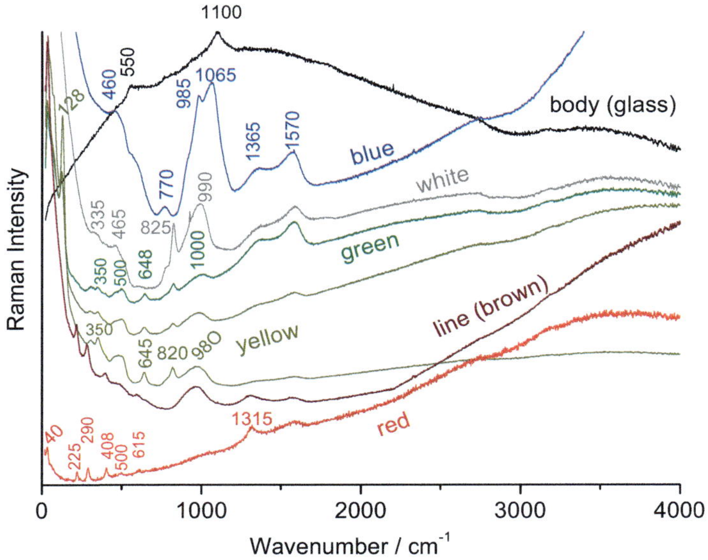

Fig. 10.7 Representative spectra of enamels and glass body of a nineteenth century mosque lamp (Fig. 10.5d). Note the absence of the signatures of lapis lazuli and cassiterite

Figure 10.7 undoubtedly shows the most beautiful use of lapis lazuli as an enamel pigment with enamelled glass objects from the Mameluk Dynasty. First made during the overlapping Ayyubid (c. 1171–1260 C.E.) and Mameluk periods (c. 1250–1517 C.E.) in Egypt and the Levant, enamelled glass mosque lamps and bottles have become representative of masterpieces of Islamic art. Generally termed a "lamp," these vessels actually functioned as "lampshades" that contained a smaller glass vessel, usually a saucer or tube, which held the oil and wick to light the lamp suspended by long metal chains hooked to the vessels' integral glass handles in mosque, *madrasa*, tomb, or *khanaqa* (hostel). Collecting, studying, and copying Islamic art from earlier centuries became popular in Europe during the nineteenth century. In particular, and the French glassmaker Philippe-Joseph Brocard (1831–1896), inspired by enamelled glass mosque lamps from the Mameluk period, collected such lamps and was especially gifted at creating successful facsimiles of Islamic enamelled glass [14, 89].

Comparison of Tables 10.5a and 10.5b shows that the most significant differences concern the solutions used for green and blue enamels and to a certain extent for white opacification. These are also some of these criteria which differentiate these medieval productions from Roman productions [14]. The white of Mameluk enamels is cassiterite and rarely calcium phosphate (bone opacification) while the nineteenth century favours lead-calcium arsenates (Fig. 10.7). Mameluk blue is based on lapis lazuli while the nineteenth century version only uses cobalt ions and/ or cobalt-based phases. (In the twentieth century this is replaced by zircon doped with vanadium [37]. The green colour is obtained by different means, such as the

Table 10.5a Summary of typical colouring agents in Mameluk original artefacts

Enamel colour	Colouring agent	Glassy matrix body
Blue	Lazurite	Soda- lime (rich) glass Some foot richer in alkali+
Yellow	Lead-tin or Lead-tin-antimony Pyrochlore	
Green	Yellow + Cu^{2+} ions or lazurite, cassiterite	
White	Cassiterite, calcium phosphate	
Red	Hematite	

Also reparation using nineteenth century technology

Table 10.5b Summary of typical colouring agents in nineteenth century artefacts with a Mameluk style

Enamel colour	Colouring agent	Glassy matrix type
Blue	Co^{2+}	Soda-lime glass
Yellow	Lead antimonite, uranyl ion Or lead chromate	
Green	Lead chromate Or chromium-based phases	
White	Lead arsenates	
Red	Hematite	

association of techniques used for blue and yellow mixtures for the medieval and earlier periods while chromates are characteristic of the nineteenth and twentieth centuries [89]. There are no precise rules for yellows, the proportions Sn/Sb and other elements (Fe, Si, Zn) being quite variable except that calcium antimoniate appears to be characteristic of the oldest productions (Roman for example). The quality of enamel production can be very variable (Fig. 10.5) with no established correlation with the choice of colouring technology.

Obtaining the Raman spectrum of the transparent glass constituting the body of the objects is difficult to obtain if the laser power on the sample is less than several mW, or even 15 or 20 mW, especially if the object is old. Easily obtaining the spectrum is a strong indication that the object is of a recent manufacture. This is a general characteristic in the study of glasses. Absolute intensity is the first criterion used to assess for the age of an object [39, 74] (Fig. 10.6).

A classic use of blue colouring agents is the brightening of materials whose whiteness is desired. This is why most laundry powders sold in supermarkets contain blue colouring matter. This practice is old and thus one of the main uses of smalt (see Chap. 4) was the whitening of paper and linen sheets! One of the difficulties in obtaining the first hard paste porcelain in Europe, that is to say using kaolin, was the availability of raw materials firing perfectly white and not reddish (due to an insufficiently reducing atmosphere) or greyish (due to impure raw materials). This was the case for the first porcelains made by J.F. Böttger in Meissen, the enamel being only opacified by the presence of micrometric bubbles [38]. Unlike opacification, solutions by addition/precipitation of cassiterite or calcium saturation obtained from the precipitation of wollastonite or calcium phosphate were adopted by the Parisian soft porcelain producers [40, 43, 75], Böttger, who had previous training as a lapidary, added grains of lapis lazuli to brighten his enamel [38] and this can be easily detected using a green excitation laser (Fig. 10.5c).

Modern fakes use the dyeing of cheaper, more porous minerals such as howlite, a calcium borosilicate, $Ca_2B_5SiO_9(OH)_5$, which in various forms simulates lapis lazuli and turquoise and has also then been termed *turquenite* [1]. Recently, the appearance of a commercial synthetic alternative to lazurite based on a resin formulation which has been coloured in simulation of the genuine article has generated much interest in the lapidary and gemstone communities; this substitute is considered so good that the visual identification of genuine lapis lazuli is no longer appropriate nor successfully accomplished. A further variant of fake lapis lazuli is the so-called Swiss/German lapis, which is a type of jasper (chalcedony, an opaque silica) that has been stained blue by use of potassium ferrocyanide and ferrous sulfate. To our knowledge these fakes are not sufficiently thermostable to be used as pigment for glass.

References

1. Ali, E.M.A., Edwards, H.G.M.: Analytical Raman spectroscopy in a forensic art context: the non-destructive discrimination of genuine and fake lapis lazuli. Spectrochim. Acta A Mol. Biomol. Spectrosc. **121**, 415–419 (2014). https://doi.org/10.1016/j.saa.2013.11.005
2. Angelici, D.: Petrographic and mineral-chemical characterization of lapis lazuli: a provenance study of rocks and artefacts from cultural heritage, Plinius, 41 (2015). https://www.socminpet.it/dwl.php?file=files/download/vol-41/dott_angelici_PM_LF.pdf
3. Angelici, D., Borghi, A., Chiarelli, F., Cossio, R., Gariani, G., Lo Giudice, A., Re, A., Pratesi, G., Vaggelli, G.: μ-XRF analysis of trace elements in lapis lazuli-forming minerals for a provenance study. Microsc. Microanal. **21**(2), 526–533 (2015). https://doi.org/10.1017/S143192761500015X
4. Ariai, J., Smith, S.R.P.: The Raman spectrum and analysis of phonon modes in sodalite. J. Physics C: Solid State Physics. **14**(8), 1193–1202 (1981). https://doi.org/10.1088/0022-3719/14/8/015
5. Ballirano, P., Maras, A.: Mineralogical characterization of the blue pigment of Michelangelo's fresco "the last judgment". Am. Mineral. **91**(7), 997–1005 (2006). https://doi.org/10.2138/am.2006.2117
6. Barsan, M.M., Butler, I.S., Gilson, D.F.R.: High-pressure resonance Raman spectroscopic study of ultramarine blue pigment. Spectrochim. Acta A Mol. Biomol. Spectrosc. **08**, 457–459 (2012). https://doi.org/10.1016/j.saa.2012.08.078
7. Best, S.P., Clark, R.J., Withnall, R.: Non-destructive pigment analysis of artefacts by Raman microscopy. Endeavour. **16**(2), 66–73 (1992). https://doi.org/10.1016/0160-9327(92)90004-9
8. Boulnois, L.: La Route de la Soie, Arthaud (coll. Signes des Temps), Paris (1963)
9. Buffon (Leclerc, G.L., Comte de).: Histoire Naturelle des Minéraux, Tome 7. Imprimerie Royale: Paris (1787)
10. Bulia, M.: Medieval art and modern approaches: a new look at the Akhtala paintings. Convivium. **3**(Supplementum), 106–123 (2016). https://www.brepols.net/series/convisup
11. Bulgarelli, G.M.; Tosi, M.: La lavorazione ed il commercio delle pietre semipreziose nelle −1800 a.c., Geo-Archeologia,1-2, 37–50 (1977)

12. Caggiani, M.C., Colomban, P., Valotteau, C., Mangone, A., Cambon, P.: Mobile Raman spectroscopy analysis of ancient enamelled glass masterpieces. Anal. Methods. **5**(17), 4345–4354 (2013). https://doi.org/10.1039/C3AY40648B
13. Caggiani, M.C., Acquafredda, P., Colomban, P., Mangone, A.: The source of blue colour of archaeological glass and glazes: the Raman spectroscopy/SEM-EDS answers. J. Raman Spectroscopy. **45**(11–12), 1251–1259 (2014a). https://doi.org/10.1002/jrs.4492
14. Caggiani, M.C., Valotteau, C., Colomban, P.: Inside the glassmaker technology: search of Raman criteria to discriminate between Emile Gallé and Philippe-Joseph Brocard enamels and pigment signatures. J. Raman Spectrosc. **45**(6), 456–464 (2014b). https://doi.org/10.1002/jrs.4481
15. Caggiani, M.C., Ditaranto, N., Guascito, M.R., Acquafredda, P., Laviano, R., Giannossa, L.C., Mutino, S., Mangone, A.: Combined analysis of enamelled and gilded glassware from Frederick II Castle at Melfi (Italy) to identify technology and raw materials. X-Ray Spectrom. **44**(4), 191–200 (2015). https://doi.org/10.1002/xrs.2594
16. Caggiani, M.C., Colomban, P.: Raman microspectroscopy for cultural heritage studies. Phys Sci Rev. **3**(11), 20180007 (2018). https://doi.org/10.1515/psr-2018-0007
17. Casanova, M.: The sources of the lapis lazuli found in Iran. South Asian Archaeol, 49–56 (1989)
18. Catalano, I.M., Genga, A., Laganara, C., Laviano, R., Mangone, A., Marano, D., Traini, A.: Lapis lazuli usage for blue decoration of polychrome painted glazed pottery: a recurrent technology during the middle ages in Apulia (southern Italy). J. Archaeol. Sci. **34**(4), 503–511 (2007). https://doi.org/10.1016/j.jas.2006.05.010
19. Chukanov, N.V., Vigasina, M.F., Zubkova, N.V., Pekov, I.V., Schäfer, C., Kasatkin, A.V., Yapaskurt, V.O., Pushcharovsky, D.Y.: Extra-framework content in sodalite-group minerals: complexity and new aspects of its study using infrared and Raman spectroscopy. Fortschr. Mineral. **10**, 363 (2020). https://doi.org/10.3390/min10040363
20. Clark, R.J.: Electronic, Raman, and resonance-Raman spectroscopic studies of mixed-valence complexes. In: Mixed-Valence Compounds: Theory and Applications in Chemistry, Physics, Geology, and Biology, pp. 271–292. Springer Netherlands, Dordrecht (1980)
21. Clark, R.J.: Spectroscopic aspects of sulfur chemistry: electronic and resonance Raman spectroscopy of sulfur containing complexes, ions and radicals. In: Studies in Inorganic Chemistry, vol. 5, pp. 221–234. Elsevier (1984)
22. Clark, R.J.: Raman microscopy: application to the identification of pigments on medieval manuscripts. Chem. Soc. Rev. **24**(3), 187–196 (1995). https://doi.org/10.1039/CS9952400187
23. Clark, R.J., Cobbold, D.G.: Characterization of sulfur radical anions in solutions of alkali polysulfides in dimethylformamide and hexamethylphosphoramide and in the solid state in ultramarine blue, green, and red. Inorg. Chem. **17**(11), 3169–3174 (1978). https://pubs.acs.org/doi/pdf/10.1021/ic50189a042
24. Clark, R.J., Dines, T.J., Ferris, L.T.: Resonance Raman spectroscopic studies on the group 6B polyatomic cations S 4 2+, Se 4 2+, and Te 4 2+. J. Chem Soc Dalton Trans. **11**, 2237–2242 (1982). https://doi.org/10.1039/DT9820002237
25. Clark, R.J., Dines, T.J., Kurmoo, M.: On the nature of the sulfur chromophores in ultramarine blue, green, violet, and pink and of the selenium chromophore in ultramarine selenium: characterization of radical anions by electronic and resonance Raman spectroscopy and the determination of their excited-state geometries. Inorg. Chem. **22**(19), 2766–2772 (1983). https://pubs.acs.org/doi/pdf/10.1021/ic00161a024
26. Clark, R.J., Curri, M.L., Laganara, C.: Raman microscopy: the identification of lapis lazuli on medieval pottery fragments from the south of Italy. Spectrochim. Acta A Mol. Biomol. Spectrosc. **53**(4), 597–603 (1997). https://doi.org/10.1016/S1386-1425(96)01768-4
27. Clark, R.J.H.: Pigment identification by spectroscopic means: an arts/science interface. Compte Rendus Chimie. 5, 7–20 (2002)
28. Clement, P.-L.: Les Montgolfières, 2nd edn, p. 20. Tardy, Paris (1982)

29. Coenraads, R.R., Canut de Bon, C.: Lapis lazuli from the Coquimbo region, Chile. *Gems Gemol.* **36**(1), 28–41 (2000). https://www.gia.edu/gems-gemology/spring-2000-lapis-lazuli-chile-coenraads
30. Collongues, R., Gourier, D., Kahn, A., Boilot, J.P., Colomban, P., Wicker, A.: β alumina, a typical solid electrolyte: latest developments in fundamental approach and in battery utilization. J. Phys. Chem. Solids. **45**(10), 981–1013 (1984). https://doi.org/10.1016/0022-3697(84)90045-3
31. Colomban, P.: Une nouvelle technique d'analyse: la microsonde Raman. L'Industrie Céramique. **734**, 815–820 (1979)
32. Colomban, P.: Lapis lazuli as unexpected blue pigment in Iranian Lâjvardina ceramics. J. Raman Spectrosc. **34**(6), 420–423 (2003a)
33. Colomban, P.: Polymerization degree and Raman identification of ancient glasses used for jewelry, ceramic enamels and mosaics. J. Non-Crystall. Solids. **323**(1–3), 180–187 (2003b). https://doi.org/10.1016/S0022-3093(03)00303-X
34. Colomban, P.: Routes du lapis lazuli, lâjvardina et échanges entre arts du verre, de la céramique et de l'enluminure. In: Chine-Méditerranée, Routes et échanges de la céramique avant le XVIè siècle, Taoci 4, pp. 145–152. Editions Findakly (2005a)
35. Colomban, P.: Raman spectrometry, a unique tool for on-site analysis and identification of ancient ceramics and glasses. Mater. Res. Soc. Symp. Proc. **852**., OO8.3.1 (2005b). https://doi.org/10.1557/PROC-852-OO8.3
36. Colomban, P.: Pigment identification of a rare 18th century wallpaper from Buffon library. J. Raman Spectrosc. **42**(2), 192–194 (2011). https://doi.org/10.1002/jrs.2667
37. Colomban, P.: Full spectral range Raman signatures related to changes in enameling technologies from the 18th to the 20th century: guidelines, effectiveness and limitations of the Raman analysis. Materials. **15**, 3158 (2022). https://doi.org/10.3390/ma15093158
38. Colomban, P., Milande, V.: On-site Raman analysis of the earliest known Meissen porcelain and stoneware. J. Raman Spectrosc. **37**(5), 606–613 (2006). https://doi.org/10.1002/jrs.1494
39. Colomban, P., Tournié, A.: On-site Raman identification and dating of ancient/modern stained glasses at the Sainte-Chapelle, Paris. *J. Cult Herit.* **8**(3), 242–256 (2007). https://doi.org/10.1016/j.culher.2007.04.002
40. Colomban, P.; Robert, I.; Roche C., Sagon G.; Milande, V. 2004. Identification des porcelaines «tendres» du 18e siècle par spectroscopie Raman: Saint-Cloud, Chantilly, Mennecy et Vincennes/Sèvres. Archeosciences, revue d'Archéométrie, 28, 153–167. https://www.persee.fr/doc/arsci_0399-1237_2004_num_28_1_1070
41. Colomban, P., Tournie, A., Bellot-Gurlet, L.: Raman identification of glassy silicates used in ceramics, glass and jewellery: a tentative differentiation guide. J. Raman Spectrosc. **37**(8), 841–852 (2006). https://doi.org/10.1002/jrs.1515
42. Colomban, P., Tournié, A., Caggiani, M.C., Paris, C.: Pigments and enamelling/gilding technology of Mamluk mosque lamps and bottle. J. Raman Spectrosc. **43**(12), 1975–1984 (2012). https://doi.org/10.1002/jrs.4101
43. Colomban, P., Simsek Franci, G., Gerken, M., Gironda, M., Mesqui, V.: Non-invasive on-site XRF and Raman classification and dating of ancient ceramics: application to 18th and 19th century Meissen porcelain (Saxony) and comparison with Chinese porcelain. Ceramics. **6**(4), 2178–2212 (2023). https://doi.org/10.3390/ceramics6040134
44. Colomban, P.: Unpublished results (2024)
45. Coupry, C.: Application of Raman microspectrometry to arts objects. Analusis. **28**(1), 39–45 (2000)
46. Coupry, C.: D'une science à l'autre: Chimie et manuscrits médiévaux. Étapes d'une évolution. Bulletin du centre d'études médiévales d'Auxerrel BUCEMA. **10** (2006). https://doi.org/10.4000/cem.462
47. Coupry, C.**:** A la recherche des pigments. Revista de Historia da Arte, série W, (1), 127–136 (2011). https://www.dcr.fct.unl.pt/sites/www.dcr.fct.unl.pt/files/documentos/medieval_colours/apresentacoes/coupry.pdf. Accessed 30 Mar 2024

48. Coupry, C., Lautié, A., Sagon, G., Froment, F.: Case study: fifteen years of artefact investigations by Raman microscopy at the CNRS Laboratoire de Dynamique, interactions et Réactivité, ch. 14. In: Edwards, H.G.M., Chalmers, J. (eds.) *Raman Spectroscopy in Archaeology and Art History*, Vol. 1, pp. 207–216. RSC, UK (2005)
49. De Benedetto, G.E., Fico, D., Margapoti, E., Pennetta, A., Cassiano, A., Minerva, B.: The study of the mural painting in the 12th century monastery of Santa Maria delle Cerrate (Puglia-Italy): characterization of materials and techniques used. J. Raman Spectrosc. **44**(6), 899–904 (2013). https://doi.org/10.1002/jrs.4298
50. Delacour, C.: Redécouvrir les verres du trésor de Begram. Arts asiatiques. **48**(1), 53–71 (1993)
51. Delhaye, M., Bridoux, M., Wallart, F.: Time and/or space resolved laser Raman spectroscopy. J. Mol Struct. **79**, 51–66 (1982). https://doi.org/10.1016/0022-2860(82)85032-1
52. Delmas, A.B., Casanova, M.: The Lapis Lazuli sources in the ancient east. South Asian Archaeol. **1987**, 493–505 (1990)
53. Dhamelincourt, P., Wallart, F., Leclercq, M., Nguyen, A.T., Landon, D.O.: Laser Raman molecular microprobe (MOLE). Anal. Chem. **51**(3), 414A–420A (1979). https://doi.org/10.1021/ac50039a002
54. Dunlop, A.: Materials, origins and the nature of early Italian painting. In: *Crossing cultures: conflict, migration and convergence: the proceedings of the 32nd international congress of the history of art*, pp. 472–476 (2009)
55. Dussubieux, L., Gratuze, B.: Nature et origine des objets en verre retrouvés à Begram (Afghanistan) et à Bara (Pakistan). *De l'Indus á l'Oxus: Archéologie de l'Asie Centrale*, pp. 315–323. Association Imago, Musée de Lattes, Lattes (2003)
56. Elsner, L.: Chemical investigation of the blue color of ultramarine. J für Praktische Chemie (in German). **24**, 385–397 (1841). https://doi.org/10.1002/prac.18410240157
57. Faurel, X., Vanderperre, A., Colomban, P.: Pink pigment optimization by resonance Raman spectroscopy. J. Raman Spectrosc. **34**(4), 290–294 (2003). https://doi.org/10.1002/jrs.996
58. Favaro, M., Guastoni, A., Marini, F., Bianchin, S., Gambirasi, A.: Characterization of lapis lazuli and corresponding purified pigments for a provenance study of ultramarine pigments used in works of art. Anal. Bioanal. Chem. **402**, 2195–2208 (2012). https://doi.org/10.1007/s00216-011-5645-4
59. Frison, G., Brun, G.: Lapis lazuli, lazurite, ultramarine 'blue', and the colour term 'azure'up to the 13th century. J. Int Colour Assoc. **16**, 41–55 (2016)
60. Guichard, V., Guineau, B.: Une nouvelle technique d'analyse des colorants. Premières applications à l'étude des tranchefiles de manuscrits médiévaux. Gazette du livre médiéval. **8**(1), 1–5 (1986). https://www.persee.fr/doc/galim_0753-5015_1986_num_8_1_994
61. Guineau, B., Coupry, C., Gousset, M.T., Forgerit, J.P., Vezin, J.: Identification de bleu de lapis-lazuli dans six manuscrits à peintures du XIIe siècle provenant de l'abbaye de Corbie. Scr. Theol. **40**(2), 157–171 (1986). https://www.persee.fr/doc/scrip_0036-9772_1986_num_40_2_1443
62. Gmelin, C.G: *Ueber die künstliche Darstellung einer dem Ultramarin ähnlichen Farbe* [On the artificial preparation of a pigment similar to ultramarine]. Naturwissenschaftliche Abhandlungen. Herausgeben von Einer Gesellschaft in Würtemberg (Scientific Essays. Published by a Society in Würtemberg) (in German). 2 (10): 191–224 (1828). https://doi.org/10.1002/andp.18280901022
63. Herrmann, G.: Lapis Lazuli: the early phases of its trade. *Iraq. 30(1)* Spring 1968, 21–57 (1968). https://doi.org/10.2307/4199836
64. Herrmann, G., Moorey, P.R.S.: Lapis Lazuli. In: Reallexikon der Assyriologie und Vorderasiatischen Archäologie, vol. VI, pp. 489–492, Berlin (1983)
65. Hogarth, D.D., Griffin, W.L.: Lapis lazuli from Baffin Island—a Precambrian meta-evaporite. Lithos. **11**(1), 37–60 (1978). https://doi.org/10.1016/0024-4937(78)90030-0
66. Holzer, W., Murphy, W.F., Bernstein, H.J.: Raman spectra of negative molecular ions doped in alkali halide crystals. J. Molecular Spectrosc. **32**(1), 13–23 (1969). https://doi.org/10.1016/0022-2852(69)90139-8

67. Kamura, S., Tani, T., Matsuo, H., Onaka, Y., Fujisawa, T., Unno, M.: New probe for porcelain glazes y luminescence at near-Infraredexcitation. ACS Omega. **6**, 7829–7833 (2021). https://doi.org/10.1021/acsomega.1c00322
68. Kamyshny, A., Gun, J., Rizkov, D., Voitsekovski, T., Lev, O.: Equilibrium distribution of polysulfide ions in aqueous solutions at different temperatures by rapid single phase derivatization. Environ. Sci. Technol. **41**(7), 2395–2400 (2007). https://doi.org/10.1021/es062637+
69. Knops-Gerrits, P.P., De Vos, D.E., Feijen, E.J., Jacobs, P.A.: Raman spectroscopy on zeolites. Microporous Mater. **8**(1–2), 3–17 (1997). https://doi.org/10.1016/S0927-6513(96)00088-0
70. Ledé, B., Demortier, A., Gobeltz-Hautecoeur, N., Lelieur, J.P., Picquenard, E., Duhayon, C.: Observation of the $\nu 3$ Raman band of S3− inserted into sodalite cages. J. Raman Spectrosc. **38**(11), 1461–1468 (2007). https://doi.org/10.1002/jrs1795
71. Li, J., Wan, X., Bu, Y., Li, C., Liang, J., Liu, Q.: In situ identification of pigment composition and particle size on wall paintings using visible spectroscopy as a noninvasive measurement method. Appl. Spectrosc. **70**(11), 1900–1909 (2016). https://doi.org/10.1177/0003702816645608
72. Liu, Z., Han, Y., Han, L., Cheng, Y., Ma, Y., Fang, L.: Micro-Raman analysis of the pigments on painted pottery figurines from two tombs of the northern Wei dynasty in Luoyang. Spectrochim. Acta A Mol. Biomol. Spectrosc. **109**, 42–46 (2013). https://doi.org/10.1016/j.saa.2013.02.015
73. Lo Giudice, A., Re, A., Calusi, S., Giuntini, L., Massi, M., Olivero, P., Pratesi, G., Albonico, M., Conz, E.: Multitechnique characterization of lapis lazuli for provenance study. Anal. Bioanal. Chem. **395**, 2211–2217 (2009). https://doi.org/10.1007/s00216-009-3039-7
74. Mancini, D., Tournié, A., Caggiani, M.C., Colomban, P.: Testing of Raman spectroscopy as a non-invasive tool for the investigation of glass-protected miniature portraits. J. Raman Spectrosc. **43**(2), 294–302 (2012). https://doi.org/10.1002/jrs.3010
75. Mancini, D., Dupont-Logié, C., Colomban, P.: On-site identification of Sceaux porcelain and faience using a portable Raman instrument. Ceram. Int. **42**(13), 14918–14927 (2016). https://doi.org/10.1016/j.ceramint.2016.06.132
76. Mairs, R.: Glassware from Roman Egypt at Begram (Afghanistan) and the Red Sea trade. Brit Mus Stud Anc Egypt Sudan. **18**, 61–74 (2012)
77. Mangone, A., De Benedetto, G.E., Fico, D., Giannossa, L.C., Laviano, R., Sabbatini, L., van der Werf, I.D., Traini, A.: A multianalytical study of archaeological faience from the Vesuvian area as a valid tool to investigate provenance and technological features. New J. Chem. **35**(12), 2860–2868 (2011). https://doi.org/10.1039/C1NJ20626E
78. Mangone, A., Caggiani, M.C., Forleo, T., Giannossa, L.C., Acquafredda, P.: A possible natural and inexpensive substitute for lapis lazuli in the Frederick II era: the finding of Haüyne in blue Lead-tin glazed pottery from Melfi Castle (Italy). Molecules. **28**(4), 1546 (2023). https://doi.org/10.3390/molecules28041546
79. Mofrad, A.M., Peixoto, C., Blumeyer, J., Liu, J., Hunt, H.K., Hammond, K.D.: Vibrational spectroscopy of sodalite: theory and experiments. J. Phys. Chem. C. **122**(43), 24765–24779 (2018). https://doi.org/10.1021/acs.jpcc.8b07633
80. Morris, L. 2021. The Begram Hoard and its Context. (Doctoral dissertation, München, Ludwig-Maximilians-Universität). https://scholar.google.com/scholar?hl=en&as_sdt=0%2C5&q=Begram+hoard&btnG=
81. Morris, L.: Revised dates for the deposition of the Begram hoard and occupation at the New Royal City. Parthica. **19**, 75–104 (2017). https://www.torrossa.com/en/resources/an/2193281#
82. Nibbi, A.: Ancient Egypt and some Eastern Neightbours. Noyes Press, Park Ridge (1981)
83. Nikbakht, T., Kakuee, O., Lamehi-Rachti, M.: Ionoluminescence spectroscopy and microscopy of lapis lazuli. J. Luminescence. **181**, 246–251 (2017). https://doi.org/10.1016/j.jlumin.2016.09.023
84. Osticioli, I., Mendes, N.F.C., Nevin, A., Gil, F.P.S.C., Becucci, M., Castelluci, E.: Analysis of natural and artificial ultramarine blue pigments using laser induced breakdown and pulsed Raman spectroscopy, statistical analysis and light microscopy. Spectrochim. Acta A Mol. Biomol. Spectrosc. **73**(3), 525–531 (2009). https://doi.org/10.1016/j.saa.2008.11.028

85. Pagán, E.A., de Ágredos-Pascual, M.V., Juan, M.M., Iranzo, L.R.: Red resins, lapis lazuli and vermilion: high symbolism luxury pigments identified in a Bifaz from the late fifteenth century preserved in the camp del Turia (Valencia). In: Science and Digital Technology for Cultural Heritage-Interdisciplinary Approach to Diagnosis, Vulnerability, Risk Assessment and Graphic Information Models, pp. 302–306. CRC Press (2019)
86. Pan, H., Cheng, Z., He, P., Zhou, H.: A review of solid-state lithium–sulfur battery: ion transport and polysulfide chemistry. Energy Fuel. **34**(10), 11942–11961 (2020). https://doi.org/10.1021/acs.energyfuels.0c02647
87. Re, A., Giudice, A.L., Angelici, D., Calusi, S., Giuntini, L., Massi, M., Pratesi, G.: Lapis lazuli provenance study by means of micro-PIXE. Nucl. Instrum. Methods Phys. Res., Sect. B. **269**(20), 2373–2377 (2011)
88. Reinen, D., Lindner, G.-G.: The nature of the chalcogen colour centres in ultramarine-type solids. Chem. Soc. Rev. **28**, 75–84 (1999). https://doi.org/10.1039/A704920J
89. Salvant, J., Schussler, V., McKenna, C., Bruno, L., Ganio, M., Walton, M.: Investigation of an enameled glass mosque lamp: a 13th–14th-century Mamluk example or 19th-century European version? Herit. Sci. **4**, 5 (2016). https://doi.org/10.1186/s40494-016-0073-6
90. Schmidt, C.M., Walton, M.S., Trentelman, K.: Characterization of lapis lazuli pigments using a multitechnique analytical approach: implications for identification and geological provenancing. Anal. Chem. **81**(20), 8513–8518 (2009). https://doi.org/10.1021/ac901436g
91. Seccaroni, C., Michael, J.U.N.G.: The provenance of Lapis Lazuli, the written sources, and its distribution as pigment in the wall paintings of Central Asia and the Middle East. Acta Via Serica. **7**(2), 89–107 (2022). https://www.dbpia.co.kr/Journal/articleDetail?nodeId=NODE11174765
92. Sendova, M., Zhelyaskov, V., Scalera, M., Gulliford, C.: Micro-Raman spectroscopy characterization of Della Robbia glazes. Archaeometry. **49**(4), 655–664 (2007)
93. Skhirtladze, Z.: Newly discovered early paintings in the Gareja Desert. In: Eastern Approaches to Byzantium. Papers from the Thirty-Third Spring Symposium of Byzantine Studies, pp. 149–167. University of Warwick, Coventry (1999)
94. Steudel, R.: Inorganic Polysulfides S N 2− and Radical Anions S N. In: Steufel, R. (ed.) *Elemental Sulfur und Sulfur-Rich Compounds II*, pp. 127–152. Springer, Berlin (2003)
95. Steudel, R., Chivers, T.: The role of polysulfide dianions and radical anions in the chemical, physical and biological sciences, including sulfur-based batteries. Chem. Soc. Rev. **48**(12), 3279–3319 (2019). https://pubs.rsc.org/en/content/articlelanding/2019/cs/c8cs00826d
96. Thierry, N., Thierry, M.: Peintures murales de caractère occidentale en Arménie: Eglise Saint-Pierre et Saint-Paul de Tatev. Byzantion. **38**, 180–242 (1968). https://www.jstor.org/stable/44169294
97. Tosi, M.: The Lapis Lazuli trade across the Iranian plateau in the 3rd millennium B.C., pp. 3–22 (1974) Studi in onore di G. Tucci, Istituto Universitario Orientale, 1974, Napoli
98. Tsai, Y.L., Huang, E., Li, Y.H., Hung, H.T., Jiang, J.H., Liu, T.C., Fang, J.N., Chen, H.F.: Raman spectroscopic characteristics of zeolite group minerals. Fortschr. Mineral. 11(2), 167 (2021)
99. Vauquelin, L. : *Note sur une couleur bleue artificielle analogue à l'outremer* [Note on an artificial blue color similar to ultramarine]. Annales de Chimie et de Physique (in French). **89**, 88–91 (1814). https://babel.hathitrust.org/cgi/pt?id=nyp.33433062722727&view=1up&seq=98. Accessed 30 Mar 2024
100. Wainwright, I.N., Moffatt, E.A., Sirois, P.J., Young, G.S.: Analysis of wall painting fragments from the Mogao and the Bingling Temple grottoes. In: *Proceedings of the international conference on the conservation of grotto sites, Mogao grottoes, Dunhuang, October 1993*, pp. 334–340 (2004). https://en.wikipedia.org/wiki/Ultramarine#19th_century_(invention_of_synthetic_ultramarine
101. Wyart, J., Bariand, P., Filippi, J.: Lapis lazuli from Sar-i-Sang, Badakhshan, Afghanistan. Gems Gemmol. **17**, 184–190 (1981). https://www.gia.edu/gems-gemology/winter-1981-lapis-afghanistan-wyart

Chapter 11
Cobalt Blues, Smalts

Abstract The cobalt blues, cobalt sources and their mining, European, Middle Eastern and Chinese origins of cobalt blue pigments on porcelains from analytical studies and presence of minor components such as manganese, bismuth, zinc and arsenic. Extraction of cobalt from mined ores. Smalt, zaffre, natural and synthetic ultramarine. Modern equivalents.

Keywords Cobalt blue · Zaffre · Smalt · Ultramarine · Presence of minor elements · Geographical sourcing of cobalt ores

11.1 From Egyptian Blue to Phases Coloured by Co^{2+} Ions

The previous chapter presented the uses of the calcium copper silicate, cuprorivaite, in its crystalline ($CaO\ CuO\ 4SiO_2$ or $CaCuSi_4O_{10}$) and amorphous ($0.5CaO\ 0.3\text{-}1Na_2O\ 0.5CuO\ 0.5Fe_2O_3\ 4SiO_2$) forms, a pigment widely used since the Pharaonic (second millennium BCE) and Mesopotamian periods (first to second millennium BCE) until the end of the Roman Empire (fourth-Century CE) [43]. Centuries after, Chinese Han potters developed a barium homologue [4, 18, 19, 67] in particular for the decoration of the famous terra cotta army of the Qui Shi Huang Xiang mausoleum at Xi'an in the late third-Century BCE [102]. But the most used technique for colouring solid or layered glass (glaze and enamel) is the addition of a small quantity of Co^{2+} ion (0.5% wt CoO is sufficient to colour strongly a glass) dissolved in the vitreous silicate network or hosted by a crystalline phase (a pigment) dispersed in the glassy network, a technique used intermittently since ca. fifteenth-Century BCE in Egypt, using cobalt-alums extracted from Dakhla oasis while the other source from the fifth-Century BCE to the end of the Ptolemaic period (first-Century BCE) was expected to come from Iran [64]. On the contrary, the cobalt used for the Mycenaean (twentieth-Century BCE), Phoenician (tenth-Century BCE) and Hallstadt (eighth-Century BCE) periods is supposed to come from the Erzgebirge (Saxony-Bohemia) [33]. The only alternative until the

H. G. M. Edwards, P. Colomban, *Blue by Fire: A Marker of the Technical History of Glass and Ceramics*, Cultural Heritage Science,
https://doi.org/10.1007/978-3-031-77122-4_11

development of new pigments such as zircon doped with vanadium in the 1970s [20, 42] was the use of lapis lazuli grains [28], a rock rich in lazurite, a natural blue silicate hosting S_2^- and S_3^- ions (the chromophores), giving the mineral the ultramarine colour [18, 19]. This natural pigment and its synthetic equivalent will be discussed in the following chapter. This chapter focuses on the use of cobalt but alternative blue colouring techniques will also be mentioned.

11.2 Saffre and Smalt: Definitions

From at least the Middle-Ages in the West, the source of cobalt in Fire Arts as in other uses is *smalt* (or rather as we will see later, *smalts*) [5]. *Smalt* is a potassium silicate glass rich in cobalt (up to ~20% wt CoO [59, 60, 81, 98]) initially prepared by mixing potash glass with *saffre*, the slag obtained during the treatment of silver (and/or bismuth) ores [28, 38, 59, 60, 77, 79]. Indeed, in the extraction of silver or even bismuth, the remainder after recovery of the metal by fusion, namely the slag, contains cobalt and metallic or non-metallic elements such as sulphur, arsenic, etc.); it is also called *speisses* or *matte* in German. With lapis lazuli [18, 19] and azurite [54] *smalt* was the pigment used in painting (and in enamelling as lapis lazuli) before the development of nineteenth-Century organic (Prussian blue, phthalocyanine blue) and inorganic (cobalt silicate and aluminate) pigments [10, 18, 19]. The typical composition of *smalt* is: 66–72 SiO_2, 10–21 K_2O, 2–18 CoO and 0–8 As_2O_3 [81]. Recently, much evidence of the use of *smalt* outside the Ancient World, such as in early Qing Dynasty Chinese painting [61, 69, 104] and enamelled objects [27, 30, 31, 41, 47] and in Japanese Edo period paintings [62] and porcelain (see Table 12.2 in ch. 12), was demonstrated.

11.3 Cobalt Geology: Sites and Associated Features

Cobalt is the 30th most abundant element on earth, like lithium, niobium, scandium and lead, but sites rich in cobalt are rare. Cobalt is an element existing exceptionally in its native state in certain meteorites [50] and in the placers of the Aidyrlya River, south of the Urals (Russia) [82]. It is only exploited as a by-product from the exploitation of other metals (98% of cobalt is today recovered from copper exploitation; the remaining part is a by-product of nickel and arsenic extraction; in the past cobalt was a by-product of arsenic, silver and bismuth production [35].

Cobalt ores are classified into three types of deposit, primary, secondary and tertiary, according to the geological conditions of formation [28]. The last type of site corresponds to the geological corrosion of primary and secondary sites to produce sedimentary rocks, laterites and evaporites. If this type of site has been little used in the past (except in Egypt for salt lake deposits), undoubtedly due to the low concentrations, these sites are important today [35]. In Europe none of the old mines are productive except in Scandinavia and the Balkans. On the other hand, it is possible that from Persia to the Urals, Tibet and the Philippines, ancient sites remain productive. The Bou Azzer deposit in Morocco (erhytrite cobalt-nickel-iron arsenide) produced only cobalt [46] and can be considered as representative of ancient mining technology (crushing, flotation, roasting and acid attack) [35].

Cobalt is a transition metal and therefore its chemical properties are close to those of other transition metals, namely iron, manganese, copper and nickel mainly. It was isolated by Georg Brandt in 1735 [55] but already identified at the end of the Middle Ages, as noted in Paracelsus' and Georgius Agricola's books (1522 and 1550, respectively) [2]. It follows that many transition metals are associated together in natural deposits. For instance, cobalt-rich crusts (nodules) are formed on the deep seafloor, typically in areas with volcanic activity [48, 56]. These crusts are composed of concentric layers of cobalt-rich minerals (oxyhydroxides), such as manganese and iron oxides that grow over time. Cobalt-rich crusts are considered a potential future source of cobalt, although they are currently not economically viable for commercial extraction due to the technical and environmental challenges associated with deep-sea mining [57]. Continental drift and associated orogenesis cause these deposits formed at the seabed to be found in little-transformed emerged lands in the form of oxyhydroxides of Fe, Mn, Ni, Cu and Co, in deposits that are economically exploited for those metals (in which cobalt is present as a byproduct). This type of 'primary' deposit is observed in association with recent massifs—in the geological sense of the term—such as those around the Himalayas, and is characteristic of many Asian sources of cobalt.

Until the middle of the nineteenth-Century [59, 60, 80] the classification of different minerals containing cobalt was made mainly on visual criteria (ashy, specular, vitreous appearance, efflorescence, earthy, etc.) and the presence of arsenic in the ores. Confusion between arsenic and cobalt was common.

If these deposits encounter hydrothermal circulations the primary minerals are dissolved. Aqueous fluids are often associated with magmatic or volcanic activity, and they can precipitate cobalt minerals in veins, fractures, or other rock formations in association with a group of five metals: Ni-Co-As-Ag-Bi [66]. These hydrothermal deposits—the secondary type—are found in association with sulphide and arsenide minerals [9, 28]. They are typical of ancient mountains like the Hercynian massifs of Europe. The classification of cobalt ores based on the elements associated with the cobalt element was therefore sought from the availability of elementary composition data. Thus, Kaczmarczyk identified the use of different sources of cobalt by the Egyptians [64]. Dayton linked the association of nickel and bismuth with cobalt in ancient blue Mediterranean glass to *Erzgebirge* (namely Ore Mountain in English, *Monts métallifères* or *Misnie* in French) ores [33]. Gratuze et al. was interested in enamelled glass and ceramics from the Mediterranean basin and showed the importance of 3 elements as main classification tools (arsenic, nickel and zinc) and incidentally manganese, copper, bismuth, and indium [51, 52].

The re-crystallization into arsenides (and sulphides) of transition metals in Ni-Co-As-Ag-Bi veins is at the origin of the word cobalt [28, 76]. This term cobalt comes from the name of an evil mountain demon, the *kobolt (or Kobelet*, *Kobolt* or *Kobelt* and in Latin *cobaltum*). In fact, the toxicity—and the volatility of arsenic during the roasting of ores—have been well understood since the Middle-Ages, as described in the famous 1556 book of Georgius Bauer, alias Georgius Agricola [68]. Kobalt is already identified as such in Paracelsus' book (1522).

11.4 Tentative Distinguishing of the Cobalt Ore Sources

The limitations of these first-made summary classifications based on non-quantitative detection, or even on ratios or biplots related to the presence of some elements were shown by Colomban et al. [28]. The use of ternary diagrams based on the compositions when the measurement allows it (the local analysis of strongly coloured and homogenous 'thick' regions by SEM-EDS or by XRF) or from the intensities of the normalized characteristic XRF signals if the hypotheses allowing the calculation of composition are not fulfilled, is more reliable especially when the measurements can only be done from the surface (non-invasive analysis). We will illustrate the advantages of the ternary diagram visualization as the basis of the study and understanding of the densification-reaction mechanisms in Fire Arts.

The ternary diagrams presented in Fig. 11.1, adapted from Colomban et al. [28] illustrate the interest of these visualizations for differentiating cobalt sources from the XRF analyses of artefacts: diagrams show the differentiation between non-Asian (Timurids, Ottoman and related fritware tiles/shards, thirteenth-Century to eighteenth-Century CE) and Asian shards/artefacts (blue-and-white decoration of Yuan, Ming and Qing Dynasty periods (thirteenth-Century to nineteenth-Century CE) porcelain produced in China and Vietnam).

It is obvious that overall the clustering of the data differs clearly, most of the cobalt used in Timurids and Ottoman production coming from secondary deposits (hydrothermal veins in Persia or perhaps Europe, free of manganese) while the cobalt used in Asia mainly comes from primary deposits rich in manganese. However, some data differ from the main group.

Classifications based on the presence of certain elements or simple biplots are insufficient for unambiguous conclusions to be made. The distribution along certain axes—whether we use a metallic element or oxide representation-, binary (the sides of the diagrams) or according to defined relationships (lines, e.g. along the axes, for example CoO-Fe_2O_3 or CoO-MnO), can have several origins. As indicated above, transition metals have similar properties and minerals are generally solid solutions. The same goes for the Ni-Zn lines. The dispersion of the distributions of the points can also come from the variable contribution of the characteristic chromophoric elements due to their variable concentrations in the glaze or the enamel in order to create the decoration. For example, a colourless coating can contain more iron than cobalt ore: in that case, a light blue colour will contain more Fe than a dark blue. The distributions then form a straight line. Finally, if the coloured layer has a thickness less than that probed by the measurement technique [30, 31, 39], the energy of the XRF peaks characteristic of light elements being low, the fluorescence of the X-rays probing only the surface layer (a few μm) for these elements, while for

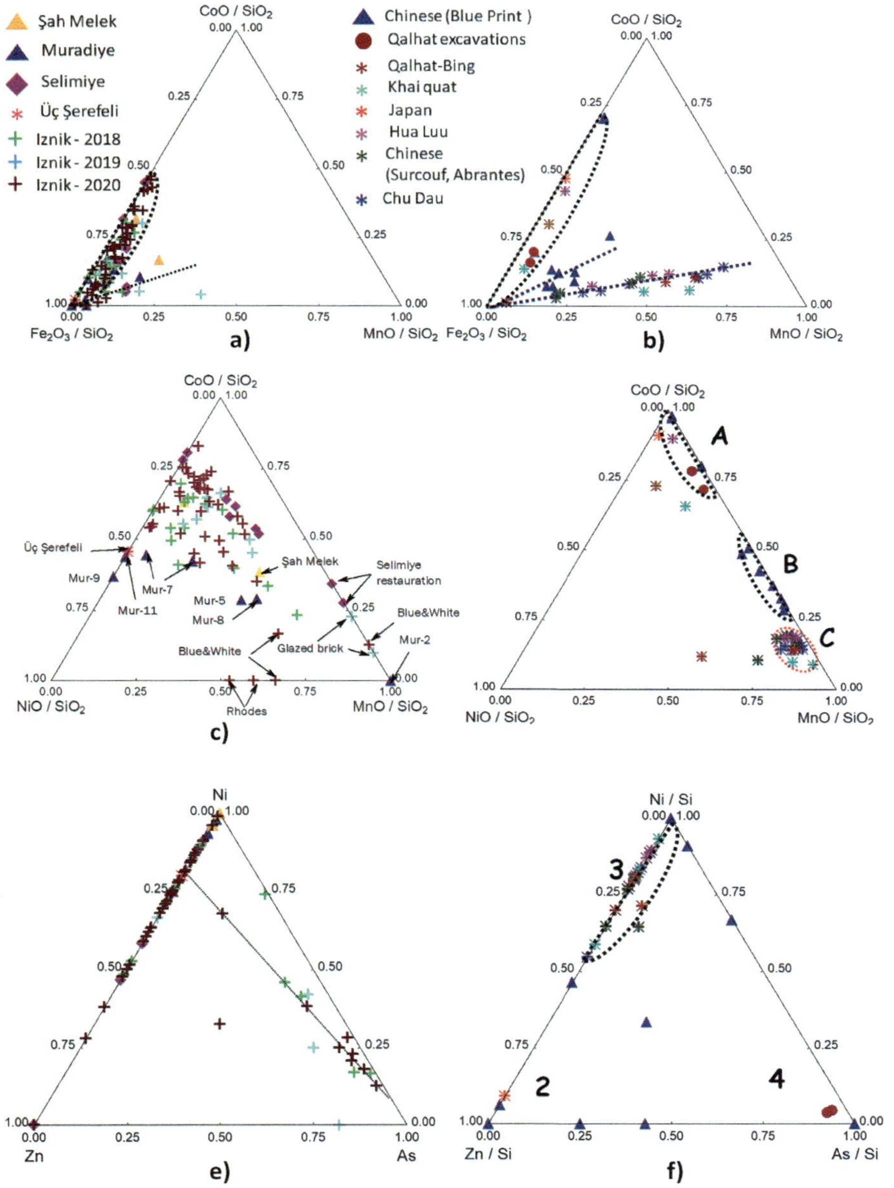

Fig. 11.1 Ternary diagrams comparing the composition ratios in the blue glazed areas of different Ottoman tiles (**a**, **c** and **e**, left side: Sah Melek, Muradiye Üç Serefeli and Selimiye Edirne mosques and shards from Iznik kiln excavation) and shards from excavations or Asian porcelain objects (**b**, **d** and **f**, right side: Chinese, Vietnamese and Japanese shards and artefacts, contemporary periods of the Chinese Yuan and Ming Dynasties). Diagrams constructed from data from [28]

heavy elements (Pb, Sn, Ba, Bi, etc.) the signal comes from a larger thickness (up to mm): the case of transition metals (chromophores) [11, 29, 39] is intermediate in that the depth probed in XRF is 200 to 300 μm. Devices using very energetic ion beams (PIXE) or electron beams (SEM-EDS) for the excitation of X-ray fluorescence lines are less sensitive to thickness effects because the absorption mainly concerns the X-rays that issue and not the exciting beam.

The CoO-Fe_2O_3-MnO and CoO-NiO-MnO diagrams (Fig. 11.1a and c) highlight the very different nature of the cobalt ores used for Ottoman tiles (secondary deposits) and for Asian porcelain from contemporary periods to the Yuan and Ming Dynasties (primary type deposits). Similar things are observed for the Ni-Zn-As and CoO-NiO-Cr_2O_3 diagrams. In Fig. 11.1 the visualization is made from the intensities of the specific signal of certain elements converted into their oxide equivalent, the usual visualization of ceramicists. Members of the elemental analysis community preferentially use a visualization in elemental content (wt %) or in the intensity of the signal characteristic of an element. This is what is used here in Fig. 11.2.

For comparisons between different measurement campaigns, in addition to the standard calibration in relation to the sensitivity of the instrument (e.g. rhodium or silver anti-cathode peak according the instrument) and the associated corrections, normalization in relation to a major element which varies little from one object to another, e.g. silicon, is recommended. When the thickness of the enamel (1–2 mm) and its homogeneity are sufficient to consider the hypothesis of a homogeneous and infinite material, the element (or oxide) content can be calculated. This is not the case if the enamelling stratigraphy of the setting is complex—as found in multilayers of a thickness less than ~500 μm -; then, only the intensities are comparable. The diagrams are constructed with the compositions—or intensities—of the XRF signals characteristic of a different element exhibiting a significant signal.

The clustering of data corresponding to similar manufactures (nearly similar compositions, and the use of the same raw materials, therefore similar quantities of associated elements or their impurities) is preserved for the two procedures. This is shown in diagrams (a) and (b) of Fig. 11.2. These Co-Fe-Mn and Co-Mn-As diagrams (normalized relative to Si) compare the data measured by XRF for porcelain with blue-and-white decoration from different locations (excavations in the Qalhat, Sultanate of Oman [92], excavations at Chu Đau, Hoa Luu and Khal quat in Vietnam [70], specimens from the wreck of the Corsaire Surcouf in France and other collections (Palais D'Abrantès, Lisbon, Portugal, Blue Print Collection) [21]). Most of these porcelains correspond to the periods of the Yuan and Ming Dynasties. Some shards from the nineteenth and twentieth century (among the shards from Hoa Luu, the Vietnamese medieval capital) and a Japanese shard from the twentieth century have been included. For objects from the Blue Print collection (Yuan and Ming dynasties) [91] the data calculated in the compositions and those of the intensities are indicated on the diagrams. The positions are shifted but the clustering is preserved. These diagrams are very educational: thus the modern Japanese shard and those modern specimens from Hoa Lu are located at the Co/Si pole (Fig. 11.2b) because the cobalt used is pure, free of arsenic and manganese. On the other hand, as the vitreous matrix of the surface contains a little iron, in the Co-Fe-Mn diagram

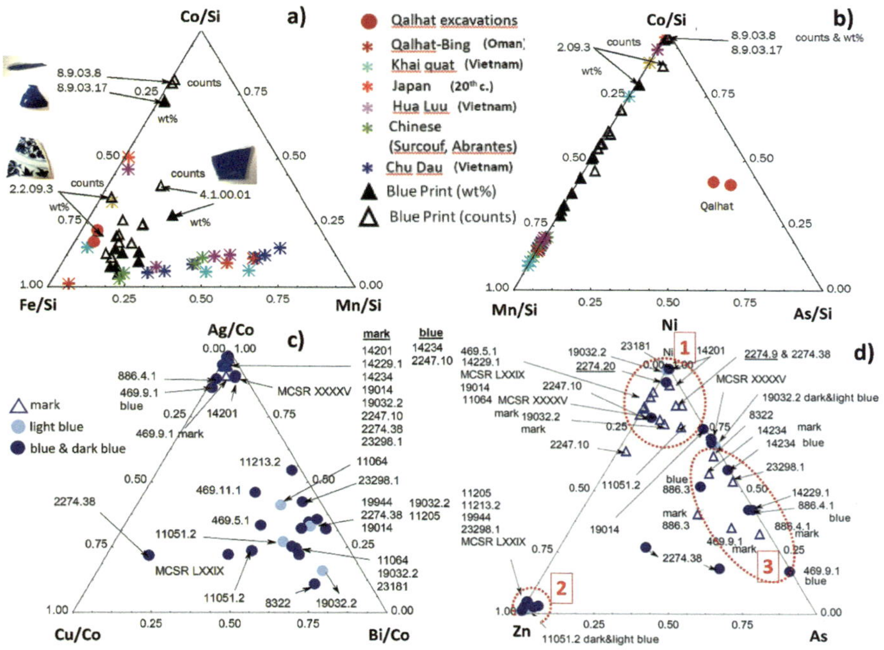

Fig. 11.2 Comparison of the intensities of the XRF peaks characteristic of the cobalt chromophore and elements associated with this element (Fe, Mn, As, Cu, Bi, Ni, and Zn), normalized by the signal of si (**a**, **b**) or Co (**c**), for blue decoration of Asian (China, Vietnam and Japan, a and b) and European (Meissen, Saxony, **c** and **d**) porcelain. See references in the text for complete information regarding the shards and objects analysed. The numbers in the data in c and d are the inventory numbers of the collection of the French national ceramic museum, Cité de la Céramique, Sèvres

(Fig. 11.2a) these shards are in the middle of the Fe-Co line, which confirms the determining role of the contribution of the matrix in the study of the chromophores and the associated elements.

Diagrams show that the data always present a dispersion whose origins are multiple: ii) the ore selection is crude (visual), therefore with contamination by the associated minerals which can be large; ii) natural minerals are generally solid solutions of which the content of certain elements is very variable and certain elements not contributing to the colour can be associated in significant proportion with the chromophore element; iii) the treatment of the minerals is a long-time summary (roasting, acid attack, washing, etc.); iv) as noted above, certain analysis techniques or procedures lead to a 'pollution' of the signal by the material surrounding the coloured zone; v) finally, different grades are produced at different cost according the purification procedure. As a conclusion, it is difficult to specify whether the distribution comes from the disturbance of the signal by the non-coloured materials, from the variability of the colouring element source or from the low efficiency of the refining process [28]. The comparison, for example, between the measurement on an uncoloured zone, a moderately coloured zone and another very strongly coloured zone contributes to the understanding of the spectrum.

11.5 Cobalt Mining and Purification

The 'purification' treatment of ores that changes with time, can lead to a variable elimination of certain elements. Knowledge about the cobalt mines, their mineralogical and chemical characteristics and their exploitation procedures, the treatment and purification processes and the economic and commercial structures are beginning to be better known. Mimoso [76] in a detailed study shows the determining role of Dutch entrepreneurs and their quasi-monopoly in the marketing of *smalt*, glass rich in cobalt extracted from the Erzgebirge on the borders of Saxony and Bohemia and from other places in Europe. Information is available regarding certain mines such as those in the Erzgebirge. For example, it is well established that the elements characterizing ores are different from one mine to another, located in the same region. For instance, the Schneeberg mine (Saxony), exploited since the early Middle-Ages, and known to also produce bismuth [88]. On the other hand, the neighbouring Freiberg mine (60 km away) produced silver and cobalt but with zinc and uranium [89, 103].

Knowledge of the characteristics of old mines, and even their location, remain little studied. Attempts to characterize the mines of Iran [73, 74] mentioned in ancient texts [3] have been made but the period of exploitation of the mines is old and there is no guarantee that the minerals still present are representative of those exploited several centuries or even tens of centuries ago. A recent review paper [28] summarizes the information available on ancient cobalt ores in Europe and Asia.

The Swedish chemist Georg Brandt, specialist in arsenic minerals, was the first to recognize the specific element underlying his ores or mining by-products between 1733 and 1738. We can consider that it will only be in the nineteenth century, particularly in the second part of the century, that 'refined' salts and then cobalt oxide will be available for use in fire art, the tradition of using natural products—at low cost compared to the refined 'chemical' products (carbonates, sulphates, nitrates, oxalates, oxides…) being those used in the arts of fire. In addition, it is well known to ceramicists and solid state chemists, but also to metallurgists, that certain impurities are favourable to densification and the achievement of optimal properties. Using products that are 'too' pure leads to inconvenience and can be economic nonsense.

Colomban et al. [28] lists the main uses of cobalt in the seventeenth and eighteenth century: the brightening of linen sheets, the brightening of paper, in fire arts, and in painting. According to the Encyclopedia, the roasting of the ore (secondary site, rich in arsenic and sulphur) lasts from between 4 and 9 h [59, 60]. After grinding, the powder obtained is mixed with quartz or flint crushed by astonishment (heating to red heat and then immersed in water) and potash or potash glass (frit) in a kiln heated with very dry wood for 6 h (i.e. to avoid formation of hydrogen), the molten mixture is mixed with an iron rod and the refining lasts another 2–6 h depending on the grades required. The hot, liquid glass is poured into water to obtain a frit, called *smalt* [59, 60]. It is reported that the bottom of the crucible is made of bismuth with pure *speiss* on top, believed to be purified cobalt (which assumes reducing atmospheric conditions apply). The frit is then washed, then crushed in an aqueous medium, sieved and decanted. Ten grades of different quality (and undoubtedly of different composition) are indicated by Baron d'Holbach in the Encyclopedia [59, 60], from 'perfect' quality to 'fine colour' quality (the roughest). We understand that these grade varieties lead to distributions of compositions of coloured zones depending on the grade used. Valmont de Bomaré [98] says that 6 grades exist depending on the uses. It is likely that these products differ depending on their cobalt concentration, their particle size but also their purity, i.e. the level of other metallic elements associated with the cobalt. A study of the clustering of compositional data will classify the cobalt according to their grade and not only their origin (mining sites).

Figure 11.3 compares the data concerning the production of a single hard porcelain factory, at Meissen (near Dresden, Saxony), the first and undoubtedly the main hard paste porcelain factory ('china') for the period of the eighteenth and nineteenth centuries. It is obvious that several types of cobalts are used simultaneously. Consider the Ag-Cu-Bi ternary diagram (normalized by the Co signal to be able to compare different decorations and marks, Fig. 11.3a) and the Ni-Zn-As ternary diagram, all of these elements being present in significant quantities in the cobalt ores from the veins, for example, from the Erzgebirge mines close to the factory. The XRF signal of bismuth classifies cobalt into 2 groups; zinc and arsenic signals identify 3 groups (Fig. 11.3b); on the other hand, the nickel and chromium signals (Fig. 11.3d) are not very effective in classifying cobalt but they do separate other colours. The study of the types of object shows that if a certain type of cobalt is associated with the first porcelains of the eighteenth century, the use of this 'luxury *smalt*' continues for pieces of great sophistication whilst another 'cobalt' is used for simpler objects [30, 31]. There is therefore not a one-to-one relationship between 'associated impurities' and geological origin of the ores; the grade used, which depends on the 'purification-selection' process, is also a determining factor. Once the characteristic elements of a cobalt grade have been identified, multivariate analysis tools (Euclidean hierarchical classification (ECH) or Principal Component Analysis (PCA)) can be used to classify the equivalent objects (Fig. 11.3e).

Somes association of impurities can be characteristic of the silicate matrix constituting the glazing: for example yttrium—a rare earth often associated with zircons present in quartz (of sand, pegmatite, etc.) -, rubidium—an impurity associated with sodium and potassium—and strontium—an associated impurity with calcium -, present relatively intense XRF signals for low contents, less than 0.1% by weight. They therefore constitute good indicators for differentiating the raw materials used. Figure 11.3c thus shows 5 groups of glazes, those using enamels for the creation of blue decorations and marks (3 groups), those for the pink decoration and apart from the golden areas [30, 31]. Although these tools do not allow the identification of provenance in a direct and unambiguous manner, they do contribute to the identification of products made using the same raw materials.

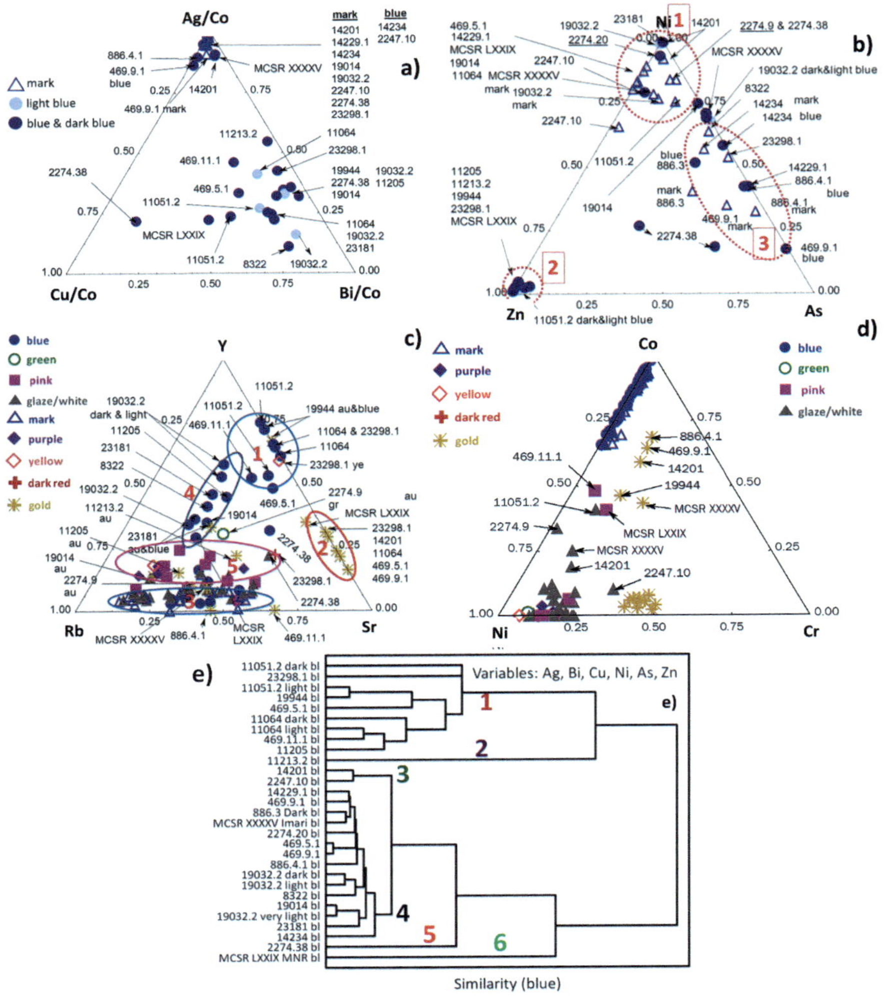

Fig. 11.3 Comparison of the intensities of the XRF peaks characteristic of the cobalt chromophore and elements associated with this element (Ag-Cu-Bi (**a**), Ni-Zn-As (**b**), and Ni-Cr (**d**)), normalized by the signal of Co (**a**), for18th and nineteenth century Meissen porcelain from Sèvres French National Museum collection. Inventory numbers are given (see [[30, 31]); for more details about the studied artefacts). Colours of analysed areas and marks (from the surface) are given. Diagram related to impurities of the silicate matrix (Y-Rb-Sr) is also presented in (**c**). An example of Euclidian hierarchical classification for a selection of variables is given in (**e**)

The different European cobalt mines are listed by different authors [28, 34, 59, 60, 66]. The main places in Europe are, as indicated above, the Erzgebirge (Saxony-Bohemia with Johanngeorgenstadt, Annaberg, Schneeberg, Marienberg, Freiberg and Joachimsthal) [66, 105, 106], the Harz massif (Saxony-Anhalt and Thuringia, now Germany), the Wirtemberg (Germany), the Vosges (Sainte-Marie-aux-Mines,

France), the Pyrenees (Giftain Valley, Aragon, Spain [83]), Norway, Russia (Ural Mountains), England (Cornwall), Slovakia (Dobsina) and Portugal [56].

If cobalt does not have isotopes measurable by LA-ICP-MS, certain associated elements such as nickel; copper and silver have measurable 'stable' isotopes. The isotopic distribution being very specific to the geology of formation of a site, it is possible that in the near future more precise characterizations will make it possible to identify mineralogical sites as has been done for lead [36].

11.6 Glasses Coloured by Cobalt and Their Characterization

Co ion is divalent in glass and the resulting glass colour depends on its overall chemical composition. Potassium glasses tend to be bluer than sodium glasses [79]. We recall that although *smalt* was used as a blue pigment as early as the eleventh century in stained glass (see ch. 17), as a chemical element Co itself was discovered by Georg Brandt in 1735 [55].

The production of blue glass has enjoyed a long tradition since the Late Bronze Age and is related to areas such as Egypt or Mesopotamia [53, 64] and large amounts of blue glass were produced—and recycled—during the Roman period [24, 45]. In Europe, the occurrence of small glassware objects started to increase significantly in the sixth–fifth-Century BCE with blue beads frequently appearing in this production. In the period between the middle of the fourth century and the beginning of the sixth century CE, a change occurred in the type of the cobalt raw material applied (this is mainly true for the Eastern Mediterranean area) [53]. In the Western Mediterranean, this trend was less pronounced and the continuous recycling of old glass shards is suggested for this period [6], as confirmed by analyses of the window glass found in Gothic cathedrals. While fragments of blue tesserae were used as a colorant in the twelfth century glass, sources of cobalt coming from Central Europe are assumed to be the colourants applied in the thirteenth century glass. In the Western Old World, the major turning point in the history of glassmaking is represented by the period of the so-called Carolingian Empire, when significant changes occurred in the raw materials applied. Ash as an alkaline raw material replaced original natron that had been used thus far. For the regions of Central and Northwestern Europe, these were mainly ashes of local plants — beech, oak, fern, etc.— [101]. According to the alkalis with a major representation, this glass is then referred to as a potassium–calcium glass. Producing this glass only required mixing a quartz raw material and ash in a suitable ratio. However, the chemical composition of ashes applied is highly variable [1, 25, 37, 63, 93–96] and depends on a host of parameters, such as the subsoil where the resource plants grow, what plant parts are processed into ash (trunk, branches, bark, leaves), as well as the type of the plants given (beech, oak, fern, etc.; [25, 95]). This is why while the variability of the composition of Roman glass is very low (a few primary sites for the production of large glass ingots), the variability of the compositions of stained glass is very wide: the travelling master glassmakers set up their production workshops near the cathedrals under construction and used local raw materials [17].

The identification of cobalt ions is possible from the colour as summarized in Table 11.1. The colours formed by the 2^+ and 3^+ ions are quite different, the first absorbing in the green and red (therefore allowing a blue colour to pass through as the latter can have a broad absorption from UV to infrared (Table 11.1).

11.7 Cobalt Oxides and Related Synthetic Blue Pigments: Raman Identification

Due to the very great coloring power of cobalt ions, their quantity does not reach the saturation thresholds leading to the precipitation of particular phases and it is only with the production and marketing of blue pigments in the second part of the nineteenth century that new blue inorganic pigments were prepared (Table 11.2). The objective was initially the production of new patented pigments for oil paint marketed 'ready to use' in individual tubes for each shade allowing artists to practice oil painting outdoors. Then some of these pigments were able to find applications in the arts of Fire. Table 11.2 lists the main pigments which fall into 2 main structural groups:

Table 11.1 Characteristic absorptions of cobalt ions in the solid state

Ion chromophore	Site	Absorption (nm)	References
Co^{2+}	Tetrahedral	524, 596, 650	[100] [99] [90] [65]
Co^{3+}	Octahedral	270, 344, 470, 506, 544, 600, 636, 644, 1196	

Table 11.2 Cobalt oxides and co-based pigments and their Raman signatures

Compound (colour)	Structural type formula	Peak wavenumber (cm^{-1})	References
Co_3O_4	Spinel	195,476,516,615,**685**	[84, 85] and [44]
CoO	Rock salt	Forbidden (second order band: 1060)	[85]
$CoSiO_3$	Perovskite Pyroxene	Low intensity, broad (poor documented)	
Co_2SiO_4	Olivine	**810–835** doublet	[40], [78]
$(Co,Zn)_2SiO_4$	Phenacite		
$CoAl_2O_4$	Spinel	**~680** **~850**	[8]
$(Co,Zn)Al_2O_4$			
$Zn(Al,Cr)_2O_4$			

Strongest band in bold (457 nm excitation)

– the spinel structure of cobalt oxide Co_3O_4 and all possible partial substitutions by aluminum, zinc, chromium and iron. $CoA_{12}O_4$ is the pigment in midnight blue porcelain enamels from the Sèvres Imperial/National Factory founded as a Royal Factory, well known for their use as official gifts from the French state since the 1880s. It is a pigment in cutlery fired at high temperatures. The Raman signature of the spinel phases is very similar whatever the exact composition with an intense peak between 650 and 710 cm^{-1}. [8, 23, 32, 85].
– the olivine $CoSiO_4$ structure is rarely detected. The signature of this structure is an intense doublet around 820 cm^{-1}. [40].

The other pigments are not documented as common glaze pigment.

11.8 Alternatives to Cobalt

Cobalt being a rare, costly chemical element and with its complex chemistry, other methods have been developed (Table 11.3).

11.8.1 The Lapis Lazuli

The oldest alternative to the use of cobalt producing a beautiful blue is the dispersion of lapis lazuli powder, a rock made mainly of minerals called haüyne or lazurite whose blue colour results from the presence of S_2^- polysulfide ions in large sites called cages of a plagioclase-type structure. [7, 97] Used widely in Mesopotamia (~thirtieth-Century BCE) for the creation of objects [58], it was under the Lagids, also called the Ptolemies (third -first-Century BCE), that the grains were used as pigments for the blue enamels on glass [12] and glazes [71]. The technique is

Table 11.3 Main alternative pigments to cobalt-based one

Compound	Structural type formula	Peak wavenumber (cm^{-1})	References
Hauyne, lazurite, nosean (lapis lazuli)	$(Na,ca)_8(Al_6Si_6O_{24})$ $[S,SO4)_2$	542,1090	[16] Natural pigment website
Zeolite (synthetic ultramarine)	$Na_8(Al_8Si_8O_{32})S_n$		[15] [13] [75]
Zircon	$ZrSiO_4$:V	1009	[42] [86] [87] [20]

observed in Sicily and southern Italy during the Norman period (thirteenth-Century) [12] and in Mameluke glass production (fourteenth–fifteenth century) [26], and as a pottery glaze [12, 18] both for blue and green as well as in Persian pottery glazes (Lajvardina) [16]. Note that the use of a neighbouring mineral (nausean) that is not coloured blue but becomes blue upon heating is also potentially considered as a pigment [72]. In the latter case, as in certain enamels on glass, cobalt is sometimes used in association with lapis grains while each of the techniques is sufficient to colour powerfully blue! Refer to the appropriate chapter for detailed information. Lapis lazuli powder is also widely used as a paint pigment and to brighten paper, textiles [28] and for the decoration of the first Meissen porcelain made by J.F. Böttger. [22].

11.8.2 Synthetic Ultramarine

Jean-Baptiste Guimet is considered by most historians, namely French, as the inventor of the ultramarine blue pigment synthesis process in 1828. For that a mixture of metakaolin, sodium carbonate, sulfur and a reducing agent is fired in several steps up to 800 ° C, first in the absence of air, followed by an oxidation step at a lower temperature to obtain a zeolith with sulphur ion chromophores in the material. Actually, between 1826 and 1828 three ways of synthesising artificial ultramarine were found independently in Toulouse, Tübingen and Meissen by Jean-Baptiste Guimet, Christian Gottlob Gmelin, Friedrich August Köttig, Carl Leverkus and Thomas Leykauf [75]. However, recent studies demonstrates that the discovery is older and dates from the 1770s with the work of Joseph de Montgolfier, one of the inventors of hot air balloons, the Montgolfières, the first aircraft [15]. The improvements concerning the synthesis of ultramarine were numerous and even recently a series of patents were filed [13, 14] Calcination in the absence of air of a mixture comprised of zeolite, sodium sulfide and sulfur, the ratios between the ingredients produced blue, pink, yellow and green pigments [49].

11.8.3 Twentieth Century Pigments

For economic reasons ancients most of the ancient pigments were replaced in the 1970S by new pigments based on the doping or partial substitution of cassiterite (SnO_2) or zircon ($ZrSiO_4$) with chromophore ions such as vanadium to obtain yellow and blue colours, respectively [20, 42, 86, 87]. Researches on stable, low cost, versions that are free of hazardous chemical elements remain active.

References

1. Adlington, L.W., Freestone, I.C., Kunicki-Goldfinger, J.J., Ayers, T., Gilderdale Scott, H., Eavis, A.: Regional patterns in medieval European glass composition as a provenancing tool. J. Archaeol. Sci. **110**, 189104991 (2019). https://doi.org/10.1016/j.jas.2019.104991
2. Agricola, G.: *De Re Metallica*. Froben, Basel (1556)
3. Allan, J.W.: Abū'l-Qāsim's treatise on ceramics. Iran. **11**, 111–120 (1973). https://doi.org/10.2307/4300488
4. Berke, H., Wiedemann, H.G.: The chemistry and fabrication of anthropogenic pigments Chinese blue and purple in ancient China. EASTM, 94–120 (2000) https://www.jstor.org/stable/43150591
5. Berrie, B.H.: Mining for color: new blues, yellows, and translucent paint. Early Sci. Med. **20**(4–6), 308–334 (2015). https://doi.org/10.1163/15733823-02046p02
6. Bidegaray, A.-I., Pollard, A.M.: Tesserae recycling in the production of medieval blue window Glass. Archaeometry. **60**(4), 784–796 (2018). https://doi.org/10.1111/arcm.12350
7. Bolotina, N.B., Rastsvetaeva, R.K., Sapozhnikov, A.N.: Average structure of incommensurately modulated monoclinic lazurite. Crystallogr Rep. **51**, 589–595 (2006). https://doi.org/10.1134/S1063774506040080
8. Bouchard, M., Gambardella, A.: Raman microscopy study of synthetic cobalt blue spinels used in the field of art. J. Raman Spectrosc. **41**(11), 1477–1485 (2010). https://doi.org/10.1002/jrs.2645
9. Breiter, K.: 800 years of mining activity and 450 years of geological research in the Krušné Hory/Erzgebirge Mountains, Central Europe. Boletim do Museu Paraense Emílio Goeldi-Ciências Naturais. **9**(1), 105–134 (2014). https://doi.org/10.46357/bcnaturais.v9i1.540
10. Burgio, L., Clark, R.J.H.: Library of FT-Raman spectra of pigments, minerals, pigment media and varnishes, and supplement to existing library of Raman spectra of pigments with visible excitation. Spectrochim. Acta Part A, **57**(7), 1491–1521 (2001). https://doi.org/10.1016/S1386-1425(00)00495-9 https://xrfcheck.bruker.com/InfoDepth. Accessed 6 July 2022
11. Bruker. https://xrfcheck.bruker.com/InfoDepth. Accessed 6 July 2022
12. Caggiani, M.C., Colomban, P., Valotteau, C., Mangone, A., Cambon, P.: Mobile Raman spectroscopy analysis of ancient enamelled glass masterpieces. Anal. Methods. **5**(17), 4345–4354 (2013). https://doi.org/10.1039/C3AY40648B
13. Carr, S.W., Hollway, F.: Pigments zeolitiques synthetiques, Patent WO1995019396A1. (1995)
14. Carr, S.W., Hollway, F., Szostak, R.: Pigments zeolitiques synthetiques, Patent WO 1995019397A1. (1995)
15. Clement, P.-L.: Les Montgolfières, 2nd edn, p. 20. Tardy, Paris (1982)
16. Colomban, P.: Lapis lazuli as unexpected blue pigment in Iranian Lâjvardina ceramics. J. Raman Spectrosc. **34**, 420–423 (2003). https://doi.org/10.1002/jrs.1014
17. Colomban, P.: On-site Raman identification and dating of ancient glasses: a review of procedures and tools. J. Cult. Herit. **9**(Suppl), e55–e60 (2008). https://doi.org/10.1016/j.culher.2008.06.005
18. Colomban, P.: Rocks as blue, green and black pigments/dyes of glazed pottery and enamelled glass artefacts–a review. Eur. J. Mineral. **25**(5), 863–879 (2013a). https://doi.org/10.1127/0935-1221/2013/0025-2305
19. Colomban, P.: The destructive/non-destructive identification of enamelled pottery and glass artifacts and associated pigments—a brief overview. Arts. **2**, 77–110 (2013b). https://doi.org/10.3390/arts2030077
20. Colomban, P.: Full spectral range Raman signatures related to changes in enameling technologies from the 18th to the 20th century: guidelines, effectiveness and limitations of the Raman analysis. Materials. **15**(9), 3158 (2022). https://doi.org/10.3390/ma15093158
21. Colomban, P., Treppoz, F.: Identification and differentiation of ancient and modern European porcelains by Raman macro- and microspectroscopy. J. Raman Spectr. **32**(2), 93–102 (2001). https://doi.org/10.1002/jrs.678

22. Colomban, P., Milande, V.: On-site Raman analysis of the earliest known Meissen porcelain and stoneware. J. Raman Spectrosc. **37**(5), 606–613 (2006). https://doi.org/10.1002/jrs.1494
23. Colomban, P., Jullian, S., Parlier, M., Monge-Cadet, P.: Identification of the high-temperature impact/friction of aeroengine blades and cases by micro-Raman spectroscopy. Aerospace Sci. Technol. **3**(7), 447–459 (1999). https://doi.org/10.1016/S1270-9638(99)00102-9
24. Colomban, P., March, G., Mazerolles, L., Karmous, T., Ayed, N., Ennabli, A., Slim, H.: Raman identification of materials used for jewellery and mosaics in Ifriqiya. J. Raman Spectrosc. **34**(3), 205–213 (2003). https://doi.org/10.1002/jrs.977
25. Colomban, P., Tournié, A., de Montmollin, D., Brother; Krainhoefner, L. Brother.: Vegetable Ash as Raw Material in the Production of Glasses and Enamels, for Example the Contemporary Vegetable Ashes from Burgundy, France (2010) https://arxiv.org/abs/1012.1504
26. Colomban, P., Tournié, A., Caggiani, M.C., Paris, C.: Pigments and enamelling/gilding technology of Mamluk mosque lamps and bottle. J. Raman Spectrosc. **43**(12), 1975–1984 (2012). https://doi.org/10.1002/jrs.4101
27. Colomban, P., Zhang, Y., Zhao, B.: Non-invasive Raman analyses of huafalang and related porcelain wares. Searching for evidence for innovative pigment technologies. Ceram. Int. **43**, 12079–12088 (2017). https://doi.org/10.1016/j.ceramint.2017.06.063
28. Colomban, P., Kırmızı, B., Simsek Franci, G.: Cobalt and associated impurities in blue (and green) Glass, glaze and enamel: relationships between raw materials, processing, composition, Phases and International Trade. *Minerals*. **11**, 633 (2021). https://doi.org/10.3390/min11060633
29. Colomban, P., Simsek Franci, G., Gironda, M., d'Abrigeon, P., Schuhmacher, A.-C.: pXRF data evaluation methodology for on-site analysis of precious artifacts: cobalt used in the blue decoration of Qing dynasty Overglazed porcelain enameled at Customs District (Guangzhou), Jingdezhen and Zaobanchu (Beijing) workshops. Heritage. **5**, 1752–1778 (2022). https://doi.org/10.3390/heritage5030091
30. Colomban, P., Simsek Franci, G., Burlot, J., Gallet, X., Zhao, B., Clais, J.B.: Non-invasive on-site pXRF analysis of coloring agents, Marks and enamels of Qing Imperial and non-Imperial porcelain. Ceramics. **6**, 447–474 (2023a). https://doi.org/10.3390/ceramics6010026
31. Colomban, P., Franci, S., Gerken, M., Gironda, M., Mesqui, V.: Non-invasive on-site XRF and Raman classification and dating of ancient ceramics. Application to 18th and 19th century Meissen porcelain (Saxony) and comparison with Chinese porcelain. Ceramics. **6**, 2178–2213 (2023b). https://doi.org/10.3390/ceramics604013
32. Cvejic, Z., Rakic, S., Kremenovic, A., Antic, B., Jovalekic, C., Colomban, P.: Nanosize ferrites obtained by ball milling. Crystal structure, cation distribution, size-strain analysis and Raman investigations. Solid State Sci. **8**(8), 908–915 (2006). https://doi.org/10.1016/j.solidstatesciences.2006.02.041
33. Dayton, J.E.: Geological evidence for the discovery of cobalt blue glass in Mycenaean times as a by-product of silver smelting in the Schneeberg area of the bohemian Erzgebirge. ArchéoSciences, revue d'Archéométrie. **1**(1), 57–61 (1981) https://www.persee.fr/doc/arsci_0399-1237_1981_sup_1_1_1131
34. De Dietrich, Baron: Des Gites de Minerai, des forges et des Salines des Pyrénées, Didot & Cuchet Libraires, Paris (1786). https://books.google.fr/books?id=CoorShRq7NQC&pg=PA296&lpg=PA296&dq=cobalt+Pyren%C3%A9es+Giftain&source=bl&ots=MuVP12wwUb&sig=ACfU3U3GELh9vVflbFQPaa6U8rUk_nPu_Q&hl=en&sa=X&ved=2ahUKEwi2opWu8OSDAxXLcKQEHX2LCiEQ6AF6BAgJEAM#v=onepage&q=cobalt%20Pyren%C3%A9es%20Giftain&f=false. Accessed 17 Jan 2024
35. Dehaine, Q., Tijsseling, L.T., Glass, H.J., Törmänen, T., Butcher, A.R.: Geometallurgy of cobalt ores: a review. Miner. Eng. **160**, 106656 (2021). https://doi.org/10.1016/j.mineng.2020.106656
36. Degryse, P., Henderson, J., Hodgins, G.: Isotopes in Vitreous Materials. Leuven University Press, Leuven (2009) ISBN: 9789058676900

37. Degryse, P., Shortland, A.J.: Interpreting elements and isotopes in glass: a review. Archaeometry. **62**(S1), 117–133 (2020). https://doi.org/10.1111/arcm.12531
38. Delamare, F.: Aux origines des bleus de cobalt: les débuts de la fabrication du saffre et du smalt en Europe occidentale. Compt.-rendus séances Acad. Inscriptions et Belles-Lettres. **153**(1), 297–315 (2009). https://doi.org/10.3406/crai.2009.92472
39. Demirsar Arli, B., Simsek Franci, G., Kaya, S., Arli, H., Colomban, P.: Portable X-ray fluorescence (p-XRF) uncertainty estimation for glazed ceramic analysis: case of Iznik tiles. Heritage. **3**, 1302–1329 (2020). https://doi.org/10.3390/heritage.3040072
40. Do Nascimento-Dias, B.L.: Overview about Raman spectroscopy of types of olivine group minerals: a brief review. J. Raman Spectrosc. **53**(11), 1942–1946 (2022). https://doi.org/10.1002/jrs.6412
41. Duan, H., Zhang, X., Kang, B.: Non-destructive analysis and deterioration study of a decorated Famille rose porcelain bowl of Qianlong reign from the Forbidden City. Stud. Conserv. **64**(6), 311–322 (2019). https://doi.org/10.1080/00393630.2019.1617589
42. Epler, R.A., Epler, D.R.: *Glazes and Glass Coatings*. The American Ceramic Society, Westerville (2000) ISBN 9781574980547
43. Etcheverry, M.-P., Schvoerer, M., Bechtel, F.: Bleu égyptien: mise en évidence de deux processus de formation de la cuprorivaïte. Revue d'Archéométrie. **25**, 87–100 (2001). https://doi.org/10.3406/arsci.2001.1004
44. Farhadi, S., Safabakhsh, J.: Solid-state thermal decomposition of the [CO (NH3) 5CO3] NO3·0.5 H2O complex: a simple, rapid and low-temperature synthetic route to Co3O4 nanoparticles. J Alloys Compd. **515**, 180–185 (2012). https://doi.org/10.1016/j.jallcom.2011.11.135
45. Fontaine, S.D., Foy, D.: L'épave Ouest-Embiez 1, Var: Le commerce maritime du verre brut et manufacture en Méditerranée occidentale dans l'Antiquité. Rev. Archéol. Narbonnaise. **40**, 235–265 (2007). https://doi.org/10.3406/ran.2007.1182
46. Formanek, V., Lauvernier, J.: Beneficiation of cobalt arsenides of Bou-Azzer (Morocco) by gravity concentration and flotation. In: Proceedings of the 6th international mineral processing congress, pp. 333–352. Cannes (1963)
47. Giannini, R., Freestone, I.C., Shortland, A.J.: European cobalt sources identified in the production of Chinese famille rose porcelain. J. Archaeol. Sci. **80**, 27–36 (2017). https://doi.org/10.1016/j.jas.2017.01.011
48. Glasby, G.P., Li, J., Sun, Z.: Deep-sea nodules and co-rich Mn crusts. Mar Georesources Geotechnol. **33**(1), 72–78 (2015). https://doi.org/10.1080/1064119X.2013.784838
49. Gobeltz, N., Ledé, B., Raulin, K., Demortier, A., Lelieur, J.P.: Synthesis of yellow and green pigments of zeolite LTA structure: identification of their chromophores. Microporous Mesoporous Mater. **141**(1–3), 214–221 (2011). https://doi.org/10.1016/j.micromeso.2010.11.016
50. Gooley, R., Moore, C.B.: Native metal in diogenite meteorites. Am. Mineral. **61**(5–6), 373–378 (1976). https://pubs.geoscienceworld.org/msa/ammin/article-abstract/61/5-6/373/40613/Native-metal-in-diogenite-meteorites?redirectedFrom=fulltext. Accessed 4 Mar 2024
51. Gratuze, B., Soulier, I., Barrandon, J.N., Foy, D.: De l'origine du cobalt dans les verres. Revue d'Archéométrie. **16**, 97–108 (1992) https://www.persee.fr/doc/arsci_0399-1237_1992_num_16_1_895
52. Gratuze, B., Soulier, I., Blet, M., Vallauri, L.: De l'origine du cobalt: Du verre à la céramique. Revue d'Archéométrie. **20**, 77–94 (1996) https://www.persee.fr/doc/arsci_0399-1237_1996_num_20_1_939
53. Gratuze, B., Pactat, I., Schibile, N.: Changes in the signature of cobalt colorants in late antique and early Islamic glass production. Fortschr. Mineral. **8**, 225 (2018). https://doi.org/10.3390/min8060225
54. Guineau, B.: Analyse non destructive des pigments par microsonde Raman laser: Exemples de l'azurite et de la malachite. *Stud. Conserv.* **29**, 35–41 (1984). https://doi.org/10.1179/sic.1984.29.1.35

55. Gusenius, E.M.: Beginnings of greatness in Swedish chemistry: Geog Brandt (1694–1768). Trans. Kans. Acad. Sci. **70**(4), 413–425 (1967) https://www.jstor.org/stable/3627593
56. Hein, J.R., Koschinsk, A., Bau, M., Manheim, F.T., Kang, J.K., Roberts, L.: Cobalt-rich ferromanganese crusts in the Pacific. In: Spencer Cronan, D. (ed.) Handbook of Marine Mineral Deposits, pp. 239–279. Routledge (2017) ISBN 9780849384295
57. Hein, J.R., Koschinsky, A., Kuhn, T.: Deep-ocean polymetallic nodules as a resource for critical materials. Nat Rev Earth Environ. **1**(3), 158–169 (2020). https://doi.org/10.1038/s43017-020-0027-0
58. Herrmann, G.: Lapis lazuli: the early phases of its trade. Iraq. **30**(1), 21–57 (1968). https://doi.org/10.2307/4199836
59. Baron d'Holbach: Safre, Saffre, Zaffre ou Smalte, L'Encyclopédie, 1re édition, Tome 14, Paris, pp. 490–493 (1751a). https://fr.wikisource.org/wiki/L%E2%80%99Encyclop%C3%A9die/1re_%C3%A9dition/SAFRE,_SAFFRE,_ZAFFRE_ou_SMALTE
60. Baron d'Holbach: Cobalt, cobolt ou kobold, L'Encyclopédie, 1re édition, Tome 3, Paris, pp. 556–557 (1751b). https://fr.wikisource.org/wiki/L%E2%80%99Encyclop%C3%A9die/1re_%C3%A9dition/COBALT,_COBOLT_ou_KOBOLD
61. Hu, K., Bai, C., Ma, L., Bai, K., Liu, D., Fan, B.: A study on the painting techniques and materials of the murals in the five northern provinces' assembly hall, Ziyang, China. Herit Sci. **1**(1), 1–14 (2013). https://doi.org/10.1186/2050-7445-1-18
62. Ichimiya, Y., Winter, J.: Blue pigments in paintings from the Ming and Qing Dynasties. Sci. Stud. Pigment. Chinese Paint. **37** (2021)
63. Jackson, C.M., Booth, C.A., Smedley, J.W.: Glass by design? Raw materials, recipes and compositional data. Archaeometry. **47**(4), 781–795 (2005). https://doi.org/10.1111/j.1475-4754.2005.00232.x
64. Kaczmarczyk, A.: The source of cobalt in ancient Egyptian pigments. In: Olin, J.S., Blackman, M.J. (eds.) *Proceedings of the 24th international Archaeometry symposium*, pp. 369–376. Smithsonian Institution Press, Washington (1986)
65. Kime, M.B., Makgoale, D.: Characterization of copper–cobalt ores and quantification of Cu2+, Co2+, Co3+, and Fe3+ in aqueous leachates using UV/visible spectrophotometry. Chem. Engn Comm. **203**, 1648–1655 (2016). https://doi.org/10.1080/00986445
66. Kissin, S.A.: Five elements (Ni-Co-As-Ag-Bi) veins, Geosci. Can. 19(3), 113–124 (1992). https://journals.lib.unb.ca/index.php/gc/article/view/3768/4282. Accessed 15 Dec 2023
67. Knight, K.S., Henderson, C.M.B., Clark, S.M.: Structural variations in the wesselsite-effenbergerite (Sr1-xBaxCuSi4O10) solid solution. Eur. J. Mineral. **22**, 411–423 (2010). https://doi.org/10.1127/0935-1221/2010/0022-2025
68. Langård, S.: Georgius Agricola, a Pioneer in the toxic hazards of mining, and his influence. In: Toxicology in the Middle Ages and Renaissance, pp. 83–90. Academic (2017). https://doi.org/10.1016/B978-0-12-809554-6.00008-1
69. Lei, Y., Yang, H., Qu, L., Wang, S.: Technical study of architectural paintings (Caihua) in the Forbidden City, Beijing, China. *Stud Conserv*. **59**(sup1), S242–S243 (2014). https://doi.org/10.1179/204705814X13975704320279
70. Liem, N.Q., Colomban, P., Sagon, G., Tinh, H.X., Hoanh, T.B.: Microstructure, composition and processing of the 15th century Vietnamese porcelains and celadons. J. Cult. Herit. **4**(3), 187–197 (2003). https://doi.org/10.1016/S1296-2074(03)00045-1
71. Mangone, A., De Benedetto, G.E., Fico, D., Giannossa, L.C., Laviano, R., Sabbatini, L., van der Werf, I.D., Traini, A.: A multianalytical study of archaeological faience from the Vesuvian area as a valid tool to investigate provenance and technological features. New J. Chem. **35**(12), 2860–2868 (2011). https://doi.org/10.1039/C1NJ20626E
72. Mangone, A., Caggiani, M.C., Forleo, T., Giannossa, L.C., Acquafredda, P.: A possible natural and inexpensive substitute for lapis lazuli in the Frederick II era: the finding of Haüyne in blue Lead-tin glazed pottery from Melfi Castle (Italy). Molecules. **28**(4), 1546 (2023). https://doi.org/10.3390/molecules28041546

73. Matin, M., Pollard, A.M.: From ore to pigment: a description of the minerals and experimental study of cobalt ore processing from the Kâshân mine, Iran. Archaeometry. **59**, 731–746 (2017). https://doi.org/10.1111/arcm.12272
74. Matin, M., Pollard, A.M.: Historical accounts of cobalt ore processing from the Kāshān mine, Iran. *Iran*. **53**, 171–183 (2015) https://www.jstor.org/stable/24595724
75. Mertens, J.: The history of artificial ultramarine (1787–1844): science, industry and secrecy. Ambix. **51**(3), 219–244 (2004). https://doi.org/10.1179/amb.2004.51.3.219
76. Mimoso, J.M.: Origin, early history and technology of the blue pigment in azulejos. In Proceedings of the GlazeArch2015, Lisbon, Portugal, 2–3 July 2015; pp. 357–376 (2015). Available online: http://azulejos.lnec.pt/AzuRe/links/07%20Origin_of_blue_pigment.pdf. Accessed on 20 Jan 2024
77. Morral, F.R.: From Cabul to cobalt: a historical view of the "mischievous metal". J. Chem Educ. **34**(4), 185 (1957). https://doi.org/10.1021/ed034p185
78. Mouri, T., Enami, M.: Raman spectroscopic study of olivine-group minerals. J. Mineral. Petrolog. Sci. **103**(2), 100–104 (2008). https://doi.org/10.2465/jmps.071015
79. Mühlethaler, B., Thissen, J.: Smalt. *Stud. Conserv.* **14**(2), 47–61 (1969). https://doi.org/10.1179/sic.1969.005
80. Natural pigments website., https://www.naturalpigments.com/artist-materials/ultramarine-blue-pigment-paint. Accessed 23 Feb 2024
81. Panighello, S., Kavcic, A., Vogel-Mikus, K., Tennent, N.H., Wallert, A., Hocevar, S.B., van Elteren, J.T.: Investigation of smalt in cross-section of 17th century paintings using elemental mapping by laser ablation ICP-MS. Microchem. J. **125**, 105–115 (2016). https://doi.org/10.1016/j.microc.2015.11.015
82. Prokin, V.A., Buslaev, F.P.: Massive copper–zinc sulphide deposits in the Urals. Ore Geol. Rev. **14**(1), 1–69 (1998). https://doi.org/10.1016/S0169-1368(98)00014-6
83. Ramond, L.-F. Observations faites dans les Pyrénées, pour servir de suite à des observations sur les Alpes, Belin, Paris, 1789. https://books.google.fr/books?id=vDWAsoINDT4C&pg=PA405&lpg=PA405&dq=%22giftain%22++pyr%C3%A9n%C3%A9es+mines&source=bl&ots=Bk9oFztDl_&sig=ACfU3U1vFY4xvLGkvNHfWq-sBgm7fpn0ow&hl=fr&sa=X&ved=2ahUKEwj8m_ehzdKEAxVZVqQEHZzNBUwQ6AF6BAgIEAM#v=onepage&q=%22giftain%22%20%20pyr%C3%A9n%C3%A9es%20mines&f=false. Accessed 1 Mar 2024
84. Rivas-Murias, B., Salgueirino, V.: Thermodynamic CoO–Co_3O_4 crossover using Raman spectroscopy in magnetic octahedron-shaped nanocrystals. J. Raman Spectrosc. **48**(6), 837–841 (2017). https://doi.org/10.1002/jrs.5129
85. Li, Y., Qiu, W.L., Qin, F., Fan, H., Fang, H., Hadjiev, V.G., Li, Y., Litvinov, D., Bao, J.: Identification of cobalt oxides with Raman scattering and Fourier transform infrared spectroscopy. J. Phys. Chem. **120**(8), 4511–4516 (2016). https://doi.org/10.1021/acs.jpcc.5b11185
86. Schabbach, L.M., Bondioli, F., Ferrari, A.M., Manfredini, T., Petter, C.O., Fredel, M.C.: Influence of firing temperature on the color developed by a (Zr,V)SiO_4 pigmented opaque ceramic glaze. J. Eur. Ceram. Soc. **27**, 179–184 (2007). https://doi.org/10.1016/j.jeurceramsoc.2006.04.180
87. Schabbach, L.M., Bondioli, F., Fredel, M.C.: Colouring of opaque ceramic glaze with zircon pigments: formulation with simplified Kubelka–Munk model. J. Eur. Ceram. Soc. **31**(5), 659–664 (2011). https://doi.org/10.1016/j.jeurceramsoc.2010.11.039
88. Schneeberg.: online: https://www.mindat.org/locentry-627563.html. Accessed on 1 Aug 2023
89. Scheinert, M., Kupsch, H., Bletz, B.: Geochemical investigations of slags from the historical smelting in Freiberg, Erzgebirge (Germany). Geochemistry. **69**, 81–90 (2009). https://doi.org/10.1016/j.chemer.2008.03.001
90. Shim, J., Venkata Reddy, C., Sarma, G., Narayana Murthy, V.S.S.P., Ravikumar, R.V.S.S.N.: Spectrochim. Acta Part A-Mol. Biomol. Spectrosc. **142**, 279–285 (2015). https://doi.org/10.1016/j.saa.2015.02.008

91. Simsek Franci, G.: Handheld X-ray fluorescence (XRF) versus wavelength dispersive XRF: characterization of Chinese blue-and-white porcelain sherds using handheld and laboratory-type XRF instruments. Appl. Spectrosc. **74**(3), 314–322 (2007). https://doi.org/10.1177/0003702819890645
92. Simsek, G., Colomban, P., Wong, S., Zhao, B., Rougeulle, R., Liem, N.Q.: Toward a fast non-destructive identification of pottery: the sourcing of 14th–16th century Vietnamese and Chinese ceramic shards. J. Cult. Herit. **16**(2), 159–172 (2015). https://doi.org/10.1016/j.culher.2014.03.003
93. Smedley, J.W., Jackson, C.M.: Medieval and post-medieval glass technology: batch measuring practices. Glass Technol. **43**, 22–27 (2002a). https://www.researchgate.net/publication/233699868_Medieval_and_post-medieval_glass_technology_Batch_measuring_practices. Accessed 4 Mar 2024
94. Smedley, J.W., Jackson, C.M.: Medieval and post-medieval glass technology: a review of bracken in glassmaking. Glass Technol. 43, 221–224 (2002b). https://www.researchgate.net/publication/233502355_Medieval_and_post-medieval_glass_technology_A_review_of_bracken_in_glassmaking. Accessed 4 Mar 2024
95. Stern, W.B., Gerber, Y.: Potassium-calcium glass: new data and experiments. Archaeometry. **46**(1), 137–156 (2004). https://doi.org/10.1111/j.1475-4754.2004.00149.x
96. Stern, W.B., Gerber, Y.: Ancient potassium-calcium Glass and its raw materials (wood-ash, Fern-ash, potash) in Central Europe. Mitt. Der Nat. Ges. **Beider** Basel **11**, 107–122 (2009)
97. Tauson, V.L., Goettlicher, J., Sapozhnikov, A.N., Mangold, S., Lustenberg, E.E.: Sulphur speciation in lazurite-type minerals (Na, Ca) 8 [Al6Si6O24](SO4, S) 2 and their annealing products: a comparative XPS and XAS study. Eur. J. Mineral. **24**(1), 133–152 (2012). https://doi.org/10.1127/0935-1221/2011/0023-2132
98. de Bomaré, V.: M. *Minéralogie, ou Nouvelle Exposition du Règne Minéral*, 2nd edn, pp. 76–96. Vincent Imprimeur-Libraire, Paris (1774) https://gallica.bnf.fr/ark:/12148/bpt6k486073x
99. Vieira Ferreira, L.F., Ferreira Machado, I., Ferraria, A.M., Casimiro, T.M., Colomban, P.: Portuguese tin-glazed earthenware from the 16th century: a spectroscopic characterization of pigments, glazes and pastes. Appl. Surf. Sci. Part B. **285**, 144–152 (2013). https://doi.org/10.1016/j.apsusc.2013.08.016
100. Visinescu, D., Paraschiv, C., Ianculescu, A., Jurca, B., Vasile, B., Carp, O.: The environmentally benign synthesis of Nanosized $Co_xZn_{1-x}Al_2O_4$ blue pigments. Dyes Pig. **87**, 125–131 (2010). https://doi.org/10.1016/j.dyepig.2010.03.006
101. Wedepohl, K.H., Simon, K.: The chemical composition of medieval wood ash glass from Central Europe. Geochemistry. **70**(1), 89–97 (2010). https://doi.org/10.1016/j.chemer.2009.12.006
102. Wei, S., Ma, Q., Schreiner, M.: Scientific investigation of the paint and adhesive materials used in the Western Han dynasty polychromy terracotta army, Qingzhou, China. *J. Archaeol. Sci.* **39**(5), 1628–1633 (2012). https://doi.org/10.1016/j.jas.2012.01.011
103. Wesch, H., Wiethege, T., Spiethoff, A., Wegener, K., Müller, K.M., Mehlhorn, J.: German uranium miner study: historical background and available histopathological material. Radiat. Res. **152**, S48–S51 (1999). https://doi.org/10.2307/3580113
104. Xia, Y., Xi, N., Huang, J., Wang, N., Lei, Y., Fu, Q., Wang, W.: Smalt: an under-recognized pigment commonly used in historical period China. J. Archaeol. Sci. **101**, 89–98 (2019). https://doi.org/10.1016/j.jas.2018.11.008
105. Zlamalova Cilova, Z., Gelnar, M., Randakova, S.: Smalt production in the ore mountains: characterization of samples related to the production of blue pigment in Bohemia. Archaeometry. **62**, 1202–1215 (2020). https://doi.org/10.1111/arcm.12584
106. Zlamalova Cilova, Z., Gelnar, M., Randakova, S.: Trends in Colouring blue Glass in Central Europe in relation to changes in chemical composition of Glass from the middle ages to modern age. Fortschr. Mineral. **11**, 1001 (2021). https://doi.org/10.3390/min11091001

Chapter 12
Islamic Glass and Glaze

Abstract Blue glazed pottery from Roman heritage through Islamic, Abbasid Dynasty period in the Middle East to its adoption in Europe in the Renaissance and afterwards. Blue glasses, Islamic blue tiles and bricks.

Keywords Roman blue glass · Islamic blue tiles · Islamic glass · Abbasid dynasty · Blue glazed bricks

We present here the development path towards European pottery with blue (and green because green is often achieved by dispersion of yellow pigment in a blue glassy matrix) decoration from the Roman heritage and Mesopotamia and the contributions and/or dialogues with Chinese productions.

12.1 Roman Heritage

Roman glasses coloured blue in the mass have been widely studied with regard to compositions [37–40, 60, 86] using especially Raman spectroscopy [6, 24, 75, 76]. A few studies deal with those decorated with enamels of different colours, including blue, [10, 25, 44].

The sources of cobalt being geologically rare [7], it is also well established that the recycling of blue glass was practiced in Roman times [36]. The Middle-Ages period is less well documented except in the Islamic region [11, 48, 53, 70, 85, 89, 95] where the production of blue glass, enamelled glass and glazed pottery using blue decoration is clearly attested and studied [1, 12, 42, 43, 80, 94]. A recent review was published by Gratuze et al. [43] related to glass. Their results are summarized in Table 12.1. They are based on the detection of a main element(s) associated with the element cobalt, namely nickel, zinc or indium, elements already used in the first classification made several decades ago by Gratuze et al. [42]. The main results are

H. G. M. Edwards, P. Colomban, *Blue by Fire: A Marker of the Technical History of Glass and Ceramics*, Cultural Heritage Science,
https://doi.org/10.1007/978-3-031-77122-4_12

Table 12.1 Cobalt type used to colour glass between first and fourteenth centuries in Mediterranean area (after [43])

Artifact	Period (centuries CE)	Glass type	Ratios CoO vs.			
			NiO	CuO	SnO	ZnO
Mosaic tesserae	1st–3rd	Low Ni Roman Co	~25 to 55	~0.5 to 1.2	~10 to 100	
Cup	2nd–3rd					
Workshop scum	2nd–3rd					
Merovingian glass beads	5th–7th	Low Ni Roman Co & High Ni Co				
Byzantine glass weights	6th–7th		~3 to 10	~0.5	~10	
Beirut workshop scum	7th	High Ni Co				
Grozon workshop scum	8th	High Ni Co				
Islamic glass	9th–10th	Zn Co	~10			~0.5–0.1
Samarra glass	9th					
Islamic glass weights	9th–12th					
Saint-Savin glass	10th–11th	Low Ni Roman Co				~0.5
Saint-Denis stained glass	12th					
Islamic Komani & Lehza glass	7th–13th	Low Ni Roman Co High Ni Co Zn Co (As) Co-In-Zn (As)	~5 to 35			~0.9
French glass	13th–14th	Co-In-Zn				~0.4

summarized in Fig. 12.1. Three groups N (CoO/NiO > 25), A (CoO/ZnO ~0.5) and R (CoO/NiO ~0.5; low ZnO) are identified [26]. This use of quantitative criteria is more relevant and effective than the simple identification of elements associated with cobalt. As mentioned in previous chapters the study of the 'isotope fingerprint' of elements associated with cobalt having stable isotopes, necessary to truly identify common sources and their source mine, will be possible in the near future with the development of high performance LA-ICP-MS instrumentation. The selection, 'purification' and grinding procedure also modify the amount of elements associated to cobalt [21, 63].

If the Gratuze' classification allows a first categorization, the finer analyses carried out for more recent periods and more sophisticated materials, for instance the Ottoman glazed tiles, show that a greater number of parameters must be taken into account due to the great variability of ingredients related to the use of poorly purified raw materials reflecting the variability of minerals [26, 77], of their selection and processing and the variety of the commercially available colouring agent. Indeed, it is reported that a variety of grades of smalt [98], numbering about 10, at different costs, were available during the eighteenth century [26]. We can

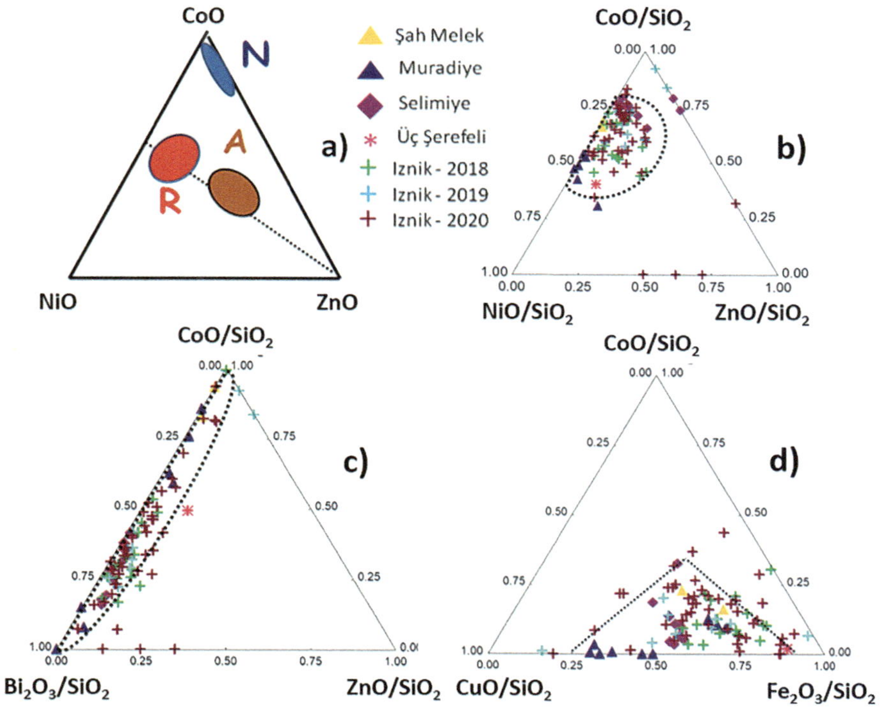

Fig. 12.1 Comparison of the distributions of composition data concerning the elements associated with cobalt in glasses (solids and enamels): (**a**) according to [42] for glasses from the Mediterranean area; (**b-d**) according to [26] for Ottoman tile enamels from the Şah Melek Paşa (built in 1429), Muradiye (built in 1435–1436), Üç Şerefeli (built from 1410 to 1447) and Selimiye (built from 1569 to 1575) mosques of Edirne, the Ottoman capital in the fifteenth century and from excavations of the Iznik kilns

reasonably assume that this varied offer existed before. The studies of mining places remain too limited to have an effective description which makes it impossible to unambiguously link the composition and origin of raw materials, even for rare elements such as cobalt, reliable data concerning old mines being almost non-existent. However, it is possible to exclude certain sources and association with historical reports and this helps formulating reasonable hypotheses.

The importance of the composition distribution visible in Fig. 12.1 for very similar objects (such as the Ottoman tiles from similar periods) highlights the variability of the raw materials which certainly given the period of the order of a century between most of the tiles being studied must have undergone a fairly comparable selection-purification process. However, we see that for certain buildings, e.g. the Muradiye mosque in Edirne built very rapidly between 1435–1436, the dispersion can be low [26, 82].

Another Roman heritage concerning blue decorations is the use of lapis lazuli both in the enamelling of glass and in that of pottery (see Chap. 9) as evidenced by the excavations of Pompeii or the treasure of Begram for the Ptolemy period (first century BCE) followed by certain Frederician sites in southern Italy in the thirteenth century. Intermediate data is lacking.

12.2 Blue Enamels on Ceramics: An Abbasid (750–1261) or Tang Dynasty (618–906) Innovation?

It has long been debated whether the potters of the Tang Dynasty or the Abbasid potters were the first to widely use cobalt as a colouring agent in enamelled decorations, in particular to produce blue-and-white wares [48]. The green-blue-white Abbasid tricolours, more common than blue-and-white wares, made of a terracotta body assumed to be covered with a layer opacified with cassiterite (SnO_2) were considered imitations of Tang porcelain/stoneware by many scholars. This simplified presentation has been called into question.

Firstly, Colomban and Truong [19, 22] have shown that in the first Abbasid tricoloured wares the precipitation of cassiterite is accidental: it only occurs, as proven by Raman analysis, when it encounters green (colouring by Cu^{2+} ions) or blue (colouring by Co^{2+} ions) overglazes rich in lead with the main alkaline-based glaze [18] while elemental analysis (SEM-EDS) detects the element Sn everywhere. The creation of a blue decoration therefore predates the imitation of the white background of porcelain!

Wood & Tite [93], who clearly did not read the spectroscopic literature, wrote in 2009: '*One of the first surprises that came from study of the seven Hira shards with white glazes and blue ornament was that only four of these glazes contained significant amounts of tin oxide. In those with negligible amounts of tin oxide (that is less than 2% SnO_2) a degree of whiteness was managed through a combination of undissolved quartz, glaze-bubbles and secondary crystallisation of calcium- and magnesium-rich silicate materials within the glazes during firing and cooling. Thus the term "tin glazed earthenware", as a technical description for Abbasid blue-and-white ware in general, may not always be an accurate one*. This once again demonstrates the need not to conclude on the simple observation of microstructure and average elemental composition data as is made by many archaeologists.

The blue areas of Tang *sancai* wares (from the Gongxian kilns, Henan Province, north of China) was coloured with a high-iron cobalt ore that usually contained copper, but was still a different material from that used in Iraq. A common assumption of the literature was, based on the view that all 'native' Chinese cobalt ores are rich in manganese, so low-manganese cobalt ores must have been imported. But in 1979 Watt showed that this was not always the case, and that low-manganese, high-iron cobalt ores are not too rare in China [90]. Chen et al. [14] suggest that traces of sulphur arise from cobalt sulphide ores such as cattierite (CoS_2) and linnaeite (Co_3S_4) intergrown with their copper and iron-sulphide analogues, as likely sources for these cobalt-rich pigments. This does not prove that the cobalt materials used at in the Tang dynasty were not imported from the Middle East (or even from Central Asia)—but rather that the argument conventionally used to make this case is not entirely sound. It may be significant, however, that the use of this particular style of cobalt ore in China ended with the Tang dynasty, and with the consequent decline in international trade. When underglaze blue resumed again on a large scale in China it was during the fourteenth century (southern China kilns), at the time of *Pax Mongolica*, and with a different type of cobalt ore, probably imported from Kashan in Persia (Part of Persia belong to Yuan Mongol Empire) [14, 51, 52].

Another decisive piece of evidence came from the cargo of the wreck of a dhow, an Arabic ship typical of the maritime trade of the Persian-Arabian Gulf discovered in 1998 on the Indonesian coast of the Java Sea, near the island Belitung, from where the name of this wreck was derived [35, 46, 47]. The combination of green-splashed decoration with painted cobalt blue as illustrated Fig. 12.2 appears to have been the invention of the potters of Basra, an Iraq city alongside the Shatt al-Arab and Persian Gulf. It is assumed that Chinese potters used cobalt imported from Iran that is consistent with the large use of cobalt in the Persian area.

Production of blue-glazed pottery during the Sassanid period is well documented [49, 66] as well as the wide use of blue glazes in Mesopotamia prior to Sassanid: for instance, the blue bricks of the Ishtar Gate (Babylon, 586 BCE) [27, 78]. The cobalt in these blue enamels is associated with Ni, Mn, Fe, Cu and Zn [2, 78]. The wreck of the Arab-style dhow contained three high-fired Chinese white dishes painted in cobalt blue with typical Abbasid *palmette* designs. The dishes have been attributed to the Gongxian kilns and dated provisionally to ~850.

Imitations of Abbasid blue and white were also made at Gongxian, using this new high-firing material, and these have been found in export contexts in China and also wrecks found in Indonesia, but not in the Middle-East itself. All the cobalt blue

Fig. 12.2 Example of tricolour pottery from Basra (ninth c.) where the cassiterite only precipitates when the green and blue overglazes rich in lead meet the colourless alkaline enamel. This is an example where the elemental analysis detecting the element Sn and not the phase is misleading (bar = 1 cm)

pigments examined on the Abbasid wares [93] were of the same high-iron type with significant amounts of zinc. Again, it is not known where this material came from, but a similar cobalt ore was used in the Abbasid blue glass. Mason reports the first Iraqi tin-containing glazes as appearing in pre-Abbasid times (Sassanids, Table 12.2), before about 700 to 750—with cobalt-blue painting and lead-glazed splashed wares arriving over this same period. Mason and Tite [59] have also supplied a thorough analytical framework for the development of tin-opacified glazes in Iraq and Egypt, from the early eighth to the later twelfth century. The development of Islamic tin opacified glazes has traditionally been attributed to the need to copy imports of Chinese whitewares into the Islamic world. However, the link between the first tin-opacified wares and earlier white opaque glazed wares of pre-Islamic Iraq would question the need for this interpretation. Many scholars ignore that tin opacification was –at least—an innovation made towards the end of Roman Empire [69], perhaps coming from Celtic recipes.

In conclusion, the continuous use of cobalt in Mesopotamia-Persia area since Antiquity as well as the Belitung cargo are consistent with the need to consider blue-and-white decoration as an Abbasid innovation.

12.3 Blue Glazed Islamic and Related Pottery

Here we will consider the different blue colouring techniques, chemical such as colouring by Co^{2+} ions dissolved in the vitreous silicate matrix and the use of synthetic ($CoAlO_4$ and more complex spinels, $CoSiO_4$ olivine) or natural blue pigments (lapis lazuli) but also physical by diffraction, as in lustre pottery. Table 12.2 list representative studies listed chronologically.

From Mesopotamia (now Iraq), the tin-opacified glaze technology was transferred to Fatimid Egypt where it was used in conjunction with lead-alkali glazes during the tenth–eleventh centuries. Subsequently, tin-opacified glazes spread in the Islamic world, eastwards to Syria, Iran, and Arabia [41] and westwards through north Africa to Sicily and Spain [76, 84]. The technology was then introduced into Italy during the first half of the thirteenth century with Spain (e.g., the Paterna workshops) influencing the development of "Archaic majolica" found in northern and central Italy, and the Maghreb (i.e., north Africa) or possibly the eastern Mediterranean influencing the development of the polychrome "Proto-majolica" found mainly in southern Italy and Sicily [92]. With respect to Archaic majolica, the earliest identified examples are probably the dishes (*bacini*) on the façade of the church of S. Lanfranco at Pavia which date to about 1200 AD [83]. A cobalt blue decoration was introduced during the first half of the fourteenth century AD, one of the earliest uses of cobalt blue being on the majolica tiles attached to Giotto's bell tower for Florence cathedral.

Thus, in Table 12.2 we can follow in a non-exhaustive way, the continuity of the creation of blue decorations in the ancient area of the Roman World from Mesopotamia, via the Sassanid then the Persian area and the deployment towards

Table 12.2 Représentative studies of pottery with blue decoration

Period (century)	Place/type	Analyses	Remarks	References
5th BCE	Mesopotamia (Babylon)	Composition	Co^{2+} ion Co associated to Ni, Mn, Fe, Cu and Zn	[2, 78]
1st BCE	Egypt	Composition Raman	Lapis lazuli	[10]
3rd–7th	Persia	Composition	Co^{2+} ion	[49, 66]
7th–13th	China (Tang & Song)	Composition	Co^{2+} ions	[97]
8th–9th	Mesopotamia & China (Tang)	Composition Microstructure	Co^{2+} ion	[93]
8th–13th	Egypt & Syria	Composition Microstructure	Lustre	[20]
11th–14th	Egypt & Mesopotamia	Composition Microstructure Raman	Lustre Co^{2+} ion	[22]
10th–14th	Spain (Al-Andalous)	Composition Microstructure	Co^{2+} ion	[15, 62, 73, 84]
13th	Iran/Lajvardina	Composition Microstructure Raman	Co^{2+} ion Lapis lazuli	[17]
13th–14th	Italy (south)	Composition Raman	Co^{2+} ion Lapis lazuli	[10]
12th–13th	Iran/minai	Composition	Co^{2+} ion	[91, 95]
13th–15th	China (Yuan & Ming)	Composition	Co^{2+} ion	[28, 50]
11th–18th	Ifriqiya (Dougga)	Raman	Co^{2+} ion Lustre	[23]
13th–17th	Italy (Della Robbia)	Composition Microstructure		[83]
15th–16th	China (Ming)	Composition Microstructure TEM, XANES	Co^{2+} ion $CoAl_2O_4$	[88]
15th	China (Ming) Vietnam	Composition Microstructure Raman		[81]
14th–17th	China	Composition XANES	$CoAl_2O_4$	[34]
14th–18th	Spain	Composition XRD	Co^{2+} ion As, Bi, Ni associated with Co Lustre	[13]
14th–18th	Italy (Albissola)	Composition	Co^{2+} ion As associated with Co	[61]

(continued)

Table 12.2 (continued)

Period (century)	Place/type	Analyses	Remarks	References
15th–17th	Italy	Composition	"Berettino", "turchino"	[79]
15th–18th	Spain (Aragon)	Composition Microstructure Reflectance	Co^{2+} ion	[74]
15th	Spain (Valencia)			[68]
15th–16th	Italy (Urbino)	Composition Microstructure, XRD	Co^{2+} ion	[3]
15th	Italy (Faenza, Venice, Genova...)	Composition	"Berettino" wares Co^{2+} ion As, Ni, Cu associated with Co	
15th–16th	Spain (Valencia, Seville)	Composition Raman,	Co^{2+} ion Fe, Ni associated with Co	[16]
16th	Italy (Sicily, Caltagirone)	Composition	Co^{2+} ion (smalt) As associated with Co	[5]
16th	Italy (Gubbio, Deruta)	Composition Microstructure, TEM	Co^{2+} ion As, Cu, Bi, Ni associated with Co Lustre	[72]
16th	Spain (Seville)	Composition		[67]
16th	Japan (Arita)	Composition	Co^{2+} ion –smalt)	[64]
16th	Italy (Casteldurante)	Composition Reflectance	Co^{2+} ion	[71]
16th–18th	Italy (Castelli)	Composition Reflectance	Co^{2+} ion	[45]
16th–17th	Portugal (Lisbon)	Composition Raman, reflectance	Co^{2+} ion	[87]
17th	Portugal	Composition Raman, reflectance	Co^{2+} ion	[31]
17th	Portugal (Lisbon, Coimbra, Vila Nova)	Composition Raman, XRD	Co^{2+} ion	[32]
16th–18th	Portugal	Composition	Co^{2+} ion Green obtained with Naples yellow mixing	[4]
16th–18th	Portugal (Lisbon) Spain (Seville)	Composition XRD		[33]
16th–18th	France (Nevers, ...)	Composition Microstructure		[54, 55]
17th–19th	France	Composition		[9, 8]

(continued)

Table 12.2 (continued)

Period (century)	Place/type	Analyses	Remarks	References
18th	Switzerland (Fribourg)	Composition	Co^{2+} ion As associated with Co	[56, 58]
18th	England (Bristol, Lancaster, Liverpool)	Composition Rama	Co^{2+} ion As associated with Co	[29]
19th	France	Composition Microstructure		[57]

the West to al-Andalus (Spain, Portugal) and Southern Europe. We can see that the studies concern Islamic sites mainly Iraq, Egypt and Spain and Italy for majolica. The sites of Ifriqiya and Magreb have been little studied. The same goes for blue lustre ceramics in which the blue colour is obtained by diffraction of the light by the glaze surface made of alternation of layers with or without metal nanoparticles The most famous example is the set of polychrome tiles of the Mihrab of Sidi Oqba (the barber of Prophet Mohamed) Kairouan mosque (Ifriqiya/Tunisia) imported in 862 from Baghdad [18, 20].

Surprisingly, publications concerning the analysis of blue decorations on European earthenware are rare. The main work was done by Maggetti et al., mainly for 18th and 19th productions. Work on porcelain is much more numerous [30, 65, 96].

References

1. Al-Saad, Z.: Chemical composition and manufacturing technology of a collection of various types of Islamic glazes excavated from Jordan. J. Archaeol. Sci. **29**(8), 803–810 (2002). https://doi.org/10.1006/jasc.2000.0576
2. Amadori, M.L., Matin, E., Poldi, G., Mengacci, V., Arduini, J., Callieri, P., Chaverdi, A.A., Holakooei, P.: Archaeometric research on decorated bricks of Tol-e Ajori monumental gate (6th century BC), Fars, Iran: new insight into the glazes. J. Cult. Herit. **60**, 63–71 (2023). https://doi.org/10.1016/j.culher.2023.01.005
3. Antonelli, F., Ermeti, A.L., Lazzarini, L., Verità, M., Raffaelli, G.: An archaeometric contribution to the characterization of R enaissance maiolica from U rbino and a comparison with coeval maiolica from P esaro (the M arches, central I taly). Archaeometry. **56**(5), 784–804 (2014). https://doi.org/10.1111/arcm.12045
4. Bandiera, M., Veronesi, U., Manso, M., Ruivo, A., Vilarigues, M., Esteves, L., Pais, A., Coentro, S.: The colour palette of 16th–18th century azulejos: a multi-analytical non-invasive study. J. Cult. Herit. **66**, 593–601 (2024). https://doi.org/10.1016/j.culher.2024.01.011
5. Barilaro, D., Crupi, V., Interdonato, S., Majolino, D., Venuti, V., Barone, G., La Russa, M.F., Bardelli, F.: Characterization of blue decorated renaissance pottery fragments from Caltagirone (Sicily, Italy). Applied Physics A. 92, 91–96 (2008)

6. Basso, E., Invernizzi, C., Malagodi, M., La Russa, M.F., Bersani, D., Lottici, P.P.: Characterization of colorants and opacifiers in roman glass mosaic tesserae through spectroscopic and spectrometric techniques. J. Raman Spectrosc. **45**(3), 238–245 (2014). https://doi.org/10.1002/jrs.4449
7. Berrie, B.H.: Mining for color: new blues, yellows, and translucent paint. Early Sci. Med. **20**(4–6), 308–334 (2015). https://doi.org/10.1163/15733823-02046p02
8. Bouquillon, A.: History of Lead in ancient ceramic materials. In: Lehuédé, P. (ed.) *Lead in Glassy Materials in Cultural Heritage* Bouquillon, A, pp. 215–248. Wiley & ISTE, London (2024)
9. Bouquillon, A., Castaing, J., Barbe, F., Crepin-Leblond, T., Tilliard, L., Paine, S.R., Christman, B., Heuer, A.H.: French decorative ceramics mass-produced during and after the 17th century: chemical analyses of the glazes. Archaeometry. **60**(5), 946–965 (2018). https://doi.org/10.1111/arcm.12349
10. Caggiani, M.C., Colomban, P., Mangone, A., Valloteau, C., Cambon, P.: Mobile Raman spectroscopy analysis of ancient enamelled glass masterpieces. Anal. Methods. **5**, 4345–4354 (2013). https://doi.org/10.1039/C3AY40648B
11. Carboni, S., Whitehouse, D.: Glass of the Sultans. Metropolitan Museum of Art, New York (2001)
12. Carmona, N., García-Heras, M., Villegas, M.A., Jiménez, P., Navarro, J.: Study of chromophores of Islamic glasses from al-Andalus (Murcia, Spain). In: *Lasers in the Conservation of Artworks*, pp. 89–94. CRC Press, Boca Raton (2008)
13. Chabanne, D., Bouquillon, A., Aucouturier, M., Dectot, X., Padeletti, G.: Physico-chemical analyses of Hispano-Moresque lustred ceramic: a precursor for Italian majolica? *Appl Phys.* **A92**, 11–18 (2008). https://doi.org/10.1007/s00339-008-4458-6
14. Chen, Y., Guo, Y., Zhang, Z.: A Study on Yuan Blue-and-White Porcelain. In: Shanghai Institute of Ceramics (ed.) *Scientific and Technological Insights on Ancient Chinese Pottery and Porcelain*, pp. 122–128. Science Press, Beijing (1986)
15. Coentro, S., Mimoso, J.M., Lima, A.M., Silva, A.S., Pais, A.N., Muralha, V.S.: Multi-analytical identification of pigments and pigment mixtures used in 17th century Portuguese azulejos. J. Eur. Ceram. Soc. **32**(1), 37–48 (2012). https://doi.org/10.1016/j.jeurceramsoc.2011.07.021
16. Coentro, S., Alves, L.C., Conesa, J.C., Ferreira, T., Mirão, J., da Silva, R.C., Trindade, R., Muralha, V.S.: White on blue: a study on underglaze-decorated ceramic tiles from 15th-16th-century Valencian and Sevillian productions. J. Archaeol. Sci. Rep. **30**, 102254 (2020). https://doi.org/10.1016/j.jasrep.2020.102254
17. Colomban, P.: Lapis lazuli as unexpected blue pigment in Iranian Lâjvardina ceramics. J. Raman Spectrosc. **34**(6), 420–423 (2003). https://doi.org/10.1002/jrs.1014
18. Colomban, P.: Secrets retrouvés du lustre Abbasside. La Revue de la Céramique et du Verre, 139, 13–19. (2004). https://www.researchgate.net/publication/260187792_Secrets_retrouves_du_Lustre_abbasside. Accessed 20 June 2024
19. Colomban, P.: Recent case studies in the Raman analysis of ancient ceramics: glaze opacification in Abbasid pottery, Medici and 18th century French porcelains, Iznik and Kûtayha ottoman Fritwares and an unexpected lapis lazuli pigment in Lajvardina ware, MRS online proceedings library (OPL) 852, OO8-4.1-8-4.8 (2005).
20. Colomban, P.: The use of metal nanoparticles to produce yellow, red and iridescent colour, from bronze age to present times in lustre pottery and glass: solid state chemistry, spectroscopy and nanostructure. J Nano Res. **8**, 109–132 (2009). https://www.scientific.net/jnanor.8.109
21. Colomban, P.: Natural nanosized raw materials and sol-gel technology: the base of pottery since millenniums. In: Dillmann, P., Bellot-Gurlet, L., Nenner, I. (eds.) Nanoscience and Cultural Heritage, pp. 59–73. Atlantis Press, Paris (2016). https://link.springer.com/chapter/10.2991/978-94-6239-198-7_3
22. Colomban, P., Truong, C.: A non-destructive Raman study of the glazing technique in lustre potteries and Faiences (9th–14th centuries): silver ions, nanoclusters, microstructure and processing. J. Raman Spectrosc. **35**(3), 195–207 (2004). https://doi.org/10.1002/jrs.1128

23. Colomban, P., Sagon, G., Louhichi, A., Binous, H., Ayed, N.: Identification par microscopie Raman des tessons et pigments de glaçures de céramiques de L'Ifriqiya (Dougga, XI-XVIIIèmes siècles). Revue d'Archéométrie. **25**, 101–112 (2001). https://doi.org/10.3406/arsci.2001.1005
24. Colomban, P., March, G., Mazerolles, L., Karmous, T., Ayed, N., Ennabli, A., Slim, H.: Raman identification of materials used for jewelry and mosaic in Ifriqiya. J. Raman Spectrosc. **34**(3), 205–215 (2003). https://doi.org/10.1002/jrs.977
25. Colomban, P., Tournié, A., Caggiani, M.C., Paris, C.: Pigments and enamelling/gilding technology of Mamluk mosque lamps and bottle. J. Raman Spectrosc. **43**(12), 1975–1984 (2012). https://doi.org/10.1002/jrs.4101
26. Colomban, P., Kırmızı, B., Simsek Franci, G.: Cobalt and associated impurities in blue (and green) glass, glaze and enamel: relationships between raw materials, processing, composition, phases and international trade. Fortschr. Mineral. **11**, 633 (2021). https://doi.org/10.3390/min11060633
27. Di Chiara, A., Tauxe, L., Gries, H., Helwing, B., Howland, M.D., Ben-Yosef, E.: An archaeomagnetic study of the Ishtar gate, Babylon. Plos one. **19**(1), e0293014 (2024). https://doi.org/10.1371/journal.pone.0293014
28. Du, F., Su, B.: Further study of sources of the imported cobalt-blue pigment used on Jingdezhen porcelain from late 13 to early 15 centuries. Science in China Series E: Technological Sciences. **51**, 249–259 (2008)
29. Edwards, H.G.M., Jay, W.H.: English delftware (c. 1770) from Bristol, Lancaster and Liverpool: a composition study using Raman spectroscopy and electron microscopy. Spectrochim. Acta A Mol. Biomol. Spectrosc. **15**(279), 121458 (2022). https://doi.org/10.1016/j.saa.2022.121458
30. Edwards, H.G.M.: 18th and 19th Century Porcelain Analysis. A Forensic Provenancing Assessment, p. 324. Springer, Cham (2020)
31. Ferreira, L.V., Casimiro, T.M., Colomban, P.: Portuguese tin-glazed earthenware from the 17th century. Part 1: pigments and glazes characterization. Spectrochim. Acta A Mol. Biomol. Spectrosc. **104**, 437–444 (2013). https://doi.org/10.1016/j.saa.2012.11.069
32. Ferreira, L.V., Ferreira, D.P., Conceição, D.S., Santos, L.F., Pereira, M.F., Casimiro, T.M., Machad, I.F.: Portuguese tin-glazed earthenware from the 17th century. Part 2: a spectroscopic characterization of pigments, glazes and pastes of the three main production centers. Spectrochim. Acta A Mol. Biomol. Spectrosc. **149**, 285–294 (2015). https://doi.org/10.1016/j.saa.2015.04.090
33. Ferreira, L.V., Machado, I.F., Pereira, M.F., Mangucci, C.: Archaeometry of tiles (azulejos) produced in the region of Lisbon–16th to 18th centuries. A comparison of Lisbon and Seville pastes for the cuerda seca and arista tiles. J. Archaeol Sci Rep. **49**, 104041 (2023). https://doi.org/10.1016/j.jasrep.2023.104041
34. Figueiredo, M.O., Silva, T.P., Veiga, J.P.: XANES study of cobalt speciation state in blue-and-white glazes from 16th to 17th century Chinese porcelains. J. Electron Spectrosc. Relat Phenomena. **185**, 97–102 (2012). https://doi.org/10.1016/j.elspec.2012.02.007
35. Flecker, M.: A 9th-century Arab or Indian shipwreck in Indonesian waters. Int. J. Naut. Archaeol. **29**(2), 199–217 (2000). https://www.tandfonline.com/doi/full/10.1111/j.1095-9270.2008.00193.x
36. Fontaine, S.D., Foy, D.: L'épave Ouest-Embiez 1, Var: Le commerce maritime du verre brut et manufacture en Méditerranée occidentale dans l'Antiquité. Rev. Archéol. Narbonnaise. **40**, 235–265 (2007). https://doi.org/10.3406/ran.2007.1182
37. Freestone, I.C.: The recycling and reuse of Roman glass: analytical approaches. J. Glass Studies, 29–40 (2015). https://www.jstor.org/stable/24726946
38. Gallo, F., Silvestri, A., Molin, G.: Glass from the archaeological museum of Adria (north-East Italy): new insights into early Roman production technologies. J. Archaeol. Sci. **40**(6), 2589–2605 (2013). https://doi.org/10.1016/j.jas.2013.01.017
39. Ganio, M., Boyen, S., Fenn, T., Scott, R., Vanhoutte, S., Gimeno, D., Degryse, P.: Roman glass across the empire: an elemental and isotopic characterization. J. Anal. Atomic Spectrom. **27**(5), 743–753 (2012). https://doi.org/10.1039/C2JA10355A

40. Gedzevičiūtė, V., Welter, N., Schüssler, U., Weiss, C.: Chemical composition and colouring agents of Roman mosaic and millefiori glass, studied by electron microprobe analysis and Raman microspectroscopy. Archaeol. Anthropol. Sci. **1**(1), 15–29 (2009). https://doi.org/10.1007/s12520-009-0005-4
41. Gianni, L., Renel, H., Kremenović, A., Colomban, P.: 'Blue' and 'Brown-speckled' pottery from Qalhât, the Sultanate of Oman (13–16th centuries): comparison with traditional Omani 19–20 century productions. Boletín de la Sociedad Española de Cerámica y Vidrio. **61**(1), 13–26 (2022). https://doi.org/10.1016/j.bsecv.2020.06.005
42. Gratuze, B., Soulier, I., Blet, M., Vallauri, L.: De l'origine du cobalt: Du verre à la céramique. Revue d'Archéométrie. **20**, 77–94 (1996). https://doi.org/10.3406/arsci.1996.939
43. Gratuze, B., Pactat, I., Schibille, N.: Changes in the signature of cobalt colorants in late antique and early Islamic glass production. Fortschr. Mineral. **8**(6), 225 (2018). https://doi.org/10.3390/min8060225
44. Greiff, S., Schuster, J.: Technological study of enamelling on Roman glass: the nature of opacifying, decolourizing and fining agents used with the glass beakers from Lübsow (Lubieszewo, Poland). J. Cult Herit. **9**, e27–e32 (2008). https://doi.org/10.1016/j.culher.2008.06.006
45. Gulmini, M., Scognamiglio, F., Roselli, G., Vaggelli, G.: Composition and microstructure of maiolica from the museum of ceramics in Ascoli Piceno (Italy): evidences by electron microscopy and microanalysis. Appl Phys A. **120**, 1643–1652 (2015). https://doi.org/10.1007/s00339-015-9376-9
46. Guy, J.: Early Asian ceramic trade and the Belitung (Tang') cargo, Trans Orient Ceram Soc 66, 13–27 (2001–2002)
47. Guy, J.: Early ninth-century Chinese export ceramics and the Persian Gulf connection: the Belitung shipwreck evidence, Taoci, 4, 9–20. (2005). https://www.researchgate.net/publication/288245350_Early_ninth-century_Chinese_export_ceramics_and_the_Persian_Gulf_connection_The_Belitung_shipwreck_evidence. Accessed 9 June 2024
48. Hallett, J.: Iraq and China: trade and innovation in the early Abbasid period. Taoci. **4**, 21–39 (2005)
49. Hojabri Nobari, A., Khorsowzadeh, A.: Trade and cultural contacts between northern and southern Persian gulf during Parthians and Sasanians: a study based on pottery from Qeshm Island. Int J Humanit. **18**(2), 89–115 (2011). http://eijh.modares.ac.ir/files/eijh/user_files_749497/archive_global-A-10-1000-8486-95904c2.pdf
50. Jiang, X., Liu, N., Xu, X., Ge, Y., Li, Z., Cui, J., Zhai, Y.: Revisiting the bleeding effect in historical cobalt porcelain pigments: mechanism, influence and technical responses. J. Archaeol. Sci. **167**, 105987 (2024). https://doi.org/10.1016/j.jas.2024.105987
51. Liu, X.: Yuan dynasty official wares from Jingdezhen, the porcelains of Jingdezhen. In: Scott, R. (ed.) Colloquies on art & Archaeology in Asia, No.e 16, pp. 33–46. Percival David Foundation of Chinese Art, London (1993)
52. Ma, W.: Research on tang dynasty blue-and-white porcelain and a discussion of certain questions concerning the cobalt materials used in Chinese blue and white. China Archaeol Art Digest. ***11***(3-4), 7–24 (1998)
53. Ma, Y., Li, Z., Zhang, L., Wu, J., Xiong, Y., Vahdati, A., Zhang, J.: A scientific study of late Islamic blue-white stonepaste wares of Iran. Archaeometry. **64**(5), 1110–1123 (2022). https://doi.org/10.1111/arcm.12765
54. Maggetti, M.: Technology and Provenancing of French faience. Seminarios de la Sociedad Espanola de Mineralogia. **9**, 41–64 (2012). https://www.academia.edu/download/26270499/Seminarios.SEM.v9web.PDF#page=42
55. Maggetti, M.: Archaeometric analyses of european 18th–20th century white earthenware—a review. Fortschr. Mineral. **8**(7), 269 (2018). https://doi.org/10.3390/min8070269
56. Maggetti, M., Serneels, V.: Chemische und technologische Analyse einiger Fayencekacheln eines JOHANN BAPTIST NUOFFER zugeschriebenen Kachelofens von ca. 1780-85. Bulletin de la Société Fribourgeoise des Sciences Naturelles= Bulletin der Naturforschenden Gesellschaft Freiburg. **105**, 32–90 (2016). https://www.academia.edu/download/61241194/188._Maggetti___Serneels_Bull._Soc.Fr._1620191117-50534-f8jqc0.pdf

57. Maggetti, M., Neururer, C., Rosen, J.: Antimonate opaque glaze colours from the faience manufacture of le bois d'Épense (19th century, northeastern France). Archaeometry. **51**(5), 791–807 (2009). https://doi.org/10.1111/j.1475-4754.2008.00442.x
58. Maggetti, M.A., Galetti, G.I., Torche-Julmy, M.T.: Analyses scientifiques de quelques catelles de poêles fribourgeois du XVIIIe s. Bulletin de la Société Fribourgeoise des Sciences Naturelles= Bulletin der Naturforschenden Gesellschaft Freiburg. **103**, 59–82 (2014). https://www.academia.edu/download/61211786/172._Bull_Frib.catelles20191114-90044-y8m48t.pdf
59. Mason, R.B., Tite, M.S.: The beginnings of tin-opacification of pottery glazes. Archaeometry. **39**(1), 41–58 (1997). https://doi.org/10.1111/j.1475-4754.1997.tb00789.x
60. Medeghini, L., Botticelli, M., Cadena-Irizar, A.C., Lepri, B., Ferrandes, A.F., Costa, M., Barrulas, P.: Blue shadows of Roman glass artefacts. Microchem. J. **179**, 107526 (2022). https://doi.org/10.1016/j.microc.2022.107526
61. Meucci, M., Carratoni, L.: Identification of the majolica polychromatic decoration by IRFC methodology. J. Archaeol. Sci. Rep. **8**, 224–234 (2016). https://doi.org/10.1016/j.jasrep.2016.06.027
62. Molera, J., Vendrell-Saz, M., Pérez-Arantegui, J.: Chemical and textural characterization of tin glazes in Islamic ceramics from eastern Spain. J. Archaeol. Sci. **28**(3), 331–340 (2001). https://doi.org/10.1006/jasc.2000.0606
63. Molera, J., Climent-Font, A., Garcia, G., Pradell, T., Vallcorba, O., Zucchiatti, A.: Experimental study of historical processing of cobalt arsenide ore for colouring glazes (15-16th century Europe). J. Archaeol Sci Rep. **36**, 102797 (2021). https://doi.org/10.1016/j.jasrep.2021.102797
64. Montanari, R., Murakami, N., Alberghina, M.F., Pelosi, C., Schiavonz, S.: The origin of overglaze-blue enameling in Japan: new discoveries and a reassessment. J. Cult Herit. **37**, 94–102 (2019). https://doi.org/10.1016/j.culher.2018.11.008
65. Montanari, R., Colomban, P., Alberghina, M.F., Schiavone, S., Pugliese, A., Pelosi, C.: European *smalt* in 17th-century Japan: porcelain decoration and sacred art. Heritage. **7**(6), 3080–3094 (2024). https://doi.org/10.3390/heritage7060145
66. Moradi, H., Karimian, H.: Classification and typology of the Islamic period potteries of Neyriz plain, Fars Province. J Archaeol Stud, 12(3), 269–291 (2020). https://jarcs.ut.ac.ir/article_79773_en.html?lang=fa. Accessed 9 June 2024
67. Myers, J.E., de Amores Carredano, F., Olin, J.S.: Compositional identification of Seville majolica at overseas sites. Hist Arch. **26**, 131–147 (1992). https://doi.org/10.1007/BF03374167
68. Nádai, Z., Kolláth, Á., Kovács, B.G.: A unique Iberian majolica fragment from the marketplace of Győr (Hungary). Hung Archaeol. **12**(3), 35–45 (2023). http://real.mtak.hu/id/eprint/182458
69. Neri, E., Morvan, C., Colomban, P., Guerra, M.F., Prigent, V.: Late Roman and byzantine mosaic opaque "glass-ceramics" tesserae (5th-9th century). Ceram. Int. **42**(16), 18859–18869 (2016). https://doi.org/10.1016/j.ceramint.2016.09.033
70. Nikbakht, T., Montazerzohouri, M.: Characterization of historical minai and colorful luster glaze ceramics, using ionoluminescence technique. J. Lumin. **231**, 117769 (2021). https://doi.org/10.1016/j.jlumin.2020.117769
71. Padeletti, G., Fermo, P., Gilardoni, S., Galli, A.: Technological study of ancient ceramics produced in Casteldurante (Central Italy) during the renaissance. Appl Phys A. **79**, 335–339 (2004)
72. Padeletti, G., Fermo, P.: Significant findings concerning the production of Italian renaissance lustred majolica. Appl Phys A. **113**, 825–833 (2013). https://doi.org/10.1007/s00339-013-7725-0
73. Pereira, M., de Lacerda-Aroso, T., Gomes, M.J., Mata, A., Alves, L.C., Colomban, P.: Ancient Portuguese ceramic wall tiles ("Azulejos"): characterization of the glaze and ceramic pigments. J. Nano Res. **8**, 79–88 (2009). https://doi.org/10.4028/www.scientific.net/JNanoR.8.79
74. Pérez-Arantegui, J., Montull, B., Resano, M., Ortega, J.M.: Materials and technological evolution of ancient cobalt-blue-decorated ceramics: pigments and work patterns in tin-glazed objects from Aragon (Spain) from the 15th to the 18th century AD. J. Eur Ceram Soc. **29**(12), 2499–2509 (2009). https://doi.org/10.1016/j.jeurceramsoc.2009.03.004

75. Ricciardi, P., Colomban, P., Tournié, A., Macchiarola, M., Ayed, N.: A non-invasive study of Roman age mosaic glass *tesserae* by means of Raman spectroscopy. J. Archaeol. Sci. **36**(11), 2551–2559 (2009a). https://doi.org/10.1016/j.jas.2009.07.008
76. Ricciardi, P., Colomban, P., Tournié, A., Milande, V.: Non-destructive on-site identification of ancient glasses: genuine artefacts, embellished pieces or forgeries? J. Raman Spectrosc. **40**, 604–617 (2009b). https://doi.org/10.1002/jrs.2165
77. Ridovics, A., May, Z., Bajnoczi, B., Toth, M.: Examination og Haban vessels with uranium-bearing blue glaze. Acta Ethnogr Hung. **60**(2), 485–515 (2015). https://doi.org/10.1556/022.2015.60.2.14
78. Rodler, A.S., Klein, S., Artioli, G., Brøns, C.: Probing the provenance of archaeological glaze colorants: polychrome faunal reliefs of the Ishtar gate and the processional way of Babylon. Archaeometry. **61**(4), 837–855 (2019). https://doi.org/10.1111/arcm.12455
79. Ruffini, A., Gualtier, I S., Fabbri, B.: Comparison between Renaissance 'berettino' glazes from some ceramic centres in northern Italy. In Proceedings of the 33rd International Symposium on Archaeometry, pp. 22–26 (2002)
80. Salehvand, N., Agha-Aligol, D., Shishegar, A., Rachti, M.L.: The study of chemical composition of Persian glass vessels of the early Islamic centuries (10th– 11th centuries AD) by micro-PIXE; case study: Islamic collection in the National Museum of Iran. J. Archaeol Sci. Rep. **29**, 102034 (2020). https://doi.org/10.1016/j.jasrep.2019.102034
81. Simsek, G., Colomban, P., Wong, S., Zhao, B., Rougeulle, A., Liem, N.Q.: Toward a fast non-destructive identification of pottery: the sourcing of 14th-16th century Vietnamese and Chinese ceramic shards. J. Cult Herit. **16**, 159–172 (2015). https://doi.org/10.1016/j.culher.2014.03.003
82. Simsek, G., Unsalan, O., Bayraktar, K., Colomban, P.: On-site pXRF analysis of glaze composition and colouring agents of "Iznik" tiles at Edirne mosques (15th and 16th-centuries). Ceram. Int. **45**(1), 595–605 (2019). https://doi.org/10.1016/j.ceramint.2018.09.213
83. Tite, M.S.: The production technology of Italian maiolica: a reassessment. J. Archaeol. Sci. **36**(10), 2065–2080 (2009). https://doi.org/10.1016/j.jas.2009.07.006
84. Tite, M.S.: The technology of glazed Islamic ceramics using data collected by the late Alexander Kaczmarczyk. Archaeometry. **53**(2), 329–339 (2011). https://doi.org/10.1111/j.1475-4754.2010.00546.x
85. Vaggelli, G., Lovera, V., Cossio, R., Mirti, P.: Islamic glass weights from Egypt: a systematic study by non-destructive μ-XRF technique. J. Non-cryst Solids. **363**, 96–102 (2013). https://doi.org/10.1016/j.jnoncrysol.2012.12.003
86. Van der Linden, V., Cosyns, P., Schalm, O., Cagno, S., Nys, K., Janssens, K., Nowak, A., Wagner, B., Bulska, E.: Deeply coloured and black glass in the northern provinces of the Roman empire: differences and similarities in chemical composition before and after AD 150. Archaeometry. **51**(5), 822–844 (2009). https://doi.org/10.1111/j.1475-4754.2008.00434.x
87. Vieira Ferreira, L.F., Ferreira Machado, I., Casimiro, T.M., Pereira, M.F., Santos, L.F.: Portuguese blue-on-blue 16th–17th century pottery. Archaeometry. **60**(4), 695–712 (2018). https://doi.org/10.1111/arcm.12336
88. Wang, T., Zhu, T.Q., Feng, Z.Y., Fayard, B., Pouyet, E., Cotte, M., De Nolf, W., Salomé, M., Sciau, P.: Synchrotron radiation-based multi-analytical approach for studying underglaze color: the microstructure of Chinese Qinghua blue decors (Ming dynasty). Anal. Chim. Acta. **928**, 20–31 (2016). https://doi.org/10.1016/j.aca.2016.04.053
89. Ward, R. (ed.): Gilded and Enamelled Glass from the Middle East. British Museum Press, London (1998)
90. Watt, J.C.: Notes on the Use of Cobalt in Later Chinese Ceramics, pp. 63–85. Ars orientalis (1979)
91. Wen, R., Pollard, A.M.: The pigments applied to Islamic Minai wares and the correlation with C hinese blue-and-white porcelain. Archaeometry. **58**(1), 1–16 (2016). https://doi.org/10.1111/arcm.12143
92. Whitehouse, D.: Roman Glass in the Corning Museum of Glass, vol. 3. Hudson Hills (1997)

93. Wood, N., Tite, M.: Blue and white—the early years: tang China and Abbasid compared. In: Person, S. (ed.) *Transfer: the influence of China on world ceramics* Colloquies on art & Archaeology in Asia no. 24, Percival David Foundation of Chinese art, pp. 21–45. School of Oriental and African Studies: London University (2009)
94. Wypyski, M.T.: Chemical analysis of early Islamic glass from Nishapur. J. Glass Stud, 121–136 (2015). https://www.jstor.org/stable/24726951
95. Yazdani, M., Ahmadi, H., Emami, S.M.A., Lamehi Rashti, M.L.R., Agha-Aligol, D.A.A., Abdillah Khan Gorji, M., Choubak, H.: Technological study of gilded Mina'i ware based on microscopic investigation and historical documents. Pazhoheshha-ye Bastan shenasi Iran. **7**(14), 161–178 (2017)
96. Zhang, R., Gethin, P.: Provenance of the cobalt pigment used for Jingdezhen minyao blue-and-white porcelain in the early Qing dynasty. Ceram. Int. **47**(18), 25763–25768 (2021). https://doi.org/10.1016/j.ceramint.2021.05.303
97. Zhang, Y., Polard, A.M.: The archaeological and scientific analysis of blue-decorated ceramics in the tang and song dynasties. Archaeometry. **64**, 1394–1410 (2022). https://doi.org/10.1111/arcm.12794
98. Zlamalova Cilova, Z., Gelnar, M., Randakova, S.: Smalt production in the Ore Mountains: characterization of samples related to the production of blue pigment in Bohemia. Archaeometry. **62**(6), 1202–1215 (2020). https://doi.org/10.1111/arcm.12584

Chapter 13
Earthenwares, Faience, Majolica and Delft

Abstract Earthenwares, faience, delftwares and majolicas—distinction between them. Translucency considerations compraison with early porcelains. Earliest terracottas and earthenwares, stonewares, glazing and decoration, celadons. Creamwares, jasperwares of Josiah Wedgwood. Analysis of several case studies.

Keywords Earthenwares · Terracotta · Majolic · Delftwares · Stonewares · Glazing techniques · Jasperware · Creamwares · Frog service · Nantgarw porcelain

13.1 Types of Fired Earthenwares, Faience, Majolica and Delftwares

Unlike porcelains, whose origin can be dated fairly accurately to the Song Dynasty in China from 960–1279 CE, with the first Emperor Taizu (960–976 CE), earthenware artefacts generically can be traced archaeologically back much earlier to about 25,000 BCE in many cultures. Following on from the Song Dynasty, the Mongol Emperors of the Yuan Dynasty (1279–1368 CE) created the idea of using their porcelains as "diplomatic gifts", which soon made them highly desirable and prepared the way for the massive exportation of Ming Dynasty (1368–1644 CE) and Qing Dynasty (1644–1912 CE) porcelains to Europe in the sixteenth, seventeenth and eighteenth Centuries. The earliest piece of Chinese porcelain still extant, dating from about 1338 CE, was presented as a diplomatic gift in Europe from the last of the Chinese Khans of the Yuan Dynasty the Emperor Togon Temur (also known as Shizu, reigned 1333–1368); this was the *qingbai* porcelain Gaignieres-Fonthill vase, now in the National Museum of Ireland in Dublin, that was taken as a gift to Pope Benedict XII by a Christian Nestorian mission from the Emperor in Peking in 1338 CE (Fig. 13.1).

Preceding the Song Dynasty, the Tang Dynasty (618–907 CE), interspersed by the interregnum of Empress Wu Zetian and the Wu Zhou Dynasty (690–705 CE), were noted for their quality of fine earthenwares, exemplified by the Tang horse

H. G. M. Edwards, P. Colomban, *Blue by Fire: A Marker of the Technical History of Glass and Ceramics*, Cultural Heritage Science,
https://doi.org/10.1007/978-3-031-77122-4_13

Fig. 13.1 The *Gaignieres—Fonthill Vase*, Yuan Dynasty Chinese porcelain, *ca.* 1338, as it is today, front view, minus its later addition of silver mounts, demonstrating the superb quality of its *Qingbai* porcelain; note that the vase itself which has a moulded floral decoration does not possess any armorial decoration, although it is recorded that silver mounts added later in the fourteenth century did have armorial devices engraved linking it with the Hungarian monarchy. *Copyright, The National Museum of Ireland, Collins Barracks, Dublin, and reproduced with permission*

figures and of course the "Terracotta Army" in the funerary mausoleum of Emperor Qin Shi Huang in Xian from 210 BCE, comprising over 8000 soldiers, 150 chariots and 700 horses in life-sized earthenware figures contained in several burial pits at the site (Fig. 13.2).

Ancient earthenwares were sometimes decorated in polychrome pigments, usually red, yellow and black, representing the iron oxide fired earth pigments, in geometric patterns and scenes from mythology and which could be protected in usage by an overglaze application. Where unglazed, earthenwares are usually assigned the name *terracotta* (from the Latin *terra cocta*, cooked earth, and the Italian *terracotta*, baked earth) but tin-glazed analogues (using a stannic oxide, SnO_2, slip) are described generically as faience, majolica (maiolica) and delft wares. All earthenwares are fired ceramics, in ancient times in pits and later in kilns, at temperatures from approximately 600 to around 900 °C, but glazing in a glost-kiln required a higher temperature approaching 1050 °C, the first examples of which are believed to originate from the Abbasid Dynasty (Iznik wares) in the Middle East. This is in contrast with porcelains, the pastes of which were fired at higher temperatures between 1100 and 1400° C, and then the glost kiln was operated at a lower temperature. Not all glazed earthenwares were tin-glazed and those from St Porchaire in France used a glaze of lead-oxide and are not strictly therefore described as faience,

Fig. 13.2 *"The Terracotta Army"*, terracotta funerary figures in the tomb of Emperor Qin Shi Huang, 210–209 BCE, Xian, China, comprising life-size figures of some 8000 soldiers, 130 chariots and 700 horses. *CC-by-SA 3.0.* Public domain

whereas the contemporary sixteenth Century glazed earthenwares from Nevers and Rouen in France are true tin-glazed faience. It should be noted in this context, however, that English delftwares from Lambeth and Southwark are essentially a hybrid in that they can be tin-glazed on the upper surface and lead glazed on the lower surface of flatwares. The name faience is derived from the early Italian production centre of glazed earthenwares at Faenza in Emilia Romagna, near Ravenna. The distinction between majolica (also known alternatively as *maiolica* for Italian examples) and delft wares is that, although both are termed faience generically, majolica is technically glazed on the upper surface only, leaving the lower surface of flatwares unglazed, whereas delft has glazing on both the upper and lower surfaces. Although faience was known for centuries before the advent into the West of Chinese porcelain there is no doubt that the ability to create a simulation of this porcelain as faience, through its fine glazed white surfaces and light coloured bodies, was seen as an immediate potential competitor economically in Europe for the expensive Chinese porcelain import.

Earthenwares can be arranged in a spectrum of fired ceramics as follows:

earthenwares (terracotta)—faience—stonewares—celadons—porcelains

For which there is an increase in kiln firing temperature range required in their production from the left to right between 600–1400° C. The other main distinction is that all earthenwares, faience, celadons and stonewares are completely opaque to transmitted light whereas porcelain exhibits a varying degree of translucency, which can depend upon the component raw materials and the heaviness of the potting:

some Chinese porcelains are not very translucent at all (for example the blue-and-white Tek Sing treasure rice bowls, Fig. 13.3) whereas Nantgarw porcelain, described as "the world's finest" in its time, and some varieties of bone china, are highly translucent and claims were made that one could read a newspaper through them, see Fig. 13.4 for a Nantgarw porcelain plate from the blue-and white *Lady Seaton* service (Edwards, *eighteenth and nineteenth Century Porcelain Analysis: A Forensic Provenancing Assessment,* [13]; Edwards, *Porcelain Analysis and Its Role in the Forensic Attribution of Ceramic Specimens,* [14]) and Fig. 13.5 which demonstrates the superb translucency of Nantgarw porcelain.

Historically, it can be somewhat confusing in that earthenware celadons are often referred to as "porcelains", but occasionally early porcelains did have a green glaze applied. The categories in the spectrum of fired ceramics listed above have several additional subdivisions, which include ironstone, Coade stone, jasperware, celadons, hard and soft paste porcelains, hybrid porcelain, phosphatic porcelain and soaprock (magnesian or steatitic porcelain). The common feature of earthenwares, celadons and faience is that they all contain clays, an alkaline additive to aid fusion and sand as component raw materials but not any vitreous material such as glass, which however does appear in the glazing component of faience with the addition of tin oxide [10]. Care should be taken in the acceptance of an absolute terminology historically for "faience" as *Egyptian faience* is basically a vitreous glassy material and is much admired artistically for its blue funerary figurines (ushabtis) and scarabs, which comprise essentially Egyptian blue, cuprorivaite, $CaCuSi_4O_{10}$, a blue calcium copper silicate glass that was known to the Romans as *caeruleum.* In contrast, several of the other high-fired ceramics listed above do have powdered glass (either crown or flint glass) as an essential raw material component, such as Coade

Fig. 13.3 The *Tek Sing* cargo, a blue and white decorated bowl with lotus leaf motifs, *ca.* 1820, photographed in transmitted light, demonstrating its correct attribution to a porcelain classification and its pale green translucency, but quite thickly potted. Private Collection

Fig. 13.4 Dinner plate from the named *Lady Seaton* service of Nantgarw porcelain, 1817–1819. The actual association of this service with Lady Seaton of Bosahan Manor, Cornwall, is unknown. Marked impressed, NANT-GARW C.W., underglaze blue hand-enamelled floral bouquets pattern, London decorated through John Mortlock's, Oxford Street, with the characteristic dentil gold edging. Private Collection

stone and some porcelains which can contain as much as 10% or more of glass in their composition, but it should be noted that Chinese hard paste porcelain does not include glass as a raw material and has only kaolin clay and *petuntse* (a form of china stone feldspar) in its recipe. Alexander Brogniart, the well-respected Director of the Sevres Porcelain Museum, hence divided porcelains into *true porcelain* for the Chinese hard paste body and *artificial porcelain* for all the others (Brogniart, *Traite des Artes Ceramiques, ou des Poteries, Considerees dans Leur Histoire, Leur Pratique et Leur Theorie*, [5]).

Celadons need a special mention since their glaze is usually a beautiful green in colour in simulation of Chinese jade, and they were highly prized for this property: indeed, they are often referred to as "green wares" or *seiji* in Chinese. The origin of the name celadon is lost in antiquity but there is a hypothesis that it is derived from the Abbasid Sultan Saladin, who in 1171 sent 40 pieces of celadon as a diplomatic gift to his neighbour, Nur ad-Din Zenghi, the Sultan of Syria. The height of their production in China was during the Song Dynasty at the Longquan site in Zhejiang province: the green colour is affected by the amount of iron oxide added to the glaze, generally about 1–2% is ideal for a jade colour, with a high firing temperature of the glazed artefacts required between 1290 and 1300 °C in a reducing kiln, which converts the red iron (III) oxide additive, Fe_2O_3, to the green iron (II) oxide, FeO. Too great a concentration of iron oxide in the glaze will darken the colour significantly to a dull olive green and eventually black, whilst too little will confer a blue colour

Fig. 13.5 Trumpet-shaped Nantgarw porcelain spill vase of beautiful translucency, *ca.* 1817–1819, with applied enamel decoration in a chrome green ground and individual pink roses contained in trellis lozenges between the fine gilded edges of a lattice gridwork pattern. Private Collection

upon the glaze, which was sometimes intended as an acceptable variation in colour. The presence of titanium (IV) oxide, TiO_2, will confer an unacceptable yellow tinge to the colour of the celadon. Chenwang Wang et al. [8] have recently analysed the glaze on several ancient celadons from the Donggou kiln in Ru, which comprise feldspar, kaolin, quartz and iron oxide as raw materials: the analyses give average values of Al_2O_3 of 11%, SiO_2 70%, K_2O 4%, CaO 9%, and Fe_2O_3 2%, with smaller amounts of Na_2O at 1% and P_2O_5 at 2% and TiO_2 less than 0.3%. Sometimes barium carbonate, witherite, was added as a flux (about 2%) to the glaze and to help develop the colour. The minerals anorthite and albite have been identified in the glaze. The glaze after firing also develops a characteristic *craquelure* pattern effect due to the release of potash and soda from the feldspar component at high temperatures, which enhances the appearance of the celadon artefacts. Although originating in China, it is claimed as early as the Iron Age around 1000 BCE, celadons were also produced in Japan, Thailand, Korea and other parts of Asia. They were exported to the Abbasid caliphate in the Middle East in the twelfth Century, whence afterwards they found their way into Europe sparsely through trade and appeared in several notable ceramic collections.

13.2 The Decoration of Early Earthenwares and Faience

The earliest earthenwares were predominantly purely functional artefacts designed for use around the home for cooking and storage vessels and the earliest glazes were simply applied for aid in their storage of oils and liquids and were produced by throwing a handful of salt crystals into the kiln at the final stages of the kiln firing process, which forms an incipient glaze of sodium silicate by reaction with silica at the surface of the artefact; although of a limited volatility, some of the salt was deposited as a vapour onto the earthenware or terracotta vessels and so giving its alternative name of "salt-glazed pottery". Early decoration comprised geometric patterns incised into the wet paste before firing and these could be further protected and enhanced by the application of an external glaze. The earliest applied decoration using pigments were polychrome and based upon a red, yellow, white and black palette of colours obtained from earth minerals and could be applied as overglaze painted or as underglaze enamel decoration. Blue-and-white decoration was comparatively more recent and the earliest example recorded of blue-and-white earthenware that is still extant is a Tang Dynasty dish (Fig. 13.6) which was salvaged from

Fig. 13.6 Rice dish, Tang Dynasty earthenware, ca. 820–830 CE, with cobalt blue decoration described as a floral lozenge on a white slip background, Giongxian kiln, Henan province. From the wreck of an Arab Dhow off Belitung Island, Sumatra. In the Museum of Asian Civilisation, Singapore, accession number 2005.1.00474. *CCO-by-SA 3.0.* Public domain

the archaeological excavation of an Arab dhow engaged in navigating the Maritime Silk Road between the China Sea and the Persian Gulf that was shipwrecked off Belitung Island, near the coast of Sumatra in the Java Sea, in 826 CE, carrying a large cargo of lead ingots, bronze mirrors and some 60,000 ceramic artefacts from several manufacturing sites in China, such as Gongxian, Chaozhou, Changsha, Yue and Xing (Krahl et al., *Shipwrecked Tang Treasures and Monsoon Winds,* [23]). This was certainly the largest hoard of Tang Dynasty ceramics ever discovered and although most of the pieces and shards were decorated earthenware in a brown and ochre palette, three complete and extremely rare pieces of blue-and-white glazed earthenware were found and are now on view in the National Museum of Asian Civilisation, Singapore, Khoo Tech Puat Gallery (Chong et al., *The Tang Shipwreck: Art and Exchange in the ninth Century,* [9]) along with many examples of the more common early Tang brown and ochre ware. The occurrence of blue-and-white decoration on Tang Dynasty ceramics was unknown before this archaeological discovery in 1985. The discovery prompted the connoisseur Nigel Wood to write *"Song blue-and-white is rare but Tang blue-and white was unheard of"* (Wood, *Chinese Glazes: Their Origins, Chemistry and Recreation,* [42]). It is interesting that analysis of the blue pigment on this Tang dish reveals that it is a cobalt silicate, formulated as Co_2SiO_4, and hails from the Middle East, in the neighbourhood of Iran, and potentially from a mine near Kashan, which indicates that an active trade in ceramics and cobalt oxide ores was already ongoing at this time between the Chinese and the Abbasid caliphate, before these ceramics had penetrated into Europe. Islamic blue-and-white tile decoration particularly was being much admired by travellers and developed into the Iznik wares (Caiger-Smith, *Tin-Glazed Pottery in Europe and the Islamic World: The Tradition of 1000 Years in Maiolica, Faience and Delftware,* [6]).

An example of early glazed Vietnamese earthenware which provided a decorative theme as well as having a functional purpose is the ewer in the shape of a rising phoenix from the fifteenth/sixteenth Century, beautifully modelled and with an attractive underglaze blue decoration shown in Fig. 13.7.

The European earthenwares developed their polychromatic and blue-and-white palettes in the southern Mediterranean with quality wares from Spain, France, Portugal and Italian factories in competition with the Chinese imports that commenced in the sixteenth and seventeenth Centuries, commencing at the height of the Ming Dynasty porcelain production (1368–1644 CE) (Caiger-Smith, *Tin-Glazed Pottery in Europe and in the Islamic World: The Tradition of 1000 years in Maiolica, Faience and Delftware,* [6]). The copying of designs on Chinese import porcelains of the period is illustrated in the dish from the faience factory at Nevers, France, dating from 1680, which depicts a typical Westernized version of a Chinese willow pattern scene in blue-and-white (Fig. 13.8) and the eighteenth Century Dutch delft plate and dish shown in Fig. 13.9, where the plate has a Chinese family scene and the dish a geometric idealised floral decoration, both in underglaze blue on a white background. Vieira Ferreira et al. [36–39, 41] have made a comprehensive analytical study of early European faience from Portugal and have made comparisons of this with the contemporary production from Spain, the Low Countries and Italy and have also studied Phoenician unglazed terracotta amphorae [40] dating from the late Iron

Fig. 13.7 Ewer in the shape of a phoenix, Vietnam, *ca.* fifteenth Century, height 29 cm, with underglaze cobalt blue decoration. Metropolitan Museum of Art, New York, gift of Betty and John R. Menke, 1992. Public domain

Fig. 13.8 Faience dish, Nevers, France, *ca.* 1680, depicting Chinese willow pattern scene in cobalt blue as used on Ming porcelains. Musee des Arts Decoratifs, Paris. *CCO-by-SA 3.0.* Public domain

Fig. 13.9 Dutch delftware plate and dish with Chinese family scene and geometric floral design, eighteenth Century. Musee Ernest Cognacq, St Martin-de-Re, Ile de Re, France. Public domain

Age, seventh and eighth Centuries BCE. Using a combination of elemental and molecular spectroscopic instrumentation comprising XRD, XRF, SEM/EDAXS and Raman spectroscopy, they demonstrated that local clays from the Tagus estuary had been used as raw materials and had been fired at a temperature of 900–950 °C, where anorthite, but not mullite (which requires a temperature of greater than 1000 °C), had been formed and detected in the fired ceramics. Cobalt blue, cobalt silicate Co_2SiO_4, was identified in the faience in the form of light blue and dark blue glazes.

13.3 English Delftwares

English delftwares were created in the mid-sixteenth Century, the earliest recorded piece being from the East Malling Pottery with a hallmarked silver mount dated to 1550 and now in the Victoria and Albert Museum, London, collection. The British delftware centres were located in London, Liverpool (Fazackerley), Bristol, Glasgow and Dublin and delftware was manufactured through the eighteenth Century until about 1840 (Dawson, *English and Irish Delftwares*, 1570–1840, [11]). John Stow, who wrote the first authoritative history of Elizabethan London (Stow, *Survey of London: Written in the Year 1598*, [33]), a tailor born in 1525 who died in Leadenhall in 1605 and was buried under a terracotta monument (!), mentioned that two potters from Antwerp established themselves in Norwich in 1567 and there made *"gally pairing wares and apothecary's vessels"*, later being granted the sole right to practice "*gallypotting*" in London by Queen Elizabeth I. At that time, the English terminology for delftware was "galliware", and gallipots were manufactured for the storage of oils, spices and medicines, and mainly used by apothecaries. Decoration of early English delftwares has been described as "*provincial and rather naïve rather than sophisticated*" (Carnegie, *Tin-Glazed Earthenwares*, [7]): chargers, or large circular platters, were very popular for the execution of more advanced artistic work with their larger surface area. Early examples of English delftwares are a sack bottle decorated with cobalt blue, dated 1645, of undefined manufacture (Fig. 13.10) and a large dish painted by Richard Irons of the Southwark Pottery in London inscribed with a painted mark IC/1638, which was probably made for a tradesman in the City of London (Fig. 13.11). The latter shows a much more ambitious style of painting of a classical scene of a troubadour playing a mandolin with a recumbent stag in a landscape in polychrome, blue, yellow and green colours applied on a white slip.

The Lambeth Pottery, London, features strongly in English delftwares and an example of a chinoiserie bowl with blue -and -white decoration dating from about 1760 is shown in Fig. 13.12. It is interesting that, although the earlier seventeenth Century earthenwares were extensively polychrome in decoration in contrast with the Chinese blue-and-white porcelain then being imported in the seventeenth Century from the Ming (Chongzhen) and Qing (Kangxi) Dynasties, the Qing Emperor Yongzhen after 1723 made the transition over to a polychrome palette for the porcelain decoration in recognition of the brilliance of the Limoges enamels that had been presented to him as diplomatic gifts. The imported Chinese *famille rose*, *famille jaune* and *famille vert* porcelains were very much favoured by clients over the classic blue-and-white versions from the second quarter to the mid-eighteenth Century. Surprisingly, however, the English earthenwares still maintained a blue-and-white decorative palette well into the late eighteenth century as evidenced by the Lambeth delftware bowl shown here. An unusual feature of Bristol and Lambeth delft earthenwares in the mid-eighteenth Century was the adoption of the Italian *bianco sopra bianco* technique, literally *white-on-white,* where the tin glaze on the upper surface was treated with a very small quantity of cobalt oxide and then a pure white slip was used in the superficial decorative application—essentially therefore being a white-on-blue decoration! The death knell of English delftwares is attributed historically to the phenomenal success of Josiah Wedgwood's fired creamware launched in the 1770s, which with its light coloured and durable body, and glossy lead glaze rivalled the imported porcelains, especially after being given the royal seal of approval and patronage by Queen Charlotte of Mecklenburg -Strelitz, Queen Consort of King George III, and by the Empress Catherine the Great of Russia (Ekaterina II), enabling Wedgwood to rename his creamware "Queen's Ware". An example of the very large service ordered through the British Ambassador to Russia, Lord Cathcart, by Empress Catherine the Great for use in her new Chesma Palace, called the *Frog* service for the green enamelled frog contained in a false heraldic escutcheon in the verge, is shown in Fig. 13.13 and a closer view of the enamelled frog is given in Fig. 13.14. Further details of the *Frog* service, its complement and its decoration, are provided in Edwards (*Welsh Armorial Porcelain,* [15]).

Fig. 13.10 Sack bottle, English delftware, manufactory undefined, with blue decoration and dated 1645. Gardiner Museum of Ceramic Art, Toronto, Canada. Public domain

13.3.1 Early English Delftware Armorials

Although early English armorials in the late seventeenth century were commissioned on Kangxi Chinese porcelain expedited through the Honourable East India Company trade with Canton, the earliest example on record being a set of blue-and-white jardinieres for Sir Edward Johnson ordered in the late 1680s, a surviving example of which is shown in Fig. 13.15, English delftwares did not seem to embark fully into this area of business (Denyer, Denyer and Edwards, *Armorial Porcelain: The Genesis*, [12]; Howard, *A Tale of Three Cities: Canton, Shanghai and Hong Kong: Three Centuries of Sino-British Trade in the Decorative Arts,* [22], p.95).

Fig. 13.11 Polychrome English delftware dish, dated 1638, probably by Richard Irons, Southwark Pottery, London. Victoria and Albert Museum, London. Public domain

Fig. 13.12 Lambeth delftware, blue-and-white chinoiserie pattern dish, *ca.* 1760. *Californian Palace of the Legion of Honor, San Francisco, California.* Public domain

Fig. 13.13 Serving platter from Josiah Wedgwood's Queen's Ware *Frog* service made for Empress Catherine the Great of Russia (Ekaterina II) in 1773 for use in her Chesma Palace—a very large, combined dinner and dessert service originally comprising 944 pieces with hand painted illustrations in sepia monochrome of individual English houses and estates, each containing a green enamelled frog motif contained within an escutcheon. This piece shows Ditchley Park in Oxfordshire, seat of the Earl of Litchfield, and is from the dinner service, which had a distinctive border of oak leaves and acorns. *City of Birmingham Museum of Art.* Public domain

Sir Edward Johnson built ships for the Honourable East India Company trade with China in his shipyards in Deptford and Blackwall. A noted exception generically is the relatively large number of royal commemoratives that were produced in the mid- to late-1600s, featuring the monarch with a crown and cipher, namely King Charles II and King Willam III and Queen Mary II. These were generally polychromatic and the decoration is rather naïve and simplistic: a typical example is shown in Fig. 13.16, which depicts King William III on horseback with the cipher W.R. and dated to 1688. William acceded to the English Stuart throne as King William III in what was termed the "Glorious Revolution" through his marriage in 1677 to Mary, elder daughter of King James II who was deposed in 1688, and this was celebrated in his native Holland with a similar release of commemorative Dutch delftwares, such as that showing William and Mary in Fig. 13.17 and an English version shown in Fig. 13.18.

Fig. 13.14 Detail of the green enamelled frog motif in an escutcheon from the *Frog* service; demonstrating that one could easily be misled into believing that this was a genuine armorial service!

13.3.2 King Charles II English Delftware Coronation Mugs

Recently, in December 2022, an early English dated tin-glazed earthenware (delftware) small tankard appeared at auction in Middle Claydon, Buckinghamshire: it was attributed to a London factory, possibly Southwark, and bears a bust length portrait of King Charles II in full armour and wearing a crown with a monogram CDR 2 and a date of 1660 (Fig. 13.19). His father, King Charles I, fought the English Civil War unsuccessfully against the Parliamentarians led by Oliver Cromwell between 1642 and 1646 and was executed in January 1649. His son, Charles, who had been declared King in Scotland, was defeated at the Battle of Worcester in 1651 and he exiled himself to France and The Netherlands. There followed the period of the Commonwealth under the Lord Protector Cromwell until he died in 1658: the Restoration of the Monarchy began in May 1660 when Charles was invited by

Fig. 13.15 Circular armorial jardiniere, Kangxi porcelain, 33 cm diameter, 19.7 cm high, ca. 1693–7, depicting the armorial bearings of Sir Henry Johnson impaled by those of his second wife, Martha Lovelace, who were married in 1692. In 1697 Lady Martha Johnson became Baroness Wentworth and Baroness Despenser *suo jure* and her arms would have reflected this change in heraldic status. Metropolitan Museum of Art, New York, the gift of Mr. and Mrs. Rafi Y. Mottahedeh, 1976, accession number 1976.112. Public domain

Parliament to return to London on his 30th birthday, which was received with much public acclaim. It is believed that this tankard represented a celebration of the restoration of the Monarchy and can be considered an armorial because of the heraldic regal crown that it bears. There are some several surviving dated examples of such armorial mugs, most of which depict King Charles II with a crown in coronation robes bearing an orb and sceptre. His depiction wearing armour on such examples is rare and this tankard is one of only four examples known, three of which show the King in armour: the others are in The Museum of London (A.4946), Colonial Williamsburg, Virginia (1956–531) and one sold at Christie's in 2002 from the Margaret Cadman Collection. This tankard is now in the collection of E. & H. Manners Ltd., London, where it has been attributed to the Southwark factory based upon its stylistic similarity to other factory products (Lipski & Archer, *Dated English Delftware,* [24]; Britton, *London Delftware,* [4]; Archer, *Delftware: The Tin-glazed Earthenware of the British Isles,* [1]; Tyler et al., *London's Delftware*

Fig. 13.16 King William III on horseback, English delftware plate, polychrome palette, *ca.* 1688, marked WMR. Gardiner Museum of Ceramic Art, Toronto, Canada, accession number G83.1.413. Public domain

Fig. 13.17 Dutch delftware dish in blue-and -white depicting the accession of William and Mary to the English throne, *ca.* 1689. Marked KWD 3 MSKG. Gardiner Museum of Ceramic Art, Toronto, Canada, gift of Robert A. Ellison, 2014, accession number 2014.712.3. Public domain

Fig. 13.18 English delftware plate depicting William and Mary on their accession to the English throne, *ca.* 1690, tin-glaze on front surface but lead glaze underneath, decorated in polychrome palette of blue, manganese purple and ochre. Gardiner Museum of Ceramic Art, Toronto, Canada. Gift of George and Helen Gardiner, accession number G83.1.419. Public domain

Fig. 13.19 English tin-glazed delftware armorial tankard, height 10 cm, diameter 7 cm, decorated in blue, ochre and purple pigments showing King Charles II in armour and bearing the monogram CDR 2, wearing the regal crown and dated 1660. Made in 1660 to celebrate the Restoration of the Monarchy. A rare example attributed to the Southwark earthenware manufactory, London. (Reproduced with permission from the Collection of E. and H. Manners Ltd., Church Street, Kensington, London)

Fig. 13.20 Lambeth delftware mug depicting King Charles II, marked C2R 1664, dated 1664. Tin-glazed to front and lead glazed to underside and inside. Gardiner Museum of Ceramic Art, Toronto, Canada. Gift of George and Helen Gardiner, accession number G83.1.542. Public domain

Industry: The Tin-Glazed Pottery Industries of Southwark and Lambeth, [35]). Another version of this type of delftware armorial mug is shown in Fig. 13.20, which has been attributed to the Lambeth delftware manufactory.

13.4 Josiah Wedgwood Jasperwares

In the 1770s, Josiah Wedgwood in Etruria, Staffordshire, brought out his jasperware, a type of unglazed biscuit stoneware with a smooth matte finish in light blue or dark blue with an applied white relief decoration of neoclassical subjects made separately in moulds in a cameo style. This was very popular and resonated with the aristocracy who were engaged in pursuing the Grand Tour of Europe, especially Italy and Greece, and returning to England with visions of having neoclassical statuary and interior furnishings embellishing their estates. Jasperware originally was a formulation of barium sulfate (57%), secondary ball clay (29%), flint (10%) and barium carbonate (4%) which contained varying amounts of cobalt blue pigment according to the colour of the ceramic desired and fired at a temperature of up to 1000 ° C. The major component, barytes or "heavy spar", was a byproduct of lead mining in nearby Derbyshire and acted as a flux for the high temperature firing process. A later version of jasperware, known as dipped ware, comprised a ceramic paste that was dipped in a dyed slip and then fired, but the earthenware paste often contained a small additive of cobalt blue to counteract the creaminess of the biscuit and effectively create a whiteness on the exposed underlying ceramic interior. Other colours could be made in addition to the commonly encountered blue cobalt (II) oxide addition to the slip, including chromium (III) oxide for a green ground jasperware, manganese (IV) oxide for a lilac colour, antimony (III) oxide for a yellow colour, and iron (III) oxide for a black colour. An example of a very fine, unglazed, light blue jasperware vase with applied white cameos is shown in Fig. 13.21.

13.5 Stonewares

These are normally treated separately from earthenwares on account of their composition being rather different and because they require a significantly higher firing temperature in the range 1100–1200 °C. The raw materials essentially comprise primary (kaolinite) and secondary ball clays, china stone or feldspar, mica, flints and sometimes waste glass and grog. The resultant product is described best as a semi-vitreous ceramic. of a grey-brown colour which is normally salt-glazed. Generically, the term stonewares historically includes a range of high-fired materials including ironstone, Cologne ware, caneware, Crouch ware, Coade stone and sometimes even jasperware, although in composition this is naturally exclusive. Coade stone, an artificial stoneware created by Eleanor Coade in 1769 to simulate carved marble statuary and for use in architecture, such as columns, pilasters, voussoirs, plaques and pediments is a stoneware subject in itself (Edwards and Brooke, *Coade Stone: A History and Analysis,* [16]).

Of particular relevance is the production in the Rhineland of the Cologne ware, so called from the major port of shipment of the finished artefacts rather than the place of manufacture, which was desirable in the fourteenth–seventeenth Centuries for its durability for household utensils such as large jugs, ewers and platters. A large saltglazed stoneware jug in Cologne ware, *ca.* 1560–1570, known as a *Bartmannkrug* from the presence of the bearded man depicted on the lip is shown in Fig. 13.22. The bearded man was a popular representation of the "wild man" in European legend, who was later likened to a representation of Cardinal Roberto Bellarmino (1542–1621), an outspoken and vehemently anti-Protestant opposer of the consumption of alcohol, who gave his name to this type of jug, called a Bellarmine, in the sixteenth Century amongst German and Low Countries

Fig. 13.21 Josiah Wedgwood blue jasperware vase and lid, *ca.* 1790, with applied white cameos of neoclassical figures. Victoria and Albert Museum, London. CCO-by-SA 3.0. Public domain

Protestants. The "wild man" in mediaeval legend, half-man and half-beast and covered in hair, was based upon the Roman god of the woods, Silvanus. Made in Cologne, or more likely in nearby Raeren or Flechen, the quality and robustness of this stoneware was much admired in Europe and in early America and was a very desirable import into England and the American Colonies (Gaimster, *German Stoneware, 1200–1900: Archaeology and Cultural History,* [18]; Reineking von Bock, *Kunstwerkenmuseum der Stadt Koln: Steinzeug*, [29]; Kerry & Hood,

Fig. 13.22 An important Cologne stoneware Cardinal Bellarmine jug (*Bartmannkrug* or "wild man jug"), displaying the Tudor coat- of -arms of Queen Elizabeth I of England and the arms of the City of Cologne, *ca.* 1560–1570. (Reproduced with permission from the Collection of E. and H. Manners Ltd., Church Street, Kensington, London)

Salt-Glazed Stonewares in Early America, [30]; Noel Hume, *If These Pots could talk: Collecting 2000 Years of British Household Pottery*, [26]). This particular jug bears the Tudor arms of Queen Elizabeth I (shown in detail in Fig. 13.23) and the arms of the City of Cologne (shown in detail in Fig. 13.24), namely the three crowns of the Magi (whose bones lay buried in a sarcophagus in Cologne Cathedral) and the ermine tails of St Ursula who was martyred by Atilla the Hun in Cologne in the fourth Century, underneath the double-headed eagle of the Holy Roman Empire and flanked by rampant lion supporters.

Fig. 13.23 Detail of the Tudor arms of Queen Elizabeth I of England on the Cologne ware jug shown in Fig. 11.22. (Reproduced with permission from the Collection of E. and H. Manners Ltd., Church Street, Kensington, London)

These stoneware jugs and household artefacts were predominantly of a grey to brown colour and the grey artefacts particularly from Raeren were often decorated with cobalt blue and manganese purple in the seventeenth Century and were superficially covered by the salt glaze: a white slip was not used on the ceramic body of these artefacts and so they are often described as having a "blue-on-grey" palette.

Fig. 13.24 Detail of the arms of the City of Cologne on the Cologne ware jug shown in Fig. 13.22. (Reproduced with permission from the Collection of E. and H. Manners Ltd., Church Street, Kensington, London)

13.5.1 The Fulham Pottery of John Dwight

The stonewares produced by John Dwight in his Fulham Pottery in London in the last quarter of the seventeenth Century are remarkable for their high quality and fineness of modelling and connoisseurs are divided in their opinion as to whether or not they should be classified as early porcelains: in fact, following composition analyses, several shards excavated at the pottery site by the Museum of London have revealed that both porcelain and stonewares were being made there from 1670 until about 1703, when John Dwight died.

John Dwight, who was born in Chester in 1637/38, established a manufactory of salt-glazed earthenware in Fulham at the junction of the New King's Road and Burlington Road, not far from Putney Bridge on the River Thames, which he called the Fulham Pottery [2]. The Fulham Pottery managed to survive the major political changes and events occurring through the early eighteenth Century and under the later ownership of William de Morgan became a driving force for architectural and functional pottery and tiles in Victorian England well into the 1880s. Excavations at the site have revealed the presence of a ceramic kiln which has now been preserved on this site. Just prior to the foundation of his pottery at Fulham, John Dwight filed a patent in April, 1671, describing his experiments in the use of clays and minerals to make stoneware and a phrase from this patent is particularly revealing, namely *"... the discovery of the mistery of transparent earthenware commonly known by the name of porcelaine in China."* Critically, this statement alerts the reader to Dwight's patent for the manufacture of a *"transparent porcelain"* rather than the opaque earthenware at Fulham with which he is normally credited. Robert Hooke FRS in February 1674 (in [19]) confirmed the production by Dwight of *"Severall little Jars of several colours all exceeding hard as flint, very light and very good shape"*. Dwight used a silicaceous- alumina-calcareous (SiAlCa) glaze for his vessels, with some lead oxide in addition, but he experienced problems with his glazing which was suspected by Ramsay & Ramsay (*The Evolution and Compositional Development of English Porcelains from the sixteenth Century to Lund's Bristol c. 1750 and Worcester c. 1752- the Golden Chain",* [27]) to arise from the silica -alumina -calcareous glaze composition which required very high firing temperatures which Dwight was unable to achieve in his kiln. This was also believed by Tite et al. [34], who attributed the defects in his manufactured products to the adoption of an intractable SiAlCa glaze. Furthermore, Spataro et al. [31] have discovered evidence for Dwight's reversion to a lower firing temperature of a lead glaze, which he then used to recover several articles which had nevertheless successfully survived his preliminary SiAlCa glaze trials. Sir John Lowther FRS in 1698 [19] visited Dwight at his Fulham Pottery and was shown "*20 or 30 varieties of more china like than is in ye world beside, nothing in Germany is like his, nor had he any help from them at setting up, but owes all to his own studies*", which clearly refers to the Cologne stonewares that were then being imported. Caroline Stanford [32] has drawn an analogy between the high quality Dwight stoneware Fulham Pottery productions and the Cologne wares form the Rhineland.

Although John Dwight is now credited with having made the first English porcelain (see the evidence presented for this in Chap. 4 of Edwards, *eighteenth and nineteenth Century Porcelain Analysis: A Forensic Provenancing Assessment,* [13]), no complete pieces have survived but two jugs have been recreated from shards excavated at the site of the pottery by the Museum of London; in contrast, several high quality artefacts in Dwight's white, high-fired stoneware are still extant, especially the bust of Prince Rupert of the Rhineland Palatinate wearing his Order of the Garter, *ca.* 1673, and the personal effigy of his daughter Lydia, inscribed

Fig. 13.25 Brown salt-glazed stoneware bottle with a thistle medallion from the Fulham Pottery of John Dwight, *ca.* 1675, emulating the Cologne Rhenish Bellarmine wares. In the Metropolitan Museum of Art, New York, Robert A. Ellison Collection, accession number 2014.712.13. Public domain

"Lydia Dwight dyed March 3rd 1673" which are now in the British Museum and the Victoria and Albert Museum, respectively. In Fig. 13.25 is a brown salt-glazed stoneware bottle, *ca.* 1675, from the Fulham Pottery made in semblance of the Cologne Bellarmine Rhenish stoneware, with an applied medallion of a thistle.

13.6 Dutch Delftware

No survey of seventeenth and eighteenth Century pottery and faience would be complete without mention of the Dutch delftware that gave its name to the eponymous generic tin-glazed faience called "delft" that was being made in many countries at that time. The Dutch East India Company, Vereeningde Oost-Indische Compagnie (with its monogram VOC), was established by the Staaten-General of the Republic of the Seven Netherlands in March 1602 after securing the government monopoly of trade in spices with China and the Far East. With the first importation of Chinese porcelain as well as spices by the Dutch East Indies Company in the early 1600s, a vigorous trade route was soon established between the Netherlands and their Far Eastern base of Batavia (Jakarta), which first acted for the receipt of porcelains carried by Chinese junks from Canton. The VOC undoubtedly had the majority of this trade by the number of ships and their tonnage carried: between 1602 and 1796, after which they became bankrupt, the VOC had 4785 ships, compared with 2690 ships of the Honourable East India Company of Great Britain (HEIC) over the same period and the VOC carried five times the tonnage of the HEIC. In contrast, the Swedish and Portuguese East India Companies adopted relatively minor roles in this trade [12], although the Portuguese did have the advantage of first opening up the trade route between Europe, India and the Far East through Vasco da Gama in the late fifteenth and early sixteenth Centuries.

It should be noted that the first Chinese porcelain to arrive in the Netherlands came from the piracy of Portuguese vessels carrying cargoes of Chinese porcelain by Dutch naval privateers: in March 1602, a Dutch fleet, comprising the *Zeelandia* and the *Langebarck,* captured the *Sao Tiago,* a Portuguese carrack, off St Helena and the artefacts and cargo were sold at auction to eager bidders at Middelburg in the Netherlands. This was followed by the capture of the Portuguese *Santa Catarina* by Admiral Jacob van Heemskerck, in the *Witte Leeuw* and *Alkmaar,* off Johore in the Straits of Singapore in February 1603, with its cargo of 100,000 pieces of porcelain, which were auctioned off in Amsterdam later that year. These auctions undoubtedly generated the desire for European ownership of these blue-and-white Chinese ceramics and contributed indirectly to the rise of interest in the acquisition of delftwares. The *Santa Catarina* incident was defended against Portuguese criticism and reprisals by Hugo de Groot (Hugo Grotius) in a lengthy treatise prepared for the VOC, entitled *De Jure Praedae Commentaris* (*On the Law of Prize and Booty*) and the incident certainly raised a *casus belli* for the war between the Netherlands and Portugal that followed [3].

Faience was being made in Antwerp in the early 1500s by Italian artisan emigres from Faenza but, following the persecution of Huguenot ceramic artisans in France from about 1585, a migration north saw them settle in Delft and establish faience manufactories there in derelict brewing warehouses and factories. In 1615 potters in Delft, and also in Amsterdam, Rotterdam, Haarlem and Gouda, were already manufacturing *Delfts blauw,* a tin-glazed faience with an underglaze cobalt blue decoration and adopting personal monograms and marks from 1640 to identify their wares in simulation of the imported Chinese porcelain with their Imperial seal marks. The impetus for this surge in production was the end of the reign of the Emperor Wanli of the Ming Dynasty after 1620, when supplies of blue-and-white Chinese porcelain started to be less reliable in production and more scarce to acquire from China: the delftwares copied the blue-and-white chinoiserie scenes with a Westernised influence, also adopting biblical scenes (*istoratio*) and setting local figures in landscapes, and they also copied the Japanese polychrome Imari wares that were then becoming popular from the mid-1600s (Figs. 13.26, 13.27 and 13.28).

A trade mark of delft faience earthenwares was the very large number of blue-and-white tiles that were created and required for use in household fireplaces, stoves and as wall decoration: it has been estimated that these would have numbered many

Fig. 13.26 Delft jar, Imari Japanese design, *ca.* 1700, Greieske A factory. Museum Geelvinck Hinlopen Huis, Amsterdam. Public domain

hundreds of millions ….. they were so popular that even the Chinese started to simulate the Netherlands delft faience tiles in their porcelain for export to Europe—a reverse trade technology! By the end of the eighteenth Century, delftwares were losing out to the higher quality British porcelains and desirable creamwares (Fourest, *Delftware: Faience and Production in Delft*, [17]). The marriage between Princess Mary, daughter of the Stuart King James II of England. Scotland and Ireland, to Prince Willem of Orange of the Netherlands occurred in 1677 and the Glorious Revolution, whereby they were invited to succeed King James in 1688 as Protestants on the throne as King William III and Queen Mary II, was celebrated in both Great Britain and the Netherlands: Queen Mary was already a devoted supporter of delftwares in her adopted Netherlands and she commissioned several exquisite pieces to decorate her palaces — including the flower pyramids (Fig. 13.29) in blue-and-white delft faience which could be used to show prized tulips especially.

Fig. 13.27 Delft, Japanese style vase, *ca.* 1680. Art *Public domain.* Art Institute, Chicago, Illinois, USA

13.6.1 A Delftware Diplomatic Gift

We have seen earlier that the diplomatic gifts of high quality European and Chinese porcelains to cement alliances was highly esteemed in the sixteenth and seventeenth Centuries and certainly delftwares were treated similarly. A pair of delftware pilgrim flasks with screw lids have recently appeared with a clear diplomatic gift provenance (Errol Manners, personal communication [25]). The pair of tin- glazed earthenware flasks are shown in Fig. 13.30, decorated in blue pigment on a white background and bearing the arms of Count Jacob Johann Hastfors, an influential member of the Swedish aristocracy in the Swedish Court: the flasks can be reliably dated to 1689 and are marked AK for Adriaen Kocks of the De Griekse A manufactory in Delft and one with the monogram JW of Jacob Wemmersz Hoppestyn of the Het Moriaenshoofd manufactory in Delft. It is most unusual to

Fig. 13.28 Delft tin-glazed faience charger of the Shiba Onko design, ca. 1730. Gardiner Museum of Ceramic Art, Toronto, Canada. Public domain

find an apparently matching pair of flasks that were made in two different manufactories: the first was made and decorated at the Het Moriaenshoofd factory but subsequent marks indicate that it was fired again at the De Griekse A factory some months later, whereas the second flask was made exclusively at the De Griekse A factory, and there are several minor differences to be observed between them in the placement of the armorial decoration design.

These flasks were a diplomatic gift from Willem of Orange in the Netherlands, who became King William III of England in February 1689, who wished to secure the support of Sweden against French aggression in the Nine Years' War which commenced in 1688 and lasted until 1697. At this time, both King William III of England and the Netherlands and King Louis XIV of France were actively seeking

Fig. 13.29 Delft flower pyramids designed for Queen Mary II, *ca.* 1690. *Credit Gerritt Schreurs*

an alliance and cementing relationships with the Scandinavian countries (Hatton, *Gratification and Foreign Policy: Anglo-French Rivalry in Sweden During the Nine Years' War,* [20]). A second point of interest is revealed in the existence of a similar pair of Dutch silver flasks given by King William III to William, the fourth Earl of Devonshire, at Chatsworth in 1688, bearing the Cavendish coat of arms. Although pilgrim flasks were usually decorative in concept, the pair illustrated here also had a functional role in being used to store tea, which was an extremely valuable commodity in the seventeenth and eighteenth Centuries. In his gift to Count Hastfors, King Willem III clearly wished to advance the appreciation and standing of Netherlands Delftware by using exemplars as a diplomatic gift.

Fig. 13.30 Pair of tin-glazed armorial earthenware pilgrim flasks, Delft, bearing the arms of Count Jacob Johann Hastfors of Sweden, 1689–1690, a diplomatic gift from King Willem III of England and the Netherlands. (Reproduced with permission from the Collection of E. and H. Manners Ltd., Church Street, Kensington, London)

13.7 Other European Faience

Many other examples of European earthenwares, faience and majolica are known dating from the fifteenth and early sixteenth Centuries: good exemplars of these are shown in Figs. 13.31 and 13.32, which are a basin in tin-glazed Manises (Valencia) earthenware with a brown lustre decoration which dates from about 1485 and bears the arms of Maria of Castile (Queen of Aragon and Naples and Princess of Asturias), and a similar period Italian tin-glazed 'maiolica dish from Nicola di Urbino bearing the arms of Isabella D'Este and Gianfranco Gonzaga dating to 1524, respectively.

Fig. 13.31 Basin in tin-glazed Manises (Valencia) earthenware with a brown lustre decoration which dates from about 1485 and bears the arms of Maria of Castile (Queen of Aragon and Naples and Princess of the Asturias). Manises was famed for its lustre decoration which was achieved using a mixture of silver, copper, cinnabar and ochre ground in vinegar and fired in a reduction furnace. Victoria and Albert Museum. London, VA 243–1853. Public domain

The latter is in an *istoriato* classical style on the theme of Hippolytus and Phaedra and includes a scene from Ovid, with a comment in Latin "*Without hope and without fear, you are defeated*".

Figure 13.33 shows a Spanish tin-glazed armorial dish which displays an escutcheon beneath a stylised helm surrounded by dense polychrome foliage in yellow ochre, brown and manganese blue on a white background in the style of the Talavera *serie tricolor* wares and dating from *ca.* 1600–1625 (Ray, *Spanish Pottery*, 1248–1898, [28]). However, this item has been established to originate from Muel in Aragon (Errol Manners, personal communication[25]).

In Fig. 13.34 is shown a Portuguese tin-glazed faience jug, *ca.* 1635–1640, made for export from Lisbon for the Northern European market in Germany and the Hanseatic States, in a blue and yellow polychrome palette in imitation of the underglaze blue Ming porcelain exports that were so much admired and desirable in Europe at that time (Houkjaer, *Tin-Glazed Earthenwares from The Netherlands, France and Germany, 1600–1800*, [21]) and in competition with the Chinese *kraak* porcelain that was then being imported by the Dutch East India Company (VOC). The remainder of the decoration is a copy of the peony floral decoration that is found on contemporary Ming export porcelains.

Fig. 13.32 Italian tin-glazed maiolica dish from Nicola di Urbino bearing the arms of Isabella D'Este and Gianfranco Gonzaga dating to 1524. This is decorated in an *istoriato* classical style on the theme of Hippolytus and Phaedra with a comment in Latin "*Without hope and without fear, you are defeated". Salting Bequest, Victoria and Albert Museum, London, C2229-1910. Public domain: credit Valerie McGlinchey*

Fig. 13.33 A Spanish faux or pseudo-armorial dish, *ca.* 1600–1625, tin-glazed earthenware from Muel, Aragon. (Reproduced with permission from the Collection of E. and H. Manners Ltd., Church Street, Kensington, London)

Fig. 13.34 A Portuguese tin-glazed faience pseudo-armorial jug, ca. 1635–1640, made for the Northern market in imitation of Ming porcelain analogues with a helm, crest, torse, lambrequins and escutcheon. However, the heraldic crest and escutcheon contain floral sprays rather than distinct heraldic achievements. The remainder of the decoration in blue enamel copies a peony design found on Ming porcelain of the same period. (Reproduced with permission from the Collection of E. and H. Manners Ltd., Church Street, Kensington, London)

References

1. Archer, M.: Delftware: the Tin-Glazed Earthenware of the British Isles, a Catalogue of the Collection in the V&A Museum, London. HMSO/V&A Museum Publications, London (1997)
2. Bimson, M.: John Dwight. English Ceram Circle Trans. **5/2**, 101 (1961)
3. Borschberg, P.: The seizure of the *Santa Catarina* revisited: the Portuguese empire in Asia, VOC politics and the origins of the Dutch -Johor Alliance, 1602- *ca.* 1616. J South East Asian Stud. **33**, 31–62 (2002)
4. Britton, F.: London Delftware. Jonathan Horne, London (1987)

5. Brogniart, A.: Traite des Artes Ceramiques, ou des Poteries, Considerees dans Leur Histoire, Leur Pratique et Leur Theorie, 2 Vols. Bechet Jeune, Libraire Editeur, Paris (1854)
6. Caiger-Smith, A.: Tin-Glazed Pottery in Europe and the Islamic World: the Tradition of 1000 Years in Maiolica, Faience and Delftware. Faber & Faber, London (1973)
7. Carnegie, D.: Tin-Glazed Earthenware. A. & C. Black/Bloomsbury Publishers, London (1993)
8. Wang, C., Zhang, W., Bo, W., Zhang, B.: Scientific study of the celadon glaze from the Donggou kiln, China. Nucl. Instrum. Methods Phys. Res., Sect. B. **504**, 1–5 (2021)
9. Chong, A., Murphy, S.A., Flecker, M., Krahl, R., Guy, J., Miksic, J., Hey, D., Shuyi, K., Louis, F., Donfang, Q. (eds.): *The Tang Shipwreck: Art and Exchange in the 9th Century*. Asian Civilisation Museum, Singapore (2017)
10. Colomban, P.: Pottery, Glass and Enamelled Artefacts: how to Extract Information on their Manufactory Technology, Origin and Age. In: Edwards, H.G.M., Vandenabeele, P. (eds.) *Analytical Archaeometry: Selected Applications*, pp. 245–267. Royal Society of Chemistry Publishing, Cambridge (2012)
11. Dawson, A.: English and Irish Delftware, 1570–1840. British Museum Publications, London (2010)
12. Denyer, R., Denyer, M.C., Edwards, H.G.M.: *Armorial Porcelain: the Genesis*. Springer-Nature, Dordrecht. to be published (2024)
13. Edwards, H.G.M.: 18th and 19th Century Porcelain Analysis: a Forensic Provenancing Assessment. Springer-Nature, Dordrecht (2020)
14. Edwards, H.G.M.: Porcelain Analysis and its Role in the Forensic Attribution of Ceramic Specimens. Springer-Nature, Dordrecht (2021)
15. Edwards, H.G.M.: Welsh Armorial Porcelain. Springer-Nature, Dordrecht (2022)
16. Edwards, H.G.M., Brooke, C.J.: Coade Stone: a History and Analysis. Springer-Nature, Dordrecht., to be published (2024)
17. Fourest, H.P.: *Delftware: Faience Production in Delft*, Transl. from the French by Katherine Watson. Thames & Hudson, London (1980)
18. Gaimster, D.: German Stoneware, 1200–1900: Archaeology and Cultural History. British Museum Press, London (1997)
19. Haselgrove, D., Murray, J.: John Dwight's Fulham pottery, 1672–1978: a collection of documentary sources. J Ceram Hist at Stoke-on-Trent City Museum. **11** (1979)
20. Hatton, R.M.: Gratification and foreign policy: Anglo-French rivalry in Sweden during the nine years' war. In: Hatton, R.M., Bromley, J.S. (eds.) William III and Louis XIV Essays, pp. 1680–1720. Liverpool University Press, Liverpool (1968)
21. Houkjaer, U.: Tin-Glazed Earthenwares from The Netherlands, France and Germany, 1600–1800, Designmuseum Danmark, Copenhagen, pp. 106–107 and 346–349, 2016
22. Howard, D.S.: A Tale of Three Cities – Canton, Shanghai and Hong Kong: Three Centuries of Sino-British Trade in the Decorative Arts. Sotheby's Publishing, London (1997)
23. Krahl, R., Guy, J., Wilson, J.K., Raby, J. (eds.): Shipwrecked Tang Treasures and Monsoon Winds, Arthur M. Sackler Gallery. Smithsonian Institution/National Heritage Board, Washington DC/Singapore (2010)
24. Lipski, L., Archer, M.: Dated English Delftware. Sotheby Publications, London (1984)
25. Manners, E..: personal communication
26. Noel Hume, I.: If These Pots Could Talk: Collecting 2000 Years of British Household Pottery. Chipstone Foundation, Milwaukee (2001)
27. Ramsay, W.R.H., Ramsay, E.G.: The Evolution and Compositional Development of English Porcelains from the 16th Century to Lund's Bristol c. 1750 and Worcester c. 1752- the Golden Chain. Invercargill Press, Invercargill (2017)
28. Ray, A.: *Spanish Pottery, 1248–1898*, A Catalogue of the Collection in the V&A Museum. V&A Museum Publications, London (2000)
29. Reineking von Bock, G.: Kunstwerkenmuseum der Stadt Koln: Steinzeug., Stadt Koln, 1986
30. Skerry, J.A., Hood, S.F.: Salt-Glazed Stoneware in Early America. University Press of New England and Colonial Williamsburg Foundation, Lebanon/Williamsburg (2009)

31. Spataro, M., Meeks, N., Bimson, M., Dawson, A., Ambers, J.: Early porcelain in 17th century England: non-destructive examination of two jars from Burghley house. Br Mus Res Bull. **3**, 37–48 (2009)
32. Stanford, C.: "The mistery of Cologne or stone wares": were John Dwight and Robert Hooke pioneers in the use of fired artificial stone in architecture? Georgian Group J. **XXVIII**, 1–20 (2020)
33. Stow, J.: *Survey of London: Written in the Year 1598, with Diuers Rare Notes on Antiquity*. J.M. Dent & Sons/E.P. Dutton & Co., London/New York (1603)
34. Tite, M.S., Bimson, M., Freestone, I.C.: A Technological Study of Fulham Stoneware. In: *Proceedings of the 24th International Archaeometry Symposium*, pp. 95–104, Washington, DC (1986)
35. Tyler, K., Betts, I., Stephenson, R.: London's Delftware Industry: the Tin-Glazed Pottery Industries of Southwark and Lambeth. Museum of London Archaeology Publications, London (2008)
36. Vierra Ferreira, L.F., Casimiro, T.M., Colomban, P.: Portuguese tin-glazed earthenware from the 17th century: part I, pigments and glazes characterisation. Spectrochim. Acta A. **104**, 437–444 (2013a)
37. Vierra Ferreira, L.F., Casimiro, T.M., Colomban, P.: Portuguese tin-glazed earthenwares from the 16th century. Appl. Surf. Sci. **285**, 141–152 (2013b)
38. Vierra Ferreira, L.F., Ferreira, D.P., Conceicao, D.S., dos Santos, L.F., Pereira, M.F.C., Casimiro, T.M., Machado, I.F.: Portuguese tin-glazed earthenware from the 17th century: part II, a spectroscopic characterisation of pigments, glazes and pastes from the three main production centres. Spectrochim. Acta A. **149**, 285–294 (2015)
39. Vierra Ferreira, L.F., Ferreira Machado, I., Casimiro, T.M., Pereira, M.F.C., Santos, L.F.: Portuguese blue-on-blue 16th–17th century pottery. Archaeometry. **60**, 695–712 (2018a)
40. Vierra Ferreira, L.F., Barros, L., Ferreira Machado, I., Gonzales, A., Pereira, M.F.C., Casimiro, T.M.: Spectroscopic characterisation of amphorae from the 8th to the 7th century BCE found at the Almaraz settlement in Almada, Portugal. J Archaeol Sci Rep. **21**, 166–174 (2018b)
41. Vierra Ferreira, L.F., Ferreira Machado, I., Gonzales, A., Pereira, M.F.C., Casimiro, T.M.: Portuguese 16th to early 18th century tin-glazed ceramics found at the Tagus Estuary saltpans. Global J Archaeol Anthropol. **11**, 156–167 (2021)
42. Wood, N.: Chinese Glazes: their Origins, Chemistry and Recreation. A & C. Black/Bloomsbury Publishing, London (1999)

Chapter 14
European Porcelains—Their Genesis and Development

Abstract The genesis of European porcelains from their first manufacture by Duke Francisco Medici in Florence in the 1570s to the exponential growth of manufactories in France and in continental Europe in the eighteenth century: through Fulham, Rouen, St Cloud and Paris to Meissen, with the first commercially successful operation from 1712. Several key factories are described in detail. The Fulham Pottery is included here chronologically between Medici and Rouen, although technically forming part of the English manufactories described separately in Chap. 15.

Keywords Medici porcelain · Florence · Fulham Pottery · John Dwight · Rouen · St Cloud · Meissen · King Augustus the Strong · Vincennes · Sèvres

14.1 The Genesis of European Porcelains

Here, we shall consider the growth of European porcelain manufactories from the rather tentative start-ups that were made in the sixteenth century to the huge increase numerically in the number of manufactories that were known in the eighteenth century: a detailed survey of these has been made by Edwards [17]. In this survey, some 62 European continental manufactories were identified in the eighteenth century, plus those of Rouen and St Cloud from the late seventeenth century, along with 30 English contemporaries, along with that of Fulham from the late seventeenth century. Clearly, it would be beyond the scope of the current text to describe all of these in detail here, but some of the earlier ones are considered to establish the potential reasons for the genesis of European porcelain and its decoration. In the past, historians have alleged that the growth of the European manufacture of porcelain in its widest context could be levelled at the establishment of a competitive business with the Chinese porcelain imports that were then flooding into Europe from the major trading activities in Far Eastern commodities such as spices, silks and tea with the British Honourable East India Company (HEIC), their Dutch and Swedish counterparts (VOC and SOIC) and the Portuguese chartered vessels of the Casa da India

H. G. M. Edwards, P. Colomban, *Blue by Fire: A Marker of the Technical History of Glass and Ceramics*, Cultural Heritage Science,
https://doi.org/10.1007/978-3-031-77122-4_14

known as the "Bombay ships". Firstly, it will transpire that the commercial business aspect was not in fact the driving force underlying this urge to manufacture porcelain in Europe and that the prime reason initially was more subtle in that it centred upon the immense desirability of its ownership and the social status that this conferred upon its owners, who were "diplomatically gifted" special creations of this wondrous new material, which King Augustus of Saxony and Poland termed his "white gold" [26]. Secondly, a close examination of the dates of foundation of the porcelain manufactories in the eighteenth century offers the presence of an intriguing chronological hiatus that existed between those founded in England and their contemporaries in continental Europe, predominantly in France, Saxony, Austria and Italy. Chronologically, as will be seen later, the first European porcelain manufactory in Florence was followed by that of Fulham in London, then those of Rouen and St Cloud, before King Augustus of Saxony enrolled Ehrenfriede von Tschirnhaus and the alchemist Johann Bottger to synthesise Meissen porcelain for him at Albrechtsburg Castle in Dresden in the first decade of the eighteenth century. Many historians claim that Meissen was the first European porcelain to be made outside of China but this will be seen to be manifestly incorrect chronologically: there is little doubt that Meissen was the first truly successful commercial European manufacturing porcelain enterprise, although there is evidence that St Cloud porcelain was offered for purchase to a clientele in limited quantities! It is less clear, however, why in the face of so many start-ups in France and Saxony in the first two decades of the eighteenth century there was a singular inactivity and unnatural quietude in England in this context after Fulham closed with the death of its proprietor, John Dwight, in 1703. Historians have attempted to attribute this hitherto to the erstwhile English manufacturers not having the wherewithal in technical competence and specific ceramics knowledge to emulate their French counterparts: according to this hypothesis, it is alleged that this apparent knowledge transfer eventually occurred some thirty years or so after the successful production of porcelain at Meissen and following the disclosure of the Chinese hard paste true porcelain recipe by Pere Xavier d'Entrecolles SJ in Peking, which was transmitted to his superior Pere Orry in Paris in 1717 and 1722, which stimulated the French manufactories [17].

It is further believed by some historians that the acquisition of the relevant body paste recipes from France therefore was necessary before English porcelain manufacture could be undertaken in the eighteenth century. However, the implication from this assertion of knowledge transfer, of course, is that the French recipes would then comprise the standard formulations for the early English porcelain and that is certainly not the case from recent analytical compositional determinations, which clearly demonstrate that the English recipes and formulations that appeared in Bow in 1744 or just before must have been developed independently from those then operating in Continental Europe, as the raw materials and paste compositions of the porcelain are sensibly different.

It is indeed also remarkable that the early success in porcelain manufacture of John Dwight at his Fulham manufactory in the late seventeenth century was not taken up by successors following his death in 1703—but already, even by that stage, Dwight had remarked at his lack of success in securing ongoing funding from sponsors for his process: it can be inferred from his remarks that the Royal Society, who had been actively sponsoring his earlier efforts at creating a silicaceous refractory material for crucibles capable of withstanding alkaline and acidic molten salts at the high temperatures necessary for the extraction of metals from their mineral ores was deemed to have been successful and that there was no further interest in the manufacture of porcelain *per se*. John Haslem has summarised the situation faced by John Dwight [23] as follows:

> In 1671 John Dwight had a patent for having by his own industry and at his own proper costs and charges invented and sett up at Fulham the mistery of Porcelain or China and Persian ware. Again in 1684 he protected the mistery of porcelain and other wares never before made in England or elsewhere. It would seem that the Doctor was in advance of the requirements of his day as his invention met with such poor encouragement that he burnt his receipts and implements in disgust and it is doubtful whether there is a specimen of his Porcelain extant. It is also said that he buried all his models, moulds and tools in some secret place on the premises at Fulham, observing that the production of such materials was expensive and unremunerating and he wished therefore to put it out of the power of his successors to perpetuate the business?

What is certainly true is that in many early porcelain manufactories there was a culture of the initial founder and driving force behind the operation taking his secrets to the grave and resulting in the premature closure of the manufactory: Francisco Medici, Duke of Tuscany, at Florence and Louis Poterat at Rouen are just two of these.

In a parallel context it is also evident that the early efforts at making porcelain in England in the eighteenth century never achieved the financial support of the aristocracy, who nevertheless still desired to acquire the finished articles for their estates, unlike their continental contemporaries in France, Saxony and elsewhere, where royal connections and privileges were extant. With the possible exception of the Chelsea manufactory in the 1750s, no English manufactory could rely upon aristocratic or royal influence to promote its products in contrast to King Louis XIV and his brother, the Duke of Orleans, in France and King Augustus of Saxony, for example. An irrefutable conclusion from the chemical analyses carried out on early English porcelains mentioned above is that they are totally different from their French and Saxon counterparts in composition and have hence been developed along a sensibly different pathway. This revelation has not been fully accepted generally by connoisseurs, who have repeatedly opted in favour of a dependence route from their French competitors for the English products such as those of Bow, Chelsea, Limehouse and Liverpool, which upon inspection actually has no credibility scientifically. It is even more apparent that the earliest syntheses of English porcelains can be reliably traced to the peripatetic movement between these early factories of a small number of skilled experimentalists and experienced potters with the relevant knowledge of ceramic clays, raw materials and additives, some of which do not feature in the French and Saxon porcelains, including calcined bone ash, which was later patented by Josiah Spode as the premier component of his highly phosphatic English bone china! In France, for example, it was realised early on that the knowledge transfer that could be accessed by the engagement of workforce from a competitor was a valuable adjunct in porcelain manufactories and King Louis XV issued a warrant at his Vincennes manufactory which forbade the departure of key employees without permission!

There now follows a summary of several key early European porcelain manufactories, some of which were successful and others not: essentially, these have been selected to illustrate the perceived reasons for their foundation and start-up and from which we can form a picture of the historical situation that then applied for their development. Not all of the manufactories will be discussed in detail but the listing supplied in Table 14.1 should be referred to as can the reference cited previously, where further exemplars are discussed [17]. Also, Fig. 14.1 indicates graphically the significant increase of porcelain manufactories in Europe during the period 1550–1850 and its exponential increase through the eighteenth and into the nineteenth century: a histogram is also supplied in Fig. 14.1 which relates the manufactories associated with modern-day European countries (and including the USA).

14.2 The First European Porcelain Manufactory: Florence

Undoubtedly, the first European porcelain manufactory is deemed to be that of Duke Francesco de Medici of Tuscany, who commenced his preparations for the synthesis of the first European porcelain in 1575 in his Casino di San Marco in Florence. Giorgio Vasari [61] recorded that Bernardo Buontalenti was experimenting with the art of manufacturing porcelain in Florence in the 1560s but without success initially being achieved. However, in 1575, the Venetian ambassador to the Court of Tuscany, Andrea Gussoni, mentioned that Francesco de Medici I, Grand Duke of Tuscany, had established a porcelain manufactory in Florence, having discovered the secret of the manufacture of the "*porcelain of India*" after 10 years of experimentation. Gussoni said that the Medici porcelain was "*transparent, hard, light and delicate*" [33]. It is interesting at this point to try to discover what catalysed this interest in the Medici family that gave rise to their groundbreaking venture in the synthesis of porcelain, which was most certainly the first successful attempt at manufacturing this novel ceramic in Europe.

It is recorded that Marco Polo, the Venetian adventurer, returned to Venice from the far East in 1295, bringing with him a small white bottle made of *qingbai* porcelain [44], probably similar in shape to that shown in Fig. 14.2, which dates from the Southern Song Dynasty (1127–1279).

Marco Polo's description of translucent fine Chinese porcelain in his classic travelogue of his journeys in China is generally accorded to be its first mention in Western literature: the text was actually transcribed by Rustichello da Pisa from Marco Polo's dictation when both were imprisoned in Genoa between 1296 and 1298 and was believed to have been further refined by Polo personally following his release later in Venice. However, several centuries earlier, Arabic texts refer to this wondrous material made of clay through which one could see their contents: Pirazzoli t'Serstevens [41] and Zhao [63] have described the extensive trade that

Table 14.1 Foundation date of the earliest European and western *factory/workshop* producing porcelain (up to 1826). For simplification, the country of production is named by its current name (region and/or state), even if the affiliation was different at the time. When the place is located near an important city, the name of this city is also indicated

City (Country), date	City, date
Florence (Italy), 1575	Venice (*Le Nove Factory*, Italy), 1762
Padua (Italy), 1627?	Berlin (*Königste Porcellan Factory*, Germany), 1763
Fulham (UK), 1671	Niderviller (Lorraine, France), 1763
Rouen (France), 1673	Sceaux – Paris (France), 1763
Saint-Cloud (France), 1690	Venice (*Cozzi Factory*, Italy), 1764
Paris (*A. Pavie Factory*, France), 1692	Wallendorf (Thuringia, Germany), 1764
Dresden (Saxony, Germany), 1694	Verbiliki - Moscow (*Gardner Factory*, Russia), 1764
Meissen (Saxony, Germany), 1708	Lunéville (Lorraine, France), 1766
Lille, (France), 1711?	Audun-le-Tiche (*Villeroy & Boch Factory*, Lorraine, France), 1766
Vienna (Austria-Hungaria), 1718	Plymouth (Devon, UK), 1768
Venice (*Vezzi workshop*, Italy), 1720	Rörstrand – Stockholm (Sweden), 1770
Chantilly – Paris (France), 1725	Philadelphia (*Bonnin & Morris Factory*, USA), 1770
Sesto Fiorentino - Doccia, (*R. Ginori Factory*, Italy) 1735	Limoges (France), 1771
Pont-Sainte-Maxence - Saint-Germain-en-Laye (Paris, France), 1738	Naples (*Real Fabbrica di Napoli*, Italy), 1771
Vincennes – Paris (France), 1740	Saint Amand – Arras (France), 1771
Savannah (*Duché workshop*, Georgia, USA), 1741?	Loosdrechts (NL), 1774
Capo di Monte – Naples (Italy), 1743	Aynsley – Longton (Staffordshire, UK), 1775
Chelsea – London (UK), 1743	Copenhagen (*New Royal Factory*, Denmark) 1775
Saint-Petersbourg (*Lomonov Imperial Factory*, Russia), 1744	Ilmenau (Graf von Henneberg Factory, Thuringia, Germany), 1777
Mennecy – Paris (France), 1745	Hollohaza (Austria-Hungary), 1777
Frankfurt (*Höchst Factory*, Hesse, Germany), 1746	Rauenstein (Thuringia, Germany), 1783
Bow – London (UK), 1747	Weimar (Thuringia, Germany), 1790
Fürstenberg (Saxony, Germany), 1747	Horni Slavkov (*Hass & Czjek Factory*, Austria-Hungary), 1792
Schloss Nymphenburg (Bavaria, Germany), 1747	Stoke-on-Trent (*Mintons Factory*, UK), 1793
Derby (*Royal Crown Factory*[a], UK) 1750	Thun - Kasterec nad Ohri, Chomutov (Austria-Hungary), 1794
Alcora (*Real Fabrica*, Spain), 1750	Tettau (Thuringia, Germany), 1794
Tournai (Hainaut, Belgium), 1751	Burslem (*Wedgwood Factory*, UK), 1795
Worcester (*Royal Factory*, UK), 1751	Stoke-on-Trent (*Spode Factory*, UK), 1800
Strasbourg (France), 1751	Bourg-la-Reine – Paris (France), 1804
Liverpool (UK), 1754	Cmielow (Poland), 1809
Frankenthal (Rhineland-Palatinat, Germany), 1755	Wedgwood (Stoke-on-Trent, UK), 1812
Copenhagen (*Royal Factory*, Denmark), 1755	New York (*Mead Factory*, USA), 1813
Hotel de Lassay/de Lauraguais – Paris (France), 1755	Hutschenreuther – Selb (Bavaria, Germany), 1814
Sèvres – Paris (*Royal Factory*, France), 1756	Madrid (*Real fabrica de la Moncloa*, Spain), 1817
Gotha (Thuringia, Germany), 1757	Swinton – Rotherham (*Rockingham Factory*, UK), 1820
Lowestoff (UK), 1757	Lichte (Thuringia, Germany), 1822
Marieberg (Stockholm, Sweden), 1758	Vista Alegre (Portugal), 1824
Ludwisburg (Baden-Württemberg, Germany), 1758	New York (*Decasse & Chanou Factory*, USA), 1824
Ansbach (Bavaria, Germany), 1758	Jersey (New Jersey, USA), 1825
Weesp (NL), 1759	Philadelphia (*Tucker & Bird Factory*, USA), 1826
Buon Retiro (Spain), 1760	
Kloster Veilsdorf (Thuringia, Germany), 1760	
Kelsterbach (Hessen, Germany), 1761	
Volkstedt (Rudolfstadt, Thuringia, Germany), 1762	

[a]Founded by a French Hugenot

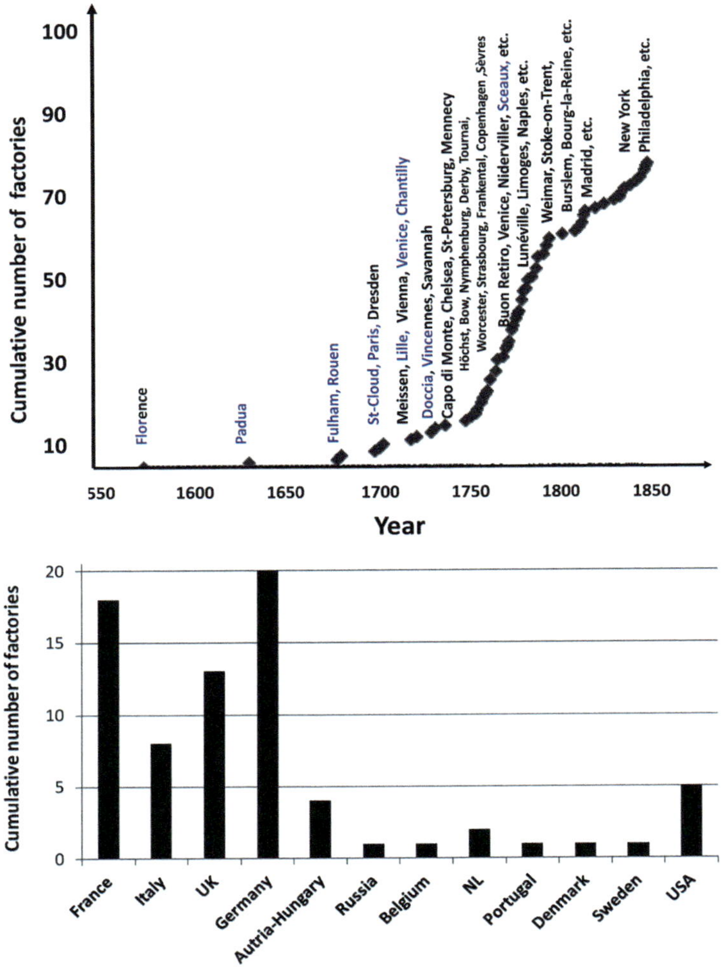

Fig. 14.1 Upper: graphical plot of the European porcelain manufactories in the period 1550–1850, including the USA for comparison. Lower: histogram of the national manufactory locations arranged according to the modern countries

was taking place between the Arab Dynasties in the Middle East and East Africa and China driven by curiosity and information gathering between the ninth and thirteenth centuries, in which *qingbai* Chinese porcelain exports featured prominently and was being eagerly acquired. Even then, envoys from Arabia, and from Chola, India, Sumatra, and Vietnam were being received in Kaifeng, China, one of the ancient capitals of China and seat of government during the Northern Song Dynasty. Frankopan [18] has stated that "Dining tables in the Middle East (author: during the 10th Century) were adorned with Chinese ceramics of such high quality that one contemporary observer wrote that you could see water sparkle through them even though they are ceramic".

Fig. 14.2 Qingbai porcelain bottle vase, with lid and with prunus blossom moulding, lotus sprays and pale blue glaze, Southern Song Dynasty, Jingdezhen, Jiangxi province, *ca.* 1127–1279. Some of the first Chinese porcelain was made as export gifts to European aristocracy. *In the Los Angeles County Museum*

In 1338, the last Mongol Emperor of the Yuan Dynasty (1206–1368), Togon Temur (Huizong), sent a Chinese delegation of Christian Nestorian monks in 1338 to Pope Benedict XII (1334–1342), who was then based in Avignon, and on the way it is maintained that they visited King Louis the Great of Hungary and presented him with a piece of *qingbai* porcelain from Jingdezhen, so named for its blue-green glaze (*yingqing*), bearing his coat of arms and those of the Kingdom of Poland. This represents the earliest piece of Chinese porcelain that was specifically made for the European market. Known as the *Gaignieres-Fonthill Vase*, it now resides in the National Museum of Ireland, arriving there via a circuitous route through Charles III Durazzo, King of Naples, then via the Kingdom of Aragon and thence to Jean, Duc de Berry and Dauphin of France, the son of King Louis XIV. This started the trend for using porcelain as "diplomatic gifts", a practice which was then carried on in Europe thereafter until later in the eighteenth century.

This intriguing story has now been adopted into the annals of early porcelain history but actually the forensic chronology requires close examination as the tale is unfortunately incorrect: firstly, King Louis the Great of Hungary (Lajos Nagy, reigned 1342–1382) did not accede to his throne until the 21st July, 1342: Louis was born in 1326 to King Charles I and Elizabeth of Poland and he would only have been twelve years old in 1338. This means chronologically that the delegation of Chinese visitors could not have appeared at the Court of King Louis in 1338 .... also, the date of their visit would have been most unlikely to have taken place as late as 1342 as Pope Benedict, the intended recipient of the diplomatic gift from this delegation, died in April that year and Pope Clement VI then assumed the papacy. Also, King Louis of Hungary did not assume the title of King of Poland until the death of his uncle in 1370, so the apparent depiction of his coat of arms as King of Hungary and King of Poland (where he then became King Ludwig) could not possibly have appeared on a piece of porcelain as early as 1338! The earliest description of this piece of porcelain is given by Barthelemy Remy, valet of Francois Roger de Gaignieres in 1713, when the *qingbai* vase was in the possession of Francois Lefebvre de Caumartin of Chalons-en-Champagne, who was Bishop of Vannes and Blois. The vase by then had been fitted with silver mounts, a silver foot and a silver handle bearing the coat of arms of King Louis of Hungary, the Angevin court at Naples and the Kingdom of Poland (Fig. 14.3). It is believed that this attachment was made either by Queen Joanna of Naples or by her successor King Charles III Durazzo in the 1380s to cement their allegiance with King Louis of Hungary and Poland and this is borne out by an inscription on the silver handle of the vase which gives the name "JEHENNA", translated as "Joanna". The vase then found its way into the Angevin line of French kings as described above and eventually into the collection of William Beckford at Fonthill Abbey, whence it was acquired by John Farquhar, the gunpowder millionaire, in a Phillips auction sale in 1823. At the sale of the effects of Alexander, the tenth Duke of Hamilton, at Hamilton Palace in 1882 the National Museum of Ireland purchased the vase for the sum of £28 with an unrecognised provenance as by that time its silver mounts had been stripped away to leave the vase in its original, unmodified condition, which does not in fact bear any motifs which relate to King Louis at all! Initial research has established that the *Gaignieres- Fonthill Vase* is certainly of Yuan Dynasty *qingbai* porcelain with floral mouldings and a characteristic blue-green glaze dating from 1300–1330, and that it probably in fact had not been presented at all to the Hungarian court of "King Louis" at that time [1, p. 132, 25, 33, 34] but had rather been given to the Angevin kingdom of Naples by the Nestorian envoys from China in 1338 *en route* to Avignon to their meeting with Pope Benedict XII. This account is a classic example of how an earlier supposition about an historical piece of porcelain, which was made inappropriately and incorrectly on the basis of a flawed chronology and in the absence of accompanying documentation, has still persisted into modern history despite compelling evidence to the contrary. Nevertheless, the *Gaignieres-Fonthill Vase* is of high importance as the earliest recorded piece of Chinese porcelain expressly made for the European market as a diplomatic gift from a Chinese delegation from a Yuan Dynasty Chinese Emperor and it can now be seen on display in its original form and without its silver mount embellishment in the National Museum of Ireland, Collins Barracks, Dublin. The *Gaignieres -Fonthill Vase* as it now appears in the National Museum of Ireland devoid of its mounts is shown in Fig. 14.4.

Other early diplomatic gifts of Chinese porcelain were made by the Chinese Emperors to the Ottoman sultans at the Topkapi Palace in Constantinople and to the Mongol Emperors in Samarkand in the early to mid-fifteenth century. Lorenzo de Medici received several precious porcelain vases as a diplomatic gift from the Mameluk Sultan of Egypt in 1478, described as being "*large vessels of porcelain,*

Fig. 14.3 The *Gaignieres-Fonthill Vase*: Yuan Dynasty, *qingbai* porcelain, *ca.* 1300–1330, the oldest piece of Chinese porcelain recorded definitively as a diplomatic gift from a Chinese mission sent from the Yuan Emperor to a European monarch. This picture shows a watercolour of the vase with additional silver mounts added later, bearing the coats of arms of King Louis the Great of Hungary, the King of Poland and the Angevin Monarch of Naples. The watercolour painting was made in 1713 by Barthelemy Remy, valet of Francois Roger de Gaignieres from the vase which was owned at that time by Francois Lefebvre Caumartin. This vase is now in the National Museum of Ireland, Dublin, minus its associated silver mounts, where it was acquired from a sale of effects in Hamilton Palace in 1882. *Copyright, The National Museum of Ireland, Collins Barracks, Dublin, and reproduced with permission*

the like of which had never been seen before" [55]. These greatly impressed him and the visitors to his Tuscan Court and this seemed to trigger a desire amongst the successive Medici Grand Dukes of Tuscany to manufacture porcelain themselves after his death in 1492 [27, 33]. Cosimo de Medici I (reigned, 1537–1574) was an avid collector of Chinese porcelain and a catalogue, dated 1553, listed over 400 pieces in his personal collection. Cosimo first initiated the effort to manufacture porcelain in his palace workshops, especially when he learned that the alchemists Camillo and

Fig. 14.4 The *Gaignieres—Fonthill Vase* as it is today, front view, minus its later silver mounts, demonstrating the superb quality of its porcelain; note that the vase itself does not possess any moulding which associates it with King Louis of Hungary. *Copyright, The National Museum of Ireland, Collins Barracks, Dublin, and reproduced with permission*

Battista Gatti working in Ferrara for his rival, Duke Alfonso II d'Este, in 1561/1562 were attempting to make porcelain there [43]. Cosimo started work on porcelain manufacture in his palace with Bernardo Buontalenti as director of operations and Giorgio Vasari, as we have already noted earlier, claimed he came near to producing porcelain in 1568, but it was also stated that he did not actually produce any. In 1574, Cosimo's son, Francesco de Medici, became Grand Duke of Tuscany and he constructed new workshops in the Boboli Gardens and in the Casino di San Marco in Florence. Francesco is recorded as working personally in these workshops alongside alchemists, potters and lapidaries to produce his porcelain. They succeeded in manufacturing good quality porcelain with the help of a "Levantine" and several large pieces of porcelain were put on show there for visitors. Recent analytical studies of Medici porcelains [6] support ambassador Gussoni's report of an Oriental influence in this porcelain production and there is a strong existence of a link with itinerant Islamic Iznik faience workers [36].

Medici porcelain is classed as a soft paste porcelain, which includes a white clay raw material component from Vicenza, which fortuitously actually contained kaolin. Francesco de Medici used his porcelain artefacts as diplomatic gifts and his manufactory was never intended to be a commercial enterprise: for example, he presented 20 pieces to the d'Estes of Ferrara and some more to King Philip II of Spain which were very highly appreciated. Around this time, there are also reports that Cipriano Piccolpasso, working for Francesco Maria II, Duke of Urbino, had succeeded in making *la vera porcellana* in 1578/9 [38]. Clearly, there was much activity in sixteenth century Italy, particularly in the Duchies of Savoy, Urbino, Ferrara and Tuscany, in the potential manufacture of porcelain.

The surviving specimens of Medici porcelain, numbering today approximately 70 in total, indicate that although they do possess a reasonable translucency, which certainly qualifies them as being classed as porcelain, the potting is undoubtedly rather heavy and the decoration comprises exclusively a simple underglaze blue palette in simulation of the Chinese porcelain imports. An example of a Medici porcelain gourd with blue and white floral decoration is shown in Fig. 14.5. Some early specimens exhibit the mark of Filippi Bruneschelli's dome of the Cattedrale di Santa Maria del Fiore, otherwise known as Il Duomo, a well-known landmark in

Fig. 14.5 Medici porcelain gourd decorated in underglaze blue with floral motifs, *ca.* 1575–1587. *Musee du Louvre, Paris. Creative Commons Attribution—Share Alike 3.0*

Florence, along with a capital *F* for Francesco de Medici. Others bear a Medici emblem known as the *alle*, several balls contained within an heraldic shield. These probably constitute the first records of manufactory marks on European porcelain, a distinction that has previously been accorded to the crossed swords underglaze blue mark of King Augustus on his Meissen porcelain some one hundred and thirty years later. Several pieces of Medici porcelain bearing the heraldic arms of the powerful King Philip II of Spain (*El Rey Felipe Segundo*, 1527–1598), a ruler who was also King of Naples and Sicily (1554–1598) and of all Spanish dominions in The Netherlands and America, King of Jerusalem (and *jure uxoris* King of England through his marriage to Mary Tudor, Queen Mary I, from 1554 until 1558) and also Duke of Milan (from 1540), were presented as diplomatic gifts and were very much admired by the recipient. The presentation of porcelain as "diplomatic gifts" was indeed highly regarded in the sixteenth century and for this reason Francesco set out to make the highest quality porcelain product that he could. Francesco had itemised in his personal collection about 310 pieces of Medici porcelain at the time of his death in 1587: these were maintained initially by his descendants at their villa in Rome, but after the due passage of time they became less well appreciated by his family in comparison with the finer porcelain pieces that were then appearing in Europe, hence, the collection was gradually dispersed but still left the 60–70 pieces of Medici porcelain that are known today [10, 54]. The Casino di San Marco manufactory ceased porcelain production immediately after Francesco's death in 1587 and it is not known whether he took the recipe for porcelain manufacture with him without ever divulging it or if there was a disinterest in the continuation with his non-commercial porcelain factory operation by his descendants. Ferdinand Cosimo II de Medici, his successor, did carry on with the factory for a little while, it is thought probably using the existing stocks of Medici porcelains for decoration without actually engaging in their manufacture.

Medici porcelain has been universally categorised historically as a *soft paste* porcelain comprising feldspar, calcium phosphate, calcium silicate (*alpha*—and *beta*—wollastonite) with quartz sand and a white earth clay from Vicenza. However, as Colomban et al. [6] point out from their chemical analyses, this mixture certainly does not qualify as a standard soft paste body and it appears that Medici porcelain should now be more strictly classified as a *hybrid* porcelain, that is between a *hard paste* and a *soft paste* body [29]. The use of a tricalcium orthophosphate component is of interest, especially since this must be the earliest recorded usage of calcined bone ash in use for a ceramic in Europe, although from the analytical evidence this appears to be localised in the enamels and glaze rather than in the porcelain body as such [6]. This usage can be ascribed to the adoption of the bone ash component as an opacifier and whitener in the glaze, as befitting the glaze used in Islamic Iznik faience and tiles, again reflecting the suggested input of Islamic ceramics knowledge (referred to as "Levantine" by contemporary observers) to the Medici operation. This bone ash component additive to porcelain raw materials does not appear again until the advent of Bow porcelain in England in the 1740–1750s, or perhaps just before, where it is used actually in the porcelain paste body in what came to be known as a characteristically English porcelain body component and this became an essential ingredient of English bone china as perfected later and patented by Josiah Spode. It has also been suggested that Francesco de Medici perhaps had obtained some detailed recipe information in his experiments from Islamic ceramicists engaged with the production of Iznik faience tiles in Turkey and that his adoption of bone ash as a component in his Medici porcelains, even if only as an opacifier in the glazes and enamelling, could probably have originated from that source. The use of bone ash by Islamic faience producers to opacify their glaze would then explain the appearance of calcium phosphate in the Medici porcelain glazes, rather than in the body components as has now been determined analytically [6].

The technical difficulty of achieving the high temperatures necessary and adequate kiln temperature control maintenance for the fusion of the paste components and their conversion into a translucent porcelain has been highlighted as a problem for Francesco de Medici but several of his surviving pieces nevertheless seem to be well formed, delicately painted in monochrome blue and glazed, although most are heavily potted and have suffered a pitting of the paste acquired in the glazing process. The style is rather eclectic and exotic, blending the Chinese and Islamic (Iznik) influences with the Italian Renaissance and employing a Westernised oriental decoration incorporating floral sprigs and scrolls. Medici porcelain is, hence, firmly established as an early European porcelain and surely can lay a fair claim to be the first successful Continental European porcelain production, even if it was never intended to be a commercial enterprise.

Based on a very rare and authenticated Medici porcelain shard preserved at the International Museum of Ceramics at Faenza, Italy, Kingery [28] has found the presence of quartz, tridymite, cristobalite, and feldspar, with some alkaline earth silicates in a glassy phase, with the presence of the minerals wollastonite and diopside held in domains in large—sized host material pores. In composition, compared with the later French soft paste porcelains which followed in the eighteenth century, Medici porcelains have a higher Al_2O_3 content (9.5% versus 2.5% for the French porcelains), a very low calcium oxide content (2.5% versus 15% for the French porcelains) and a larger potash content (6% versus 3% for the French porcelains). Clearly, Medici porcelain, however it has been classified historically, is not generically a *soft paste* porcelain (although it does contain the mineral marker signature of wollastonite that characterises a soft paste porcelain); although it does contain a very high percentage of silica at 78%, which is in the regime expected for a standard hard paste body, the alumina content is also far too low to be acceptable for a standard hard paste porcelain body. Tite [60] has estimated that the raw materials for the Medici porcelain bodies contained approximately 60% quartz, 25% white clay and 15% *marzacotta*, a sintered mixture of sand and alkali which is similar to that used in Italian majolica glazes [30]. This composition, Tite has suggested [60], is very close to that of the Islamic stonewares which typically comprised 10 parts crushed quartz sand, 1 part white clay and 1 part glass frit. Likewise, the glaze analysis carried out by Tite gives a highly silicaceous glaze (55% silica), low alumina (5% alumina) and a very large 25% lead oxide presence. It has also been estimated that the kiln firing temperature of Medici porcelains would have had to have been accomplished in the region of 1100 °C. The ternary phase diagram of key elemental oxides constructed from the compositional formulation shows a very shallow *liquidus* region, which means that a small kiln variation in this firing temperature in practice would have been disastrous for the integrity of the porcelain artefacts being fired and that this would not have been helped by the large amount of alkaline and alkaline earth flux used, approaching 13% in total, which is greater than the alumina content, namely, 9.5%. It is, hence, believed that Francesco's kiln control would not have been finely achievable enough in the 1570s to maintain the narrow *liquidus* phase in the fired pastes, so he would have had to endure a serious distortion in many of his porcelain artefacts [28]. However, despite its obvious *hybrid* nature analytically, ceramics historians still refer to Medici porcelain as a "soft paste porcelain" and that has been maintained here conventionally, although as has been determined analytically it clearly was not a *soft paste* bodied porcelain and really would better now be classified under the *hybrid* porcelain category.

In most ceramic historical accounts, Italian porcelain production ceases with the Medicis in the late sixteenth century (around 1587) and does not recommence until the foundation of Vezzi in Venice in the early eighteenth century but recently there have been several connections made to an enterprise that started in Padua and continued for just over ten years from 1627 [33, 37, 38]. It appears that two pieces in the collection (accession numbers C.377-1919 and B.41-1905) of the Victoria & Albert Museum in London have been attributed to the Padua manufactory by John Mallet, both being marked in underglaze blue with *G/I.G.P.F* and a cross with dates of 1627 and 1638 and having a floral decoration. It is alleged that perhaps the *P.F* could mean either *Pisanus fecit* or *Padovano fecit*? These have also been recorded in Regine Plinval de Guillebon's survey of early porcelains, including the manufactory founded at Fulham between 1671 and 1673, as a precursor of the Rouen and St Cloud porcelain manufactories some fifty years later [15]. The basis for the attribution to Padua is not stated but, if confirmed, it undoubtedly represents another example of an early European porcelain manufactory site. In passing, it is perhaps noteworthy that the mention of the unsuccessful attempt by Duc Alfonso II de l'Este to create porcelain earlier in the sixteenth century has a geographical link in that the Duc's ancestral seat was located near Ferrara, which is only a few kilometres distance from Padua and it is not inconceivable that after the Duc died in 1597 some lingering expertise in faience and porcelain synthesis remained in the region which perhaps generated a prospective start-up in Padua some years later? There is also a reference to another ceramics manufactory in Liguria near Genoa, which was already known for its faience production and a papal privilege was granted to Savignoni in 1633 to make *porcellana di Genoa* and also a possible further small scale attempt was made by Canon Manfredo Septalio (Settala) in Milan about 1664, which may not have been successful (Mallet 1995, 2008). A summary of these earliest Italian attempts to create porcelain is given by Wilson [62].

14.3 The First English Porcelain Manufactory at Fulham

After the closure of the Medici porcelain manufactory in Florence at the end of the sixteenth century, aside from the possible later sporadic and questionably successful ventures at manufacturing at Padua and Genoa, the next European manufactory was set up in the Fulham Pottery in London, some one hundred years later by John Dwight. Unlike Medici porcelain, extant examples of the Fulham wares are not readily identifiable, with the possible exception of the Burghley House Jars. The Museum of London have on display one small damaged, glazed garniture porcelain vase which has been reconstructed from pieces found in their archaeological excavation of the site [21] of the Fulham Pottery, and another similarly constructed damaged, unglazed vase, and these two items comprise the only known survivors of the Fulham manufactory found on site [38]. The Fulham shards were clearly affirmed to be porcelain from their translucency and dated to 1675–1680, but it should also be noted that, like the Medici porcelain from Florence, porcelain was not made commercially for sale at Fulham. John Dwight, with the technical support of eminent scientists of the Royal Society of London, succeeded in making a *hard paste* china body that was both highly siliceous and refractory, yet which possessed a satisfactory translucency which enabled it to be termed "*porcellaneous*" by several independent contemporary scientific observers. Fulham porcelain was essentially manufactured as a non-commercial operation but, unlike its Medici precursor which was made for diplomatic gifts, the Fulham product it is believed was primarily made for the scientific realisation of refractory materials that were capable of withstanding chemical attack by strong acids and fused alkalis at high temperatures and which could then be adopted for synthetic refractory crucible manufacture for the extraction of metals from their mineral ores.

John Dwight, who was born between 1633 and 1636, studied natural philosophy at Oxford University and was appointed as a laboratory assistant there to Robert Boyle FRS, the "Father of Chemistry". Recent archaeological excavations undertaken at the site of the Fulham Pottery by the Museum of London have revealed the presence of a ceramic kiln which has now been preserved on this urban site. Just prior to the foundation of his pottery at Fulham, John Dwight filed a patent in April 1671 describing his experiments in the use of clays and minerals to make a high quality "stoneware" and a phrase from this patent is particularly revealing, namely:

> ... the discovery of the mistery of **transparent earthenware** commonly known by the name of **porcelaine** in China.

The wording of this patent is critical in that Dwight has been credited by several historians with the manufacture of opaque salt-glazed stonewares only at Fulham and the granting of this patent for the manufacture of a "*porcelain*" is revealing. Sir Arthur Church [5] had some detailed knowledge of Dwight's patent and he alleged that Dwight must have achieved the manufacture of something very similar to what we now call porcelain in that "*... he produced if not porcelain then something distinctly porcellaneous*", which reflects the quality and extent of light transmission of the material. The role of the Royal Society of London in early English porcelain research and manufacture was mentioned in an earlier ceramics work by Solon [50] and has been reopened with the interpretation of several key documents [22] which demonstrate clearly that in the mid- to late-seventeenth century, within a few years after its foundation during the Restoration of King Charles II in 1660, the Royal Society was actively encouraging and sponsoring research and experimentation in refractory materials with John Dwight at the Fulham Pottery and that this occurred 40 years prior to the successful Meissen porcelain manufacturing operation undertaken by King Augustus, Elector of Saxony. Because of the key scientific link between John Dwight and Robert Boyle, Fellows of the Royal Society were monitoring closely the high temperature synthetic ceramics work being carried out at the Fulham Pottery in the 1670s. The eminent scientist Robert Hooke FRS on February 1674 (in [22]) confirmed the successful production by Dwight of his "*English china*" including some figures and:

> Severall little Jars of severall colours all exceeding hard as flint, very light and very good shape.

Later, Dr. Robert Plot FRS in 1677 [42] noted that John Dwight had "*discovered the mystery of the Hessian wares and ways to make an Earth white and transparent as Porcellane"*. He also mentioned that Dwight had experienced problems with the glazing of this *white porcellane*, which was attributed to arise from his adoption of a silica -alumina -calcareous glaze composition which required very high firing temperatures, which Dwight was unable to achieve reliably in his kiln. This had also been suggested by Tite et al. [59], who attributed the defects in Dwight's manufactured products to the adoption of an intractable SiAlCa glaze which had problems in adhesion to the ceramic body substrate. Dr. John Houghton FRS in 1693 [22] maintained that "*Mr Dwight of Fulham uses clay, the same earth China ware is made from and this is made by not lying long in the earth but in the fire .... we make as good China here as any in the world"*.

The only chemical analysis that has been carried out thus far into the Fulham Pottery products is described in the comprehensive work of Tite et al. [59], which is entitled "*A technological study of Fulham stonewares*". The title of this paper could be misinterpreted as it seems to exclude specifically any porcelains made at the site but the word *stonewares* should be regarded in this instance as being all-encompassing in this respect as this analytical paper describes the study of high temperature fired shards that were found and analysed at the Fulham Pottery site, which included both stonewares and translucent pieces of porcelain. The clay raw material was confirmed as comprising a natural secondary ball clay, sourced from Poole in Dorset, with a mineralogical composition of 55–65% quartz, 15–25% mica and 15–25% kaolinite, in admixture with sand and lime as components rather than a naturally sourced raw material [24]. The five porcelain shards that were analysed using SEM/EDAXS produced the data shown in Table 14.2 confirming a highly siliceous and aluminiferous body, low in iron oxide content: both body and glaze were analysed independently and the glaze composition is very similar to that of the

Table 14.2 Analytical data/Elemental oxide percentages /% for Fulham Pottery White Stonewares and Glazes

Specimen	SiO_2	Al_2O_3	CaO	MgO	K_2O	Na_2O	Fe_2O_3	TiO_2
Stonewares								
FP16	76.6	17.2	0.14	0.28	3.28	1.09	1.00	0.68
FP20	80.6	14.0	0.57	0.30	2.67	0.46	0.50	0.86
FP23	76.9	18.0	0.34	0.31	2.69	0.37	0.63	0.71
FP27	80.3	12.4	0.39	0.20	5.35	0.47	0.74	0.48
FP28	83.5	11.6	0.39	0.22	2.24	0.62	0.97	0.50
Average	80±3	15±3	0.4±0.2	0.26±0.05	3±2	0.6±0.3	0.8±0.3	0.6±0.2
Glazes								
FP16	72.6	14.7	0.99	0.89	7.03	1.64	1.41	0.76
FP28	73.7	13.7	1.69	0.36	7.82	0.87	0.78	1.02
Average	73.1±0.6	14.2±0.5	1.3±0.4	0.6±0.2	7.4±0.4	1.3±0.4	1.1±0.3	0.9±0.1

After Tite et al. (1986)

ceramic body in its major components except for a significantly larger contribution from the alkali metal oxides, which cumulatively comprise about 10% of the glaze mixture compared with about 4% for the ceramic body.

Tite et al. [59] also inferred from their earlier optical vitrification measurements [39] that a firing temperature of between 1200 and 1250 °C had been used for the white shards compared with 1100 °C for the salt glazed brown stonewares. The white ceramic body analyses matched a selection of those quoted by John Dwight in his notebooks [22] which detailed seven variants of a white "transparent clay porcelain" body all comprising mixtures of 30–70% Dorset ball clay and a white earth/white sand component which was made separately by sintering quartz sand with nitre (potassium nitrate) and tartar emetic (potassium hydrogen tartrate), which effectively gave a potash frit of composition 97% silica and 3% potash. Of particular comparative interest in this analytical work was the observation that these fine white porcellaneous bodies analysed fell within the same compositional range of the *yinqing* (*qingbai*) Chinese porcelain bodies of the Yuan Dynasty (1260–1368), which had been analysed in a similar fashion by the same authors two years previously [59] but they are also significantly different in composition with much lower alumina/silica ratios from the later Chinese porcelains of the Ming and Kangxi Dynasties. It is also interesting that the firing temperature adopted for the Yuan Chinese porcelains has been determined to be in the range 1200–1300 °C, which is a range analogous to that believed to have been used by John Dwight in his kilns at Fulham. The calculated mineral compositions of the Yuan and Fulham ceramic bodies is also quite similar indeed with quartz 57–64% versus 57–73%, kaolinite 8–18% versus 19–30% and mica 14–27% versus 9–14%, respectively, for the Yuan and Dwight formulations. The very close similarity between Dwight's white porcelain and the *yinqing* Chinese porcelain is seen in the ternary compositional diagram, representing extremes of $Na_2O + K_2O/Al_2O_3/SiO_2$ given in the article of Tite et al. [59], where the data for both ceramic bodies spatially overlay each other indicating that they had very similar relative elemental oxide compositions. It was also shown from an XRD analysis that the Dwight porcelain shards gave evidence of the presence of mullite and tridymite minerals, the key markers of a genuine *hard paste* porcelain which has been fired at a very high temperature. This is surely well-defined analytical evidence that John Dwight had indeed produced a genuine porcelain, which analytically is extremely similar to the Chinese *yinqing* porcelains of the Yuan Dynasty (1260–1368 CE), which preceded the Ming Dynasty and Qing Dynasty, and which together were responsible for the export of such significant quantities of Chinese porcelains into Europe in the eighteenth and nineteenth centuries. What Tite et al. [59] also found is that the Dwight porcelains, which can now be re-defined as really *proto-yinqing* porcelains, do not match the Kangxi porcelains of the early eighteenth century in composition as verified later by Spataro et al. [56, 57].

However, even though Dwight has closely reproduced the Chinese body, the applied glazes are very different for the two porcelains in that the Yuan porcelain glaze was applied conventionally to the porcelain substrates as a liquid lime-alkaline slip, which resulted in a smoother and thicker glaze layer after firing in the glost kiln, whereas Dwight's glaze appeared to have been produced in a method rather more akin to the salt-glazing process by the casting of salt mixtures for volatilisation into the kiln atmosphere towards the end of the body firing process. This latter procedure produced a much thinner and more diffuse coating than the alternative slip method and explains the difficulty noted by several contemporary observers in the ability to perfect the glazing process on Dwight's porcelain. Tite et al. [59] theorised that Dwight could have perfected his glazing procedure and achieved a much better adhesion with the body substrate if he had only increased the content of calcium oxide in his glaze mixture to about 15%, which he had not obviously considered undertaking at the time.

Furthermore, Spataro et al. [57] have discovered evidence for Dwight's later reversion to a more amenable and achievable lower glaze firing temperature using a lead glaze, which he then used to recover several articles which had nevertheless successfully survived his preliminary but less successful SiAlCa glaze trials and kiln firing.

It can fairly be concluded, therefore, that sometime between 1672 and 1680, John Dwight had succeeded in manufacturing a ceramic body that was equivalent to the translucent hard paste porcelain perfected and manufactured in China in the Yuan Dynasty and he should be credited with that achievement. It is true that he did experience severe problems with his glazing technique initially, which were partially overcome in later years by his institution of a re-glazing method involving a lead oxide flint glass additive.

Although, it is now difficult to positively identify the products of Dwight's early porcelain from the Fulham manufactory with certainty, it is believed that the *Burghley House Jars* (Fig. 14.6) could be early examples of this porcelain production at Fulham and recent analytical work has confirmed their similar highly siliceous, glassy content. The Fulham manufactory ceased to make porcelain and reverted to the manufacture of stonewares in 1703 after John Dwight had died. It is

Fig. 14.6 The *Burghley House Jars*, comprising the central larger jar known as the *Virtues Jar* and the smaller pair with lids sometimes known as the *Buckingham Jars* after the first Duke of Buckingham. The *Virtues Jar* contains panels depicting cherubs and scenes enacting virtuous behaviour in a style that matches the latter half of the seventeenth Century. *From the Ceramics Collection at Burghley House, courtesy of Lady Miranda Rock, Director of the Burghley House Trust, Stamford, Lincolnshire, UK*

still conjectural for some historians to accept that he did make true porcelain at the Fulham Pottery but there is now much analytical evidence extant that in fact he did so successfully and these are backed up by the statements from a range of contemporary observers which confirm the highly translucent nature of his pieces.

An intriguing comment has emerged from Daniels' research, namely, a reference to a visit made to the Fulham Manufactory by Ehrenfriede von Tschirnhaus, who later "*transferred to the Meissen Factory*" in Saxony, where history now tells us (seemingly incorrectly!) that European porcelain was invented! Von Tschirnhaus (1651–1708) arrived in London from Paris in May, 1675, with an introductory letter from the philosopher Spinoza and met with influential member scientists of the Royal Society where they discussed their novel experiments on refractory porcelains, which then stimulated von Tschirnhaus' recorded visit to the Fulham Pottery to view the experimental products for himself. On his return to Paris, Ehrenfriede von Tschirnhaus immediately commenced his own experiments on porcelain synthesis, using his famous burning mirror, and eventually he set up with Johann Bottger at Meissen under the patronage of King Augustus, the Elector of Saxony. It could be argued, therefore, that Dwight's manufacturing success at Fulham catalysed and stimulated the early experiments of von Tschirnhaus made after his visit there in 1675 which culminated in the synthetic porcelain manufactured at

Meissen in 1709. It must be said that this is in disagreement with opinion that Fulham would have had "*no influence at all on continental European porcelain manufacture*", whereas the evidence now suggests otherwise.

In a special case study which has been recounted fully in Edwards [17] the analytical evidence strongly suggests that the *Burghley House Jars* (Fig. 14.6), which comprise a refractory porcelain of very unusual composition [56, 57] are potential survivors of John Dwight's Fulham manufactory. There is irrefutable documentary evidence for these ceramic artefacts being in existence in the 1670s and early 1680s and it is actually difficult to attribute their origins to any source chronologically other than Fulham. Chemical analysis by analytical scientists from the British Museum Research Laboratories has revealed that their composition categorises them as very early porcelain [59] and this has further been credibly attributed to the Fulham manufactory. If this is really the case, then it firmly establishes that John Dwight must have succeeded in making a limited number of pieces that were deemed satisfactory at least for presentation to an esteemed clientele. Ramsay and Ramsay also highlight the critical mentoring role of Robert Boyle with John Dwight as analogous to that adopted between von Tschirnhaus and Bottger — both resulting in the production of what must be early and historically important translucent refractory materials, be these didactically *porcelain* as we know it today or potentially at least a noteworthy but still translucent variant.

In conclusion, a statement made by Soden-Smith in 1873 [49], is very revealing in this context:

> Fulham Porcelain: Porcelain was made at Fulham by Mr John Dwight as early as 1671, who took out patents to secure his invention, and is mentioned by contemporary writers as having produced porcelain. Mr Bailey, of Fulham, possesses the original memorandum book in which reference is made to the receipts for the composition of such **porcelain**, and it is believed that the white mugs here shown are of Dwight's manufacture. Their material is a white, **very translucent**, somewhat vitreous porcelain with Oriental ornament in relief.

It is believed that these early Fulham porcelain mugs were contributed by a Mr. Willett to the ill-fated Alexandra Palace Exhibition of 1873. In all, some 2074 pieces of porcelain, representing the finest examples of their type of production ever made in Great Britain, were displayed at the Alexandra Palace Exhibition. However, tragedy struck on Monday, June 8th, 1873 when a fire swept devastatingly through the Alexandra Palace and almost all of the porcelain exhibits on show there were destroyed completely. It is recorded that only a few pieces of Benjamin Lund's Bristol porcelain and Josiah Wedgwood's *Jasperware* survived the intense conflagration—this was ascribed at the time to their ability to withstand the heat of the blaze because of the high firing temperatures used in their production. This cannot really be a satisfactory explanation, however, as the 24 pieces of Nantgarw and Swansea porcelain also displayed in the Exhibition, described as the finest pieces ever produced at these factories and on loan from several well-known personal collections were also completely destroyed and certainly in the case of the Nantgarw porcelains, as we now know, these had been fired at the highest kiln firing temperatures of 1400–1420 °C adopted in British porcelain manufacture up to that time.

It is far more reasonable to suppose that the damage to the collections of fine china on exhibition there was caused by the thermal stresses induced in the china by the conflagration, the physical disintegration of the building and its roof collapse onto the fragile exhibits. Sir Arthur Church [5] mentioned that he managed to recover a handle from a rare piece of Brancas-Lauraguais porcelain, which was made in London around 1760, from the debris left after the conflagration and that he based his published analysis of Brancas-Lauraguais porcelain subsequently upon the destructive chemical analysis of that sole item. No other specimens of Brancas-Lauraguais porcelain have survived either intact or in pieces elsewhere to date.

It is also very clear that the composition of the early English wares from Fulham, Bow, Limehouse and Chelsea differed completely from those being produced on the Continent in the 1740s and this strongly suggests that the production of porcelains in Western Europe developed along two main and parallel fronts, namely the English and the Franco/German themed avenues, and that the manufacture of porcelain in England did not, therefore, depend upon the transfer of knowledge from the Continent [60].

14.4 The First French Porcelain Manufactory at Rouen

Rouen had been a centre for the production of faience and tin-glazed earthenwares in France since 1540 and in 1647 King Louis XIV granted Nicholas Poirel, seigneur of Grandval, and Edme Poterat of the Rouen faience manufactory a 50 years monopoly to manufacture faience. Porcelain manufacture in France started in Rouen on a different site but clearly used the ceramics background and workforce skills that existed in faience manufacture there. The porcelain manufacture was of a *soft paste* body (*pate tendre*), which again never seemed to aim ostensibly for a commercial output, and only some 10–12 believed genuine Rouen porcelain pieces have now survived to the present day to attest to its successful manufacturing operation. Cyrille Froissart has recorded details of all existing surviving specimens of Rouen porcelain but several of these are believed possibly to be of doubtful provenance and await analysis for confirmation [19]. Rouen porcelain pieces have been often misattributed to the slightly later and very much more productive St Cloud factory porcelain artefacts.

In 1674, Louis Poterat, the son of Edme, bought out Nicholas Poirel and started another family factory in Rouen to manufacture porcelain exclusively [11, 51]. Edme Poterat died in 1684 and his younger son Michel then continued with the faience arm of the Rouen business. Louis Poterat, with the authority and royal *privilege* granted by the King on the 1st October, 1673 for an initial period of 30 years [20], continued to manufacture porcelain at Rouen until his death from typhus in 1696. He was very wary of the potential loss of his manufacturing details and recipe through espionage, so he only made small quantities personally. He took his secret porcelain recipe and details of his manufacturing processes with him to the grave and porcelain manufacture finished at Rouen very soon afterwards. The *privilege* granted by the King to the Rouen manufactory specified that the manufacture should be for

> .... any kind of dishes, pots and vases in porcelain similar to the Chinese (Author: i.e. in blue and white) and the production of polychromatic or blue and white faience in the Dutch style.

On the 16th June, 1694, the Intendant of Rouen, Lefevre d'Ormessan, reported to Count Louis de Phelypeaux of Pontchartrain that Louis Poterat was now making quality porcelain and faience. In 1696, the Royal *privilege* was renewed for a further 20 years so the King must have been pleased with the quality of the porcelain production at Rouen, but recognising the secrecy surrounding Louis Poterat's process and recipe, this renewal contract was made conditional upon his making his secret known to the King at the end of that time [51]. However, Louis died of typhus the same year and the secret of the manufacture of Rouen porcelain died with him. Porcelain production effectively stopped in Rouen upon the death of Louis Poterat in 1696 and the Royal patronage was removed. Louis Poterat had always insisted upon carrying out the critical operations alone and he only ever made small quantities of porcelain at any time, so it can but be concluded that anye Rouen wares that were being issued after the death of Louis Poterat in 1696 were perhaps factory "seconds" or stock that was defective in some way [31]. An interesting corollary to the secrecy surrounding porcelain production and how the knowledge of a process could be acquired easily by a competitor through their employment of members of a rival's key workforce can be seen cited in other examples later, especially involving the Meissen and Vienna du Paquier manufactories. In an attempt to minimise this practice, the following legal injunction was issued on the 24th July, 1745 by the Vincennes porcelain manufactory, as mentioned above, to ensure an exclusivity for their manufactured products:

> Workers who left the plant without official permission and the confiscation of objects therefrom and a fine also for the factories them offered it being them also forbidden to employ previous employees from Vincennes without factory agreement.

The meaning is quite clear regarding the concern felt by the factory owners at the possibility of the transfer of working knowledge about their materials and processes to a competitor's factory through the recruitment or subversion of their key staff.

The King was apparently pleased with the Rouen product, these becoming known as "*Porcelaine francaise – la veritable porcelain de Chine*" ("French porcelain – the true porcelain of China") [40]. However, the porcelain production was always limited and an intriguing comment has been found in the records that in the kilns the relatively fragile porcelain items had to be surrounded by the more robust faience items during firing to protect them "*from the violence of the furnace*", which presumably refers to the critical inability to control the kiln temperature [13].

For a while the details of the foundation of the Rouen factory were lost but the "*privilege*" was rediscovered by Andre Pottier of the Sevres Museum in 1847 who stated:

> These precious objects are admirable for the delicacy with which they are decorated and the exquisite vitrification of the paste. The whiteness of the body is lightly coloured with green, like the faience of the same period.

The porcelain made at Rouen possessed a slight greenish-blue tinge: it was translucent and decorated in underglaze blue in a distinctive *camaieu* style and used *lambrequin* motifs, involving scalloped borders with lacework drapes and scrolls, reflecting the wall furnishings of the period and instituting a characteristic Rouen style reflecting heraldic emblems (known as the *style rayonnant*) [2, 32, 40, 46]. Much of the limited production at Rouen centred upon luxury items such as small vases, cosmetic jars, condiment vessels and cabinet cups with lids. An example of Rouen soft paste porcelain is shown in Fig. 14.7, a mustard pot or *moutardier* with lid *ca.* 1695 with the characteristic decoration in underglaze blue and bearing the coat-of-arms of a French legal advocate and counsellor to the Departement de Normandie, Jacques Asselin de Villequiers. This was identified positively as Rouen by Xavier Riocreux, curator of the Musee de Sevres, and has proved to be a highly significant comparator for other porcelain items of potential Rouen manufacture: an example of this is the potpourri jar, shown in Fig. 14.8, with raised moulded panels and characteristic Rouen blue underglaze decoration which is now in the Metropolitan Museum of Art, New York. Of the ten or a dozen known Rouen

Fig. 14.7 Rouen porcelain mustard pot and lid, *ca.* 1695, bearing the coat-of-arms of Jacques Asselin de Villequiers, decorated in underglaze blue. *Musee national Ceramique de Sevres*. CC BY- SA 3.0

Fig. 14.8 Potpourri jar, Rouen porcelain, with raised moulded panels minus its lid, gifted to the Metropolitan Museum of Art, New York, by R. Thompson Wilson in memory of Florence Ellsworth Wilson in 1950 (accession number 50.211.86). Formerly in the collection of the Comte Xavier de Chavagnac and thereafter Mrs. Helen Dupuy before its acquisition by Mr. Thompson Wilson via a Sotheby's sale in 1948. *Metropolitan Museum of Art, New York*

examples surviving it is believed that all but one are in Museum collections and that several attributed to Rouen may actually be St Cloud porcelain: one of these was recently identified analytically as being potentially of St Cloud origin from the composition of its underglaze blue enamels [9]. Colomban et al. [8] have analysed early Rouen and St Cloud porcelains and have determined that the *moutardier* mentioned here has actually a *hybrid* body between *hard* and *soft paste* porcelains; it is now in the Musee de Ceramique at Sevres and is the earliest exemplar known of Rouen porcelain.

Although Rouen closed officially for the manufacture of porcelain in 1696, Nicolas Levavasseur attempted to revive porcelain manufacture there again in 1743 but failed because his products were deemed inferior to those then being made at the other prestigious French porcelain manufactories of Vincennes, Mennecy and Chantilly [3].

14.5 St Cloud Porcelain

The soft paste porcelain manufactory at St Cloud, in a suburb of Paris, commenced in 1693 under the protection of Philippe, Duc d'Orleans (1640–1701), younger brother of King Louis XIV, and so overlapped that at Rouen for some three years and whilst Rouen was still operating under its royal privilege. St Cloud became the first French porcelain manufactory to make porcelain commercially, obtaining an official patent for this in 1702, although already in 1700 a factory inventory registered that 3146 pieces of finished and completed porcelain were held in stock at the site, so clearly the production of porcelain must have been undertaken there for some while before. Dr. Martin Lister FRS (1639–1712), an English naturalist and natural scientist, reported on his visit to the St Cloud porcelain manufactory in 1698 ([35]) that:

> They were producing already a manufacture of porcelain as white and translucid as the one that came from the East.

The brothers Francois and Claude de Reverend, late of the faience factory in Rouen, had set up a competitive faience factory in St Cloud in the 1660s under the direction of Jean-Baptiste Colbert specifically to manufacture "*the Indian style porcelain*" and employed Pierre Chicaneau to decorate their faience: Chicaneau took over the faience factory in 1674 and experimented with the production of porcelain [52, 53]. After his death, his widow, Barbe Coudray, married Henri Trou from Nevers, squire to the Duc d'Orleans: they bought the factory in 1683 and along with the Chicaneau children in 1693 they decided to install a separate factory on the same site to manufacture porcelain. There is a report that they persuaded several workers from Rouen to join them when that factory closed in 1696. In a patent granted to St Cloud in 1702, Pierre Chicaneau is named as the discoverer of the porcelain secret. A family rift soon developed between the Chicaneau family and Henri Trou which saw Marie, the widow of Pierre Chicaneau's son, Dominique Chicaneau and Marie-Louise Soisson found a new factory in the 8eme arrondissement in Paris called Ville L'Eveque in 1711 and this was operational until 1766 making a similar soft paste porcelain to St Cloud and earmarking Ville L'Eveque as one of the earliest French porcelain manufactories—but around this time the factory of Antoine Pavie had also started up in Paris. As Philippe, Duc d'Orleans, died in 1701, the royal patronage would most likely have been taken over by his elder brother, King Louis XIV.

Fig. 14.9 St Cloud porcelain vase, *ca*. 1695–1700, decorated in underglaze blue with *camaieu* decoration similar to the Rouen factory. *Musee des Artes Decoratifs*, Paris. CC BY-SA 3.0

St Cloud porcelain is very similar to that of Rouen but is of an ivory colour with a satin glaze and the decoration is of a similar *camaieu* and *lambrequin* style in underglaze blue, so it is not surprising that in the past the output from both factories has been confused. An example of St Cloud porcelain is shown in Fig. 14.9 and the following description applies:

> In 1702, letters-patent were granted to the family of Pierre Chicaneau, who were said to have improved upon the process discovered by him, and since 1693 to have made porcelain as "perfect as the Chinese". The typical, blue-painted Saint-Cloud porcelain is one of the most distinct and attractive of porcelains, and not the least part of its charm lies in the quality of the material itself. It is rarely of a pure white, but the warm yellowish or ivory tone of the best wares of the period is sympathetic and by no means a shortcoming; and while actually very soft and glassy, it has a firm texture unlike any other. The glaze often shows a fine satin-like pitting of the surface that helps to distinguish it from the brilliant shiny glaze of Mennecy, which is otherwise similar. The heavy build of the pieces is also characteristic and is saved from clumsiness by a finer sense of mass, revealed in the subtly graduated thickness of wall and a delicate shaping of edges.

St Cloud porcelain has a rather characteristic mixture of component raw materials, specifically comprising a mixture of quartz sand, clay, sandstone, limestone or marl, alum, soda ash and salt, bonded with *Marseilles black soap* and with calcium additives comprising chalk and calcareous marl [9]. This complex formulation was finely ground and heated to 950 °C for 30–35 h, then fritted and mixed with a curious additive called "*chymie*" which was described by Alexander Brogniart as an animal glue. The second firing of this frit was mixed with more limestone and an illitic aluminium—rich clay from Argenteuil and moulded into shape followed by a final glaze firing after enamelling at 1100 °C, but a lack of plasticity often resulted in the appearance of cracks and the deformation of larger objects in the firing kiln. Typical St Cloud soft paste porcelain phases corresponding to wollastonite, tridymite, cristobalite and a vitreous phase have been identified. Most surviving pieces are thus small in size, such as snuff boxes, beakers, saucers, teabowls, cane handles and cutlery handles as it proved difficult to maintain the soft paste porcelain shapes of larger pieces in the kiln upon firing. Early pieces are marked with an underglaze blue motif representing a rayed sun (representing King Louis XIV, who was known as *Le Roi Soleil,* The Sun King), whereas later pieces had a colder bluish tone and were marked with a St. C legend above a T in underglaze blue, representing Henri Trou.

By 1766, St Cloud was unable to compete with the more sophisticated products of Chantilly, and Vincennes so they filed for bankruptcy—the same year that the porcelain manufactory at Ville L'Eveque closed, probably for a similar reason.

In an early spin-off from St Cloud, in 1703, Antoine Pavie (d. 1727), son of Nicolas Pavie, a faience maker in the Faubourg Saint Antoine in Paris, engaged with Pierre Pelissie (1678–1756) a former member of the workforce at St Cloud to manufacture porcelain "*in the Chinese manner such as that being made at St Cloud*" and for this purpose he bought the recipe and manufacturing secret from Pelissie for the sum of 200 livres (equivalent to about £4000 today), setting up his factory in the Rue de la Muette. Despite this acquisition of the manufacturing recipe, it is interesting that although the underglaze decoration on Pavie porcelain is similar stylistically to those of Rouen and St Cloud, the porcelain body analysis shows that the soft paste body composition is rather different in that it contains glass frit, implying that the recipe that Pavie had procured at that time was apparently not used. The Pavie porcelain was marked "*AP*" in underglaze blue, which for a long time was

Fig. 14.10 Blue and white porcelain ewer, from the Parisian factory of Antoine Pavie, Rue de la Muette, *ca.* 1710, with exotic birds and garlands of flowers. *Metropolitan Museum of Art New York: J.P Pierpont Morgan Bequest*

incorrectly believed to refer to the very similar St Cloud factory porcelain on the basis that "*AP*" could mean "A Paris", as St Cloud is situated in the district of Passy, Paris. About eight examples of Pavie porcelains are known, all are small and based on contemporary silver models and a detailed comparative analysis of the porcelain bodies of Rouen, St Cloud and Pavie porcelains would be an interesting exercise. The Pavie porcelain "*AP*" mark, along with those of St Cloud, represent some of the earliest marks on French porcelains [12–14, 40]. An example of Pavie porcelain and the *AP* mark is shown in Figs. 14.10 and 14.11.

Fig. 14.11 Underside of Antoine Pavie porcelain ewer (Fig. 3.10.) with *AP* mark in underglaze blue of Antoine Pavie, Paris. *Metropolitan Museum of Art, New York: J.P Pierpont Morgan Bequest*

14.6 Meissen Porcelain

German porcelain in the early eighteenth century is synonymous with the foundation of the Meissen manufactory, which is generally accepted historically to be the first in continental Europe to develop porcelain commercially and it has been alleged by several historians that the resounding success in this operation at Meissen spawned the creation of many other factories later in the century as can be seen in Table 14.1. The discovery of European porcelain is credited historically to Ehrenfriede von Tschirnhaus and Johann Bottger working together in the first decade of the eighteenth century under the patronage of Augustus the Strong, Elector of Saxony and King of Poland. The factory was established by Augustus, which was entitled at that time the Royal Polish and Electoral-Saxon Porcelain Manufactory (Koningslich—Polnische und Kurfurstlich-Sachsisische Porzellan Manufaktur) at his castle in Albrechtsburg in Meissen, near Dresden. Augustus called the porcelain material "*white gold*"—this nomenclature directly reflects the original purpose of Augustus bringing together these two experimental alchemists to work for him which was to assist in the alchemical transmutation of base metals into gold [26]. Von Tschirnhaus died of dysentery in 1708 before the production went commercial and it was left to Bottger to take it forward to completion. The very first ceramic products at Meissen were a hard glazed red stoneware called "*Bottger wares*" (Bottgersteinzeugt), which appeared first at the Easter Fair in Leipzig in 1710 but the production of glazed white and painted porcelain soon followed and the full commercial production of a *hard paste* porcelain appeared in 1713 [47, 48]. To protect the goods, a crossed swords mark was introduced in 1720, which has been claimed to be not only the first factory mark to appear on porcelain (the author questions this claim chronologically in the light of the marks that had already appeared prior to this time on St Cloud porcelain and indeed on the even earlier Medici porcelains) but it is undoubtedly believed to represent correctly the first ever trade - mark on a *commercial* product. Underglaze blue pigments, termed "*Meissen blue*", were introduced by Frederick Kottig and a wide range of other overglaze pigments in 1723 were adopted by Johann Herold. In the early 1730s, china wares were sold from the factory in the white to private workshops (Hausmalerei) in Augsburg and Bayreuth for independent retail and sale as finished, decorated Meissen items [16, 45].

The first recorded setback to the successful enterprise at Meissen came in 1717 when Samuel Stoltzel, who was head of arcanism and crafts at Meissen, became disaffected with his role and status at the manufactory and took the secret recipes to the fledgling Vienna China Works—a critical revelation in this recipe was the use at Meissen of the local Saxon china clay (kaolin) from Colditz (Ebengeburg). Early identification marks at Meissen prior to the underglaze blue crossed swords trademark were underglaze blue initials, such as AR (Augustus Rex), KPM (Konigsliche Porzellan Manufaktur), MPM (Meissen Porzellan Manufaktur) and KPF (Konigsliche Porzellan Fetcrip). Augustus' passion for porcelain resulted in his construction of the Japanese Palace in time for the marriage of his son, Crown Prince Friedrich (who later became King Augustus II) to the Archduchess Maria Josephine in 1719. Augustus made some important acquisitions of early Japanese and Chinese porcelains to furnish his new Japanese Palace: in 1717, he exchanged a regiment of 600 elite soldiers for 151 choice pieces of Chinese porcelain from the Prussian Royal Collection including the famous Dragoon vases. In 1722, he acquired a further 804 pieces of porcelain for the Japanese Palace, including several pieces of St Cloud porcelain which were catalogued as Meissen in his Collection because no provision was made for porcelain categories other than Japanese, Chinese and Meissen! In fact, one of these "St Cloud" pieces, a spice box, it transpires cannot be of St Cloud origin at all as it bears a monogrammed AP mark in blue enamel underneath—which as we have seen above actually belongs to the early French porcelain manufactory of Antoine Pavie. This demonstrates the lurking danger of assuming an apparently irrefutable provenance for a piece of porcelain historically catalogued from an important Royal Collection, which unfortunately was mis-attributed in the first place [4]!

In 1756, during the Seven Years War, Prussian troops occupied Meissen and King Frederick II of Prussia discovered the recipes and formulations of the Meissen Porcelain Manufactory and used these to set up his own eighteenth century porcelain manufactory in Berlin, thereafter known as the Konigsliche Porzellan Manufaktur Berlin. Meissen experienced intense competition with the French manufactory at Sevres from 1760 to 1780, especially with the introduction by Sevres of a fine quality matte finish on their undecorated biscuit porcelain of a range of figurines which were designed to simulate white marble statuary, the possession of which was deemed to be very desirable by its clientele and which were immediately copied at Meissen.

In the eighteenth and nineteenth centuries, Meissen was famous for its figures modelled in their eighteenth century rococo and classical poses and involving a wide menagerie of animals. The "*Meissen Temples*" were created as porcelain table decorations (Tafelaufsatz) around 1727–1728 for court dinner parties at King Augustus' Japanese Palace and there is a record of 6 "*Venus Tempel*" being made for this location in the 1720s. These porcelain temples were of complex construction, modelled by Johann Joachim Kaendler, and very few have survived: one of these comprised 24 pillars in eight groups of three with an arched cover supporting an oval pierced ornament on which stand eight torsos carrying trophies of arms on an entablature of eight putti holding sheaves of corn and flowers. Grape vines and fruit curl up the pillars and in the centre a figure of Fame blowing a trumpet is suspended from the cover. This temple was later purchased by the Dresden Porzellansammlung where it is described now as the "*Temple of Honour*" (*Ehrentempel)*. A similar "*Temple of Love*" also exists in Frankfurt and a third *Temple* is at Longleat House in Warminster, Wiltshire, which was purchased by John Thynne, the Marquess of Bath, in 1881 for a royal visit made there that year. These temples were used at Meissen as table centrepieces for the royal weddings of the children of King Augustus III in the mid -eighteenth century from 1747 onwards.

Large Meissen armorial dinner services were very much in vogue for local consumption and for export from the 1730s: 17 Meissen dinner services given as part of her dowry to Princess Maria Amalia of Saxony upon her wedding to Charles, King of the Three Sicilies, who later became King Carlos III of Spain. Other notable and very large services made at Meissen were the *Swan* service (*Schwanenservice*), 1737–1743, made for Count Heinrich von Bruhl, which numbered more than 1000 pieces, and the *Mollendorf* service of over 960 pieces made in 1762 and presented by King Frederick II of Prussia to General Wichard von Mollendorf to celebrate his victories in the Seven Years War. Much Meissen blue and white porcelain copied the contemporary successful Chinese imports into Europe, such as the famous "*Blue Onion*" pattern (*Zwiebelmuster*), the Chinese motif actually representing a pomegranate—a fruit unknown in Europe at that time. This was adopted by Horoldt in 1729 at Meissen on the basis of a Kangxi design but, never having seen such a fruit in the West, it was assumed to represent an onion in underglaze blue pigment (Fig. 14.12.). In 1830, the ownership of Meissen reverted to the state of Saxony.

Fig. 14.12 Meissen porcelain, *ca.* 1725–1730, the famous *Zweibelmuster* pattern, or *Blue Onion,* imitating the Chinese pomegranate pattern (which at that time was believed to represent an onion) on imported Kangxi porcelains. *Public domain*

Detailed workbooks of recipes and pigments used at various stages in the factory production history have facilitated analytical interpretations of the data from specimens of Meissen porcelains from some of the earliest products to later variants [7, 58].

References

1. Arnold, L.: Princely gifts and Papal treasures: the Franciscan mission to China and its influence on the art of the West, 1250–1350. In: Stephen, M. (ed.) Princely Gifts and Papal Treasures; the Artistic Legacy of China in Europe, pp. 119–134. Desiderata Press, London (1999)
2. Baghdiantz- McCabe, I.: Orientalism in Early Modern France: Eurasian Trade, Exoticism and the Ancien Regime. Bloomsbury Publishing, London (2008)
3. Battie, D. (ed.): Sotheby's Concise Encyclopedia of Porcelain. Conran Octopus, London (1990)
4. Cassidy-Geiger, M.: Meissen and St Cloud, Dresden and Paris. In: Rondo, B. (ed.) Discovering the Secrets of Soft Paste Porcelain at the Sat cloud Manufactory, ca. 1690–1766, pp. 97–111. Bard Graduate Center for Studies in the Decorative Arts/Yale University Press, New York/New Haven Connecticut/London (1999)
5. Sir A.H. Church: English Porcelain: A Handbook to the China Made in England During the 18th Century as Illustrated by Specimens Chiefly in the National Collection, A South Kensington Museum Handbook. Chapman & Hall Ltd., London (1885 and 1894)

6. Colomban, P., Milande, V., Lucas, H.: On-site Raman analysis of Medici porcelain. J. Raman Spectrosc. **35**, 68–72 (2004)
7. Colomban, P., Milande, V.: On-site analysis of the earliest known Meissen porcelain and stonewares. J. Raman Spectrosc. **37**, 606–613 (2006)
8. Colomban, P., Milande, V., Lu, T.-A.: Non-invasive on-site Raman study of blue decorated early soft paste porcelains. The use of arsenic-rich (European) ores compared with huafalang Chinese porcelains. J. Eur. Ceram. Soc. **38**, 5228–5233 (2018)
9. Colomban, P., Gironda, M., Edwards, H.G.M., Mesqui, V.: The enamels of the first soft paste European porcelains: the Rouen, St Cloud and Paris. Complementarity of Raman and X-Ray fluorescence analysis with mobile instrumentation, in preparation for submission (2021)
10. Cora, G., Fanfani, A.: La Porcellana dei Medici, pp. 70–71. Fabbri Publishing, Milan (1986)
11. Dawson, A.: French Porcelain – A Catalogue of the British Museum Collection. British Museum Publications, London (1994)
12. de Plinval de Guillebon, R.: La Marque AP Indentifiee. L'Estampille/L'Objet d'Art. **271**, Juillet/Aout, 72–77 (1993)
13. de Plinval de Guillebon, R.: Faience et Porcelaine de Paris: XVIIIe -XIXe Siecles. Faton, Paris (1995)
14. de Plinval de Guillebon, R.: Les Ceramistes du Faubourg Saint Antoine Avant 1750. Bulletin de la Societe de L'Historie de París et de L'Ile de France, 6 (2002)
15. de Plinval de Guillebon, R.: L'Eveil de la porcelaine à Paris. Revue de la Société des Amis du Musée National de Céramique, 55–68 (2005)
16. Ducret, S.: German Porcelain and Faience, With Vienna, Zurich and Nyon. Oldbourne Press, London (1962)
17. Edwards, H.G.M.: Porcelain Analysis and Its Role in the Forensic Attribution of Ceramic Specimens. Springer-Nature, Cham (2022)
18. Frankopan, P.: The Earth Transformed: An Untold History. Bloomsbury Publishing, London/Dublin (2023)
19. Froissart, C.: Collection d'un Amateur, Rouen 1680–1740: De la Premier Porcelaine de l'Age d'Or de la Faience, Sotheby's Paris, 18th June 2008
20. Grandjean, G.: The porcelain of Rouen. In: Rondot, B. (ed.) Discovering the Secrets of Soft Paste Porcelain at the St Cloud Manufactory, ca. 1690–1696. The Bard Graduate Center for the Studies in the Decorative Arts, New York (1999)
21. Green, C.: John Dwight's Fulham Pottery: Excavations, 1971–1979 Archaeological Report 6. English Heritage, London (1999)
22. Haselgrove, D., Murray, J.: John Dwight's Fulham Pottery, 1672–1978: A Collection of Documentary Sources. J. Ceram. Hist. **11**, Stoke-on-Trent City Museum (1979)
23. Haslem, J.: The Old Derby China Factory: The Workmen and Their Productions, Containing Biographical Sketches of the Chief Artist Workmen, the Various Marks Used, Facsimiles Copied from the Old Derby Pattern Books and the Original Price List of More than 400 Figures and Groups, George Bell & Sons, London, (1876)
24. Holdridge, D.A.: Ball clays and their properties. Trans. Br. Ceram. Soc. **55**, 369–440 (1956)
25. Horvath, J.: The pedigree of Louis the Great's ewer. In: Vardy, S.B., Grosschmidt, G., Domonkos, L.S. (eds.) Louis the Great, King of Hungary and Poland, pp. 325–345, New York (1986)
26. Keating, M.: The Hunt for White Gold. Hodder & Stoughton, London (2011)
27. Kerr, R., Wood, N.: Science and Civilisation in China, Volume 5, Part 12, Ceramic Technology. Cambridge University Press, Cambridge (2004)
28. Kingery, W.D.: The development of European porcelain. In: Kingery, W.D. (ed.) Ceramics and Civilisation, High-Technology Ceramics, Past, Present and Future, vol. III, pp. 153–180. Ohio Press (1986)
29. Kingery, W.D., Smith, D.: The development of European soft paste porcelain. In: Kingery, W.D., Lense, E. (eds.) Ancient Technology to Modern Science, pp. 273–292. American Ceramic Society, Columbus (1984)
30. Kingery, W.D., Vandiver, P.A.: Medici porcelains. Faenza. **70**, 441–453 (1984)

31. Lacombe, C.: L'Apparition de la Porcelaine Tendré a Rouen Chez les Poterat: L'Hypothese Protestante? Sevres Museum Publications No. 15. Sevres Museum Publications, Sevres (2006)
32. Lane, A.: French Faience. Faber and Faber, London (1948)
33. Lane, A.: Italian Porcelain. Faber & Faber, London (1954)
34. Lane, A.: The Gaignieres-Fonthill vase- a Chinese porcelain of about 1300. Burlingt. Mag. **103**, 124–132 (1961)
35. M. Lister: Journey to Paris in the Year 1698, ed. R. P. Stearns, Southern Illinois University Press, Carbondale, 1967
36. Mack, R.E.: Bazaar to Piazza: Islamic Trade and Italian Art, 1300–1600, pp. 109–111. University of California Press, Berkeley/London (2002)
37. Mallet, J.V.G.: Porcellana tenera del Seicento in Liguria? Due ciotole at Victoria & Albert. In Atti XLV Courgno Internazionale della Ceramica Albisola, 337–345 (1995)
38. Mallett, J.V.G.: The earliest English porcelains? Part 6: The place of Buckingham porcelain in ceramic history. Trans. Engl. Ceram. Circle. **20**, 211–230 (2008)
39. Maniatis, Y., Tite, M.S.: Technological examination of Neolithic -Bronze Age pottery from central and South Eastern Europe and from the Near East. J. Archaeol. Sci. **8**, 50–76 (1981)
40. Munger, J., Sullivan, E.: European Porcelain in the Metropolitan Museum of Art: Highlights of the Collection. Metropolitan Museum of Art, New York (2018)
41. Pirazzoli t'Serstevens, M.: Une Denree Recherche: La Ceramique Chinoise Importee dans le Golfe Arabo-Persique, IXe – XIVe Siecles. Mirabilia Asiatica. **2**, 69–88 (2005)
42. Plot, R.: The Natural History of Oxfordshire Being an Essay Towards the Natural History of England. The Theatre, Oxford (1677)
43. Pollard, A.M.: Letter from China: a history of the origins of the chemical analysis of ceramics. Ambix. **62**, 50–71 (2015)
44. Polo, M.: Il Milione, cap. CLVIII, dell edizione a cuva di L.F. Benedetto, translated (1928)
45. Roentgen, R.E.: The Book of Meissen, 2nd edn. Schiffer Publishing, Altglen (1996)
46. Savage, G., Newman, H.: An Illustrated History of Ceramics. Thames and Hudson, London (1985)
47. Schulle, W., Goder, W.: Die erfinding des europanischen porzellans durch Bottger – eine systematische schopferisch entwicklung. Keram. Z. **34**, 598–600 (1982)
48. Schulle, W., Ulrich, B.: Ergebnisse gefugeanalytischer untersuchungen an Bottgerporzellan. Silikatechnik. **33**, 44–47 (1982)
49. Soden-Smith, R.: Catalogue of the Collection of English Pottery and Porcelain Exhibited on Loan at the Alexandra Palace, 1873. R.K. Burt & Co., London (1873)
50. Solon, M.L.: Art of the Old English Potter, pp. 32–35. Bemrose & Son, London (1883)
51. Solon, M.L.: The Rouen porcelain. Burlingt. Mag. Connoiss. **7**(26), 116–124 (1905)
52. Solon, M.L.: "The St. Cloud porcelain", Part I. Burlingt. Mag. Connoiss. **10**(43), 24–25. and 27–28 (1906a)
53. Solon, M.L.: "The St. Cloud porcelain", Part II. Burlingt. Mag. Connoiss. **10**(44), 89–93. and 95–96 (1906b)
54. Spallanzani, M.: Medici porcelain in the collection of the last Grand Duke. Burlingt. Mag. **132**(1046), 316–320 (1990)
55. Spallanzani, M.: Ceramiche alla Corte dei Medici nel Cinquecento, pp. 55–56. Franco Cosimo Panini, Modena (1994)
56. Spataro, M., Bimson, M., Meeks, N.: Part 3: a scientific investigation of three pieces of porcelain presumed seventeenth century, from Burghley House. Trans. Engl. Ceram. Circle. **20**, 187–194 (2008)
57. Spataro, M., Meeks, N., Bimson, M., Dawson, A., Ambers, J.: Early porcelain in 17th century England: non-destructive examination of two jars from Burghley House. Br. Mus. Res. Bull. **3**, 37–48 (2009)
58. Stiegelschmitt, A., Tomandl, G.: Untersuchung der materialzusammensetsung aller porzellan im hublick auf ihre kunsthistorische zuordnung. Sprechsaal. **118**, 974–978 (1985)

59. Tite, M.S., Bimson, M., Freestone, I.C.: A technological study of Fulham Stoneware. In: Proceedings of the 24th Archaeology Symposium. Washington DC, pp. 95–104 (1986)
60. Tite, M.S.: Ceramic production, provenance and use – a review. Archaeometry. **50**, 216–231 (2008)
61. Vasari, G.: Le Vite de piu eccelenti pittori, scultori e architettori, Lorenzo Torrentino, Florence, 1568, transl. into English by E.L. Seeley (1908)
62. Wilson, T.: Western Decorative Arts, Part I, pp. 234–237. National Gallery of Art, Washington, DC (1993)
63. Zhao, B.: Chinese style ceramics in East Africa from the 9th–16th centuries: a case of changing value and significance in multi-partner global trade. Afriques-Debate, Methodes et Terrains d'Histoire. **6**, not paginated (2015)

Chapter 15
English, Welsh, Irish and Scottish Porcelains

Abstract The rise of British porcelain manufactories (excluding Fulham which has been considered in Chap. 14) chronologically with the mainland European manufactories, and growth of early English porcelains from Pomona, Bovey Tracey, Bow, Chelsea, Limehouse and Vauxhall in the 1740s to Derby, Worcester in the 1750s and afterwards. The information derived from analytical science of early English porcelains and the need to re-adjust the concept of an inheritance of production stemming from continental Europe. Also, a brief description of Welsh, Scottish and Irish porcelains and ceramics is given.

Keywords English porcelains · inheritance of Fulham Pottery · Bow · A-marked porcelains · Vauxhall · Chelsea · Limehouse · Derby · Worcester · Welsh ceramics · Nantgarw · Swansea · Scottish ceramics · Irish ceramics

15.1 Introduction

The historian of early porcelains is aware of a gap in the chronology of succession between the initial production of porcelain in Europe in Florence, Tuscany, by Duke Francisco de Medici from the 1570s to about 1600, with possible start-up ventures in Padua and Genoa in the early to mid-seventeenth century, until the efforts of John Dwight at the Fulham manufactory in the 1670s and thereafter at Rouen and St Cloud in France to 1703 (Chap. 14). At this time the French porcelain manufactories at Rouen and St Cloud were in the forefront of European manufacture, generating other spin-offs at Paris such as Pavie and Ville d'Eveque, until the porcelain manufactory of King Frederick II at Meissen became established commercially in Saxony from 1712, which itself spawned early competitors in Vienna, Sevres, Vincennes, Mennecy and Chantilly. In Chap. 14 we have discussed the early porcelain manufactories in Europe, which have included that at Fulham Pottery in England for its chronological position lying between those of Florence and Rouen towards the end of the seventeenth century, and here it is appropriate we shall expand upon the

H. G. M. Edwards, P. Colomban, *Blue by Fire: A Marker of the Technical History of Glass and Ceramics*, Cultural Heritage Science,
https://doi.org/10.1007/978-3-031-77122-4_15

early English manufactories and others which appeared later on in the eighteenth century and afterwards.

Unlike the continental European upsurge in porcelain manufacture, there seems to be a hiatus in English porcelain production after John Dwight of the Fulham Pottery died in 1703 and his manufacturing business closed down, until the fledgling factories of Bovey Tracey and Pomona opened in 1743/4 and the potential reason for this has been open to much speculation historically. Certainly, from the mid- to late- 1740s and through into the 1750s, several English porcelain manufactories were established such a those at Bow, Limehouse, Vauxhall, Chelsea, Derby and Worcester with other minor ones such as that at Kentish Town. The output initially was predominantly in blue and white wares, often of a utilitarian design, but artefacts of a more decorative nature soon were appreciated, especially when polychrome decoration became desirable and armorial motifs of clients could be reproduced appropriately and faithfully on china. Early armorial porcelains were commissioned by English armigerous families and the aristocracy firstly in Chinese hard paste porcelains, where the arms and decoration were applied in cobalt blue enamel in Canton or Jingdezhen from templates provided by the clients to the supercargoes of the Honourable East India Company ships [13]: the earliest examples of such blue and white armorials are a set of octagonal and circular jardinieres (Fig. 15.1) that were supplied to Edward Johnson of London, who built East Indiamen ships for the HEIC in the Deptford and Blackwall dockyard on the River Thames, and these have been firmly dated to 1678. The Sir Thomas Pitt armorial service supplied in 1708, also of Chinese porcelain, is the first dinner service that is still extant and is remarkable in its very early and ground-breaking use of a polychrome decoration, which is nevertheless contained in an otherwise predominantly blue and white decorative palette (Fig. 15.2). By the 1750s, the atelier of James Giles in London was actively using polychrome decoration on English and Chinese porcelains, especially armorials, and at one time his stock in-house of Worcester hybrid porcelain was actually greater than that held in the Worcester China Works itself, according to historical insurance documents.

15.2 Compositional Characteristics

What clearly differentiates early English eighteenth century porcelains from their mainland continental European analogues is that they are predominantly soft-paste in composition, or hybrid equivalent variations of these such as that at Worcester, and that they have been developed along a completely different route to their continental contemporaries: this is confirmed analytically with the detection of components such as calcined bone ash present in some of the earliest Bow porcelains [51] in 1744/1745, which eventually became the staple characteristic component of English porcelains, culminating in Josiah Spode's "English bone

Fig. 15.1 Circular armorial jardiniere, Kangxi porcelain, 33 cm diameter, 19.7 cm high, ca. 1693–7, depicting the armorial bearings of Sir Henry Johnson impaled by those of his second wife Martha Lovelace, who were married in 1692. In 1697 Lady Martha Johnson became Baroness Wentworth and Baroness Despenser *suo jure* and her arms would have reflected this. Now in the Metropolitan Museum of Art, New York, the gift of Mr. and Mrs. Rafi Y. Mottahedeh, 1976, accession number 1976.112. (*Public domain*)

china" in the 1790s, which was the envy of other porcelain manufacturers in England and in Europe on account of its lightness, translucency, thinness of potting and robustness regarding survival and damage in use. Despite this compositional difference between English and continental European porcelains, which must arise from an independent and distinct contemporary development procedure in porcelain manufacture, several historians and connoisseurs still maintain that the English manufactories required an input of technology, technical advice and recipe formulations from the continental manufactories to get started, which is nonsense as, analytically, the two types of porcelain are quite distinctive in their different compositions.

Firstly, we shall consider the early English manufactories whose foundations will be described briefly and this will be followed by an analytical comparison between these and the major French and other continental manufactories.

Fig. 15.2 Kangxi porcelain, Qing Dynasty, *ca.* 1705, Chinese armorial porcelain plate from the *Pitt* service, made for Sir Thomas Pitt of Ewerne - Stapleton, Dorset, the earliest surviving example of a Chinese armorial service ordered for the British aristocracy. Note the curious placement of the over-large crest below the escutcheon and the upwards curvature to the torse matching the curve of the cavetto of the plate, which should properly fit over the helm and require a downwards curvature. The oval escutcheon displays three bezants—a Byzantine gold coin discovered by the Crusaders *en route* to Jerusalem and in Latin called "*Bizantius aureus*", a gold bezant. The bezants depicted here are more akin to elliptical golden roses and should properly be shown as being plain and circular heraldically. (*Private Collection. Reproduced with the permission of Kingschina Ltd, Mayfair, London*)

15.3 The Early English Porcelain Manufactories: Pomona, Chelsea, Limehouse and Bow

Many historians and ceramic writers have identified Bow as the earliest English porcelain manufactory of note, founded in 1744 [44], and closely followed by Limehouse and Chelsea within the same year, whilst discounting Pomona, which was also founded in 1744, but for which no extant and recognised complete example of its products has survived and only a very few badly damaged pieces now exist, these having been professionally reconstructed from porcelain shards found in the factory waste pit [5].

15.3.1 Pomona, Newcastle-Under-Lyme

Pomona is arguably the premier foundation of the English porcelain manufactories mentioned here chronologically, being instituted in Newcastle-under-Lyme in Staffordshire, in 1744, by William Steers. Lasting for only three years duration, Pomona is now regarded as an experimental venture, which although it produced many wasters, it is thought never produced porcelain in sufficient quantities to achieve a commercial sales success. Not one piece of finished Pomona porcelain has been identified as surviving into the modern era although several pieces now exist in the Brampton Museum in Newcastle-under-Lyme which have been reconstructed from fragments and wasters found on the site of the factory [5]. Functional wares such as teapots, tankards, teabowls, mugs, jugs, a mustard pot and a spittoon, decorated in a blue pigment, have been identified from these excavated fragments. An example, dating from 1744 to 1747, is a mug decorated in underglaze blue with a recognisable *chinoiserie* scene in an imitation of the imported Chinese porcelains of the time (Fig. 15.3).

15.3.2 Bow, Stratford, London

Bow, in common with the earlier French factories of Rouen and St Cloud, is credited with being a soft paste/glassy porcelain manufactory initially, unlike Meissen and Vienna, which were exclusively hard paste porcelain manufactories [1]. Bow was founded by Thomas Frye and Edward Heylyn in 1744/1745 on a site just outside London in the borough of Stratford at Grove Hill on the western bank of the River Lea near the church of St Mary's, Bow, which put it just within the boundary of the county of Middlesex ([11])—hence, its mention in Page [40]. Frye and Heylyn applied for a patent to manufacture porcelain, firstly in 1744 and again in 1748, each of which was granted the following year [12]. The first patent was for "*A Method for the Improvement of the English Earthen Ware,and have at last invented and brought to perfection a new Method of manufacturing a certain Material, whereby a ware may be made of the same nature or kind, and equal to, if not exceeding in goodness and Beauty China, or porcelain ware imported from abroad*".

Heylyn had family connections through his brother John and his estates in North Carolina, where he had noticed the fine white earthenwares made by the Cherokee Indians locally [61]. Edward Heylyn conceived the idea of importing this fine white earth called "*unaker*" from the Cherokee lands [45, 46] to make porcelain in England, it is believed being encouraged to do this by Andrew Duché, whom it is

Fig. 15.3 A Pomona porcelain mug, *ca.* 1744–1727, reconstructed from shards excavated at the site, with underglaze blue decoration in the Chinese style. (*Courtesy of Brampton Museum and Art Gallery, Newcastle-under-Lyme Borough Council, Newcastle-under-Lyme, Staffordshire, United Kingdom*)

thought may have tried to undertake porcelain synthesis himself in the Carolinas in the 1730s using this "*unaker*"clay [43]. In the mid-1740s there was a shortage of quality iron-free white earths in Europe and the Chinese, who originally used kaolin as a ballast for the transportation of their export porcelains by the English and Dutch East Indiaman vessels from China to Europe soon realised that they were presenting such a commodity *gratis* to their potential competitors in Europe and then immediately ceased to do this. It would be several years before native kaolin and china stone deposits would be found in Tregonning, St Austell, Cornwall, by William Cookworthy in quantities enough to sustain the fledgling British porcelain industry following his own patent for Plymouth porcelain manufacture in the 1760s.

In the late 1740s, the Bow manufactory moved its site east across the River Lea and into the borough of Newham, west of Stratford High Street in the county of Essex, as indicated in a document dated the 7th July, 1749. The new premises were called New Canton and further invoices from 1750 appeared for the Bow Porcelain Company bearing that address. The location of a large slaughterhouse nearby at Three Mills enabled a ready supply of bones for the provision of the calcined bone ash raw material and glass cullet would be readily obtained from Heylyn's own local glass works nearby. Bow wares started off with a simple *Blanc de Chine* in imitation of the Dehua porcelains then being imported from Fujian, China. These differed from the main importation source of Chinese porcelains at Jingdezhen in being undecorated or very simply decorated in underglaze blue on a warm ivory/

cream substrate body. Bow designs were modelled on the Chinese imports and also imitating its rival porcelain manufactory at Chelsea, situated across London. The use of calcined bone ash, claimed by Josiah Spode as his invention for his famous staple "bone china" later in the century, was actually a novel phosphatic component additive to early Bow porcelains some fifty years before [51].

Bow rapidly became the largest porcelain manufactory in England and by 1758 was employing more than 300 people, of whom 90 were "painters of china". The production output, like that of Chelsea, was mainly luxury items, including blue and white tablewares in imitation of the Chinese and Japanese Kakiemon styles, and many figures, taking the lead from Meissen in this respect. They adopted transfer printing in the 1750s and also used cold paint decoration for some figures, i.e. an unfired, overglaze decoration which, of course, was prone to flaking off and liable to experience extensive superficial damage with use. A Bow landmark in 1750 was the production of the earliest full length figures in English porcelain namely the *Fine Lady* and *Fine Gentleman.*

After 1760, the fortunes of the Bow Porcelain Company went into recession as several partners died or became bankrupt with other failed business ventures: Thomas Frye died at this time—his epitaph stating "… *this was the inventor and first manufacturer of porcelain in England*". John Crowther continued to run the depleted business at Bow by selling off existing stock but eventually the factory formally closed in 1764 and was bought by William Duesbury of Derby in 1776, a former enameller at Bow, who then transferred the moulds and equipment to his own Derby Porcelain Works. Edward Heylyn apparently survived the closure of the Bow Porcelain Company with his other business interests but died in 1765.

Bow porcelains [1] have been the subject of several experimental analytical studies in which the analyses have not always concurred with the established and avowed expert ceramic opinion based historically on purely stylistic considerations. Tite and Bimson [56] analysed four specimens of Bow shards from three production periods, namely, 1750–1755, 1755–1760 and 1765–1770. All were found to be phosphatic in type with an average percentage of phosphorus pentoxide of 16.2%, representing a significant bone ash content in the original recipe of approximately 38–40%. Corresponding percentages of silica of 49.2%, alumina of 6.4%, magnesia of 0.4%, lime of 23.8% and a combined soda and potash content of 1.2% are supportive of an assignment to a predominantly highly phosphatic and soft paste porcelain body. On the other hand, some recent studies of first patent Bow porcelains by Ramsay and Ramsay [44] give a totally different body composition with 60.4% silica, 26.8% alumina, 6.3% lime, and 6.5% combined soda and potash, with zero percent phosphorus pentoxide, which is more typical of a true or hard paste porcelain. It is clear that from the first patent Bow body registered by Edward Heylyn and Thomas Frye in 1745 [28], several major body variants were being experimentally evaluated and this complex story has been unravelled in a recent and comprehensively detailed scholastic work published by Ramsay and Ramsay [49]. In their ground-breaking research on Bow porcelains, Ramsay and Ramsay have analysed data pertaining to the mysterious early English "*Factory A*" porcelains for which it has long been postulated belonged perhaps to a completely unassigned and unrecognised porcelain factory from the mid-eighteenth century but for which a cogent scientific argument based upon analytical data has now been presented for their re-assignment to early Bow factory porcelain body variants (for further details, see [16]). An interesting conclusion from this work is that it seems that porcelain production at Bow actually commenced several years earlier than 1744 and that some compositions included hard paste variants: a recent paper by Ramsay and Edwards [51], tracing the activities of William Reid from Pomona, through Bow, to Staffordshire and Liverpool informs us analytically that indeed he was a major player in the creation of various compositions of early English porcelains and that several early Bow variants can be credited to his work. Additionally, this same paper, following on from an earlier study of a polychrome snuff box [50] has also now credibly established that examples surviving of the so-called "A-marked" porcelains can now be assigned to the earliest Bow recipes [47, 48], without invoking a new as yet-unidentified factory location as was also demonstrated in an earlier paper [17]. The nature of the meaning of the letter A inscribed into A-marked porcelains has generated some discussion; some alternative hypotheses include the Duke of Argyll's sponsorship at Gorgie, Edinburgh (see below) and "athanor", an alchemical furnace used for chemical digestion at high temperatures [11] and at one time were believed to be porcelains of Italian origin [32].

An important point emerges from this particular study in that, unlike some later manufactories such as Derby or Rockingham, where the successful soft paste porcelain recipes seemed to have been maintained unchanged or at least very little changed over relatively larger periods of time, other factories of an admittedly earlier foundation experimented with several body variants which could possibly then have been retained in parallel and a concurrent commercial production was established over a significant production period. Forensically, therefore, it might then be foreseen to be rather difficult to establish a characteristic set of analytical data which would unambiguously define a factory porcelain body for Bow with a single set of unique characteristic elemental oxide compositions which could perhaps be utilised to unequivocally identify the potential origin for an unknown piece, unless examples of these body variants had been well-characterised analytically and separately for known periods of production at the factory.

15.3.3 Chelsea Porcelain

The establishment of the Chelsea China Works by Nicholas Sprimont, a Huguenot silversmith [26], in 1745 gives Chelsea the credit of being one of the first three English china manufactories to be founded within a year or so of each other, the others being Pomona and Bow. The Chelsea manufactory, was founded across London from Bow, on a site near the socially fashionable Ranelagh Gardens. Documentation relating to the early Chelsea factory is sparse. The earliest products of the factory were tablewares and figurines, tending to imitate those emanating

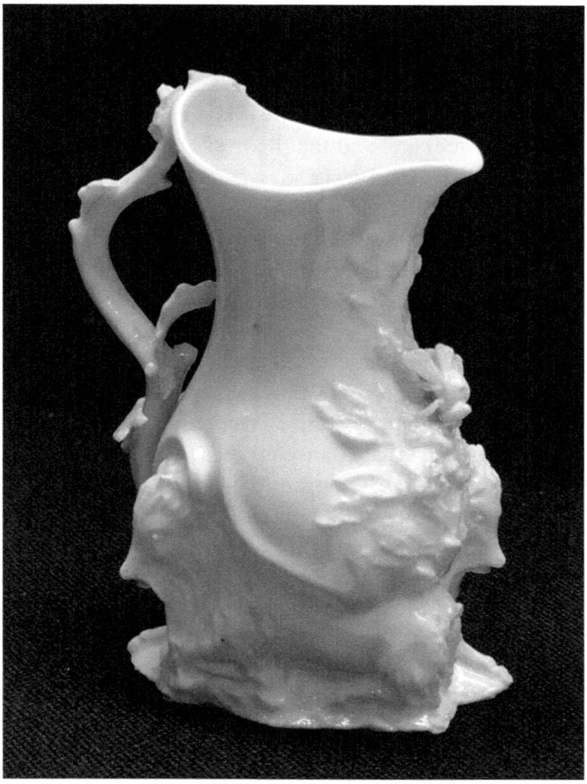

Fig. 15.4 Chelsea porcelain, *ca.* 1745, Goat and bee jug, glassy soft paste porcelain. Modelled by Nicholas Sprimont with the characteristic depiction of two goats at the base placed parallel but facing in opposite directions: Sprimont was previously a silversmith and silver versions of this jug appear with his mark. (*Gardiner Museum, Toronto, Canada*)

from Meissen and one of the most celebrated Chelsea shapes from this period is the "goat and bee" jug (Fig. 15.4) based on its silver variant which appeared in 1745. Some versions of this bear an incised triangle and a "Chelsea 1745" script mark. The incised triangle is characteristic of early Chelsea wares up until 1749 [53]. A rare mark for this period is a crown and trident in underglaze blue.

It is believed by some historians that Charles Gouyn, also a Huguenot silversmith and a porcelain dealer, was involved as a partner with Sprimont in the setting up of the porcelain factory at Chelsea. By 1749 Gouyn had departed and Sprimont took over control of the factory in March, 1749 and he moved to a neighbouring site where they started to produce more tablewares based on Meissen designs such as salt cellars and cream jugs. He introduced the new mark of a *raised anchor* (1749–1752) [53] which is believed to be related to the location of the new site closer to the River Thames. Gouyn is also believed to be the Chelsea "arcanist", the person responsible for deriving the formulations for the body paste, glaze, pigment colours and kiln operating characteristics. It has also been suggested that Sir Everard Fawkener and Prince William, Duke of Cumberland, son of King George II, were early investors in the Chelsea porcelain manufactory from 1746. If true, this is indeed a noteworthy event as there had been no confirmation recorded historically of any investment in English porcelain manufacturing by royalty, unlike the situation that was clearly documented in France and Germany at that time. A contemporary observer commented that Prince William was "a great encourager of the Chelsea porcelain" and that he possessed several services for his own use. In 1751 Fawkener, who died in 1758, was materially involved in the personal borrowing of several items from the Meissen factory for copying and reproduction at Chelsea.

This second phase at Chelsea was the most successful period for the factory and the new composition body and glaze with their clear white surfaces were an ideal canvas for artistic developments in polychrome enamelling. Chelsea was still noted for its small figures, sometimes only 2″ (5 cm) high, and delicate artefacts such as snuff boxes, bonbonnieres and etuis, collectively known as "toys" [8], but now they were also making inroads into the production of larger figures in classical poses and decorating them in the style of Vincennes, at a pre-Sevres amalgamation period [19].

The third phase at Chelsea is characterised by a red anchor mark, *ca.* 1752–1756 [54] and again the repertoire of the factory was extended into copies of Japanese Kakiemon export wares and tablewares; especially desirable were the botanically illustrated pieces based upon Sir Hans Sloane's Chelsea Physic Garden, which had been set up nearby in 1722. Much Chelsea porcelain was sold through "*chinamen*", the name accorded to porcelain dealers in London warehouses, and through china auctions rather than through the works directly as occurred in other contemporary china factory establishments.

The gold anchor period, phase four of the Chelsea porcelain factory production, occupied the period between 1756 until the closing year, 1769 [55]. Now the influence of Sevres became more evident and the French rococo style of rich colours and heavy gilding produced garnitures of elaborate vases with lavish decoration. The Chelsea body paste now included a bone ash component for the first time, the glaze was thicker and subject to pooling in greenish coloured areas. King George III and Queen Charlotte commissioned a lavish and large Chelsea service for the Queen's elder brother, Adolphus Frederick IV the Duke of Mecklenburg-Strelitz, that was much admired and cost the fabulous sum at that time of £1200, which is equivalent to the sum of £ 250,000 today. Some 137 pieces of this service have survived and are now in the Royal Collection of Her Majesty Queen Elizabeth II. This was also the height of the Japanese Imari style production and an observer commented that "*the Chelsea copies greatly surpass in beauty their dull originals*". After 1763, the production from the factory diminished as Sprimont approached retirement in a state of ill health and died in 1771: a sale occurred in 1763 of his moulds and some equipment prior to the factory winding down, it being sold in 1769 to James Cox who immediately sold it on to William Duesbury of the Derby China Works the following year on the 14th February, 1770. Duesbury continued to operate the factory at Chelsea until 1784, during the so-called Chelsea-Derby period, and the porcelain produced there became effectively indistinguishable from that produced at the parent Derby works.

Early Chelsea products undoubtedly had a glassy porcelain body and the use of glass cullet was certainly prevalent in their recipe and composition: analytically, this is not as easy to define as it was with later factories as it seems that the incorporation of lead-free crown glass (or soda glass) was the norm at Chelsea, perhaps in admixture occasionally with a lead-rich flint glass additive, and this would of course render only a rather small presence of lead oxide to be detectable analytically in the body. The analytical data for Chelsea porcelain from the work of Tite and Bimson [56] revealed that the earliest Chelsea shards from the marked raised triangle (*ca.* 1745–1749) and red anchor mark (*ca.* 1755–1760) periods both show lead oxide percentages of 5.1 and 1.9% only, along with high silica and low alumina percentages and higher percentages of alkaline oxide potash and lime cumulatively amounting to about 18–20% in total. In contrast, however, a shift in the Chelsea body composition seems to have occurred in the later gold anchor period between 1760 and 1770, since this body is totally different in composition to the earlier glassy variety and analysis now records some 46% silica, 11% alumina, 22% lime and 17% phosphorus pentoxide—the latter being equivalent to a significantly high 44% calcined bone ash raw material component. A similar situation can be described for the Vauxhall manufactory (*ca.* 1753–1764), where the analysis of shards [56] showed that very small lead oxide percentages of approximately 0.3% could not be correlated with the recorded use of ground flint glass cullet in the paste, which can now therefore be confirmed as originating from a crown glass component rich in potash but also being virtually lead-free [35]. The Vauxhall porcelain differs from Chelsea in that it has a significant percentage of magnesia in the shards, which approximates to about 25–30% of a soapstone raw material used in the original recipe, although the silica percentage is still high at approximately 75%. A phosphatic porcelain from Vauxhall has also been identified more recently from shards excavated at the site [6].

15.3.4 Limehouse

The Limehouse Porcelain Manufactory was among the first of the English porcelain manufactories, being founded just a year after Bow and along with Chelsea in 1745, by William Ball, John Wilson and Joseph Wilson within two miles of the Bow manufactory site at Fore Street near Duke Shore at the Limehouse reach on the River Thames in London (conforming to the location of the present-day Narrow Street) [57]. In 1990, the site was excavated archaeologically and remains were found of a kiln and a large quantity of glazed and unglazed china wasters were recovered, amounting altogether to 1402 shards [42]. Jay and Cashion [29] report a recent Raman spectroscopic analytical study of some of these wasters and confirm that the body is comprised mainly of silica and alumina with magnesia and lead oxide also present and that a silicaceous and lead glaze has been applied where appropriate. Generally, the porcelain quality was considered to be not very good which contributed to the closure of the factory in 1748 but much interest has been stimulated in Limehouse porcelain from analytical studies of the shards by Freestone [21, 22], Owen [38, 39] and by Ramsay, Daniels and Ramsay [46]. From the analytical data obtained from the relevant factory shards, Ramsay, Daniels and Ramsay [45] have identified a time-line chronological pathway that existed from Bow-Limehouse-Lund's Bristol-Worcester which spanned just 8 years in time between 1744 and 1752. In a scholastic piece of detective work, the analytical data from the two separate and independent studies of Ramsay et al. [46] and Owen [38, 39] reveals that there were actually up to three distinct phases of porcelain production at Limehouse. Ramsay et al. [46] could ascribe the following chronology to their shards—for the period 1745-early 1746: a silica /alumina body with a silica/alumina/lime glaze, for the period early 1746–June 1747: a silica/alumina/lime body with a moderate lead oxide glaze, and from June 1747 until early 1748: a magnesia/phosphate body with a moderate lead oxide glaze. All of these were compositionally derived from Bow porcelain. Freestone [21] defined two Limehouse bodies, both silicaceous and aluminaceous, respectively, the first being two parts of alkaline lime/bottle glass and one part of a silica / alumina body similar in composition to the Bow first patent body, followed by a second group with lower silica levels but including higher percentages of lime, potash and soda. Both bodies had a low-lead content glaze. Finally, Owen [38, 39] also found two types of porcelain body, the earliest comprising mainly silica and alumina with a low lime content and a lime-alkaline glaze and the later period shards with a similar body but with a lead oxide glaze. Bernard Watney [59] confirmed the presence of a magnesian-phosphatic body in his collection of Limehouse porcelain shards. Ian Freestone [21] noted that glass frit had been added to the china clays in the formulation and that this suggested a link between the three factories of Pomona, Bow and Limehouse. The poorer quality of the porcelain body at Limehouse has been attributed to the inability of the proprietors to source the premier grade china clay which their Bow rivals had obtained from the Cherokee "*unaker*" deposits in America and as a result were forced to use the inferior sedimentary ball clays, probably sourced from Cornwall. The analyses are summarised in Ramsay et al. [46] and demonstrate a highly refractory porcelain which for the *hard paste* silica-alumina-calcia type body comprises 73% silica, 11% alumina, 7% calcia, 6% soda and potash, whereas the *soft paste* magnesia-phosphatic body contains 65% silica, 3% alumina, 11% magnesia and 6% phosphorus pentoxide and the mineralogical signatures of mullite, sanidine, diopside and bytownite were identified in the *hard paste* body. The raw materials were confirmed as ball clay, crushed quartz sand, saltpetre, and lime-alkali glass frit for the *hard paste* body and steatite m calcined bine ash, crushed quartz sand and a lead-free alkaline glass frit.

The eventual demise of the Limehouse operation can be ascribed to, firstly, a lack of demand for their porcelain product, which was deemed to be inferior to Bow and, secondly, the depletion of their source of soapstone raw material deposits obtained from the mine at Kynance Cove in Cornwall which became worked out in 1747. The intriguing link between Bow, Limehouse and then Lund's Bristol and eventually Worcester proposed by Ross Ramsay is extremely relevant to the wider consideration of the transferability of knowledge and technologies from one factory to another at this time, particularly when they were close neighbours to each other as they were especially at Bow and Limehouse. It is even more revealing when we consider the compositions of the formulations for Bow, Limehouse and Lund's Bristol: these comprise three ceramic types of soft paste porcelain: a high clay silica/ ball clay /saltpetre composition at Limehouse, which was switched to the magnesian based bodies in June 1747, then comprising a magnesia/lead oxide porcelain made from soapstone, silica and lead oxide glass frit, followed by a magnesia / phosphate /lead oxide porcelain made from soapstone, bone ash, silica and a lead oxide flint glass frit. The composition of Lund's soft paste Bristol porcelain produced between 1749 until late 1750 was a magnesia/phosphate/lead glass body with a moderate lead oxide glaze followed by a late 1750 until early 1751 body variant with a magnesia/lead glass body plus a moderate lead oxide glaze which clearly pointed to information being acquired from Bow and Limehouse. It is also known from historical correspondence that Edward Heylyn, Thomas Frye, William Ball and Benjamin Lund frequently visited each other's factories and it can be surmised that the transfer of empirical compositional and technological information about their experimental products would have been undertaken in a supportive way by these four leaders in English porcelain experimentation in its earliest days.

William Ball, the founder of the Limehouse China Works, emigrated to the American Colonies with his family in 1769, being recruited as a "master potter" from William Reid's Brownlow Hill manufactory in Liverpool by representatives of Bonnin and Morris, who were intent upon establishing their own porcelain manufactory in Philadelphia, Pennsylvania. [51] Ball and about ten key skilled workers from Liverpool landed in America in October 1769 and they soon helped set up the American Porcelain Manufactory in Philadelphia, which produced blue and white decorated porcelains over the short period between 1770 and 1772.

Other early English porcelain manufactories such as Vauxhall, Isleworth, Worcester, Derby and Longton Hall which were all founded around 1750 or just before were contemporary with those of Bow, Limehouse and Chelsea and will not be discussed further here. Reference has also been made to another short-lived porcelain manufacturing enterprise at Lambeth, initiated by a Thomas Briand in the early 1740s [7, 37].

Another rather strange phenomenonhat deserves some comment is the rare occurrence of blue and white decoration on Chelsea wares compared with those of its contemporaries at Bow land Limehouse. Figure 15.5 shows a rare Chelsea plate, dated to 1754/55, decorated in underglaze cobalt blue and marked with an underglaze blue anchor; it imitates the Japanese Kakie4mon style and shows two birds in a landscape with five vignettes of flowers in the reserve. Figure 15.6 shows a Bow plate from 1755/60 decorated in the chinoise style in underglaze blue with the *Golfer and Caddy* pattern.

Fig. 15.5 Chelsea, porcelain plate decorated in underglaze blue in the Japanese Kakiemon style, *ca.* 1754/1755, blue anchor mark. (*Private collection*)

15.4 Analytical Evidence for the Independent Development of Early English Porcelain Manufactories from Mainland European Influence

A major discrepancy arises between the affirmed conclusions made by earlier historians on the basis of an apparent mis-interpretation of historical documents and the recent results of analytical science on artefacts and shards from the archaeological excavation of several early English porcelain sites, such as Bow, Limehouse and Liverpool. It has always been a matter of conjecture as to the underlying reason that English porcelain manufacture seemed to halt after John Dwight's Fulham Pottery closed soon after his death in 1703 until the start-up of Pomona, Bow, Chelsea and Limehouse in the mid- to late 1740s, which it has been argued required the input of mainland European manufactories with advice, expertise and recipes for its production. This is tantamount to inferring that the English manufactories were, therefore, reliant completely upon Continental support from places such as Meissen, St Cloud, Paris, Sevres, Vienna and Vincennes for their products: the overriding conclusion would have to be made that the English porcelain recipes, formulations and manufacture could not have developed independently but would hence be similar if not identical to those of their European continental counterparts! Analysis has now shown that nothing could be further from the truth and that, in fact, early English porcelain manufactories were ground-breaking in their empirical recipes and formulations and that their compositional characteristics were very different

Fig. 15.6 Bow, porcelain plate decorated in underglaze blue, *ca.* 1755/60, in the *Golfer and Caddy* pattern. (*Private collection*)

indeed from those in Europe. John Mallet typifies the misconception by alleging in a recent paper [34] that "English factories can be compared with Mennecy, St Cloud and Vincennes".

A critically important historical document is the patent filed by Edward Heylyn and Thomas Fry of the Bow China Works in 1744, which described their procedures and recipe for the manufacture of porcelains; whereas some historians feel that this is really the marker for the commencement of English porcelain manufacture, there is in existence a raft of documentation which indicates that experimental trials were proceeding at Bow some years before in the 1730s, catalysed by the bringing of specimens of the white Cherokee clay "unaker" to London by Andrew Duche. Several key articles have been published in recent years which highlight the anomaly especially those of Daniels ([11, 12, 44, 47–49]). What is clear from the analytical data obtained from the early "A-marked" porcelain wares, which have been categorically redefined to an early Bow manufactory production around 1744 on the basis of analyses [17, 43] and especially that of a Bow "A-marked" snuff box [50]. It appears that far from concentrating upon a sole selected composition, the Bow

Manufactory in its New Experimental & Development phase up to 1744, prior to the New Canton period,actually made five or more bodies, an SiAl hard paste body akin to stonewares, an SiAlCa body, a Bone ash phosphatic body and a magnesian body. These were found to also exist at Limehouse (1746–1748) and at Liverpool (Brownlow Hill) (1755–1761) and this can be attributed to the movement in employment of William Reid and William Ball. Analytically, these bodies are of a significantly different composition: the SiAl body has alumina at 20% and potash at 3%, the SiAlCa body has alumina at 7–85, calcia at 4–6%, whereas the bone ash phosphatic body has alumina at 6–11%, phosphorus pentoxide at 11–16% and calcia at 14–23%. The calcia: phosphorus pentoxide ratio at 3.3 is in the range expected for calcined bone ash as used in bone china. The SiAl body is clearly a hard paste porcelain in type. Greater details of the compositional changes in early Bow and Liverpool and relevant discussion can be found in Ramsay & Edwards [51], which traces the movements of William Reid alluded to above to eth establishment of the William Reid factory in Liverpool in the late 1750s.

This means that Bow was making English bone china with alumina at 12% and phosphorus pentoxide from 18% to 20% in 1744 to 46, some 50 years before Josiah Spode claimed its invention (which he termed *Stoke china*) with his patent in 1790. The formulation for English bone china is not unique: for example, one Spode formulation is 36% bone ash, 24% primary china clay, 18% china stone, 12% secondary blue clay and 12% flint, whereas another one was 50% bone ash, 25% kaolin and 25% china stone, which can be compared with the recipe given in the literature of 6 parts bone ash, 4 parts china stone and 3.5 parts china clay, corresponding to 44% bone ash, 30% china stone and 26% china clay [9]. However, Whiter [60] in his history of the Spode China Works does point out that the production of bone ash porcelain and bone china can be traced back to Bow in the 1740s.

Another interesting fact emerges from these analyses: that Bow was making true hard paste porcelain in England some years before Cookworthy's venture materialised firstly at Plymouth, and later at Bristol, in 1768. Clearly, the message is that analytical data must be considered as a fundamental component of the attribution of unknown porcelains and that one can reasonably expect that one's perception that the established historical view of a particular factory as a result may well be compromised.

15.5 Welsh, Scottish and Irish Porcelains

Although there were many small ceramic potteries in Wales, Scotland and Ireland which were run as family businesses making earthenwares for local sale, the incidence of porcelain manufactories is much lower.

15.5.1 Wales

As an example of the previous statement, the village of Ewenny in Glamorganshire, near Bridgend, is recorded as having approximately 15 separate families making ceramics locally in 1427, several of which were grouped together in 1610 by Evan Jenkins into a small pottery called the Ewenny Pottery, which in 1883 attracted the attention of Horace Elliott, an Arts and Crafts designer who designed artefacts there which carried his fleur-de-lys mark and inscriptions in Esperanto! Still running today, the antique Ewenny Pottery items command high prices from collectors for their sometimes whimsical decoration on plain glazed brown earthenwares,

The more well-known Cambrian Pottery, founded by John Coles of Neath in 1767, in partnership with George Haynes from 1789, and taken over by Lewis Weston Dillwyn in 1802 became the first porcelain manufactory in Wales when Dillwyn opened his Swansea China Works on the same site in 1812. The engagement of William Billingsley and Samuel Walker in 1814 gave Dillwyn the specific expertise and kiln control he needed to manufacture his finest "duck-egg" phosphatic porcelain from 1815 to 1820 with its characteristic blue-green translucency that was a suitable vehicle for the exquisitely fine decoration by Billingsley that became admired locally and by his London clients. In his efforts to create a more robust porcelain, Dillwyn undertook some planned empirical experiments with his composition and his inclusion of steatite in place of other raw materials produced successfully a toughened soaprock porcelain in 1817, called "trident" ware from its impressed mark, which unfortunately did not appeal to his clientele, who regarded the pigskin-like texture and lack of translucency as inferior to his "duck-egg" analogue. The Swansea China Works fell into decline and was sold in 1820, but Dillwyn retained control of the Cambrian Pottery on the same site, which was managed by his son Lewis Llewellyn Dillwyn until that too fell into decline and passed into a succession of owners until its closure in the 1870s. The Glamorgan Pottery under Baker, Bevans and Irwin opened in Swansea in competition but did not survive beyond the 1840s. A third pottery, the Ynysmeudwy Brickworks and Pottery Co., opened in the Swansea Valley in 1848 and manufactured garden ornaments bricks, chimney pots and tiles in red terracotta and cream biscuit clay resembling the stone china output of Mason's ironstone. The YMP closed in 1877 [25, 31, 36].

In the meantime, Billingsley and Walker had left Swansea in 1816/17 to set up their own porcelain manufactory at Nantgarw in the Taff Valley alongside the Glamorgan canal linking Merthyr Tydfil and Cardiff, providing excellent access to high quality anthracite coal supplies and a deep water port. At Nantgarw what has been described as the world's finest porcelain was made between 1817 and 1820, with undoubtedly the finest clear white translucency but again the extremely high kiln losses of 90% meant that it was not a viable economic proposition despite it being the most desirable acquisition of clients in London who would purchase all of Billingsley's output at a very high premium for decoration the London ateliers of John Mortlock, their agent in the capital. Nantgarw fell into bankruptcy and was taken over by the Quaker William Weston Young, who will feature again in

Chap. 16, who engaged Thomas Pardoe to decorate stockpiled porcelain in the white locally [14, 15]. Young became disillusioned when he failed to recreate the Nantgarw porcelain body as Billingsley and Walker had taken the recipe with them when they left for Coalport in 1820. The death of Pardoe in June 1823 meant the final closure of the Nantgarw China Works, which reverted to the manufacture of earthenwares and clay pipes under William Pardoe and his family successors until the pottery finally closed in the 1920s [23, 24]. A fine example of Nantgarw porcelain decorated with floral sprays in in cobalt blue enamel is shown in Fig. 15.7: this is a dinner plate from the *Lady Seaton* service, decorated in a London atelier between 1817 and 1820, and is unusual for its blue and white decoration and simple dentil-edge gilding in contrast to the polychrome examples that are more prevalent.

A Swansea porcelain version of the *Lady Seaton* pattern also exists from the Bevington (post-Dillwyn) period, *ca.* 1820 [18], where the floral sprays are transfer printed in a pale blue pigment on a plain glazed body without any additional gilding: according to Morton Nance [36] engravings were made of sketches of the flowers by Henry Morris, an accomplished Swansea China Works artist, and were sued in this transfer printing. The "*Lady Seaton*" terminology for this service is interesting since this service was sold from the estate of Arthur Jones of Bryn Newydd, Swansea, in 1864 whence it was purchased by Lady Seaton, so it was

Fig. 15.7 Nantgarw porcelain dinner plate from the *Lady Seaton* service, *ca.* 1817–1820, decorated in London with sprays of flowers in cobalt blue and the characteristic dentil-edge gilding. (*Private Collection*)

certainly not an original commission of hers with eth Swansea China Works and therefore comes under the classification of a "named service by association".

The final Welsh pottery that deserves mention here is the Llanelly Pottery, which was contemporary with the later years of the Cambrian Pottery from 1840 onwards, first known as the South Wales Pottery, which made simple utilitarian artefacts, described as "simple cottage ornaments with an unconscious artistry" by Herbert Read of the Victoria and Albert Museum. An example of their wares is shown in Fig. 15.8, being a locally painted "cockerel" plate with cobalt blue edging (as seen in the Carmarthen Museum, Abergwili). Noted decorators in the early 1900s were Samual Shufflebotham and Sarah Jane Roberts (known as "Aunty Sal"). The Llanelly Pottery eventually closed in 1922 [27, 30, 41].

Fig. 15.8 Llanelly pottery, "cockerel" plate, *ca.* 1900, painted in a free flowing style by Sarah Jane Roberts (Auntie Sal) with a cobalt blue border: the cockerel designs were multivarious as were the accompanying borders. (*Private Collection*)

15.5.2 Scotland

As in Wales and England, there were many small local Scottish potteries operating from the late seventeenth century onwards such as Watson's Pottery in Prestonpans, the Rathbone Pottery in Portobello, the Glenochar Pottery, the Newbygging Pottery in Musselburgh and the Delftfield Pottery in Glasgow, which used local clays and manufactured earthenwares for utilitarian usage. Robert and George Gordon's Pottery at Morison's Haven, Prestongrange, East Lothian, provides an example of the siting together of glass manufacture and pottery manufacture in the eighteenth century [10]. The Delftfield Pottery in Glasgow is said to be the earliest Scottish pottery of note, founded in 1748, on th Dienwiddie estate, which found local clays to be rather unsatisfactory for the production of earthenwares and started importing clays for their creamware production and were noted for their association with James Watt in the early nineteenth century. Perhaps, the most famous Scottish pottery is Wemyss, Kirkcaldy, Fife, originally the Fife or Gallatown Pottery,which was founded by Archibald and Andrew Grey in 1817 under the patronage of the local family at Wemyss Castle. It developed its artistic reputation for the freeflowing and naturalistic representation of flowers in 1882 when Robert Heron of the Fife Pottery engaged Karel Nekolar as decorator. In 1932 the pottery closed and Joseph Nekolar, Karel's son, moved to Bovey Tracey Pottery, where he continued to make Wemyss ware analogues in Devon. Joseph Nekolar died in 1952 and the Bovey Tracey Pottery closed in 1957. Wemyss ware is still produced today at Ceres, Fife, by Griselda Hill since 1985.

Scottish porcelain is exemplified in the eighteenth century by the West Pans Manufactory at Musselburgh, East Lothian, which produced the first china in Scotland [20, 33] after the initial manufacture of red terracotta pottery there. The main production was tablewares and figures, decorated in cobalt blue and bearing an underglaze blue script mark of crossed L's, representing William and Jane Littler. Littler moved to West Pans after the failure and bankruptcy of his manufactory at Longton Hall in Staffordshire around 1760 and he established the West Pans manufactory in 1764. After closure of eth West pans paottery in 1777, Littler returned to Staffordshire to work with Ralph Baddeley and make the Baddeley -Littler porcelain there until 1785.

Occasionally, one comes across reference to a Scottish porcelain venture at Gorgie, near Edinburgh, that was founded by an Alexander Lind (1695–1756) FRSE, who was a wealthy landowner and Deputy Sheriff, sponsored by the 3rd Duke of Argyll: definitive examples of Gorgie porcelain are rare and several historians have attempted to link the A-Marked porcelain f early Bow to Gorgie, reasoning that the A would refer to the Duke of Argyll [52, 58]!

15.5.3 Ireland

The first name in Irish porcelain that springs to mind is Belleek: the Belleek Pottery Works was founded in Belleek, Fermanagh, in 1849 by John Bloomfield who conducted geological surveys in the local area and found they were rich in minerals: he started to produce a pale cream Parian ware in 1862. Then a local group of investors acquired the premises and engaged with a master craftsman called Frederick Slater to produce the now characteristic cream porcelain which contained Irish Celtic motifs, shamrocks and basket weave patterns, very thinly potted and slightly iridescent [2]. Still produced today, Belleek porcelain exemplifies Irish porcelain.

Parian ware is essentially a biscuit porcelain that is designed to simulate marble and it acquired its term from Paros, a Greek island noted in antiquity for its fine-textured and white marble that was desirable for sculpture [3, 4]. It was developed in Staffordshire by Minton's, as a version of their ironstone wares, and was called "statuary porcelain" by Thomas Battam at Copeland; at the Great Exhibition of 1851, items of parian ware were submitted for assessment by both Minton's and Belleek Pottery and the judges could not decide who was the prime inventor of parian ware [3].

References

1. Adams, E., Redstone, D.: Bow Porcelain. Faber & Faber, London (1981)
2. Archer, M.: Irish Pottery and Porcelain Irish Heritage Series. Eason and Son, Swords (1979)
3. Atterbury, P., Batkin, M., Greenwood, M., Smith, R., Ward-Johnson, P., Glynn, G.D.V.: The Parian Phenomenon: A Survey of Victorian Parian Porcelain Statuary and Busts. R. Dennis Publications, Shepton Beauchamp, Somerset (1989)
4. Barker, D.: Parian Ware. Osprey Publications, Oxford (2008)
5. Bemrose, P.: The *Pomona* potworks, Newcastle, staffs. Trans. Engl. Ceram. Circle. **9**, 1–18 (1975)
6. Bimson, M., Bimson, I., Freestone, I.: Phosphatic porcelain from the Vauxhall site. Trans. Engl. Ceram. Circle. **18**, 91–95 (2002)
7. Bridge, A., Bundock, N.: A newly-found reference to porcelain making in the early 1740s. Trans. Engl. Ceram. Circle. **26**, 68–81 (2015)
8. Bryant, G.E.: The Chelsea Porcelain Toys, Scent-Bottles, Bonbonierres, Etuis, Seals and Statuettes, Made at the Chelsea Factory, 1745–1769 & at Derby-Chelsea, 1770–1784. Medici. Soc., London (1925)
9. Burton, W.: Porcelain: a Sketch of its Nature, Art and Manufacture. Cassell & Co., London (1906)
10. Cressay, M., Johnson, M., Haggarty, G., Turnbull, J., Wilmott, H.: 18th century glass and pottery manufacture at Morison's haven, Prestongrange, East Lothian. Post-Mediae. Archaeol. **46**, 36–55 (2012)
11. Daniels, P.: The Origin and Development of Bow Porcelain, 1730–1747, Including the Participation of the Royal Society, Andrew Duche and the American Contribution. Resurgat Publishers, Oxford (2007)

12. Daniels, P., Ramsay, W.R.H.: Bow porcelain. New primary source documents and evidence pertaining to the early years of the manufactory between 1730–1745 and the John Campbell's letter to Arthur Dobbs. South. Inst. Technol. J. Appl. Research. **1**, 1–30 (2009)
13. Denyer, R., Denyer, M., Edwards, H.G.M.: Armorial Porcelain: The Genesis. Springer-Nature, Cham (2024)
14. Edwards, H.G.M.: Swansea and Nantgarw Porcelains. Springer -Nature, Cham (2017)
15. Edwards, H.G.M.: Nantgarw and Swansea Porcelains. Springer -Nature, Cham (2018)
16. Edwards, H.G.M.: Porcelain Analysis and Its Role in the Forensic Attribution of Ceramic Specimens. Springer-Nature, Cham (2022)
17. Edwards, H.G.M., Jay, W.H., Ramsay, W.R.H.: High-fired early english porcelains of the "a-marked" group, east London (*ca.* 1744): A Raman spectroscopic and electron microscopy confirmational study. J. Raman Spectrosc. **53**, 785–809 (2022)
18. Fairclough, O.: The use of set pattern decoration at the Swansea China works. In: Gray, J. (ed.) Welsh Ceramics in Context, Part II, pp. 193–208. Royal Institution of South Wales, Swansea (2005)
19. Fisher, S.: The Decoration of English Porcelain: A Description of the Painting and Printing on English Porcelain of the Period, 1750–1850. D. Verschoyle, London (1954)
20. Forbes, S., Haggerty, G.: The ceramic history of west pans, part 1. North. Ceram. Soci. J. **21**, 29–47 (2005)
21. Freestone, I.C.: A technical study of Limehouse. In: Drakard, D. (ed.) Limehouse Ware Revisited, pp. 68–73. London, English Ceramic Circle (1993)
22. Freestone, I.C.: A-marked porcelain: some recent scientific work. Trans. Engl. Ceram. Circle. **16**, 76–84 (1996)
23. Gray, J. (ed.): Welsh Ceramics in Context, Part I. Royal Institution of South Wales, Swansea (2003)
24. Gray, J. (ed.): Welsh Ceramics in Context, Part II. Royal Institution of South Wales, Swansea (2005)
25. Hallesy, H.: The Ynysmeudwy Pottery, 1845–1877. In: Gray, J. (ed.) Welsh Ceramics in Context, Part II, pp. 272–303. Royal Institution of South Wales, Swansea (2005)
26. Honey, W.B.: Old English Porcelain, 3rd edn. Faber & Faber, London (1977)
27. Hughes, G., Pugh, R.: Llanelly Pottery. Llanelli Borough Council, Llanelli (1990)
28. Hurlbutt, F.: Bow Porcelain. G. Bell & Sons Ltd., London (1926)
29. Jay, W., Cashion, J.D.: Raman spectroscopy of Limehouse porcelain sherds supported by Mossbauer spectroscopy and scanning electron microscopy. J. Raman Spectrosc. **44**, 1718–1732 (2013)
30. Jenkins, D.: Llanelly Pottery. DEB Books/Watkins, Swansea (1968)
31. Jewitt, L.L.: The Ceramic Art of Great Britain, J.S. Virtue & Co., London (1883)
32. Lane, A.: Italian or English porcelains; the A-marked cup. Keramikfreunde der Schweiz, Mittelungsblatt. **43**, 15–18 (1958)
33. Lane, A.: William littler of Longton hall and west pans, Scotland. Engl. Ceram Circle Transactions. **5**, 82–94 (1961)
34. Mallet, J.: Part 6: The place of Buckingham porcelain in ceramic history. Trans. Engl. Ceram. Circle. **20**, p. 220 and ff., (2008)
35. Massey, R.: Ceramics of Vauxhall: 18th Century Pottery and Porcelain. English Ceramics Circle Publications, London (2018)
36. Morton Nance, E.: The Pottery and Porcelain of Swansea and Nantgarw. Batsford & Co. Books, London (1942)
37. Mountford, A.: Thomas Briand: a stranger. Trans. Engl. Ceram. Circle. **7**, pp. 91 and ff (1969)
38. J.V. Owen, "A preliminary assessment of the geochemistry of porcelain sherds from the Limehouse factory site". In Limehouse Ware Revisited, eds. K. Tyler and R. Stephenson, MoLAS Monographs, No. 6, Museum of London, London, pp. 61–63, 2000a

39. Owen, J.V.: A preliminary assessment of the geochemistry of porcelain sherds from the Limehouse site. In: Tyler, K., Stephenson, R. (eds.) The Limehouse Porcelain Manufactory: Excavations at 108–116, Narrow Street, London,1990 Museum of London Archaeology Service, Monograph No. 6. Museum of london (2000b)
40. Page, W.: A History of the County of Middlesex: General – Ashford, East Bedfont with Hatton, Feltham, Hampton with Hampton Wick, Hanworth, Laleham, Littleton, Volume 2, pp. 142–146. Victoria County History, London (1911)
41. Pugh, R.: Lanelly pottery, 1840–1922. In: Gray, J. (ed.) Welsh Ceramics in Context, Part II, pp. 257–271. Royal Institution of South Wales, Swansea (2005)
42. Potter, J.: The Limehouse story. In: Barker, D., Cole, S. (eds.) Digging for Early Porcelain, pp. 40–53. City Museum and Art Gallery, Stoke-on Trent (1998)
43. Ramsay, W.R.H., Gabszewicz, A., Ramsay, E.G.: The chemistry of A-marked porcelains and their relationship to the Heylyn and Frye patent of 1744. Trans. Engl. Ceram. Circle. **18**, 264–283 (2003)
44. Ramsay, W.R.H., Ramsay, E.G.: A case for the production of the earliest commercial hard paste porcelains in the English-speaking world by Edward Heylyn and Thomas Frye in about 1743. Proc. R. Soc. Vic. **120**, 236–256 (2008)
45. Ramsay, W.R.H., Daniels, P., Ramsay, E.G.: The Limehouse Porcelain Factory, its Output, Antecedents and the Influence of the Royal Society of London on the Evolution of English Porcelains Based on Composition and Technology, Invercargill, New Zealand (2013)
46. Ramsay, W.R.H., Daniels, P., Ramsay, E.G.: Limehouse Porcelain: Are Limehouse Porcelains in Fact All Limehouse ? Evidence from Archaeology, Science, and Historical Documents. Resurgat Publishers, Oxford (2015)
47. Ramsay, W.H.R., Hansen, J.A., Ramsay, E.G.: An A-marked covered porcelain bowl, Cherokee clay and the colonial American contribution to the English porcelain industry. In: *Ceramics in America*, pp. 60–77. Chipstone Foundation (2004)
48. Ramsay, W.H.R., Hill, G.R., Ramsay, E.G.: Re-creation of the 1744 Heylyn and fry ceramic patent wares using Cherokee clay: implications for raw materials, kiln conditions and the earliest English porcelain production. Geoarchaeology. **19**, 635–655 (2004)
49. Ramsay, W.R.H., Ramsay, E.G.: The Evolution and Compositional Development of English Porcelains from the 16th Century to Lund's Bristol c. 1750 and Worcester c. 1752- the Golden Chain. Invercargill Press, Invercargill (2017)
50. Ramsay, W.R.H., Edwards, H.G.M., Manners, E., Howkins, A.: Geochemical investigation of a ceramic snuff box - body, glaze and enamels: A-marked porcelain attribution confirmed. In: Ceramics in America, pp. 91–112. Chipstone Foundation, Milwaukee (2023)
51. Ramsay, W.R.H., Edwards, H.G.M.: New insights into the porcelains of William Reid & Co., Liverpool, and their prior links to London, *ca.* 1744–1748. North. Ceram. Soc. J., 101–135 (2024)
52. Rock, J.: A-marked porcelain: further evidence for the Scottish option. Trans. Engl. Ceram. Circle. **17**(1), 474–499 (2001)
53. Severne McKenna, F.: Chelsea Porcelain: the Triangle and Raised Anchor Wares. F. Lewis Publishers, Leigh-on-Sea (1948)
54. Severne McKenna, F.: Chelsea Porcelain: The Red Anchor Wares. F. Lewis Publishers, Leigh-on-Sea (1951)
55. Severne McKenna, F.: Chelsea Porcelain: The Gold Anchor Wares. F. Lewis Publishers, Leigh-on Sea (1952)
56. Tite, M.S., Bimson, M.: A technological study of English porcelains. Archaeometry. **33**, 3–27 (1991)
57. Tyler, K., Stephenson, R., Owen, J.V., Philpotts, C.: The Limehouse Porcelain Manufactory: Excavations at 108-116 Narrow Street, London, 1990 Museum of London Archaeological Service, Monograph No. 6. Museum of London Archaeological Service (2000)

58. Valpy, N.: A-marked porcelain—A for Argyll? Trans. Engl. Ceram. Circle. **13**, 98. and ff. (1987)
59. Watney, B.M.: Limehouse, its relationship to Newcastle-under-Lyme (Pomona) and other manufactories. In: Dakard, D. (ed.) Limehouse Ware Revealed. English Ceramic Circle, London (1993)
60. Whiter, L.: Spode: A History of the Family, Factory and Wares from 1733–1833. Barrie & Jenkins, London (1970)
61. Wills, G.: The Bow China factory and Edward Heylyn. Connoisseur. **132**, 100–102 (1954)

Chapter 16
The Quaker Influence on the Development of Industries Related to Ceramic Production

Abstract A history of Quakerism and the role of the Quakers in British industrial development from the seventeenth Century onwards in many fields of activity such as heavy metal refining, chocolate, foods, and especially porcelain manufacture. Key personalities such as George Fox (the founder of Quakerism), Andrew Duche, William Weston Young, William Cookworthy, Benjamin Lund, Richard Champion and their roles in the development of early English and Welsh porcelains are described.

Keywords Quakers · Andrew Duche · Richard champion · William Weston young · William Cookworthy · Benjamin Lund

16.1 Introduction

Unlike mainland Europe, porcelain manufacturers in Britain did not receive particular support from the King. Thus, whereas King Augustus of Saxony, Empress Catherine II of Russia, and King Louis XIV and Louis XV all gave significant financial support and encouragement to their fledgling porcelain industries, this was not the case in England and all manufactories required the raising of capital from wealthy investors and families to survive. As will be shown, the Quakers played a vital role in providing financial and technical assistance to both the porcelain industry and related industries at large.

16.2 Quaker History

Quakers, like Puritans and other Nonconformists, also known as the Society of Friends, had their origins in the hotbed of European religious dissent in the seventeenth Century and grew directly out of the English Civil War (1640–49)

H. G. M. Edwards, P. Colomban, *Blue by Fire: A Marker of the Technical History of Glass and Ceramics*, Cultural Heritage Science,
https://doi.org/10.1007/978-3-031-77122-4_16

(Coffey, *The Oxford History of Protestant Dissenting Tradition, Volume 1: The Post-Reformation Era, 1559–1689*, [2]). This Protestant Revolution freed up numerous 'ordinary' men and women from the constraints and control of the old feudal monarchical state. The Protestant philosophy of 'individualism', namely that each person could communicate directly with God, without the need for an intermediary State-authorised person, meant that those who accepted this way of thinking were empowered to challenge local or national authority.

Following the Civil War, people in Britain formed into numerous radical Protestant sects and one such radical was George Fox who was born in Fenny Drayton, Leicestershire in 1624, to a Puritan family. When Fox was 19 years old he believed he had received a revelation from God and understood this revelation to be a sign that everyone should be guided by his or her "Inner Light", a concept earlier known as "Christ within". It was in 1647 that Fox began to preach within the Anglican Church and attracted many converts. Fox did not intend to start a new religion, but his followers soon began to group together with such names as Children of Light, Friends of Truth, and later, Society of Friends. In 1669 at the age of 45, George Fox married Margaret Fell who then assisted him in his work. During this period, Fox had been imprisoned in Derby on a charge of blasphemy, and again imprisoned at Nottingham for interrupting a church service. He preached that if you claimed to be acting under the direction of God, "doing God's will" then you could claim to be answerable to a higher power than that represented by your local Lord, Corporation, Magistrate or General. The outcome being, that in 1650, Fox was convicted of blasphemy. At the sentencing Fox advised the judge to tremble at the word of the Lord. The judge then called Fox and his followers "quakers", bestowing upon them the name Quakers by which they are known today. Fox was imprisoned on no less than five occasions (Williams, *The Rich Heritage of Quakerism*, [14]).

Fox harnessed the popular resentment across the country towards ecclesiastical authority and in 1652 the first Quaker community was formed in northern England. Their challenge to the state authority under the Commonwealth of Oliver Cromwell was to lead to numerous Quakers being persecuted because they would not attend places of established worship, pay tithes, or bear arms. Quakers refused to swear oaths, saying that their word alone was good enough and this led to them being open to punitive fines and their members were extremely political and confrontational. With the restoration of the monarchy and King Charles II ascending the throne in 1660 this radicalism began to decline. With the return of the British monarchy in 1660, attempts were made to suppress the changes that had taken place during Cromwell's reign. A series of Acts were passed known as the Clarendon Code which had the effect of severely limiting the freedom of non-conformists, and particularly Quakers. Quakers were now required to take an oath and many Quakers were imprisoned for refusing to comply. Charles Lloyd, for instance, was imprisoned on these grounds for 10 years. In 1661, 4230 or over 10% of the Quaker population, were imprisoned with in a few weeks. When Quakers refused to pay tithes (the religious tax to support the established church) the persecution became even more extensive. Vast numbers had their household goods, farm animals and equipment, and trade tools were confiscated or broken.

Clergy in the Anglican Church of England denounced Quakerism. Gradually over the following decades, Quakers became less political. Increasingly, Quaker meetings came under the 'guidance' of their more conservative, wealthier members. Instead of holding a philosophy of direct action and involvement in political affairs the concentration upon 'Inner Light' and reflection came to the fore. Then, in 1673 the Test Act was passed which obliged all people of England to take Oath and Communion according to the Church of England and this was then followed by a mass round-up of all members of all non-conformist sects.—their meeting rooms closed down and they were pursued and harassed by government officers. Because of this persecution many Quakers migrated to the West Indies and to America. The initial American colonies were formed with the help of William Penn (1644–1718) in the 1660's, particularly in New Jersey, Rhode Island and in Pennsylvania.

Quakers were still barred from professions where oath taking was a condition of entry because Quakers believe that taking oaths sets a double standard of integrity and so they refused to swear. The death of George Fox in 1691 was essentially the end of the "Founding Quakers" and there was no longer the need for pioneers to take on leadership and to inject fervour as in the early days. Now that active persecution had ceased, it was actually an offence to interrupt and interfere with the public worship of others. So Quakers began the process of consolidating and developing their own particular faith and way of life and in this they excelled in support of each other and communities as a whole: it was in this climate that they made an impact upon the industrialisation of Britain.

16.3 Quakers in British Science and Industry

Quakers soon developed a system of apprenticeship where Quaker children stayed with other Quaker families and learned to acquire their technical and business skills. Bankruptcy was considered a reason for disownment or expulsion from the Society, so Quaker Meetings organised assistance for Quakers in financial difficulties to try and avoid failure. If a fellow Quaker was in financial difficulties, finance and partners could be introduced to assist. Thus, a lot of new enterprises were started out of operating surpluses rather than speculative money borrowing. However, as will be seen in the porcelain industry, many Quakers were still made bankrupt but they could still be assisted by other solvent Quakers.

Because they could not secure employment in the public service or attend university, Quakers turned their talents to the world of industry, banking and commerce with great success and some prominent Quakers were at the forefront of the Industrial Revolution which began in Britain in the late eighteenth and early nineteenth Centuries.

Quaker families became the leading entrepreneurs in cotton (John Bright, Samuel Hunter), wool (Barclay, Gurney), safety matches (Bryant, May) coal, brass (Champion), banking and finance (Price Waterhouse,Barclay, Baring, Gurney, Lloyd), railways (Stephenson, Bradshaw) brewing (Burton, Truman), iron (Darby, Lloyds, Tregelles Price), steel (Huntsman, Ransome), clock manmufcature (Tompion, Quare), food (Cadbury, Fry, Huntley & Palmer, Rowntree), porcelain (Cookworthy, Champion, Dillwyn, Young), pottery (Wedgwood), shipping, mining and chemicals (Allen & Hanbury, Crosfield, Reckitt Coleman, Lever). Notable scientists were Young, Eddington, Dalton and Lister. Through what they called their 'Atlantic Community', the Quakers became active traders between England and America and played a significant part in the business enterprises of the fledgling United States. In the United States steel industry, for example, both Bethlehem Steel in Baltimore, Maryland, and Lukens Steel in Coatsville, Pennsylvania, were founded by Quakers. Joseph Wharton, after making his fortune in steel, went on to endow a business school at the University of Pennsylvania. Herbert Hoover (who translated *De Re Metallica* and was the Manager of the Sons of Gwalia gold mine at Leonora, Western Australia) became the thirty-first president of the United States and, Richard Nixon was the thirty-seventh president, both Quakers.

Quakers demonstrated that business relations founded upon honesty and trust encouraged the support of business customers and clients and were an important factor in long term success and stability. The concept of the role of business being the pursuit of profit for the owners to the exclusion of other stakeholders was considered foreign to Quakers. The trades that Quakers pursued had to be "innocent" and not contributing to war or its preparations in any way. Curiously, many were indirectly involved in the slave trade. Although Quakers actively traded with Africa they were at the forefront of the movement to abolish slavery. Quakers saw it as their life's mission to carry out God's work of capitalist expansionism. The new colonies needed commodities in huge quantities and Quakers produced commodities and traded in them. Bristol soon became the second most important port city in England and was a thriving centre for Quakerism.

A notable Quaker ironmaster was Joseph Tregelles Price (1784–1854), whose father helped found the Neath Abbey Ironworks in 1799 and who took over the managing of the operation in 1818, absorbing the Mines Royal Neath Abbey copper smelting works that had moved there in the eighteenth Century from Aberdulais, along with the manufacture of steam engines, locomotives and iron ships: he was motivated to start the first free school for poor children in Neath Abbey (in Welsh, Abaty Nedd) in 1801. The Neath Abbey Ironworks trained many early engineers such as Sir Benjamin Baker, who designed the Forth Bridge, and David Thomas, father of the American iron industry. Two of the impressive, charcoal and coal-fired blast furnaces from the 1790s still remain at the Neath Abbey Ironworks site, some 20 m tall (65 feet), adjacent to Brunel's viaduct that carries the Neath -Swansea main railway line. When Tregelles Price died in 1854 the *Cambrian* newspaper described him as "an indefatigable philanthropist" (Rhys Phillips, *A Romantic Valley in Wales—History of the Vale of Neath,* [12]).

16.4 Quakers in the Porcelain Industry

By the end of the Seven Years War in 1763, Britain was the foremost colonial power in the world, which brought with it the desire to become a manufacturing powerhouse and by the end of the century it was a net exporter of manufactured goods. Early English attempts to manufacture porcelain were centred on the understanding of glass production. Glass (as cullet), silica (sand), alkalis (lime), lead oxide, amongst other raw materials, formed the basis for these early ceramic trials. These glassy bodies were difficult to control during the firing process and items were prone to buckle, sag and melt.

An important aspect of early English porcelain is the part played by Quakers in early porcelain manufacture. Andrew Duché, Edward Heylyn, William Cookworthy, Benjamin Lund, Richard and Joseph Holdship and Josiah Wedgwood, were all members of the Quaker fraternity and they all knew each other. Typically, they were merchants (Heylyn and Lund—copper and brass), a chemist (Cookworthy), and a glover (Richard Holdship). It is reasonable to assume that they would have discussed their common interest in porcelain manufacture with each other. Nicholas Crisp, a jeweller and merchant, is known to have had a working relationship with William Cookworthy and is also reported to have known Emmanuel Mendes de Costa the author of "*A Natural History of Fossils*" [11]. Sometime prior to 1753 he gave Thomas Pennant a complete collection of steatites and soapstone specimens (R. [10].)

16.4.1 Bow

Edward Heylyn has been described as a Quaker, copper merchant, clothier, and a burgess of Bristol. He was one of the founding partners of the Bow manufactory along with Thomas Frye, an Irish portrait painter and engraver. Heylyn is said to have had a glass works at Bow, Middlesex. The Bow manufactory was originally supported financially by George Arnold, a wealthy Quaker haberdasher and London alderman. In 1744 a patent was filed for porcelain manufacture using "an earth found on the land of the Chirokee nation in America and called by the Chirokee 'unaker'" and this was granted to Thomas Frye and Edward Heylyn for a porcelain formula based upon a white clay shipped from America, probably by Andrew Duché. It has been reported that Edward Heylyn's brother had interests in North Carolina and this would quite possibly have been the link between Heylyn and Duché. It has been established that wares made under this patent were the "A" marked group of wares, the "A" being, possibly named after their financial backer, George Arnold This led to the foundation of the Bow porcelain factory at Stratford-le-Bow in East London.

Later, in November 1749 a second patent was applied for by Frye, who was then the manager of the Bow works and who remained in this position until 1759, for a porcelain formula containing calcined bone ash (phosphatic porcelain). It can definitively be ascribed to Bow that they were the first factory in England to produce porcelain comparable with that being produced in China. Recent evidence indicates that Bow also experimented with soaprock in their experimental trials which clearly began in the 1730s, much earlier than the original patent of Heylyn and Frye would apparently indicate.

16.4.2 Chelsea

Chelsea, founded in 1743/4 by Nicholas Sprimont at Church Lane, London, manufactured porcelain from the late 1740s (*ca.* 1745–1770). Along with Limehouse (London, *ca.* 1745–1748) came porcelain works at Pomona Inn, Newcastle-under-Lyme (*ca.* 1744–54), the 'Girl-in-a-Swing' factory, St. James, London, run by Charles Gouyn, then Lund's Bristol in 1748/9 (*ca.* 1748–1751).

16.4.3 Bristol and Worcester

Benjamin Lund, a Quaker, married Christobel Ingram, daughter of Robert Ingram of London, at the Friends Meeting House, Bristol. The wedding was witnessed and the marriage register was signed by some of the Champion, Hartford, Lloyd and Scandrett prominent Quaker families. On February 3, 1728, Benjamin Lund and Francis Hawksbee, of London took out a patent for "manufacturing copper, extracting silver therefrom, making brass to be cast into plates, etc., and a new way of mixing calamy and charcoal and making thereof brass without pans or pots". The Lund's Bristol porcelain works was established in about 1749, the proprietors being Benjamin Lund and William Miller: they manufactured soft paste porcelain until 1752 when the operation was acquired by Dr. John Wall and William Davis of the Worcester porcelain manufactory. Miller, who was a wealthy merchant, banker and grocer probably funded the operation because Lund at that time was bankrupt since the eighth December 1741. Lund had been granted a licence to mine Cornish soaprock on seventh March 1749, at Gewgraze in the parish of Mullion. The site of this factory is unknown, but is assumed to be in the Redcliffe Backs district of Bristol. This early soft paste porcelain is not to be confused with the later factory manufacturing hard paste porcelain established in Briston by William Cookworthy and Richard Champion, who moved the manufactory there from Plymouth in 1770.

The most commonly held view is that Benjamin Lund and William Miller (a banker and a grocer) commenced manufacturing porcelain in 1749. Blue and white wares were almost exclusively produced by this manufactory. Lund was declared bankrupt in 1740 and was still bankrupt in 1753, clearly indicating that he would require the services of a financial partner to enable him to conduct the business. In bankruptcy proceedings dated 23rd February 1755, Benjamin Lund is described as a 'china maker now of the City of Worcester'. A series of advertisements were placed in The Bristol Intelligencer between November 1750 and July 1751 indicating that for some time, porcelain wares of high quality had been produced and the proprietors wished to extend their works by employing boys over the age of 14 to become apprentices and "that they shall learn the Art of Pottery as practiced in Staffordshire .... And in particular, Children of either sex, not under the above Age, may be learned to draw and paint by Persons appointed for that Purpose, that they may be qualified to paint the said Ware ...".

The Richard Holdship bankruptcy proceedings indicate that the sale of the Bristol factory took place on 21st February 1752. These papers confirm that William Miller and Benjamin Lund were the owners of the Bristol works. Possibly, it was the financial pressure associated with being a bankrupt that led Lund to readily dispose of the successful Bristol factory to the Worcester partners when an offer was received from Richard Holdship probably around 1750–51. Richard and Joseph Holdship were both Quakers and were two of the 15 original partners forming the Worcester porcelain manufactory in 1751.

On 24th July 1752 and again on 1st August of that year, it was announced in the *Bristol Intelligencer* that the proprietors of the manufactory "... are now united with the Worcester Porcelain Company where for the future the whole Business will be carried on; therefore the said Proprietors are determined to Sell their Remaining Stock of Ware, very cheap at their Warehouse in Castle Green until the whole is disposed of." Although this notice indicated that the factory was to be transferred to Worcester in 1752, it would appear that much Bristol ware still remained unsold. In January 1755 a further announcement was made announcing that the remaining stock would be sold from the Castle Green warehouse. And, on 31st January 1756, the following notice was placed in the Bath Advertiser: "Bristol, Jan. 24. This is to give Notice, That I have lately purchased a large Assortment of Enamel'd and Blue and White Useful Worcester China Ware, which will be sold, at the Warehouse in Castle-Green, cheaper, better in Quality, and in greater Variety than has hitherto been expos'd to Sale there." The advertisement also indicated that Robert Carpenter was in charge of the Castle Green premises. Later advertisements indicated that the business moved from Castle Green to the Key opposite the Drawbridge, and it is reported that he also had a shop in Temple Street, Bristol on certain occasions. It would appear that items emanating from both Bristol and Worcester production were being sold from the same premises at the same time. For this reason, it is quite possible that the Bristol warehouse would have requested the Worcester factory to mark their blue and white production to differentiate it from Lund's production for ease of identification or stocktaking. The Worcester factory could have initiated the marking of their blue and white porcelain for their own identification purposes, particularly as the moulds produced in Lund's Bristol factory were later used at Worcester.

From the Worcester factory, a number of other enterprises which also included soaprock in their formulations were established, some of which were later acquired by or amalgamated with the Worcester operation. These included the porcelains produced at the Liverpool works established by Richard Chaffers and Philip Christian who relied on the information received from Robert Podmore (an original Worcester employee) in 1755 to incorporate soapstone into their porcelain formulation. It was reported by Josiah Wedgwood that Podmore was his foreman and "he left with the intention of setting up a potworks in America. Mr. Chaffers, a Liverpool manufacturer, had caught the intending emigrant during his pre-embarking stay in Liverpool, and finding that Podmore showed so much intelligence and practical knowledge of the business, had made him sufficiently liberal offers to induce him to remain in England".

16.4.4 Vauxhall

The Vauxhall porcelains produced in London by Nicholas Crisp (*ca.* 1751–1764), a London jeweller who was known to William Cookworthy, and John Sanders who were granted a licence to quarry Cornish soapstone in 1752. Nicholas and his brother Thomas were financially involved in the East India Company and its trading with the Orient and therefore Nicholas was probably the financier of the Vauxhall factory early in its existence.

The English soft paste porcelain, porcelain which is fired at 1100—1150 °C, the formulation of which was originally based on either bone ash or soaprock (steatite) and which contains ingredients otherwise used in glass manufacture is considered by some to not be a 'true' porcelain as defined by the Chinese hard paste version (Brogniart). Steatite was certainly the basis for the porcelain produced at the short-lived production by Lund's Bristol in 1748/49 and then by Worcester for the next 90 years. The use of steatite in porcelain manufacturing was first reported in the 1720s when 'several trials' were attempted. Thus, soaprock had been reported to be able to be used on the production of porcelain in Dr. J. Woodward's "*Natural History of the Fossils of England*"[15]:

> From several Tryals that have been lately made for the baking and making this earth into Pots, I am satisfy'd that 'tis not much inferior to that of which Porcellain is made …

The Rev. William Borlase in his "*Natural History of Cornwall, 1758*" [1], in a number of references concludes: "The principal property of our soap-rock is, that it withstands the fire in a wonderful manner; …[it] prevents vitrification, and makes porcelain ware more tough and tenacious…".

On breakage, these steatitic soft paste porcelains give a granular appearance. When correctly fired, they are generally translucent with a strong greenish tinge when viewed by transmitted light. Unlike Chinese porcelain which was glazed and fired in one operation, these soft paste formulations required separate firings for the frit, the biscuit (body) and for the glaze. Poor temperature-control of the kilns existing at that time resulted in a considerable loss of production—often upwards of 25% or more of the fired artefacts. Soft paste formulations were fired at a lower temperature than those of the hard paste formulae and then were subsequently glazed in an additional firing at an even lower temperature using a lead-based glaze. These glazes tended to be thicker than the feldspar-based glaze used on hard paste porcelain and often resulted in a blurring and softening of surface detail. It did however, provide an excellent base on which coloured enamels could be applied and then subsequently fired, the glaze allowing the enamels to fuse into the surface to provide a coherent finish to the object.

16.4.5 Caughley (Salopian)

Another manufactory emanating from the Worcester pottery was the Caughley or 'Salopian' works established by Ambrose Gallimore further up the River Severn near the town of Broseley in Shropshire. An existing pottery was previously located at the site on the estate of Edward Browne from about 1751. In 1772 Gallimore entered into a partnership with Thomas Turner who is reputed to have worked at Worcester, although this cannot be definitively established. However, Robert Hancock who was the most important early engraver to work for Worcester and who was a partner in the Worcester factory between 1772 and 1774, was at Caughley in 1775. The Caughley factory survived until 1799 when it was sold to the Coalport China Works partners led by John Rose. During its time in production it manufactured what may be described as close copies of the Worcester wares. Worcester could also be accused of copying some of the items produced by these factories. Caughley was the main supplier of china in the white to Robert Chamberlain founder of Chamberlain's Worcester porcelain works, after he left Worcester to commence decoration in his own right. It is believed that Chamberlain's criticism of the diminished quality of Caughley's production in the early 1790s was instrumental in the demise of the factory.

16.4.6 William Cookworthy: Bristol and Plymouth

William Cookworthy was born at Kingsbridge, Devon, on April 11th 1705, the son of a weaver. His father died in 1718 and the family was quickly reduced to a state of poverty. At the age of 14 he was apprenticed to Timothy and Silvanus Bevan, both Quakers, who were chemists and druggists in Cheapside, London. It seems to have been a good apprenticeship, for he was taught Latin, French and Greek in addition to drug dispensing.

When he had completed his training in 1726, the Bevans' offered him a position in a new wholesale pharmacy business that they were setting up in Plymouth. By 1735 he had become a partner and the business was renamed Bevan and Cookworthy. In that year he married Sarah Berry. Ten years later he bought out his partners but Sarah's unexpected death that year left him with a family to bring up on his own.

His brother, Philip, joined him as a partner and the business became Messrs. William Cookworthy and Company. His youngest brother, Benjamin, married into another Plymouth family, the Colliers. William continued to practice as a Quaker and became a minister in the Plymouth Society of Friends. Because Cookworthy spoke several languages he met with many foreign dignitaries when they visited Plymouth. He entertained Captain Cook, Sir Joseph Banks and Daniel Solander to discuss with them their plans for the voyage in 1769 in the Endeavour to Otaheite in the South Seas for the observation of the transit of Venus. He similarly entertained John Smeaton and give him practical assistance in the erection of the Eddystone Lighthouse.

Cookworthy is said to have read a copy of Pere Du Halde's translation of the Jesuit missionary Francois Xavier d'Entrecolles relating to the methods for the manufacture of Chinese porcelain which was published in London between 1738 and 1741 [6]. Andrew Duché, a Quaker from America, having visited Heylyn and Frye at Bow in 1745 also later in the same year visited William Cookworthy at Plymouth. Cookworthy then wrote to a Mr. Hingston on May 30, 1745, saying that kaolin and petuntse had been discovered in America, and that he had seen specimens said to have been manufactured from the American materials. One letter of his on the subject states: "I had lately with me the person who hath discovered the china-earth. He had with him several samples of the china ware of their making which I think were equal to the Asiatic. 'Twas found on the back of Virginia, where he was in quest of mines, and having read Du Halde, he discovered both the petuntse and the kaolin. 'Twas this latter earth which he says is essential to the success of the porcelain manufacture. He is going for a cargo of it, having bought from the Indians the whole country where it rises. They can import it for £13 per ton, and by that means afford their china as cheap as common stone ware. The man is a Quaker by profession, but seems to be as thorough a Deist as I ever met with".

When Cookworthy was visiting the Great Work Mine in Cornwall, he noticed that the miners were repairing furnaces with clay. He enquired where it came from and took some samples back to Plymouth with him. With this information, Cookworthy looked for suitable clays and in 1746 he discovered china clay close by at Tregonning Hill in the parish of Germo, Cornwall, where it was known as "moorstone". He found it was good for making porcelain and leased several clay pits on the Hill. The clay was shipped from Porthleven to Plymouth, where he carried out experiments to find the best way to process, glaze and fire the artefacts. It was not until 1756 that Cookworthy also discovered nearby china stone or petuntse, a feldspar-like material which is mixed with kaolin for the manufacture of true hard paste porcelain. By December 1766 he had set up a small factory in Plymouth in which he manufactured items as a trial. On March 17th 1768 Cookworthy obtained a patent for "Making porcelain from Moorstone, Growan and Growan Clay". The patent gave him the exclusive right to use china clay and china stone for porcelain manufacture. Better quality deposits were later found by Cookworthy on land owned by Mr. Thomas Pitt in the parish of St Stephen's, near St Austell, and this quickly replaced that which was acquired from Tregonning Hill. Mr. Pitt, who was later to become Lord Camelford in 1784, invested money in the process and with William Cookworthy set up the Plymouth China Works in Plymouth. The business was not very profitable and it was soon amalgamated with one at Bristol in the name of Messrs. William Cookworthy and Company [7]. William Cookworthy died on Tuesday October 17th 1780 and was buried in the family vault in the Westwell Street Burial Ground. The earliest known extant piece of Cookworthy's hard-paste porcelain is now in the British Museum; a blue decorated mug bearing the Arms of Plymouth and the inscription "14 March 1768 C.F."—presumably the initials mean "Cookworthy fecit" (Cookworthy made this).

16.4.7 Richard Champion: Plymouth and New Hall

Richard Champion (1743–1791) was born into an influential Bristol Quaker family on 6th November 1743. A distant relative, William Champion, invented a new method of smelting zinc. He was treasurer of the infirmary from 1768–79, a position which other family members had occupied from 1740–65. Richard became a merchant, trading with America and the West Indies, joining the Society of Merchants in 1766, and becoming a warden of the Society in 1772. He became a burgess of the city of Bristol on 20th April 1765. In 1764 he also met a fellow Quaker, William Cookworthy, and later they worked together in the manufacture of porcelain. A partnership with James Brice and Joseph Harford in shipping was dissolved in 1769 and after that he traded alone. By 1775 he probably owned at least six ships. In 1765, his brother-in-law Caleb Lloyd, a Quaker from Charleston, South Carolina, sent him a "box of porcelain-earth" from the internal parts of the Cherokee nation, four hundred miles from Charleston on mountains scarcely "accessible" together with another box of clay for a relative of Richard Champion, the Earl of Hyndford, who desired Champion to make experiments with it. Champion and Cookworthy are said to have experimented unsuccessfully "on the principle of Chinese porcelain" having decided to use the clay that they had discovered in Cornwall, which was "not so fine as the Cherokee; however, there can be no chance of introducing the latter as a manufacture when it can be so easily procured from Cornwall". Cookworthy also supplied Cornish materials to the Bristol venture which led to nothing due to discolouration upon firing.

In 1768 Richard Champion invested in Cookworthy's Plymouth china venture, with borrowed money. After the works were transferred to Bristol, in 1770, he had some involvement, but it would be wrong to call him the manager, which was a post held by John Britain. The first apprentices, in 1772, were assigned to Richard and Judith Champion. Cookworthy transferred the patent to him in September 1773, although the deed was not executed until May 1774, leaving Champion as the proprietor. This agreement, however, left Champion tied to a 99 year lease for the raw materials at twice their true cost, although these costs were only a small part of manufacture. It was to Richard Champion that Cookworthy sold his interest in the

Plymouth porcelain company when he decided to retire in 1773/74, the Company then changing its name to Richard Champion & Co. There is some evidence that the Quaker, Joseph Fry, founder of Fry's chocolates, had a financial interest in the business. After the sale of the business to Richard Champion, Cookworthy continued to receive a royalty on every ton of clay sold. When Richard tried to renew Cookworthy's patent in 1777, Josiah Wedgwood and other potters in Staffordshire raised objections and although the patent formula was upheld, the actual use of the china clay was released to enable the manufacture of other ceramic products provided that the formula was not infringed. The cost of the legal battle unfortunately crippled the company and Richard sold the formula in 1782 to the New Hall Porcelain Company, which had been formed by the Staffordshire potters. Champion left Bristol on 5th November 1781 to reside at Merril Street, Newcastle, Staffordshire where he acted as manager of the New Hall porcelain manufactory. The porcelain is not true hard paste, but of a hybrid type and cheaper to produce.. By the late 1790s other potters, such as Coalport, began making hybrid hard paste porcelain. About 1800 Josiah Spode developed an improved hybrid paste incorporating bone ash which he called his "Staffordshire china"— and later became known as English bone china. Champion left Staffordshire on 8th April 1782 and settled in Camden, South Carolina, where he had a plantation. He died on 7th October 1791, his wife having passed away in the previous year.

16.4.8 William "Quaker" Pegg and Derby

William "Quaker" Pegg is regarded as being one of the finest flower painters to have ever worked in the ceramics industry (Twitchett, *Derby Porcelain, 1748–1848*, [13]). Born near Newcastle- under -Lyme in 1775, the son of a gardener, William Pegg worked in a pottery from the age of ten and by the age of 13, was already an accomplished self- taught flower painter on earthenware bodies. At the age of 15 William Pegg started an apprenticeship as a china painter at one of the Staffordshire potteries. The hours were long and hard. He wrote that he worked up to 15 h a day at the china factory in addition to his studies and wrote that he found it a punishing schedule but his efforts served him well later.

In 1796, at the age of age 21, William Duesbury II hired him at the Nottingham Road, Derby China Works to succeed the innovative and well-respected William Billingsley as chief flower painter who had just departed to set up the Pinxton China Works with John Coke (Twitchett, *Derby Porcelain, 1748–1848*, [13]; John, *William Billingsley, 1758–1828,* [8]). William Pegg spent two periods at the factory, the first from 1796–1801. After being at the factory for 7 years on this second occasion in 1820, at the age of 45, William left the profession and opened a small shop at his home where he worked with his wife for the rest of his life living with deteriorating sight. He died on December 27, 1851 aged 76.

16.4.9 *Josiah Wedgwood*

Josiah Wedgwood was born into a family of potters on 12 July 1730 at Burslem, Staffordshire. His father's death in 1739 led him to an early start working as a 'thrower' in the pottery of his eldest brother, Thomas, to whom he was later apprenticed. Thomas refused Josiah a partnership in the business, so the younger man moved first to a small pottery run by John Harrison, then more happily to the firm of Thomas Wheildon of Fenton. From there, he opened works of his own, first at his cousin's Ivy House and later at the Brick House factory. At these works, Wedgwood made many models himself, and also prepared clay mixes. In June 1769, he opened a new factory at Etruria, near Stoke-on-Trent, in partnership with Thomas Bentley. Attached to the factory was a village where his workmen and their families could live in decent surroundings. Wedgwood greatly improved the ordinary earthenwares of the day, introducing durable, simple and regular artefacts His cream coloured earthenware was named 'Queen's Ware' after Queen Charlotte, who appointed him Queen's potter in 1762. Other eminent patrons included Empress Catherine II of Russia, who ordered 952 such pieces in 1774 for her Chesma Palace *"Frog"* service (Denyer, Denyer & Edwards, *Armorial Porcelain: The Genesis*, [3]). Wedgwood experimented with the incorporation of barium sulphate into his paste and from it produced his famous jasperwares in 1773. Jasperware, is used for a host of ornaments, blends metallic oxides, often with different shades of blue with separately moulded reliefs in white cameo. Wedgwood realised the importance of preserving secrecy about his recipe and materials so he adopted a handwritten code for his trial experiments which is now preserved in the V&A Museum, London (MS E26-19117): Rebecca Klarner has written a compelling account and reproduced this code [9] in which in a letter to Thomas Bentley in 1776 Wedgwood refers to the ingredient barium sulfate for his jasperware as "cawk" and numbered it 74 in code. In all there are 97 such numbered codes ranging frm 1, which was borax, to 97, which was Newhaven clay. Some such cameo reliefs were designed for Wedgwood by John Flaxman. Other wares included black basalts, frequently enhanced by 'encaustic' colours such as red, to imitate Greek terracotta vases. Wedgwood was elected a Fellow of the Royal Society in 1783, primarily for inventing a much-needed pyrometer to measure oven and kiln temperatures. He took a keen interest in efficient factory organisation, and in improving the transport of raw materials and finished wares by canals, such as the Grand Trunk Canal, and by road. His daughter Susannah was the mother of Charles Darwin.

16.4.10 *William Weston Young and Lewis Weston Dillwyn: Swansea and Nantgarw*

William Weston Young, born on 20 April 1776, Lewin's Mead, Bristol, England, into a devout Quaker family, the third son of Edward Young, a Bristol merchant and Sarah (Sally) Young (née Weston). He was educated at Gildersome Quaker boarding school in Yorkshire, which among other things gave him a rudimentary knowledge of science which he was to apply effectively later in his invention of the silica fire-brick which revolutionised industrial high temperature smelting furnaces. He was also an accomplished artist, botanist, wreck-raiser, surveyor and ceramicist.

After an abortive attempt to migrate to America, William Weston Young settled in Bristol, found employment and married fellow Quaker, Elizabeth Davis in April 1795. In 1798, Young had acquired the financial backing to lease a farm and water mill from John Llewellyn of Ynysygerwn, at Aberdulais, in the Neath Valley, Glamorganshire. After a lucrative start to his new venture as a miller, corn-factor and farmer, he made a large purchase of corn, beans and grist which coincided with the Treaty of Amiens that severely diminished the value of his goods and he was made bankrupt in May 1802.Young's early bankruptcy had a lasting impact upon his career, being no longer independent—he sought employment and so put his artistic skills to use. On 23 January 1803, Young and his wife moved to new lodgings in Swansea, Glamorganshire, where he had gained employment under fellow Quaker, Lewis Weston Dillwyn, as a draftsman at Dillwyn's Cambrian Pottery, where he remained until August 1806. Dillwyn and Young, both in their mid- twenties, struck up a close friendship due to their common interest in natural history. Many of Young's painted wares feature accurately depicted flora and fauna as well as the taxonomic names of the illustrated species (Edwards, *From Porcelain to Silica Bricks: The Extreme Ceramics of William Weston Young, 1776–1847,* [5]).

During this period, Young shared his time between decorating the Cambrian Pottery's best creamwares and assisting with Dillwyn's botanical field work. Between 1802–1814, Dillwyn worked on a groundbreaking study of British algae species, The British Confervae. Young's illustrative plates are to be found in Dillwyn's publications (Dillwyn, *The British Confervae*, [4]). In these publications, Dillwyn credits Young with the discovery of *Conferva dissiliens* in Crymlyn Bog, Swansea, and *Conferva youngana* in the limestone rock pools near Dunraven Castle, Glamorganshire.

Young's discoveries and collaboration on the work with Dillwyn earned him Associate Membership of the Linnaean Society. In 1806 Young conceived of an improved "grab" or "forceps" mechanism to be used in wreck-raising and set about a wreck-raising business, retrieving sunken vessels in the Bristol Channel. His first commission to raise the freight ship '*Anne and Teresa*', salvaged a cargo of copper therefrom, making him enough money to establish himself comfortably in the village of Newton Nottage, Glamorganshire as a wreck-raiser, merchant and farmer. In 1811, the death of established local surveyor John Williams of Newland, near Margam, Glamorgan, enabled Young to add the profession of land surveyor to his list of occupations to fill the niche that Newland left in the region.

It was during his work as a land surveyor, that Young, an amateur geologist, discovered the potential of a limestone found at Mumbles, Swansea, Glamorganshire being fashioned as marble. In 1814, Thomas Mansel Talbot (1747–1814), father of Christopher Rice Mansel Talbot (1803–1890), died at Penrice, Gower, Glamorganshire and Young was commissioned to design Talbot's tomb using locally-sourced materials. The tomb is a large and elaborate edifice, deploying

Penrice alabaster and Mumbles marble, and took 6 years of design and modelling before its completion in February 1820 in the nave of Margam Abbey Church, Margam, Glamorganshire. In early 1814, Young became the major investor in William Billingsley and Samuel Walker's venture at the Nantgarw Pottery, Nantgarw, Glamorganshire. Billingsley and Walker, through their prior ties with Flight, Barr & Barr at Worcester had signed an agreement not to disclose their new porcelain recipe to a third party, but there was no clause preventing them from using that recipe formulation themselves. The manufactory was set up and produced beautiful quality translucent porcelain but 90% of the of the artefacts were ruined upon firing. The resources of the three associates soon ran out and the group approached the British Government's Committee of Trade and Plantations asking for a grant of £500, referring to the subsidy the French Government had given the contemporary Sèvres porcelain factory. They were not successful, but one member of the committee and a porcelain enthusiast, Sir Joseph Banks, suggested to his friend and ceramicist Lewis Weston Dillwyn of the Cambrian Pottery of Swansea, Glamorganshire, that he should investigate and report on the matter.

Dillwyn made the inspection, and evaluated the extent of the firm's losses, but was so impressed with the quality of the surviving pieces that he offered Billingsley and Walker employment at the Cambrian Pottery. An annexe was built for porcelain production at the Cambrian Pottery, namely the Swansea China Works, where Walker and Billingsley were based from late 1814. Billingsley's porcelain recipe was modified and improved, and the Swansea duck-egg china was produced to great acclaim. Dillwyn still tried to improve the robustness of his process and output with a series of experimental trials and modifications, realising his trident soaprock paste version in 1817. That year Billingsley and Walker returned to start up again Phase II at Nantgarw wherein Young reinvested £1100 from his wreck-raising business in the Bristol Channel, He additionally became an art teacher at Cowbridge Free School to supplement his income He also succeeded in gaining the financial support of "ten gentlemen of Glamorgan" whom he had met during his land surveying activities for an additional £1000. Billingsley and Walker continued to fire their porcelain, at a significantly heavy loss however, until April 1820, when they departed for Coalport, leaving behind them the lease to the pottery and several thousand pieces of undecorated porcelain in various stages of production which were taken up by Young.

Young put the Nantgarw Pottery and its contents up for sale via public auction in October 1820, enabling him to buy-out his minor partners and become sole proprietor. He invited his friend and former co-worker from the Cambrian Pottery, Thomas Pardoe, to aid in the decoration of the porcelain stock. Young and Pardoe experimented to perfect a glaze for the biscuit ware, but were unable to ever add to Billingsley's stockpile of porcelain, having no knowledge of the recipe that Billingsley had kept secret. Despite this, Young still attempted to recreate Nantgarw porcelain but never realised the vital role played by Billingsley's calcined bone ash constituent and in this he was commercially unsuccessful and the Nantgarw China Works eventually closed down. The final sales of the finished porcelain that was sold between 1821 and 1822 paid Pardoe and his staff's salaries in arrears, but failed to recoup Young's total losses overall, leaving Young narrowly avoiding a further bankruptcy. Thomas Pardoe attempted to keep it going after Young's departure in 1822 but he died in 1823. (Edwards, *Porcelain to Silica Bricks: the Extreme Ceramics of William Weston Young*, 1776–1856, [5]). In 1833 the Nantgarw estate was sold to William Henry Pardoe, son of Thomas Pardoe. Young's experience of firing ceramics, together with his familiarity with the region as a local surveyor and his amateur interests in geology enabled him to conceive of a heat-proof, blast-furnace brick, using silica found in large deposits at the head of the Neath Valley at Dinas Rock. The process of "vitrifying" the walls of a ceramic brick-built furnace had been patented by William Harry, of the Swansea Valley in 1817, but Young's solution was to build the whole furnace from a "silica firebrick," made with a 1% addition of lime, to bind the blue-grey "clay" of the Dinas Rock silica. The idea being that the interior of the blast furnace would vitrify and be vastly more durable and ultimately more economical than a mere veneer of silica within a comparatively fragile ceramic shell. Young made early experiments with the recipe and fired his trial bricks at the Nantgarw Pottery kilns, while he and Pardoe finished the Billingsley porcelain for sale between 1820 and 1822, when he had finalised his recipe.

In 1822, Young applied to the Marquis of Bute to lease the lands near Craig-y-Dinas, Pontneddfechan, in the upper Neath Valley for a period of 21 years. Young had the lease, and the patent (No. 5047) but had no funds left to set-up the required brickworks. He sought financial backing from a number of sources, including his extended family once more and on 19 October 1822, the Dinas Fire Brick Co. was established in a partnership involving David Morgan, a Neath Ironmonger, John Player and Joseph Young (William Weston Young's older brother). W.W. Young was a party but could not be a partner in the final enterprise, owing to his previous bankruptcy in 1802 at Aberdulais watermill. A brickworks was built at Pontwalby, about a mile down river from Craig-y-Dinas. The company, which sold bricks to industry across the world, transferred through many hands, but the Young family, held their shares throughout, finally passing via Joseph Young to his son William Weston Young Junior (1798–1866). From "The Dinas Firebrick Co." to "John Player & Co." in 1825, to "Riddles, Young & Co." in 1829 and finally, becoming world famous as "Young & Allen". In 1852, the company brochure later mentions that it had supplied firebricks to Swansea's White Rock Copper Works for 40 years.

16.4.11 Quakers in Early American Trade

Admiral Sir William Penn (1621–1670), a Quaker, was born, married and buried in Bristol. He was Cromwell's Sea General who was responsible, with General Venables, for the British capture of Jamaica in 1655 which became the base for British expansion in the West Atlantic. William Penn, the Admiral's son, was to take this further with his proprietorship of Pennsylvania. William Penn (1644–1718), a devout Quaker, went on to become one of the largest private landowners that the world has ever seen when King Charles II gifted him with 40,000 square miles of Pennsylvania from which, using the Quaker West Indian network, Penn was ideally placed to take the advantages offered him and his associates by trade through Europe, Africa and America (called the "Triangular Trade").

English potters expressed concern over the American competition in the manufacture of ceramics. John Bartlam (1735–1781) a Quaker, established a pottery to manufacture "Queens Ware" in Cain Hoy, South Carolina in 1765. He was the subject of frequent tirades from Josiah Wedgwood who was infuriated over the defection of his workmen to Bartlam's works and because of Bartlam's attempts to produce English tablewares. In 1771, Bartlam moved his pottery to Camden, South Carolina, in the very heart of the clay supply. This seriously alarmed Josiah Wedgwood as he had once before lost his foreman, Robert Podmore, who had left him with the intention of establishing pot-works in America. Wedgwood wrote to his patron, Sir W. Meredith: "Permit me, Sir, to mention a circumstance of a more public nature, which greatly alarms us in this neighbourhood. The bulk of our particular manufactures are, you know, exported to foreign markets for our home consumption is very trifling in comparison to what is sent abroad and the principal of these markets are the Continent and Islands of North America. To the continent we send an amazing quantity of white stone ware and some of the finer kinds, but for the islands we cannot make anything too rich and costly. This trade to our colonies we are apprehensive of losing in a few years, as they have set on foot some pot-works there already, and are at this time amongst us hiring a number of our hands for establishing new pot-works in South Carolina, having got one of our insolvent master-potters (Bartlam) there to conduct them. They have every material there, equal if not superior to our own, for carrying on that manufacture; and as the necessaries of life and consequently the prices of labour amongst us are daily advancing, it is highly probable that more will follow them and join their brother artists and manufacturers of all classes who are from all quarters taking a rapid flight indeed the same way".

Recovering quickly from the French and Indian Wars, the American colonies had enjoyed a rising prosperity, an established society, and a high standard of living. Philadelphia in 1770 was a rich and fashionable centre ideal to support a porcelain factory. "The China-works now erecting in Southwark [1 January 1770] was compleated, and in motion" the following July, 1770, and was lauded in newspaper advertisements.The Philadelphia China Works proprietors were Gouse Bonnin (from Antigua) and a Philadelphia Quaker, George Anthony Morris. The latter retired in April/May 1771 and moved to North Carolina, where he died 2 years later while Bonnin in November 1772 was "embarking for England without the least prospect of ever returning to this continent". They were financed by a £500 advance from the father of Dr. James Mease, who got nothing in return but a blue-painted dinner service, from which one basket in the Worcester manner is all that survives and is now in the Philadelphia Museum. This piece and four others, all with a factory mark P in blue, were the 'known' output of Bonnin & Morris. In 1766 the English Society for the Encouragement of Arts, Manufactures, and Commerce gave a gold medal to Samuel Bowen, with the inscription that it was given to him "for his useful observations in china and industrious application of them in Georgia". It was doubtless the industrious Mr. Bowen's china that was referred to in Felix Farley's Bristol Journal, in the issue of November 24, 1764: "This week some pieces of porcelain manufactured in Georgia was imported; the materials appear to be good, but the workmanship is far from being admired". Though this china venture was of enough importance to provide a medal for Bowen, all traces of its location, progress, and fate have been lost.

Andrew Duché, was the third son of the stoneware potter Antoine (Anthony) Duché (1682–1762) who had established a "Pott-House" in Philadelphia by 1725. Born in 1710 in Philadelphia, he married in 1731, worked first at Charleston (1731–5) where he immediately began searching for clay and establishing a kiln. On 5 April, 1735 he advertised in Charlton's South Carolina Gazette "butter pots, milk pans, and all other sorts of Earthen ware ….. at much cheaper rate than can be imported". He later moved to New Windsor (1735–7) across the river from Augusta and finally at Savannah (1738–43), where he had been assured that "all reasonable encouragement" would be given him by General James Oglethorpe, founder in 1733 of the colony of Georgia. Indeed, he received a grant of £230 and built a pottery in 1743 where, according to local records, he "found out the secret to make as good porcelain as is made in China". It should be noted that in August 1738 samples of Chinese porcelain had been sent by the Earl of Egmont, the most active of the Trustees of the colony of Georgia, to a certain master potter in Savannah. These samples were to serve as models for Andrew Duché. Egmont, in his diary described Duché as "a worse believer even than a Deist" as did Cookworthy. As for his porcelains, Oglethorpe in 1738 had already reported to the Trustees that Duché had found "an earth" (kaolin, china-clay) and "had baked it into china". By February of the next year he had discovered "a whole mountain of stone" (petuntse) and, in 1740 Duché found "a quarry of Ironstone" on the five acre lot of William Gough. Whilst conducting his experiments to perfect porcelain, Duché supplied the vicinity with useful articles of common earthenware or ironstone, and stove tiles for the settlement forty miles inland. On 17 March 1738, he requested "two ingenious pot painters", and special supplies which were sent to him in August. Duché therefore had all the requirements for blue-decorated porcelain, and skilled helpers to finish it. Drawn into political squabbles, Duché came into disagreement with Colonel William Stephens, who was secretary to the Trustees of the colony of Georgia. Ostensibly to plead the cause of the dissatisfied settlers, he left Savannah in March 1743 and appeared in London in May 1744. He was evidently in contact with Edward Heylyn and Thomas Frye, the proprietors of the Bow factory supplying "an earth, the produce of the Chirokee nation in America, called by the natives unaker". Later in 1744, Duché visited William Cookworthy at Plymouth, who in a letter dated 30 May 1745 discusses "the person who has discovered the china earth, calling it kaulin, and saying that the finder is going for a Cargo of it". Cookworthy has seen "several samples of the china-ware of their making", and understands that the requisite earth is to be found "on the back of Virginia". No more is heard of Duché's activities as a potter and from 1750 to 1769 he is described as a 'merchant' and prosperous landowner in Norfolk, Virginia. Then, in 1769 he returned to Philadelphia, where he died in 1778.

16.5 Conclusions

The Quaker (Society of Friends) movement, after suffering persecution in Britain for many years, prospered during the late seventeenth and eighteenth centuries and fostered many industrial operations. Their endeavours in a wide range of manufacturing processes led to the founding of many businesses which continue to this day. Their influence in Britain and in America in porcelain manufacture was significant, with many of the factories initially established from the 1750s onwards relying on the skill and dedication of Quakers for their success.

References

1. Borlase, W.: The Natural History of Cornwall: the Air, Climate, Waters, Rivers, Lakes, Sea and Tides ... of the Inhabitants, Their Manners, Customs, Playes or Interludes, Exercise and Festivals. Their Common Language, Trade, Leisures and Crafts. W. Jackson, Oxford/London (1758)
2. Coffey, J. (ed.): The Oxford History of Protestant Dissenting Traditions, Volume 1: the Post-Reformation Era, 1559–1689. Oxford University Press, Oxford (2020)
3. Denyer, R., Denyer, M., Edwards, H.G.M.: Armorial Porcelain: the Genesis. Springer- Nature Publishing, Cham (2024)
4. Dillwyn, L.W.: British Confervae or Coloured Figures and Descriptions of the British Plants Referred by Botanists to the Genus Conferva. W. Phillips, Lombard Street, London (1809)
5. Edwards, H.G.M.: Porcelain to Silica Bricks: The Extreme Ceramics of William Weston Young, 1776–1847. Springer -Nature, Cham (2019)
6. du Halde, J.B.: A Description of the Empire of China and Chinese -Tartary together with the Kingdom of Korea and Tibet: Containing the Geography and History (Natural as well as Civil) of Those Countries, 2 Volumes, T. Gardner for Edward Cave, London, (1738–1741). (Translated from the French version published by Le Mercier, Paris in 1735)
7. Jackson, R., Jackson, P., Price, A.: Bristol pottery and potteries, 1600–1800. J Ceram Hist. **12**, 39–41 (1982)
8. John, W.D.: William Billingsley, 1758–1828, his Outstanding Achievements as an Artist and Porcelain Manufacturer. The Ceramic Book Company, Newport (1968)
9. Klarner, R.: More than just numbers: Josiah Wedgwood's ceramic trials. In: Brownsword, N. (ed.) Alchemy and Metamorphosis, pp. 40–59. Topographies of the Obsolete Publications, University of Bergen, and Staffordshire University Stoke-on-Trent, Norway (2021)
10. R. Massey, Paper Read at the Courtauld Institute on 13 January, 2001
11. Mendes da Costa, E.: In: Davis, L., Reymers, C. (eds.) The Natural History of Fossils. Grays Inn Gate, London (1757)
12. Rhys Phillips, D.: A Romantic Valley in Wales _ a History of the Vale of Neath. Rhys Phillips Publishing, Beili Glas, Swansea (1925)
13. Twitchett, J.: Derby Porcelain, 1748–1848, An Illustrated Guide. Antique Collectors Club, Woodbridge, Suffolk (2002)
14. Williams, W.R.: The Rich History of Quakerism. Pickle Partners Publishing, Potomac (2019)
15. Woodward, J.: An Attempt Towards a Natural History of the Fossils of England, Volume 1:In a Catalogue of the English Fossils in the Collection of J. Woodward MD. Part I. Of the Fossils that are Real and Natural, Eart hs, Stones, Marble, Talcs, Coralloids, Spars and Crystals, Containing a Description and Historical Account of Each with Observations and Explanations, Made in Order to Discover as Well the Origin and Nature of These and their Medicinal, Mechanical and other Uses, F. Fayram, J. Senex, J. Osborn and J. Longman, London, (1728–9)

Chapter 17
Blue European Glass

Abstract Blue glass in Europe and its use in stained glass windows of our great mediaeval cathedrals: the techniques of production of stained glass. Glass beads and their manufacture from an Islamic heritage. Pigments used and their acquisition.

Keywords Blue glass · Stained glass · Nanoparticles · Glass beads · Islamic glass heritage

17.1 Introduction

The production in Europe of objects based on blue coloured glass is ancient, as evidenced by Roman production [12, 34, 46, 64, 65, 69, 104, 105] but this colour remains rare due to the scarcity of cobalt sources [17, 40] that imposes the requirement of recycling of blue glass since that time [56]. Blue glass recycling continued in the Middle-Ages, in particular using older mosaic tessarae [19].

The first production of blue glass in large quantities was without context that intended for stained glass windows, mainly from the twelfth century [44]. Figure 17.1 presents a visualization of the distribution of stained glass production periods in France constructed using data from Bellet's work [14] for the Rhône-Alpes region and data accessible from Gatouillat et al. for other ones [63]. Depending on the region—and the country, but data out of France are difficult to find -, the proportions vary and the proportions indicated are subject to discussion, but we can agree on the dominance of stained glass windows from the nineteenth and twentieth centuries and the minimal share of stained glass windows from other periods and in particular from the medieval periods, the quantitative peak of 'old glass 'production corresponding to the sixteenth century [63, 99]. The inventory of stained glass windows and their study of the aesthetic and historical point of life has been conducted for decades by the Corpus Vitrearum Medii Aevi (CVMA) association but studies in terms of manufacturing techniques are much rarer. The information on this plan

H. G. M. Edwards, P. Colomban, *Blue by Fire: A Marker of the Technical History of Glass and Ceramics*, Cultural Heritage Science,
https://doi.org/10.1007/978-3-031-77122-4_17

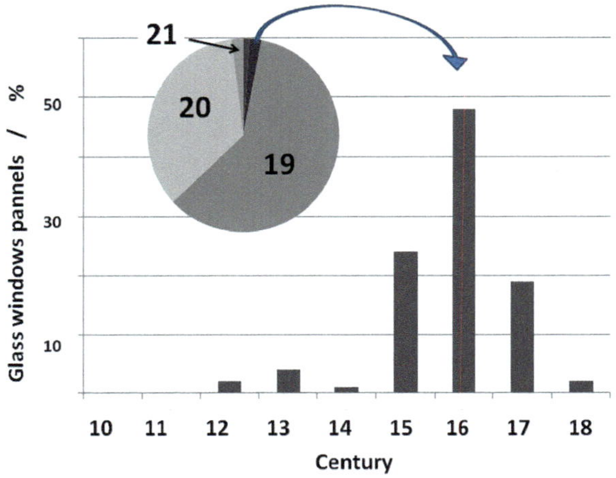

Fig. 17.1 Distribution of number of glass roofs according to their installation date for churches in France

given by the CVMA is often erroneous. However, it is only stained glass windows from the Middle-Age that have been studied, apart from some recent work on the Art Nouveau period [58] or on the production of the Franco-Spanish Society Mauměјean Frères [62, 89]. Thus, a Raman study of several panels from the Sainte-Chapelle Paris (thirteenth century) [35] showed that the classifications made by one of the great stained glass specialists, Louis Grodecki in the 1950s by observing the stained glass windows with binoculars to distinguish the authentic parts of the nineteenth century restorations were just 50%!

Significant analytical work also concern glass beads [10, 22–24, 42, 43, 60, 77–79, 83, 131–133], a minor area but of international influence through their use as a means of exchange in the various colonies and trading posts [53], whether European in Africa [31, 52, 60, 83, 133] and in America [72, 78, 79], in Asia [60, 87, 103] but also Arab/Swahili (around the Indian Ocean and Trans-Saharan caravans) [73, 120] or even the Indian Ocean islands (Madagascar, Mayotte) [40, 41].

Blue glass is used in association with gems in many objects from the Middle-Age [22–24].

The oldest stained glass window in place is considered to be that of the Prophet Daniel in the Cathedral of Our Lady of Augsburg (Germany) manufactured in 1065 at Tegernsee Abbey (Bavaria). Older stained glass windows are preserved in museums such as the Christ of Wissembourg (1060) at the Musée de l'Œuvre Notre-Dame de Strasbourg. Fragments from the sixth century were extracted from excavations in Switzerland and Italy [49, 76, 84, 99, 127]. 'In situ' recycling during renovations of old stained glass elements is also well established [2, 4, 35]. Texts attest to the presence of coloured stained glass windows at the Hagia Sophia (Byzantium) in the fourth century and fragments of large Roman blue glass (~30 × 80 cm) from the mid-third century were found in a shipwreck in Corsica [5, 99]. The continuity with Roman production is evident especially if we take into account the continuity of the use of large quantities of glass, including blue, in Roman and then Byzantine mosaics, in Ravenna (Italy, sixth century), Santa Sophie (Byzantium, tenth century), Palermo, Monreale and Céphalù (Sicily, twelfth century) [29, 104, 105]. Byzantine production influenced Islamic production [55]. For instance the famous mosaic of the Cordoba mihrab was made by Byzantine craftsmen and sent by the Emperor [68, 97].

The development of glass production in Europe benefitted from the forced transfer of Byzantine glassmakers to Venice caused by the Crusaders. The boom in the construction of churches and cathedrals in the twelfth and thirteenth century without context induced the development of stained glass [85]. There are nearly 200 churches in France which are or were cathedrals (i.e. the seat of a bishopric), half of which were built between the year 1000 and the end of the thirteenth century. Figure 17.2 shows a stained glass window from the Sainte-Chapelle built by the King of France Louis IX (Saint-Louis) from 1241 to 1248 in his Parisian castle to

Fig. 17.2 Example of stained glass windows (with view of details) from the Middle Ages (left, choir of Sainte-Chapelle Palatine church, thirteenth c., Paris) and modern (right, church of Moutier-Saint-Jean, twentieth c., Société Mauméjean Brothers Paris Hendaye). Note the hieratic character of the thirteenth century and the neo-baroque style of the Mauméjean stained glass window

house like a giant reliquary the relics of the Passion of Christ: the Crown of Thorns, the Lance, pieces of the True Cross, the Holy Sponge, etc. brought back from the East. The surface area of the stained glass windows is more than 600 m^2 and the replacement of the flying buttresses with iron hoops allows exceptional brightness. The main colours are blue, green, red and yellow.

The decline in stained glass creation in the fourteenth century resulted from the population and economic collapse following the Great Plague before the rise of the Renaissance, one of the consequences of the Crusades. The Council of Trent (1545–1563) discredited the use of stained glass windows as an illustration of the Bible and the Gospels and promoted geometric decorations, allowing light to pass widely, returning to the Cistercian tradition of simplicity, clarity and geometric patterns. The decline in the installation of new glass roofs in the seventeenth century is linked to the development of iconoclastic Protestantism and its implications for Catholicism (Counter-Reformation) which gave preference to sculpture, painting and music for celebrate the greatness of God to the detriment of stained glass windows. In fact the development of the civil use of stained glass in the seventeenth century partly compensated for the reductions in orders for religious art, with variations depending on the country. Civil stained glass windows are more common in Northern and Germanic countries [99]. Enamelling techniques are widely used.

The eighteenth century saw a complete decline in the art of stained glass with loss of technical skills. In France after the Restoration (Louis XVIII) and the expansion of the Royal Chapel of Dreux in the mid-nineteenth century, the development of Troubadour/Neo-Gothic styles, the impetus given by Napoleon III to archaeology and renovation of French Heritage (restorations by E. Violet-le-duc, J.B. Lassus etc. of the basilica of Vézelay, the Sainte-Chapelle of Paris, Notre-Dame de Paris, etc.; the installation of the workshop of G. Bontemps in Choisy and its move to London after the revolutionary days of 1848, that of A. Brongniart in Sèvres producing stained glass windows, etc.) [35, 96] caused a revival of the art of stained glass amplified by technical developments in chemistry and the arts of Fire. The succession of innovative Art Deco/Tiffany/Liberty/Secession aesthetic trends in Europe and America with a flowering of creators (J. Grüber, L. Majorelle, L.C. Tiffany, C.R. Macinstosh, etc.) led to an aesthetic and technical flowering. In the twentieth century artists like L.J.P. Barillet, G. Rouault, F. Décorchemont, M. Chagall, M. Ott, J.M. Alberola, Kim en Joong, etc.) and industrialists like the Société Mauméjean Frères (Fig. 17.2) [89] are participating in the new influence of the art of stained glass. Note that the use of coloured glass roofs is also widespread in Ottoman style mosques (Fig. 17.3) but the network supporting the glass fragments is based on stucco [121] and not on metal (lead/copper and steel) as for stained glass windows. In rare cases, concrete (we speak of claustra) serves as a support (glass claustra by Décorchemont, Perret, and Le Corbusier). Ottoman (blue) glass production starts to be studied [98].

Fig. 17.3 Examples of Ottoman stained glass. (Edirne mosque, Turkiye, sixteenth century or after)

17.2 Techniques and Compositions

17.2.1 Stained Glass

Not only the compositions but also the decoration methods and the layout evolve from century to century. The stained glass windows of the eleventh and twelfth centuries combine coloured glass in the mass (with the exception of yellow and red colours where the very strong colouring power of the nanoparticles of silver (yellow) and copper (red) requires only colouring the thin layer near the surface [37–39, 99] assembled by lead rods, the entire panel being stiffened by iron bars. The link with the aesthetic of Byzantine art of mosaics but also champlevé enamels is visible. The dimensions of the pieces of glass are limited (<~20 cm). The design is obtained by cutting pieces of glass and the network of lead rod. Only the faces and the hands of figures are made with grisaille, very porous enamel because it is fired at low temperature, based on a mixture of iron, manganese and lead oxides. The combination of high porosity and easily oxidized reduced oxides (MnO, FeO, Fe_3O_4) leads to their easy oxidative corrosion, and obscuration of the window.

The cost of the stained glass windows of a cathedral is equivalent to that of the entire stone building. In the fourteenth century the pieces reached large dimensions and the design was no longer structured solely by the lead rods (Fig. 17.2). Coloured areas are made on glass of another colour. Indeed the glass production techniques, the Norman method of *cives* (crown) which produced circles of 60 cm in diameter in the fourteenth century and 100 cm in the eighteenth century and the technique of sleeves ('*manchons*' or cylinder) which already produced from the twelfth century

100 cm plates are being improved and make available large pieces of glass. The variation in the dimensions of the fragments is therefore mainly an aesthetic choice.

Table 17.1 compares some stained glass compositions from different periods with the typical composition of modern window glass and Roman glass. These latter soda-lime compositions are quite close with ~70% SiO_2 and ~ 17% Na_2O and 6–8% CaO. During the Middle-Ages, difficulties in supplying soda from natron (extracted from the salt lakes of the Near East) or from halophytic plants (from the Near East or Muslim Spain (al-Andalus) led to the use of plant ashes (mainly wood, ferns, algae). This leads to potassium compositions (5 to 25% K_2O) but also with around 4% P_2O_5, this element being characteristic of living matter. From the fourteenth century a return to soda-lime glasses occurred, but in England, potash glasses continued to be used until the nineteenth century [99].

The great variability of stained glass compositions is due to their method of production. While the places of production of Roman glass were rare, located mainly in the Near East but produced significant quantities (ingots of several tens of kg), the workshops of shaping were numerous scattered throughout the Empire [114, 134]. For the stained glass windows, a workshop was created for each new project to produce pieces of glass and had to use local raw materials as much as possible, as transport costs were very high at that time. The master masons and their team could move from one site to another and we can assume that the same was true for the master glassmakers [7, 71, 102]. On the other hand in England glass was imported [4].

The colouring power of cobalt being very strong, 0.02% CoO is sufficient to colour glass several mm thick. Thus recycling blue glass or even adding some pieces of blue glass as colouring agent (the content can reach 2%) was easy. On the other hand, for enamels on glass, of reduced thickness (50–300 μm) 0.5 to 1% wt CoO is common. At these low cobalt concentrations, the Co^{2+} ions remain dissolved in the silicate network of SiO_4 tetrahedron connected by some of their vertices. The Raman signature is therefore insensitive to the blue colour except that the absorption in the complementary colours eliminates a large part of the continuous fluorescence background for excitation by a green laser. Figure 17.4 presents the characteristic spectra of the 3 main types, called 1, 2 (soda-lime glass) and 3 (potash-lime glass) as well as their evolution in the event of acid corrosion (K^+/H_3O^+/H^+ exchange) [37, 38, 118]. The three types of glass are easily distinguished but as shown in the Fig. 17.4, leaching causes a strong modification of the spectrum. The variability of the compositions visible in Table 17.1 and the literature [16, 37, 38, 99, 110] is a characteristic of stained glass windows that is found in spectral responses.

Figure 17.5 compares the representative spectra of the different types of composition, in the non-corroded state and having undergone artificial leaching leading to the replacement of mobile alkaline ions by oxonium ions and/or protons []. We then observe the appearance of components at lower wave numbers for both stretching and bending modes. Indeed, the protonation of certain oxygen ions leads to the weakening of neighbouring Si-O bonds and therefore to a shift towards low energies.

Figure 17.5 also compares the distributions of parameters extracted from the Raman spectra characteristic of the structure of the polymer network based on SiO4 tetrahedron constituting the glass for Roman glass samples collected in different

Table 17.1 Representative compositions of (blue) stained glasses from 12th to 19th from different places in France (Amiens cathedral, Paris Sainte-Chapelle and Troyes basilica), Germany (Cologne cathedral) and England (Canterbury cathedral and York minster). Comparison is made with the composition of a blue enamel on stained glass (Belgium) and with Roman and modern 'standard' glass. (After data of [16, 37, 38, 99, 110])

	Modern	Roman	Canterbury Cathedral	York Minster	Paris Ste-Chapelle	Amiens Cathedral	Troyes Basilica	Cologne Cathedral	Cologne Cathedral	Belgium	Amiens Cathedral	Amiens Cathedral
	20th	1st-3th	12th	12th	13th	13th	13th	14th	14th	16th–17th	19th	19th
						Type 3					Type 1	Type 2
SiO2	73.6	67	54.3	61	52	50	49.9	48.8	44.7	60	65	63
Na_2O	16	18	1.4	2.5	1.3	1–3	0.4	0.3	0.3	4	10	18
CaO	5.2	8	16.9	15.8	14.7	18	14.7	19.1	28.5	2.3	17	15
K2O	0.6	1	13.7	5.3	16.3	15	23.8	21	10.6	12	1	1.5
MgO	3.6	1	4.2	7.7	5.8	5	3.6	4	4		2.5	0.2
$A_{12}O_3$	1	2.5	1.1	2.2	1.6	2.5	2	1.9	3	2.2	2.5	2
Fe_2O_3		0.5	0.4	1.3	0.6	1	0.6	0.5	0.6	2.4	0.2	1
MnO			0.8	7.7	1.2	1	1.1	1.5	1	0.1	<1	1
P_2O_5			4.8	4.1	4.4	4	0.9	1.9	3.8		0.2	<0.05
SO3					0.2		0.3	0.2	0.1			
CoO												
PbO											0.5	
Cl					0.5		0.05	0.07		0.4		

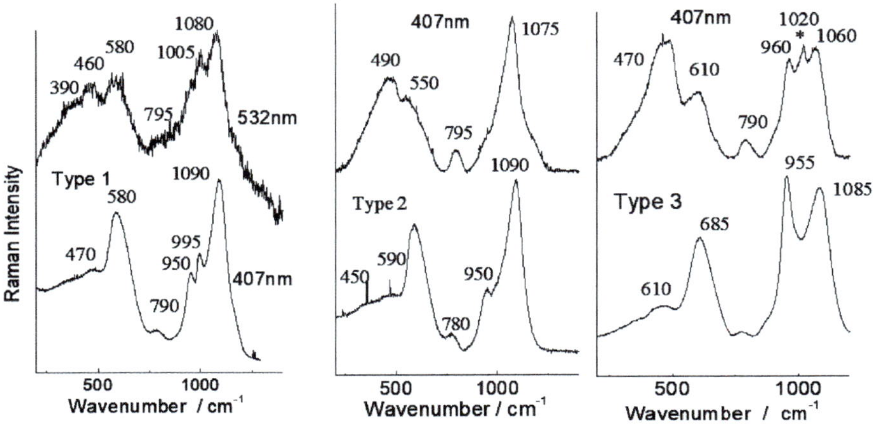

Fig. 17.4 Representative Raman spectra of the different types (1 et 2 soda-lime glass, 3 potash-lime glass) of stained glass non-corroded (bottom) and after artificial lixiviation of the alkali ions (top). See Table 17.1

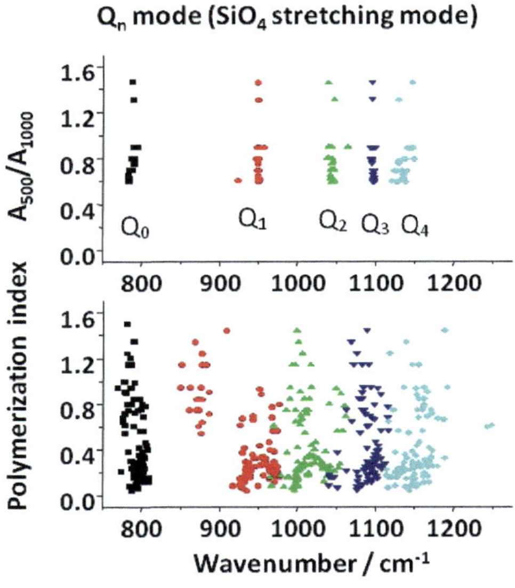

Fig. 17.5 Comparison of the distributions of the ratio of the area of the deformation band (centre of gravity around 500 cm^{-1}) to that of the elongation band of the SiO_4 vibrational unit as a function of the positions of the five Qn components of the stretching modes of Raman spectra from a set of samples from throughout the Roman Empire with a set of glass fragments from stained glass windows from the twelfth and thirteenth century from the Aube department (eastern France), dismantled before the 1914–1918 war (after data of [33])

regions of the Empire with those of stained glass samples from a very limited region of France, the department of Aube. The dispersion is low for Roman glass but very wide for stained glass, reflecting the centralized, well-defined production of Roman glass and the variability of stained glass.

17.2.2 Glass Beads

Glass beads have been produced and traded long distances for several millennia, from the Near and Middle East (Egypt, Mesopotamia) to the rest of the world either by land routes (Silk Road Caravans, Trans-Saharan Caravans), or by maritime routes (Ocean Indian, Colonial conquests). Much work has been carried out concerning the trade of beads excavated from trading posts on the African coasts of the Indian Ocean and in sites in southern Africa accessible by rivers. Indeed, these routes were used for the 'trade' of gold, ivory and slaves since at least the seventh century [13, 60, 131, 132]. One of the indirect proofs of the large scale of this trade is the rebellion of the many Zanjs (black slaves) which endangered the Abbasid Dynasty at the beginning of the tenth century [91]. This maritime trade using the Monsoon winds connected Indonesia and Malaysia to the East (themselves connected to Champa and China), to the ports of India, to those of the Arab-Persian Gulf (Hormuz, Siraf, Basra, Qahlat, Sharma, Aden etc.) and the Red Sea (Moka, Jeddah, Aqaba, Berbera, etc.) and the numerous trading posts along the African coast (from the island of Soqotra to that of Zanzibar and Kilwa outpost) and Vohémar in the north of Madagascar and Mayotte Island (Commore archipel) [13, 86, 95]. These last two sites corresponded to the maximum area that could be reached in a cycle of Mousson wind. Going further required waiting for the next one. But the routes to the gold, ivory and slaves areas, particularly via the Limpopo River, were via trading posts further south (Sofala, Chibuene). These outposts belong to the Swahili Hormuz-Qahlat Sultanate. All the sites cited have preserved thousands or even hundreds of thousands of pearls of different sizes, colours and origins. The Fig. 17.6 shows the comparison of colour proportions for four important sites in South Africa, the twin sites of Mapungubwe and K2 and those of Magoro and Schroda. These sites close to Limpopo River were occupied continuously or partially between the seventh and nineteenth centuries with beads coming in simplified chronological order from Asia (beads called Monsoon' or Indo-Pacific beads), the Islamic world, Europe (with the Portuguese then Dutch trade, mainly from Venice then Bohemia and later from France). If turquoise beads (Cu^{2+} colouring of alkaline glass) generally occupy a significant proportion, the low availability of cobalt leads to a rarity of blue beads. Some post-Renaissance beads are polychromatic (Fig. 17.6), a speciality of Venetian workshops [88] inherited from Roman glassmakers [64]. Rarely, local manufacturing of large beads is done by partial re-melting-sintering of smaller beads [83, 119].

Glass beads have been used as barter currency, hoarding, decoration of clothing and fetishes in West Africa (Ghana, Mauritania, Mali, Senegal, Burkina Faso, Niger,

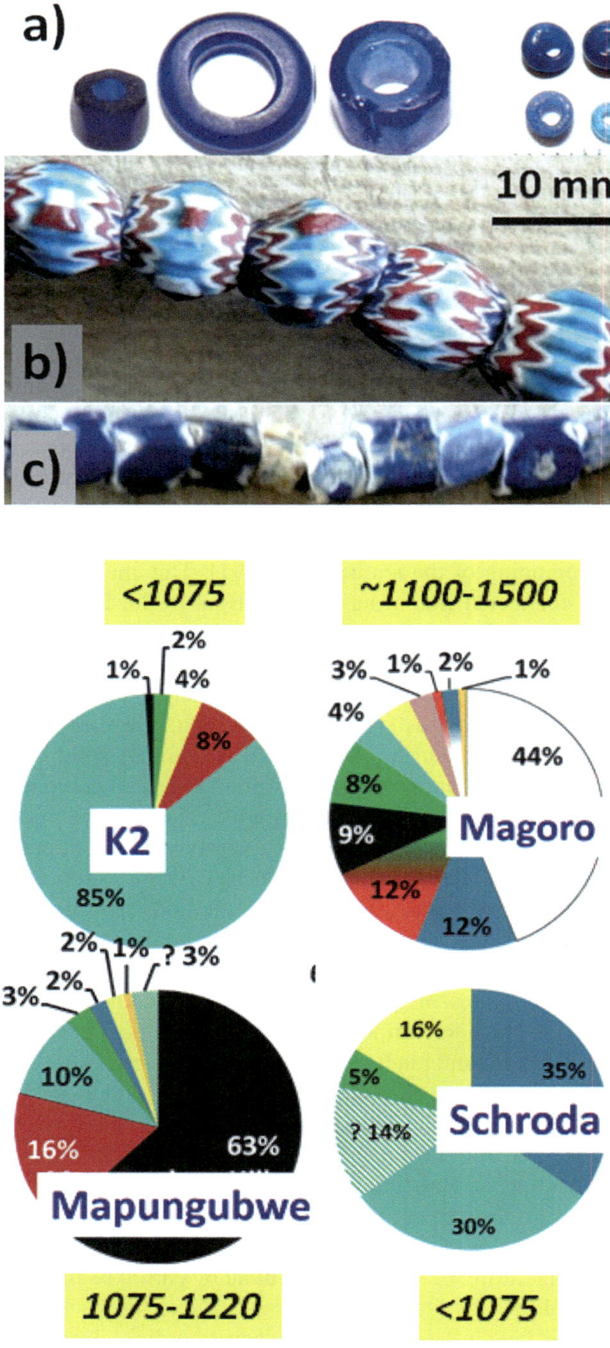

Fig. 17.6 Examples of blue pearls from sites in Southern Africa (**a**) or Madagascar (**b**, **c**, Vohémar): (**a**) the pearls in the lower range are European while the two small ones in the upper range belong to the Khami-Indopacific group, made in India from the 15th to the seventeenth century

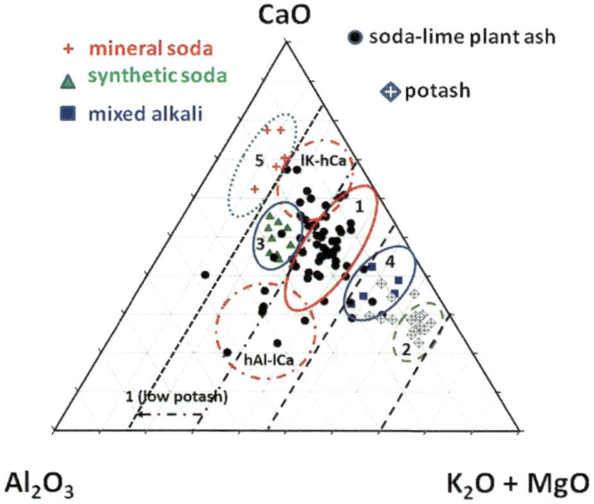

Fig. 17.7 Classification of the types of compositions of the vitreous matrices of glass beads in the ternary diagram CaO-Al_2O_3-($K_2O + MgO$) of the compositions of glass beads from sites in southern Africa for the period ~7th–19th century according to Koleini et al. [83]; hAl-lCa: high amount of Al_2O_3_low amount of CaO; lK-hCa: low K_2O-high CaO

Sierra Leone, Benin, Nigeria, Ivory Coast, Guinea), the Centre (Congo, Cameroon, Angola) [42, 43], the East (Kenya, Uganda, Zanzibar, Ethiopia, Madagascar, Sudan, Tanzania) and as mentioned above South Africa, Botswana, Namibia and Zimbabwe [83]. Glass beads have been also made in China since millenia [50, 135] and India was the main place of production, in particular Arikamedu close to Pondichéry [1, 20, 59].

Due to the very efficient coloration power of Co^{2+} ions the amount of cobalt in glass bead ranges between 0.02 (light blue) to 0.3 (very dark blue), generally 0.1 wt % CoO [83].

Numerous studies of average compositions have been carried out for several decades for sites in Africa [Dussubieux, Francis, Koleini, Wood], America [25, 79], Africa [83, 131–133] and recently Europe [51, 70, 72, 82]. The classifications (Fig. 17.7) are made to identify the source of flux, natural mineral (natron) or synthetic (prepared using Leblanc (eighteenth c.) [128] or Solvay (nineteenth c.) route [18]), or ashes of coastal plants, algae (rich in Na) or ferns, branches, plants (rich in potassium and phosphorus). Aluminium or magnesium levels provide information on the purity of the silica source (river sand, quartz pebbles, etc.). Based on these parameters, the glasses are categorized and provenances proposed. The distribution of traces, in particular of rare earths and trans-uranides, accessible by the precision of mass spectroscopy analyses gives a 'fingerprint'. Recently for certain elements (Pb, Sn, etc.) the determination of isotopic proportions allows precise and reliable attributions when the data concerning the mining sites are known [48, 93].

Fig. 17.6 (continued) (**b**) European blue-white-red-turquoise herring-bone beads, (**c**) European Nueva Cádiz beads named after their use in America by the Conquistadors, initially made in Venice. Below: comparison of the proportions of beads of different colours for the South African sites of K2, Mapungubwe and Magoro Hill based on data from Wood [130] and Koleini et al. [83]. Note the great variability in the presence of the blue colour

When the cobalt colouring bead is associated with a high amount of Mn (ratio of Co/Mn is about 0.3) production place is in Far East Asia (China and Vietnam) [40]. Raman analysis highlighted that rare phases such as malayite (a sphene silicate of calcium and tin doped with chromium) detected in soda pearls from different sites (in South Africa [83] and Antsiraka Boira necropolis on Mayotte Island (twelfth-thirteenth century) [41] confirmed the common origin of these Indo-Pacific beads.

On the contrary, when cobalt colouring the bead is associated with silver, manganese, nickel and arsenic, this is a proof of the use of cobalt from secondary deposits (hydrothermal veins) as mining place of Europe or Iran [40]. For instance, the cobalt colouring Zhizo Indo-Pacific beads with a high ratio of Co/As (>1) or cobalt associated with high iron and low manganese content (Mn/Fe < 0.1) as found in Jingdezhen glazes, using cobalt expected coming from Persia (at the time of Mongol Yuan Dynasty, Persia belong to Mongol Empire).

If elemental analysis carried out by mass spectroscopy is the dominant technique for the study of glass beads, it has the disadvantage of being destructive or micro-destructive and of averaging the information, confusing the contributions of the vitreous matrix and the pigments. Conducted in a 'too' systematic manner, this technique ignoring the microstructure missed the identification of process objects locally by partial re-fusion. The perfectly non-invasive Raman analysis not only identifies both the pigments and the chromophoric complexes and the matrices hosting these colouring agents (Fig. 17.8) and the Raman parameters make it possible to classify the types of vitreous matrices in a slightly different way from elementary analysis. The combination of the two methods is a guarantee of effectiveness in the complete characterization of the phases constituting the pearls.

17.2.3 Modern Glass Objects

At the end of the nineteenth century a revival in the Fine arts occurred with the concomitant developments of chemical industries, major Universal exhibitions presenting innovations and the discovery of Japanese fire arts with the *wabi-sabi* (faded-rusty) aesthetic, well illustrated by the Raku stonewares then at the beginning of the twentieth century by the *mingei* movement (arts without artists, the beauty of the imperfect). The research carried out for example Théodore Deck (1823–1891) and Félix Bracquemond (1833–1914) for being able to reproduce the ceramic masterpieces of non-European civilizations or antiquity are well known

Fig. 17.8 (continued) soda-lime glass with high iron content (mostly recorded for green and yellow Khami-IP series); spectrum 4: soda-lime glass recorded on European beads with high potassium; spectrum 5; lead arsenate glass recorded on European beads; spectrum 6). A baseline has been subtracted according to the procedure described in Colomban [95]. Below, the plot of Si–O bending vs. stretching vibrational modes of glass beads found in South Africa, classifying the different glass series from different sites (light blue: Basinghall Farm; dark blue: Baranda; brownish-red: Magoro Hill; yellow: Mutamba)

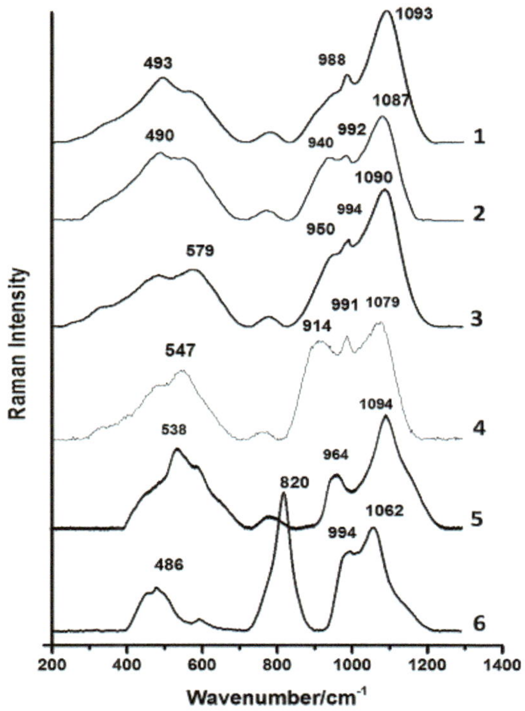

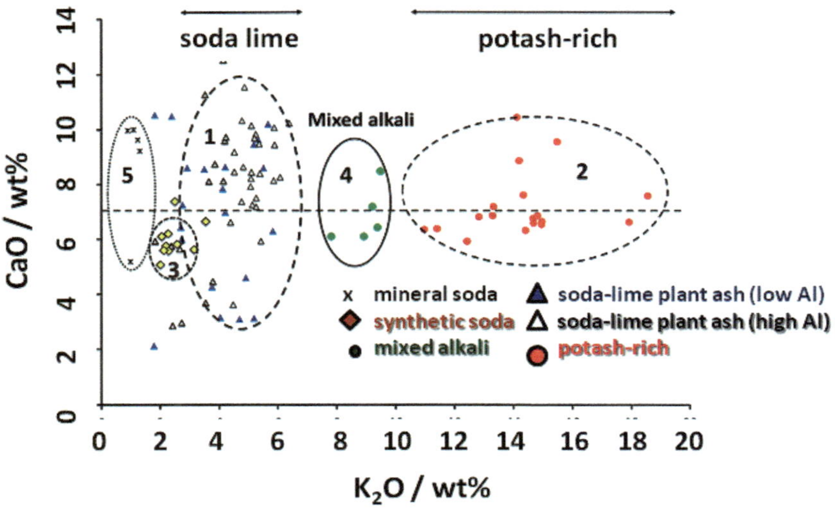

Fig. 17.8 Top, representative Raman signatures of: soda glass of the Indo-Pacific series; spectrum 1: soda glass with high iron content used as colorant; spectrum 2); soda-lime glass of the Mapungubwe Oblate, and some of the Khami-IP series and European beads; spectrum 3:

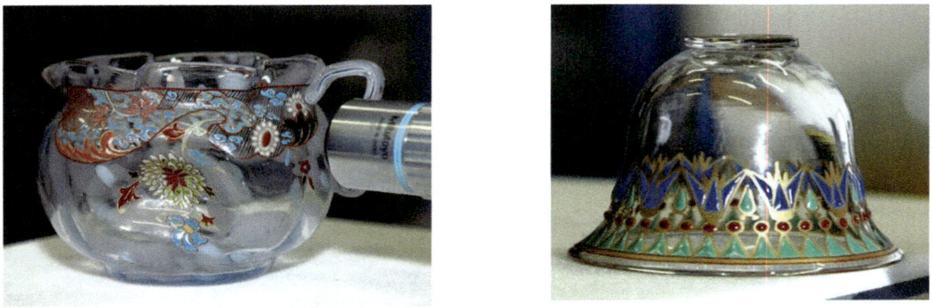

Fig. 17.9 Blown and shaped pot by Emile Gallé (ca. 1884 or before) and goblet attributed to Gallé (Coll. Musée des arts décoratifs, Paris)

[47] and the Samson factory [112] made it its specialty. In the Anglo-Saxon world William Morris also promoted the applied arts. Likewise in the glassmaking field Philippe Joseph Brocard (1831–1896) [27, 28] recreated glass masterpieces, in particular the Mamluk mosque lamps. The study of the productions of L.J. Brocard showed his technological inventiveness to be able to create objects in imitation of the masterpieces which were discovered. But it is certainly the works of Emile Gallé (1846–1904) which are the best known. Figure 17.9 shows a pot by Gallé, well representative of his inventiveness in creating new shapes and his mastery of enamelled decoration. This mastery, although the style is more classical, can be seen in the enamelled decoration of the cup. Gallé uses classic colouring techniques using Co^{2+} ions dissolved in the enamel matrix while Brocard tests new pigments such as $CoAlO_4$ [27, 28]. Analytical work on these objects remains very rare. It is the same for the large production of jewellery decorated with enamels, often blue [117].

References

1. Abraham, S.A.: Glass beads and glass production in early South India: contextualizing indo-Pacific bead manufacture. Archaeol Res Asia. **6**, 4–15 (2016). https://doi.org/10.1016/j.ara.2016.02.003
2. Adlington, L.W., Freestone, I.C., Seliger, L.: Dating Nathan: the oldest stained glass window in England? Heritage. **4**(2), 937–960 (2021)
3. Adlington, L., Freestone, I., Seliger, L., Martinón-Torres, M., Brock, F., Shortland, A.: In Situ Methodology for Compositional Grouping of Medieval Stained Glass Windows. Cambridge Scholars Publishing (2024)
4. Adlington, L.W., Freestone, I.C.: Using handheld pXRF to study medieval stained glass: a methodology using trace elements. MRS Adv. **2**(33), 1785–1800 (2017). https://link.springer.com/article/10.1557/adv.2017.233
5. Alfonsi, H.: L'épave Porticcio A. Des uvres d'art en marbre, une cargaison de vitres et une grande variété d'amphores à l'époque de Philippe Ier, dit Philippe l'Arabe. Cahiers Achéol. Subaquat. **22**, 23–114 (2014). http://journals.openedition.org/adlfi/23565
6. Almeida, T.: Study of the contemporary Portuguese stained glass: artists and works. Int J. Arch Arts Appl. **7**(2), 24–32 (2021). https://www.researchgate.net/profile/Teresa-Almeida20/

publication/354631639_A_Study_of_the_Contemporary_Portuguese_Stained_Glass_Artists_and_Works/links/6525417b2db76c669afdd28c/A-Study-of-the-Contemporary-Portuguese-Stained-Glass-Artists-and-Works.pdf
7. Aubert, M.: La construction au Moyen Age. Bull. Monum. **119**(2), 81–120 (1961). https://doi.org/10.3406/bulmo.1961.8646
8. Balakhnina, I.A., Brandt, N.N., Chikishev, A.Y., Mankova, A.A., Morozova, E.A., Shpachenko, I.G., Yuryev, V.A., Yuryeva, T.V.: Raman microspectroscopy of blue-green historical beads: comparative study of undamaged and strongly degraded samples. J. Raman Spectrosc. **49**(3), 506–512 (2018). https://doi.org/10.1002/jrs.5305
9. Barker, M.: Stained glass in France in the 1950s. J Decorative Arts Soc 1850-*the Present* **1991**, *1*(15), 5–13. https://www.jstor.org/stable/41809182
10. Barone, G., Bersani, D., Crupi, V., Longo, F., Longobardo, U., Lottici, P.P., Aliatis, I., Majolino, D., Mazzoleni, P., Raneri, S., Venuti, V.: A portable versus micro-Raman equipment comparison for gemmological purposes: the case of sapphires and their imitations. J. Raman Spectrosc. **45**(11–12), 1309–1317 (2014). https://doi.org/10.1002/jrs.4555
11. Basso, E., Riccardi, M.P., Messiga, B., Mendera, M., Gimeno, D., Garcia-Valles, M., Fernandez-Turiel, J.L., Bazzocch, I.F., Aulina, M., Tarozzi, C.: Composition of the base glass used to realize the stained glass windows by Duccio di Buoninsegna (Siena cathedral, 1288–1289 AD): a geochemical approach. Mater. Charact. **60**(12), 1545–1554 (2009). https://www.sciencedirect.com/science/article/pii/S1044580309002836
12. Basso, E., Invernizzi, C., Malagodi, M., La Russa, M.F., Bersani, D., Lottici, P.P.: Characterization of colorants and opacifiers in roman glass mosaic tesserae through spectroscopic and spectrometric techniques. J. Raman Spectrosc. **45**(3), 238–245 (2014). https://doi.org/10.1002/jrs.4449
13. Baujard, P.: The Worlds of the Indian Ocean: a Global History, 2 Vol. Cambridge University Press, Cambridge (2019)
14. Bellet, J.: Le recensement du vitrail ancien de la France, "et spécialement" en Rhône-Alpes, *Les carnets de l'Inventaire: études sur le patrimoine culturel – Rhône-Alpes & Auvergne* [en ligne]. Accessed 29 juin 2024. (2016). http://inventaire-rra.hypotheses.org/3837
15. Beltran, M., Pradell, T.: Analytical methods applied to the study of modernist stained glass. Avenços en recerca i desenvolupament del Departament d'Enginyeria Gràfica i de Disseny. **12**, 140 (2023). https://books.google.com/books?hl=en&lr=&id=o9C4EAAAQBAJ&oi=fnd&pg=PA140&dq=Beltran+M,+Pradell+T.+Analytical+methods+applied+to+the+study+of+modernist+stained+glass.+Aven%C3%A7os+en+recerca+i+desenvolupament+del+Departament+d%E2%80%99Enginyeria+Gra%CC%80fica+i+de+Disseny.+2023+Apr+12:140.&ots=9NJ3Dat50o&sig=ps_epMqD58eDqsrfdHUMjMr3AaU
16. Bernardi, A., Becherini, F., Verità, M., Ausset, P., Bellio, M., Brinkmann, U., Cachier, H., Chabas, A., Deutsch, F., Etcheverry, M.P., Bianchini, F.G.: Conservation of stained glass windows with protective glazing: Main results from the European VIDRIO research programme. J. Cult Herit. **14**(6), 527–536 (2013). https://doi.org/10.1016/j.culher.2012.11.009
17. Berrie, B.H.: Mining for color: new blues, yellows, and translucent paint. Early Sci. Med. **20**(4–6), 308–334 (2015). https://brill.com/view/journals/esm/20/4-6/article-p308_2.xml
18. Bertrams, K., Coupain, N., Homburg, E.: Solvay: History of a Multinational Family Firm. Cambridge University Press, Cambridge (2013)
19. Bidegaray, A.I., Pollard, A.M.: Tesserae recycling in the production of medieval blue window glass. Archaeometry. **60**(4), 784–796 (2018)
20. Boțan, S.P.: The road to the indies. The glass evidence. Studia Antiqua et Archaeologica. **20**(1), 153–170 (2014)
21. Brill, R.H., Pongracz, P.: Stained glass from saint-Jean-des-Vignes (Soissons) and comparisons with glass from other medieval sites. J Glass Stud. **46**, 115–144 (2004). https://www.jstor.org/stable/24190935

22. Bruni, Y., Hatert, F., George, P., Strivay, D.: The reliquary bust of Saint Lambert from the Liège cathedral, Belgium: gemstones and glass beads analysis by pXRF and Raman spectroscopy. Archaeometry. **62**(2), 297–313 (2020). https://doi.org/10.1111/arcm.12527
23. Bruni, Y., Hatert, F., Demaude, M., Delmelle, N., George, P., Maquet, J.: An Archaeometric investigation of gems and glass beads decorating the double-arm reliquary cross from Liège, Belgium. Heritage. **4**(4), 4542–4557 (2021a)
24. Bruni, Y., Hatert, F., George, P., Cambier, H., Strivay, D.: A gemmological study of the reliquary crown of Namur, Belgium. *Eur J Mineral.* **33**(2), 221–232 (2021b)
25. Burgess, L.E., Dussubieux, L.: Chemical composition of late 18th-and 19th-century glass beads from western North America: clues to sourcing beads. BEADS: J Soc Bead Res. **19**(1), 58–73 (2007). https://surface.syr.edu/beads/vol19/iss1/9
26. Caggiani, M.C., Colomban, P., Valotteau, C., Mangone, A., Cambon, P.: Mobile Raman spectroscopy analysis of ancient enamelled glass masterpieces. Anal. Methods. **5**(17), 4345–4354 (2013). https://doi.org/10.1039/C3AY40648B
27. Caggiani, M.C., Valotteau, C., Colomban, P.: Inside the glassmaker technology: search of Raman criteria to discriminate between Emile Gallé and Philippe-Joseph Brocard enamels and pigment signatures. J. Raman Spectrosc. **45**(6), 456–464 (2014a). https://doi.org/10.1002/jrs.4481
28. Caggiani, M.C., Acquafredda, P., Colomban, P., Mangone, A.: The source of blue colour of archaeological glass and glazes: the Raman spectroscopy/SEM-EDS answers. J. Raman Spectrosc. **45**(11–12), 1251–1259 (2014b). https://doi.org/10.1002/jrs.4492
29. Caggiani, M.C., Barone, G., Mazzoleni, P.: Non-invasive analysis of reference glass and historical mosaic tesserae by means of reflectance infrared spectroscopy. Spectrochim. Acta A Mol. Biomol. Spectrosc. **308**, 123675 (2024). https://doi.org/10.1016/j.saa.2023.123675
30. Capobianco, N., Hunault, M.O., Balcon-Berry, S., Galoisy, L., Sandron, D., Calas, G.: The Grande rose of the Reims cathedral: an eight-century perspective on the colour management of medieval stained glass. Sci. Rep. **9**(1), 3287 (2019). https://doi.org/10.1038/s41598-019-39740-y
31. Carter, A.K.: The production and exchange of glass and stone beads in Southeast Asia from 500 BCE to the early second millennium CE: an assessment of the work of Peter Francis in light of recent research. Archaeological Research in Asia. **6**, 16–29 (2016)
32. Ceglia, A., Meulebroeck, W., Baert, K., Wouters, H., Nys, K., Thienpont, H., Terryn, H.: Cobalt absorption bands for the differentiation of historical Na and Ca/K rich glass. Surf. Interface Anal. **44**(2), 219–226 (2012). https://doi.org/10.1002/sia.3810
33. Colomban, P.: On-site Raman identification and dating of ancient glasses: a review of procedures and tools. J. Cult. Herit. **9**, e55–e60 (2008)
34. Colomban, P., March, G., Mazerolles, L., Karmous, T., Ayed, N., Ennabli, A., Slim, H.: Raman identification of materials used for jewellery and mosaics in Ifriqiya. J. Raman Spectrosc. **34**(3), 205–213 (2003)
35. Colomban, P., Tournié, A.: On-site Raman identification and dating of ancient/modern stained glasses at the Sainte-Chapelle, Paris. *J Cult Herit.* **8**(3), 242–256 (2007). https://doi.org/10.1016/j.culher.2007.04.002
36. Colomban, P., Kırmızı, B.: Non-invasive on-site Raman study of polychrome and white enamelled glass artefacts in imitation of porcelain assigned to Bernard Perrot and his followers. J. Raman Spectrosc. **51**(1), 133–146 (2020). https://doi.org/10.1002/jrs.5745
37. Colomban, P., Tournie, A., Bellot-Gurlet, L.: Raman identification of glassy silicates used in ceramics, glass and jewellery: a tentative differentiation guide. J. Raman Spectrosc. **37**(8), 841–852 (2006a)
38. Colomban, P., Etcheverry, M.P., Asquier, M., Bounichou, M., Tournié, A.: Raman identification of ancient stained glasses and their degree of deterioration. J. Raman Spectrosc. **37**(5), 614–626 (2006b). https://doi.org/10.1002/jrs.1495
39. Colomban, P., Tournié, A., Ricciardi, P.: Raman spectroscopy of copper nanoparticle-containing glass matrices: ancient red stained-glass windows. J. Raman Spectrosc. **40**(12), 1949–1955 (2009)

40. Colomban, P., Kırmızı, B., Simsek Franci, G.: Cobalt and associated impurities in blue (and green) glass, glaze and enamel: relationships between raw materials, processing, composition, phases and international trade. Fortschr. Mineral. **11**(6), 633 (2021)
41. Fischbach, N., Ngo, A.T., Colomban, P., Pauly, M.: Beads excavated from Antsiraka Boira necropolis (Mayotte Island, 12th–13th centuries) Colouring agents and glass matrix composition comparison with contemporary southern Africa sites. Archeol Sci. **40**, 83–102 (2016)
42. Costa, M., Arruda, A.M., Dias, L., Barbosa, R., Mirão, J., Vandenabeele, P.: The combined use of Raman and micro-X-ray diffraction analysis in the study of archaeological glass beads. J. Raman Spectrosc. **50**(2), 250–261 (2019a). https://doi.org/10.1002/jrs.5446
43. Costa, M., Barrulas, P., Dias, L., da Conceição Lopes, M., Barreira, J., Clist, B., Karklins, K., de Jesus, M.D.P., da Silva Domingos, S., Vandenabeele, P., Mirao, J.: Multi-analytical approach to the study of the European glass beads found in the tombs of Kulumbimbi (Mbanza Kongo, Angola). Microchem. J. **149**, 103990 (2019b). https://doi.org/10.1016/j.microc.2019.103990
44. Cox, G.A., Gillies, K.J.: The X-ray fluorescence analysis of medieval durable blue soda glass from York Minster. Archaeometry. **28**(1), 57–68 (1986). https://doi.org/10.1111/j.1475-4754.1986.tb00374.x
45. CVMA. https://corpusvitrearum.org/. Accessed 14 July 2024.
46. Dariz, P., Schmid, T.: Raman focal point on Roman Egyptian blue elucidates disordered cuprorivaite, green glass phase and trace compounds. Sci. Rep. **12**(1), 15596 (2022). https://doi.org/10.1038/s41598-022-19923-w
47. Deck, T.: La faïence Ancienne Maison Quantin, Paris (1887)
48. Degryse, P., Vanhaecke, F.: Status and prospects for quasi-non-destructive analysis of ancient artefacts via LA–ICP–MS. Elements. **12**(5), 341–346 (2016)
49. Dell'Acqua, F.: Early history of stained glass. In: Investigations in Medieval Stained Glass, pp. 23–35. Brill (2019)
50. Dong, J., Li, Q., Hu, Y.: Multi-technique analysis of an ancient stratified glass eye bead by OCT, μ-XRF, and μ-Raman spectroscopy. Chin. Opt. Lett. **18**(9), 090001 (2020). https://doi.org/10.1364/COL.18.090001
51. Dussubieux, L., Gratuze, B.: Chemical composition of 16th-to 18th-century glass beads excavated in Paris. BEADS: J Soc Bead Res. **24**(1), 26–38 (2012). https://surface.syr.edu/beads/vol24/iss1/6/
52. Dussubieux, L., Walder, H.: The Elemental Analysis of Glass Beads: Technology, Chronology and Exchange. Leuven University Press (2022)
53. Dussubieux, L., Karklins, K.: Glass bead production in Europe during the 17th century: elemental analysis of glass material found in London and Amsterdam. J. Archaeol. Sci. Rep. **5**, 574–589 (2016)
54. Fernandes, P., Vilarigues, M., Alves, L.C., da Silv, R.C.: Stained glasses from monastery of Batalha: non-destructive characterisation of glasses and glass paintings. J. Cult. Herit. **9**, e5–e9 (2008). https://doi.org/10.1016/j.culher.2008.07.005
55. Fiorentino, S.: A tale of two legacies: byzantine and Egyptian influences in the manufacture and supply of glass tesserae under the umayyad caliphate (661–750 AD). Heritage. **4**(4), 2810–2834 (2021). https://doi.org/10.3390/heritage4040158
56. Foy, D., Jézégou, M.-P.: L'épave antique Ouest-Embiez 1. Archéologia. **407**, 22–33 (2004)
57. Fornacelli, C., Colomban, P., Turbanti Memmi, I.: Toward a Raman/FORS discrimination between art nouveau and contemporary stained glasses from CdSxSe1− x nanoparticles signatures. J. Raman Spectrosc. **46**(11), 1129–1139 (2015)
58. Fornacelli, C., Sciau, P., Colomban, P.: CdS x Se1−x quantum dots as colouring agents of art nouveau and contemporary stained glass: a combined transmission electron microscopy and Raman study. Philos. Trans. R. Soc. A Math. Phys. Eng. Sci. **374**(2082), 20160045 (2016)
59. Francis Jr., P.: Beadmaking at Arikamedu and beyond. World Archaeol. **23**(1), 28–43 (1991)
60. Francis Jr., P.: Asia's Maritime Bead Trade: 300 BC to the Present. University of Hawaii Press (2002)

61. Freestone, I., Gutjahr, M., Kunicki-Goldfinger, J.J., McDonald, I., Pike, A.: Composition, Technology and Origin of the Glass from the Workshop at 35 Basinghall Street. Museum of London Archaeology (2010). https://discovery.ucl.ac.uk/id/eprint/1496132/1/Basinghall%20 text%20Feb%202010.pdf
62. García-Heras, M., Carmona, N., Gil, C., Villegas, M.A.: Neorenaissance/Neobaroque stained glass windows from Madrid: a characterisation study on some panels signed by the Maumejean Fréres company. J. Cult. Herit. **6**(2), 91–98 (2005). https://doi.org/10.1016/j.culher.2004.12.001
63. Gatouillat, F., Hérold, M., David, V.: Des vitraux par milliers… Bilan d'un inventaire: le recensement des vitraux anciens de la France, *In Situ* [En ligne], 6. Accessed 24th June 2024 (2005). https://doi.org/10.4000/insitu.8884
64. Gedzevičiūtė, V., Welter, N., Schüssler, U., Weiss, C.: Chemical composition and colouring agents of Roman mosaic and millefiori glass, studied by electron microprobe analysis and Raman microspectroscopy. Archaeol. Anthropol. Sci. **1**(1), 15–29 (2009). https://doi.org/10.1007/s12520-009-0005-4
65. Gelzo, M., Corso, G., Vergara, A., Rossi, M., Miedico, O., Arcari, O., Chiaravalle, A.E., Piccioli, C., Arcari, P.: Archaeometric studies on a blue glass fragment from Pompeii: case study. Mediter. Archaeol. Archaeom. **21**(3), 23 (2021). https://www.maajournal.com/index.php/maa/article/view/583
66. Gibson, P.: Our heritage of stained glass and its Care in the Twentieth Century. RSA J. **136**(5379), 159–173 (1988). https://www.jstor.org/stable/41374526
67. Gimeno, D., Garcia-Valles, M., Fernandez-Turiel, J.L., Bazzocchi, F., Aulinas, M., Puges, M., Tarozzi, C., Riccardi, M.P., Basso, E., Fortina, C., Mendera, M.: From Siena to Barcelona: deciphering colour recipes of Na-rich Mediterranean stained glass windows at the XIII–XIV century transition. J. Cult. Herit. **9**, e10–e15 (2008). https://doi.org/10.1016/j.culher.2008.08.001
68. Gómez-Morón, M.A., Palomar, T., Alves, L.C., Ortiz, P., Vilarigues, M., Schibille, N.: Christian-Muslim contacts across the Mediterranean: byzantine glass mosaics in the great Umayyad mosque of Córdoba (Spain). J. Archaeol. Sci. **129**, 105370 (2021)
69. Gratuze, B., Pactat, I., Schibille, N.: Changes in the signature of cobalt colorants in late antique and early Islamic glass production. Fortschr. Mineral. **8**(6), 225 (2018). https://www.mdpi.com/2075-163X/8/6/225
70. Gratuze, B., Pion, C., Sode, T.: Indian Glass Beads in Western and North Europe in Early Middle Age. In: Ancient Glass of South Asia: Archaeology, Ethnography and Global Connections, pp. 427–450 (2021)
71. Grodecki, L.: Chronologie de la cathédrale de Chartres. Bull. Monum. **116**(2), 91–119 (1958). https://doi.org/10.3406/bulmo.1958.3831
72. Hancock, R.G., McKechnie, J., Aufreiter, S., Karklins, K., Kapches, M., Sempowski, M., Moreau, J.F., Kenyon, I.: Non-destructive analysis of European cobalt blue glass trade beads. J. Radioanal. Nucl. Chem. **244**, 567–573 (2000). https://doi.org/10.1023/A:1006705014455
73. Hitchner, R.B.: In: Mattingly, D.J., Leitch, V., Duckworth, C.N., Cuénod, A., Sterry, M., Cole, F. (eds.) Trade in the Ancient Sahara and beyond. Cambridge University Press (2017) ISBN978-1-107-19699-5
74. Hunault, M., Bauchau, F., Loisel, C., Hérold, M., Galoisy, L., Newville, M., Calas, G.: Spectroscopic investigation of the coloration and fabrication conditions of medieval blue glasses. J. Am. Ceram. Soc. **99**(1), 89–97 (2016). https://doi.org/10.1111/jace.13783
75. Hunault, M.O., Bauchau, F., Boulanger, K., Hérold, M., Calas, G., Lemasson, Q., Pichon, L., Pacheco, C., Loisel, C.: Thirteenth-century stained glass windows of the Sainte-Chapelle in Paris: an insight into medieval glazing work practices. J. Archaeol. Sci. Rep. **35**, 102753 (2021). https://doi.org/10.1016/j.jasrep.2020.102753
76. Husband, T.B.: The silver-stained roundel in northern Europe. In: InInvestigations in Medieval Stained Glass, pp. 307–318. Brill (2019). https://doi.org/10.1163/9789004395718_024

77. Karampelas, S., Kiefert, L., Bersani, D., Vandenabeele, P., Karampelas, S., Kiefert, L., Bersani, D., Vandenabeele, P.: Gem treatments, synthetics and imitations. In: Gems and Gemmology: An Introduction for Archaeologists, Art-Historians and Conservators, pp. 67–90. Springer, Cham (2020). https://link.springer.com/chapter/10.1007/978-3-030-35449-7_4
78. Karklins, K.: Glass Beads. National Historic Parks and Sites Branch, Parks Canada, Ottawa (1982)
79. Karklins, K.: Guide to the description and classification of glass beads found in the Americas. Beads: J Soc Bead Res. **24**(1), 62–90 (2012)
80. Katsaros, T., Ganetsos, T.: Raman characterization of gemstones from the collection of the byzantine & Christian Museum. Archaeology 1(2), 7–14. (2012). https://www.academia.edu/download/42022938/10.5923.j.archaeology.20120102.01.pdf. Accessed 2 June 2024
81. Koeberl, C.: Blue glass: a new impactite variety from Zhamanshin crater, USSR. *Geochimica et Cosmochimica Acta*. **52**(3), 779–784 (1988). https://doi.org/10.1016/0016-7037(88)90338-9
82. Koffi, M., Colomban, P.: (2025) to be published
83. Koleini, F., Colomban, P., Pikirayi, I., Prinsloo, L.C.: Glass beads, markers of ancient trade in sub-Saharan Africa: methodology, state of the art and perspectives. Heritage. **2**(3), 2343–2369 (2019)
84. Krueger, I., Wedepohl, K.H.: Composition and shapes of glass of the early medieval period (8th to 10th century AD) in Central Europe. Échanges et commerce du verre dans le monde antique, pp. 93–100. (2003). https://artefacts.mom.fr/Publis/Krueger_Wedepohl_2003.pdf
85. Kurmannn-Schwarz, B., Pastan, E. (eds.): Investigations in Medieval Stained Glass, Reading Medieval Sources Series, vol. 3. Leiden, Brill (2019). https://doi.org/10.1163/9789004395718
86. Kusimba, C.M., Kusimba, S.B., Dussubieux, L.: Beyond the coastalscapes: preindustrial social and political networks in East Africa. Afr. Archaeol. Rev. **30**, 399–426 (2013)
87. Lamb, A.: Some observations on stone and glass beads in early south-East Asia. J Malays Branch R Asiat Soc. **38**(208), 87–124 (1965)
88. Lima, A., Medici, T., de Matos, A.P., Verità, M.: Chemical analysis of 17th century Millefiori glasses excavated in the monastery of Sta. Clara-a-Velha, Portugal: comparison with venetian and façon-de-Venise production. J. Archaeol. Sci. **39**(5), 1238–1248 (2012). https://doi.org/10.1016/j.jas.2012.01.006
89. Manauté, B.: La manufacture Mauméjean, un acteur essentiel du renouveau de l'art sacré. Histoire de l'art. **73**(1), 117–128 (2013). https://doi.org/10.3406/hista.2013.3474
90. Meulebroeck, W., Wouters, H., Nys, K., Thienpont, H.: Authenticity screening of stained glass windows using optical spectroscopy. Sci. Rep. **6**(1), 37726 (2016). https://doi.org/10.1038/srep37726
91. Naqīb, M.Ḥ.: The political and military career of Mu'nis al-Muẓaffar at the ʿAbbāsī court (296-321/908-933) (1969). https://escholarship.mcgill.ca/concern/theses/5425kf77k. Accessed 14 July 2024
92. Niilisk, A., Ritslaid, P., Kikas, J., Haak, A., Mäesalu, A.: Elemental and Raman investigation of 13th–14th and 16th century enamelled glass beakers found in Estonia. J. Archaeol. Sci. Rep. **16**, 205–212 (2017). https://doi.org/10.1016/j.jasrep.2017.10.016
93. Norman, M., Robinson, P., Clark, D.: Major-and trace-element analysis of sulfide ores by laser-ablation ICP–MS, solution ICP–MS, and XRF: new data on international reference materials. Can. Mineral. **41**(2), 293–305 (2003)
94. Oikonomou, A., Beltsios, K., Zacharias, N.: Analytical and technological study of blue glass from Thebes, Greece: an overall assessment. Proc AIHV. **18**, 75–80 (2012). https://www.academia.edu/download/55953940/6._Analytical_and_technological_study_of_an_ancient_glass_collection_from_Thebes__Greece_An_overall_assessment_AIHV18.pdf
95. Oteyo, G., Kusimba, C.M.: The Consumption of Glass Beads in Ancient Swahili East Africa. In: China and East Africa: Ancient Ties, Contemporary Flows, p. 147 (2019)

96. Pallot-Frossard, I.: Stained glass: heritage and creation, materiality in all its states. In: InInvestigations in Medieval Stained Glass, pp. 431–444. Brill (2019). https://doi.org/10.1163/9789004395718_033
97. Palomar, T., Schibille, N., Alves, L.C., Hidalgo, R.J.D., Gomez-Morón, M.A.: Historical restorations of the Maqṣūrah glass mosaics from the great mosque of Córdoba. Boletín de la Sociedad Española de Cerámica y Vidrio. **62**(3), 204–219 (2023)
98. Palamara, E., Zacharias, N., Germanidou, S., Gerolymou, K., Palles, D., Kamitsos, E.I.: Technology and Provenance Analysis of Glass Fragments from an Ottoman Bathhouse (Hamam) in Kyparissia, Peloponnese, Greece. STAR Sci Technol Archaeol Res. **3**(2), 376–390 (2017). https://doi.org/10.1080/20548923.2018.1450130
99. Parker, J.M., Martlew, D.: Stained glass windows. In: Richet, P. (ed.) Encyclopedia og Glass Science, Technology, History and Culture, Vol. II The American ceramic society, 1st edn, pp. 1341–1359. Wiley (2021). https://lccn.loc.gov/2020055825
100. Patin, M., Nys, K., Thienpont, H., Meulebroeck, W.: Prestige markers in art: subtle stratagems in material selection for fifteenth-century stained-glass windows. Herit. Sci. **10**(1), 106 (2022). https://doi.org/10.1186/s40494-022-00698-2
101. Purowski, T., Kępa, L., Wagner, B.: Glass on the Amber road: the chemical composition of glass beads from the bronze age in Poland. Archaeol. Anthropol. Sci. **10**, 1283–1302 (2018). https://doi.org/10.1007/s12520-016-0443-8
102. Quiévreux, F.: Les vitraux du XIIIe siècle de l'abside de la cathédrale de Bourges. Bull. Monum. **101**(2), 255–275 (1942). https://doi.org/10.3406/bulmo.1942.9301
103. Ramli, Z., Rahman, N.H., Jusoh, A., Sauman, Y.: Some observations on glass beads composition in Sarawak, Singapore and peninsula Malaysia. Sari-Int J Malay Civiliz. **29**(2), 3–20 (2011)
104. Ricciardi, P., Colomban, P., Tournié, A., Macchiarola, M., Ayed, N.: A non-invasive study of Roman age mosaic glass tesserae by means of Raman spectroscopy. J. Archaeol. Sci. **36**(11), 2551–2559 (2009a). https://doi.org/10.1016/j.jas.2009.07.008
105. Ricciardi, P., Colomban, P., Tournié, A., Milande, V.: Nondestructive on-site identification of ancient glasses: genuine artefacts, embellished pieces or forgeries? J. Raman Spectrosc. **40**(6), 604–617 (2009b). https://doi.org/10.1002/jrs.2165
106. Ricciardi, P., Colomban, P.: Ramane spectroscopy of ceramics and glasses. In: Infrared and Raman Spectroscopy in Forensic Science, pp. 469–479. Wiley (2012). https://doi.org/10.1002/9781119962328
107. Pinto, A.M., Sanjad, T.A., Angélica, R.S., da Costa, M.L., Paiva, R.S., Palomar, T.: 19th century stained-glass windows from Belém do Pará (Brazil): analytical characterisation and pathology. Boletín de la sociedad española de cerámica y vidrio. **57**(4), 133–141 (2018). https://doi.org/10.1016/j.bsecv.2017.10.005
108. Robinet, L., Neff, D.D., Bouquillon, A., Pagès-Camagna, S., Verney-Carron, A., Etcheverry, M.P., Tate, J.: Raman spectrometry, a non-destructive solution to the study of glass and its alteration. In: ICOM Committee for Conservation 15th triennial meeting (2008). https://hal.science/hal-03028178/document
109. Röhrs, S., Fenoglio, G., Reiche, I., Lambacher, L.: Studying glass on metal: Raman analysis of medieval champlevé enamelled objects. J. Raman Spectrosc. **52**(1), 23–34 (2021). https://doi.org/10.1002/jrs.5934
110. Schalm, O., Van der Linden, V., Frederickx, P., Luyten, S., Van der Snickt, G., Caen, J., Schryvers, D., Janssens, K., Cornelis, E., Van Dyck, D., Schreiner, M.: Enamels in stained glass windows: preparation, chemical composition, microstructure and causes of deterioration. Spectrochim. Acta B At. Spectrosc. **64**(8), 812–820 (2009). https://doi.org/10.1016/j.sab.2009.06.005
111. Simmons, C.T., Mysak, L.A.: Transmissive properties of medieval and renaissance stained glass in European churches. Archit. Sci. Rev. **53**(2), 251–274 (2010). https://doi.org/10.3763/asre.2009.0073

112. Slitine, F.: C*'est un vrai Samson!* Sèvres. Revue de la Société des Amis du musée national de Céramique. **7**(1), 66–80. : (1998). https://doi.org/10.3406/sevre.1998.1370
113. Spencer, H.M.: Changing origins and trade routes of Scottish window glass. In: Bouwmeester, J., Patrick, L., Berryman, D. (eds.) Building Networks: Exchange of Knowledge, Ideas and Materials in Medieval and Post-Medieval Europe. Themes in Contemporary Archaeology. Springer, Cham (2024). https://doi.org/10.1007/978-3-031-51963-5_7
114. Stern, M.: Ancient Glassworking. In: Richet, P. (ed.) Encyclopedia og Glass Science, Technology, History and Culture, Vol. II, 1st edn, pp. 1285–1307. The American Ceramic Society, J. Wiley & Sons, Inc. ch. 10.5 (2021). https://doi.org/10.1002/9781118801017.ch10.5
115. Sterpenich, J., Libourel, G.: Medieval Stained Glass Windows: a Physical and Chemical Characterization of Alteration. In: Science and technologies of the materials and of the environment for the protection of stained glass and stone monuments, pp. 171–180 (2001). https://www.researchgate.net/profile/Jerome-Sterpenich/publication/236120822_Medieval_stained_glass_windows_A_physical_and_chemical_characterisation_of_alteration/links/02e7e5226ef8b4e11b000000/Medieval-stained-glass-windows-A-physical-and-chemical-characterisation-of-alteration.pdf
116. Thompson, N.M.: The creation of stained glass in Central Italy, 1250–1400. In: InInvestigations in Medieval Stained Glass, pp. 350–361. Brill (2019). https://doi.org/10.1163/9789004395718_027
117. Tissot, I., Manso, M., Guerra, M.F.: Unveiling the art of René Lalique with XRF and Raman spectroscopy–technological innovation in jewellery production. J. Cult. Herit. **33**, 83–89 (2018)
118. Tournié, A., Ricciardi, P., Colomban, P.: Glass corrosion mechanisms: a multiscale analysis. Solid State Ionics. **179**(38), 2142–2154 (2008)
119. Tournié, A., Prinsloo, L.C., Colomban, P.: Raman spectra database of the glass beads excavated on mapungubwe hill and k2, two archaeological sites in South Africa. arXiv preprint arXiv:1012.1465. (2010)
120. Truffa Giachet, M., Gratuze, B., Mayor, A., Huysecom, E.: Compositional and provenance study of glass beads from archaeological sites in Mali and Senegal at the time of the first Sahelian states. PLoS One. **15**(12), e0242027 (2020)
121. Uguryol, D., Uguryol, M.: Manufacturing Techniques, Deterioration, and Restoration Problems of Traditional Gypsum Windows. *Megaron.* **16**(2), 350–365 (2021). https://doi.org/10.14744/MEGARON.2021.87360
122. Utzig, J.: Ancient remedies for new ideas: possible sources of the Cistercian stained-glass technique and style. (2022). https://publicacions.iec.cat/repository/pdf/00000319/00000054.pdf. Accessed 29 June 2024
123. Van der Snickt, G., Schalm, O., Caen, J., Janssens, K., Schreiner, M.: Blue enamel on sixteenth-and seventeenth-century window glass. Stud. Conserv. **51**(3), 212–222 (2006). https://doi.org/10.1179/sic.2006.51.3.212
124. Van der Snickt, G., Legrand, S., Caen, J., Vanmeert, F., Alfeld, M., Janssens, K.: Chemical imaging of stained-glass windows by means of macro X-ray fluorescence (MA-XRF) scanning. Microchem. J. **124**, 615–622 (2016). https://doi.org/10.1016/j.microc.2015.10.010
125. Verita, M., Biron, I.: Enamels on venetian renaissance glass. An analytical and technical approach. Glass Technol-Eur J Glass Sci Technol Part A. **56**(*6*), 177–190 (2015). https://doi.org/10.13036/17533546.56.6.177
126. Verità, M., Bracci, S., Porcinai, S.: Analytical investigation of 14th century stained glass windows from Santa Croce basilica in Florence. Int. J. Appl. Glas. Sci. **10**(4), 546–557 (2019). https://doi.org/10.1111/ijag.13446
127. Virdis, A.: Fragmentation as a visual principle. From Cloisonné to early stained glass. Convivium Supplementum. A radical turn? Re-appropriation, fragmentation, and variety in the post-classical world (3rd-8th c.), 8-97. (2022). https://digilib2.phil.muni.cz/_flysystem/fedora/pdf/Convivium_09_2022_supplementum2_07.pdf. Accessed 29 June 2024

128. Wisniak, J.: Nicolas Leblanc. Chemical revolution and social injustice. Revista CENIC Ciencias Químicas. **35**(3), 163–171 (2004)
129. Wypyski, M.T.: Technical study of renaissance venetian enamelled glass. In: *Proceedings of the 17th congress of the International Association for the History of glass, Antwerp, Belgium*, pp. 4–8 (2006)
130. Wood, M.: Glass Beads and Pre-European Trade in the Shashe-Limpopo Region. Unpublished Masters Thesis. University of the Witwatersrand, Johannesburg (2005)
131. Wood, M.: A glass bead sequence for southern Africa from the 8th to the 16th century AD. J Afr Archaeol. **9**(1), 67–84 (2011). https://doi.org/10.3213/1612-1651-10184
132. Wood, M.: Glass beads from pre-European contact sub-Saharan Africa: Peter Francis's work revisited and updated. Archaeol Res Asia 6, 65–80. (2016). https://www.scimagojr.com/journalsearch.php?q=21100393521&tip=sid&clean=0. Accessed 15 July 2024
133. Wood, M.: Glass beads and Indian Ocean trade. In: The Swahili World, pp. 458–471. Routledge (2017)
134. Zanini, R., Moro, G., Orsega, E.F., Panighello, S., Šelih, V.S., Jaćimović, R., van Elteren, J.T., Mandruzzato, L., Moretto, L.M., Traviglia, A.: Insights into the secondary glass production in Roman Aquileia: a preliminary study. J. Archaeol. Sci. Rep. **50**, 104067 (2023). https://doi.org/10.1016/j.jasrep.2023.104067
135. Zhao, H.X., Li, Q.H., Liu, S., Gan, F.X.: Characterization of microcrystals in some ancient glass beads from China by means of confocal Raman microspectroscopy. J. Raman Spectrosc. **44**(4), 643–649 (2013). https://doi.org/10.1002/jrs.4239
136. Zlámalová Cílová, Z., Gelnar, M., Randáková, S.: Trends in colouring blue glass in Central Europe in relation to changes in chemical composition of glass from the middle ages to modern age. Fortschr. Mineral. **11**(9), 1001 (2021). https://doi.org/10.3390/min11091001

Chapter 18
Japanese and Chinese Porcelains and the Role of Jesuit Missionaries

Abstract The role of Jesuit missionaries from Europe (predominantly Portugal, Spain and France) to the Imperial Courts of Japan and China in the seventeenth and eighteenth centuries and their influence upon the production and decoration of Japanese and Chinese porcelains imported into Europe. The critical revelation of the composition and production processes of Chinese porcelains at Jingdezhen by Father Xavier Dentrecolles to his superior Father Orry in Paris in the first two decades of the eighteenth century and its impetus on European porcelain manufacture thereafter. The Jesuits and their diplomatic gifts of Limoges, Blois and Paris enamelled metalwares to the Emperor Kangxi and the resultant upsurge in the Chinese polychrome porcelain decoration and trade with Europe in pigments, especially cobalt blue, gold rose and Naples yellow: the rise of *famille verte and famille rose* decorated porcelains.

Keywords Jesuits · Seminario · Kangxi · Xavier Dentrecolles · Arita wares · Limoges enamels · Famille verte · Famille rose · Porcelains

18.1 The Context and the Questions to Be Resolved

18.1.1 Historical Context

To discuss and compare Asian and European enamelling technologies since the seventeenth century, it is essential to know the principles of the manufacturing processes and the physical and chemical phenomena involved. These processes are dependent on the raw materials and the production tools (for grinding and firing) as well as the practices and know-how of the artisans. The subject of the transfer of enamelling technologies from Europe to China and vice versa is vast, this chapter will condense the information obtained recently and details can be found in already

H. G. M. Edwards, P. Colomban, *Blue by Fire: A Marker of the Technical History of Glass and Ceramics*, Cultural Heritage Science,
https://doi.org/10.1007/978-3-031-77122-4_18

published articles and the specialized literature, these references will be cited gradually. As we will indicate later, most of the publications are based on collections of porcelain and enamelled metal objects, Chinese enamelled glasses have not generally been analysed and and hence will not be discussed further.

It is common to date the first mention of Chinese porcelain by Europeans from Marco Polo's story written in 1298, "*Le devisement du Monde*" [138]. A bowl preserved in the treasury of the church of Saint-Marc in Venice, considered to have been brought back by Marco Polo, would be one of the first and oldest imports, but its attribution still remains in dispute [27, 156]. The growing attraction for these rarities is attested by their representations in sixteenth, seventeenth and eighteenth century paintings and by some objects preserved from ancient royal collections [156]. It was only with the establishment of the Europe-Asia maritime trade networks by the Portuguese from the second half of the sixteenth century that the importation of porcelain became significant, especially blue-and-white porcelain although this was limited to elite clients, with maritime routes remaining very dangerous and the imported products being very expensive. Exportation from China during the Yuan and Ming Dynasties and after the 2nd Ming Dynasty ban on exportation (1521–1529) from Vietnam was significant towards the Indonesian Sultanates and in the Arabic-Persian Gulf. Some of these artefacts diffused towards the Ottoman Empire and are now preserved at the Topkapi Palace Museum (Istanbul, Turkiye). Portugal (Fig. 18.1) was supplanted in the seventeenth century by England (with the founding of the *Honourable East India Company* (HEIC) in 1600) and Holland (*Verenigde Oostindische Compagnie* (VOC) founded in 1602), and to a lesser extent by France (*Compagnie des Indes Orientales*, founded in 1664), and by Sweden (*Svenska Ost-Indiska Companiet* (SOIC), founded in 1731) which created groups of shareholders equipping ships for the Orient [5, 91, 151, 160] in order to mainly import tea (~90% of the tonnage, the rest being spices and dye products, and a little white or blue-and-white porcelain). Ceramics, a heavy material, provided the essential ballast for the ships on the way back to Europe, ballast often being made up on the outward journey by the flint accompanying the arquebus trade [78, 79].

The war between Portugal and the United Provinces (Holland) led to the decline of Portuguese traffic in favour of that of the Northern Companies. England (HEIC) supplanted Holland (VOC) from ~1760 [185] as shown in Fig. 1. The expansion of the number of connections from Europe led to a reduction in costs as well as the development of workshops in Canton specifically for exportation [193].

In the beginning of the eighteenth century, porcelain from China and Japan was present not only in the European royal and princely courts but also among the nobles and the bourgeoisie [5, 17, 160, 188] which led the authorities to send emissaries to gather information on porcelain pastes and their preparation and to promote alternative porcelain production in Europe [40, 80, 82, 83, 86–88, 103]. A very limited production took place from 1575 to 1587 at the Court of the Medici (Florence, from which about 60 pieces have been preserved) [24, 126]. We suspect that other productions in Padua (~1625, Italy) and Fulham (~1675, England) were made (see Fig. 14.1) but only a few pieces produced in Rouen (~1673) have survived and have been analysed [20, 53, 58, 59]. Noted productions were made in Paris and at the

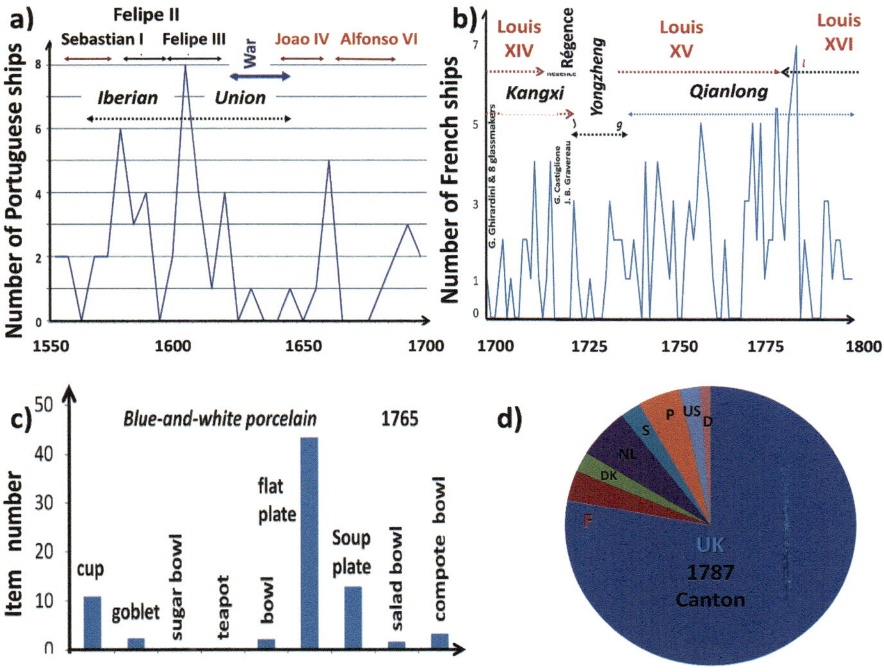

Fig. 18.1 Comparison of the numbers of Portuguese (**a**) and French (**b**) ships making the voyage to China between 1550 and 1700 and 1700 and 1800, respectively; (**c**) example of the distribution of boxes of blue-and-white porcelain from a shipment from 1765, the number of boxes of polychrome porcelain is very small; (**d**) the distribution of nationalities of ships arriving in Canton in 1787. (According to [76, 124])

neighbouring castle of Saint-Cloud belonging to the brother of King Louis XIV [25, 67, 112]. In fact all these productions used a technique different from that of Chinese artisans, a technique close to that used for the Ottoman Iznik paste made of sand bound by a vitreous 'cement' based on lead oxide, called fritware [113]. This first European 'porcelain' is called 'soft-paste porcelain' because the lead enamel covering it is 'soft', scratchable with a simple knife unlike that of Chinese porcelain, and has been fired with its enamel at a higher temperature which allows for a highly aluminous and therefore much harder alkaline glaze. The white colour of soft-pastes is warmer than that of hard- paste Chinese porcelain but the soft–paste containing very little clay is not very plastic, which makes difficult the creation of large objects, as achieved by the Chinese potters.

With the aim of independence and economic gain, the European authorities favoured obtaining the 'secrets' of Chinese porcelain paste (the letters of Father D'Entrecolles to Father Orry, Prosecutor of the Society of Jesus, brother of one of the founders of the Vincennes-Sèvres factory, are one of the best-known examples [72, 105, 162]) and from the 1740s the number of factories in Europe doubled every

5 years until the 1770s (with more than 50 significant factory being created) and then growth slows down (Fig. 14.1) [58, 59].

Three periods can be expected, therefore, a period from ~1680 to 1730 where only a few workshops hold the 'secrets' of manufacturing, a second between ~1730 and 1780 corresponding to a dissemination of know-how and then a period of consolidation after 1780.

18.1.2 Technical Context: Porcelain

In China, porcelain is obtained by combining a white rock (a phyllosilicate incorrectly called clay, '*gaoling*' in Chinese, a hydrated aluminosilicate of the kaolin or halloysite type, formula: ($Al_2Si_2O_5(OH)_4$)), or in the south of the country, sericite (another type of hydrated alkali aluminosilicate ($KAl(AlSi_3O_{10})(OH)_2$), resulting from an incomplete alteration of micas, often mixed with quartz, or sodium feldspar (albite)) and a pegmatite (called '*petuntsé*' or porcelain stone, or '*shi*' in Chinese), a rock combining different feldspars or plagioclases (alkali and alkaline-earth aluminosilicates with quartz) [73, 189, 195]. The paste prepared from crushed rocks is shaped and covered with a powder mixing the crushed rocks above with limestone or grass and wood ashes providing the elements (liquid phase former called flux) for lowering the melting point (the composition medium is an amorphous aluminosilicate of potassium and calcium with a little sodium) called 'glaze' [42, 47]. The whole is fired at high temperature between ~1300 and 1450 °C, in a reducing atmosphere (firing with wood or charcoal depending on the region) to obtain a white paste [189]. The reducing atmosphere is mandatory to reduce iron to Fe^{2+} ions to obtain a white body. The final phases are mullite (composition 3 Al_2O_3-$2SiO_2$) for ~30–40% of the volume (mullite forms long needles that forms a network that retains the molten glass in between, preserving the shape of the object during the firing) and a vitreous phase (sodium, potassium and calcium aluminosilicate, 60–70% vol.) and some residues of raw materials which have not reacted completely, mainly grains of quartz.

By comparison, a soft-paste is fired at a temperature below 1200 °C in a oxidizing atmosphere to burn '*Chymie*', the organic compounds added to improve the mechanical strength in the green state. The paste is much more siliceous (~75% of quartz grains) with a lower proportion of glass and is slightly porous [69], i.e. with much lower mechanical strength than hard-paste porcelain. The temperature range for a 'good' firing is rather limited.

The blue decoration of hard-paste porcelain is generally painted directly onto the 'green' paste of the object (dried, but nor fired, thus porous and therefore absorbing the 'blue aqueous ink' like paper) before depositing the layer which will constitute the final surface ('underglaze decoration'). Then the decorated body is fired at a high temperature. Sometimes, in southern China and Vietnam, the blue decoration is made on the glaze powder already deposited (not fired and porous, and called 'in glaze decor'). The ink is an aqueous suspension of a fine powder obtained from a more or less purified cobalt ore (the associated elements are a 'signature' of the geological origin). Adding the polychromatic decoration is usually done during a second series of firing cycles. The surface of the object is no longer porous (it is then a glass), so the decoration must then be applied (painted) with a syrupy medium (a mixture of turpentine, oil and or glues/resins) containing both the ion chromophore (a transition metal ion), namely Cu^{2+} for green, Co^{2+} or Fe^{2+} for blue, Fe^{3+} for yellow (the colour thus depends on the oxidation state of the ions, and therefore on the heating atmosphere), Mn^{2+} for purple (eggplant) and the material which will form the enamel matrix, an aluminosilicate powder and flux, namely compounds rich in sodium and/or potassium and/or calcium and/or lead and boron and lithium [47, 85]. This procedure is used for soft-paste porcelain: the decor is applied on the already fired body which is then fired again between 600 and 1000 °C.

The composition of raw materials is complex and a chemical element is always provided by several raw materials: for instance silica is present in kaolin, in pegmatite and feldspar but also in smalt [48, 49]. Ceramic art consists of balancing the mixture of several raw materials to achieve the desired aesthetic and functional characteristics. Each new decoration/colour is fired at a lower temperature of around 100 °C or more less than that of the previous firing cycle. The enamel is homogenized by fusion and reacts partially with the already fired glaze. Several firings, each time at a lower temperature, may be necessary in controlled atmospheres when the decoration is complex.

18.1.3 Technical Context: Enamels on Metal

The situation is similar for creating an enamelled decoration on metal, also a non-porous support like a glass or a glaze, but the coefficient of thermal expansion of the metal substrate is very different: that of a metal is double that of a hard-paste porcelain but only a little higher than that of soft-paste porcelain due to its high quartz content [47, 85]. As the volume after fusion is approximately half that of the original powder, several filling-firing sequences of the cells delimited by the metal walls (thin foils) of *cloisonné* enamel are necessary. At the end the enamelled object is then polished and often the metal surface is cold gilded (with a Au-Hg amalgam). To adjust the colour, different compositions or colours of glass can be deposited successively in a cell. As polishing destroys the perfectly smooth surface created by fusion therefore *cloisonné* enamels have numerous pores and, due to their high thickness (>mm), numerous cracks. They are very sensitive to shock and their intact preservation is often problematic.

A ceramic material such as glass is poorly resistant to tensile stress but much better suited to compression. When cooling, the enamel (its thickness varies between approximately 100 µm and 2–3 mm) must therefore be placed in (slight) compression and not in tension: this implies a coefficient of expansion of the enamel greater than that of the substrate; then the much thicker substrate is little disturbed by the enamel layer. The composition of the enamels must therefore be adapted to that of the support. The enamel must also react with the metal support to assure a good adhesion. The presence of an element such as fluorine facilitates this bond.

Europe, whether Celtic or Mediterranean, has produced polychrome enamelled objects on different alloys for several millennia [93, 119, 170]. In the Middle Ages, numerous masterpieces were produced in Byzantium, Italy, Spain and France, particularly in Limoges, with mainly alkaline enamels [8, 37, 93, 115, 159, 192]. Production in Limoges developed during the Renaissance and the seventeenth century. It lost its aura in the eighteenth century before a certain return to favour came at the end of the nineteenth century with the Troubadour/Neo-Gothic period. These enamel decorations on metals, alloys of gold, copper, and more rarely iron, combine coloured areas separated by 'bare' metal, from large (several mm in *champlevé* decorations) to fine (less than1 mm in the '*cloisonné*') or to form small paintings ('painted enamels').

In China, the production of *cloisonné* enamels developed under the Ming Dynasty [116, 184] and it is considered that the Silk Road was at the origin of the transfer of know-how from the Byzantine and via Islamic Worlds to China. With the exception of a few cases to which we will return, Chinese *cloisonné* enamels use a lead-based flux unlike Limoges enamels, which were almost lead-free before the nineteenth century [116, 184].

18.1.4 Missionaries in Japan and China

European missionaries (Franciscans, Jesuits, Dominicans, Lazaristes/Vincentians) arrived in the sixteenth century although some were present—episodically—well before, under the Tang and the Yuan Dynasties, with the establishment of the first bishoprics.

Japan was the preliminary stage and at the late sixteenth and beginning of the eighteenth century, Macao depended on the Prosecutor of the Japanese Mission (more than 250 churches, 300,000 Christians!), which was the made up of Portuguese (~60%), Spanish (20%) and Italian (20%) missionaries. The contribution of the Jesuits to the appearance of the first polychrome enamels on porcelain in the kilns of Yanbeta (>1640) and Nabeshima (>1690) at Arita (Kyushu) has been recently established [147] by the demonstration of the use of cobalt imported from Europe and the first use in Asia of lead-tin yellows. European smalt was also used for painted sacred art (paintings, sculptures).

The first ceramic productions of Arita were used for the recently established Tea ceremony and it is symptomatic that five of the first seven Tea Masters established by *Sen no Rikyu* (also known as *Soeki* or *Reiki*) were Christian *daimyo* (Lords) [149], which testifies to the close links between the arts of Fire, European missionaries and Tea Ceremony. It is also reported that Chinese craftsmen participated in the *Seminario*, the training 'school' for European painting techniques on all media established by Giovanni Cola in 1583 in Arie, near Nagasaki [143, 146, 148].

The reigns of Kangxi (1661–1722), Yongzheng (1722–1735) and Qianlong (1735–1796) (Fig. 17.1), the first 'great' emperors of the Qing Dynasty (1644–1912), are contemporary respectively with those of Louis XIV (1651–1715), La Régence (1715–1723), Louis XV (1723–1774) and Louis XVI (1774–1789) and in private workshops (Jingdezhen and the Customs sector of Guangzhou/Canton), with recipes and/or ingredients from Europe, as reported both in the correspondence of French missionaries or diplomats and in the archives of the Manchu Imperial Court [64, 66, 122, 123, 178, 179, 193]. Indeed, the administration of the Imperial Palace noted in detail the appreciation and advice of the Emperor on the gifts offered (the archives report that the Kangxi Emperor appreciated painted blue enamels very much), the requests for manufacturing, the assessment of the results, etc. while the missionaries regularly sent reports on their activities asking what could enable them to develop their apostolic mission, to satisfy imperial demands and thus guarantee their possibility of remaining at Court and converting members of the Chinese elite [120–123]. An action most representative of the quality of the links established with the Court is that the French Jesuits taught mathematics in Manchu to the Kangxi Emperor himself [120].

It was customary for representatives of foreign countries or 'vassals' to offer gifts to the emperor, who accepted them or not, as well as to dignitaries of all ranks. This is how Europeans offered enamelled watches and clocks, scientific instruments, and other productions from European manufacturers, some of these objects were decorated with scenes painted in enamels [180, 198]. The long duration of the Europe-China travel (~2 years at least for a round trip) led missionaries, in particular the Jesuits (the Society of Jesus was founded in 1540, the other main missionary orders present were the Franciscans, then later from 1885 the Dominicans and the Lazarists/Vincentians), from their installation in Japan as in China, created the possibilities of producing on site all the objects necessary for their apostolic mission: objects of worship and the decoration of places of worship, portraits of authorities, etc. In addition, they naturally brought with them religious objects that could withstand the conditions of transport, which is the case with enamelled objects. In addition, the mission had to be self-financing through donations or participation in commercial activities (smuggling) [148, 149].

Table 18.1 lists the most significant events reported in the archives concerning the arrival of missionaries in relation to the arts of Fire in Japan and China. If the missionaries' primary objective was China, events led to a prior establishment in the island of Kyushu (Japan) and in Macao (San Juan/Sancian island) [19] due to unexpected circumstances. The initiative of Louis XIV to send an official French delegation of Jesuit scholars (particularly mathematicians, astronomers and geographers in relation to the Cassini project of accurately mapping the earth) gave a decisive role to the French missionary in technical innovation transfer from Europe to China between 1688 and 1703.

All the above mentioned data testify to the artisanal activities developed by the missionaries upon their arrival in a foreign country and the connections between the initial settings in Japan then, after the expulsion from Japan in the first quarter of

Table 18.1 Facts and people linked to the development of artisanal activities in the art of fire by missionaries in Asia [19, 77, 120, 123, 143–147, 163, 165, 198]

Date	Place	Fact	Personnage(s)	Remarks
1541–1550	Rome	Foundation of the company of Jesuits	Ignacius de Loyola François-Xavirus	
1543	Nagasaki, Satsuma (Kyushu, Japan)	Arrival	Portuguese ship	Import of arquebus
1549	Kagoshima (Kyushu, Japan)	Arrival	François Xavirus	
1551	Sancian (San Juan in Portuguese) island (Macao, China)	Departed for and died in 1552	François Xavirus	
1559	Kyushu	Permission to teach the Catholic faith	Daimyos of Kyushu Island	Importation of firearms
1575	Macau	Diocese attached to Japan	Procurator Valignano SJ	
1582	China	Arrival of the Jesuits		
1583	Arie, Arima (Nagasaki, Kyushu, Japan)	Establishment of the seminar and workshops	Giovanni Cola SJ	Production of paintings, drawings, then glass, and watches. Links with the Kyoto school of painting
1592–1598	Kyushu	Capture, transfer and installation of Korean potters	Toyotomi Hideyoshi	
1595	Arie (Nagasaki, Kyushu, Japan)	Glass workshop		Protection of Naito Tokuen Johan and Takayama Ukon
1614–1639		Persecutions		
1640–1650	Yanbeta kiln, Arita (Kyushu, Japan)	Production of porcelain with polychrome enamels		
1644–1661		Defeat of the Ming Dynasty		
1650		End of the Christian period. Closure of the country		With the exception of Dejima Island in Nagasaki for trade with the VOC

(continued)

Table 18.1 (continued)

Date	Place	Fact	Personnage(s)	Remarks
1655		Arrival	VOC representatives	Official gifts
1685		Arrival	Jesuit missionaries sent by Louis XIV	Official gifts
1686		Arrival	VOC representatives	Official gifts
1687		Correspondence	Fontenay SJ	Request for glazing ingredients
1690	Nabeshima kiln, Arita (Kyushu, Japan)	Production of porcelain with polychrome enamel		
1691		Correspondence	Gerbillon SJ	Request for glazing ingredients
1693	Beitang, Beijing	Establishment of the enamelling workshop		Contacts with the workshops in Jingdezhen and Guangzhou
1696		Management of the glass workshop	Kilian Stumpf SJ	
1698	Guanzhou et Beijing	Arrival of the French ship The Amphitrite	Arrival of the painter Gherardini SJ, master of *trompe l'oei*l, of 3 enamellers and 8 glassmakers (coadjutor brothers)	Coming from French Brittany and Normandy
1706		Gifts from Kangxi to the Papal Nuncio (glass produced on the palate)	T. Maillard de Tournon SJ	
1715	Beitang, Beijing	Arrival	Giuseppe Castiglione SJ	Famous painter
1716		Correspondence	Matteo Ripa SJ Giuseppe Castiglione SJ	Request for ingredients
1719	Beitang, Beijing	Arrival	Jean Baptiste Gravereau SJ	Enameller-painter
1720		Correspondence	Joseph de Mailla SJ	
1722		Request	Empereur Yongzheng	Documentation

the seventeenth century due to persecution, their redeployment towards China. In Beijing (Peking) the workshops are installed on the very site of the place granted to the Jesuits by the Kangxi Emperor for the establishment of their church, Beitang (old writing: Pétang) [63]. Archives indicate that the first of the workshops was devoted to the production of glasses and their enamelling [63, 134], in particular with the production of new types of glass for astronomy, the Jesuits having taken charge of the development of the observatory created in Beijing under the Ming Dynasty [107, 139]. It was only then that the enamelling was carried out on metal and then on porcelain in dedicated workshops. A Jesuit of German origin, with links to the world of glass production in German lands, Kilian Stumpf first directed the glass workshop [165]. Several painters were Italian (Table 17.1). Potential links with the three main areas of glass and ceramic production, Italy, France and the Germanic area are therefore possible. However, as shown in Fig. 17.1, the contribution to trade of French ships will quickly be reduced, becoming much lower in number and tonnage than that of the other companies, Dutch and especially English, the Portuguese contribution having almost disappeared in the eighteenth century. Also, most of the '*Chine de Commande*' porcelain glazed in Guangdong (Canton) will be produced for Northern Europe (Holland/United Provinces, England) [82, 83].

18.1.5 Experimental Constraints

If a large part of the collections of the Imperial Palaces transferred to Nanking around 1930 left on the initiative of Chiang Kai-shek in 1948–1949 for Taipei (the current museum opened in 1965 to the public) [18, 62], and that another part is visible in the museums of mainland China, the military events of the end of the nineteenth century (many objects from Summer Palace were shared between the Empresses Victoria (UK) and Eugénie (France)) and the attraction of European collectors for these objects led Western museums, especially London and Paris, more particularly in France with the national museum of Asian arts-Guimet (mnaa-Guimet), the museum of Empress Eugénie-Château de Fontainebleau [155, 172], the Museum of Decorative Arts (MAD), the Louvre Museum and the National Ceramic Museum-Sèvres, to hold significant collections of Chinese enamelled objects, including pieces produced for the emperor and his court. The value of these pieces is very high and growing regularly, which means that the costs of packaging and insurance of these objects prohibit their transport to a laboratory and of course prohibit any sampling. The rarity of these pieces also means that the fragments are extremely rare. The analyses must therefore be carried out on site with non-destructive, contactless techniques (non-invasive) and since these objects are for the most part exposed, the time that can be devoted to their analysis will therefore be limited. Since the 2000s experience in the development of methods and conceptual tools for interpreting non-contact analyses (using electromagnetic radiation), perfectly non-invasive, has enabled us to know how to collect a lot of information contained in the material of objects and thus to compare the raw materials and manufacturing processes of objects [20, 21, 40, 41, 53, 84]. These analytical techniques use the interaction of two types of electromagnetic radiation, lasers emitting coherent monochromatic light and X-rays. In the first technique the interaction takes place with the external electronic layers (this is Raman scattering) and in the second one with the internal (this is X-ray fluorescence spectroscopy) atoms of the volume illuminated by the excitation beam, laser or X-ray beam. Depending on the accessible zone this will be the body of the object (porcelain, glass or metal alloy) or the decoration (enamels).

Non-invasive Phase Identification

Raman scattering spectroscopy resulting from the inelastic interaction of the laser beam with the external electronic layers (those which form chemical bonds) modified by atomic vibrations is often described, like absorption or infrared reflection spectroscopy, as 'molecular spectroscopy' because it identifies chemical entities very precisely: in the simplest case, for example a gas, these are true molecules (or ions sometimes) but in the solid state these entities are more complex but they remain the vibrational units which determine the Raman spectrum also called the 'vibrational spectrum'. Indeed, Raman spectrometry affords a double approach to the analysis of matter, mechanical (the signal is a function of the mass of the atoms and the chemical bond strength) and electrical (the signal is a function of the electrical coupling between atoms, in other words the electrical conductivity) [11]. In fact, if the analysis is done across the entire spectral range, the information is much richer with the possibility of identifying crystallized or amorphous phases (glass), their symmetry and certain other physical characteristics. It is thus possible to classify crystalline and glassy silicates according to their structure and their melting temperature [22, 33]. The choice of analysis optics makes it possible to modify the surface analysed from approximately 1 μm^2 to several cm^2, and even more by carrying out maps. The in-depth analysis can reach a few mm that is to say from the surface of enamel to the support paste.

The main weakness, which can also be a strength, is that the Raman signal of a phase (that is to say a form of matter defined by a particular composition and structure) is extremely variable. In a mixture the spectrum of a minority phase (<1%) can dominate and make that of the other phases barely visible, or even invisible. Indeed, the Raman response depends on the nature of the chemical bonds: it is non-existent for metals, extremely weak for very ionic compounds ('salts' such as chlorides, fluorides, etc.) but strong or even very strong for compounds with covalent bonds (oxides, nitrides, and carbides). The signal depends on both the colour of the laser excitation and that of the material. A particular combination leads to the phenomenon of 'Resonance Raman' scattering which maximizes the sensitivity of the technique. The impossibility of moving most objects requires the use of mobile instruments [34, 84] whose performance remains to date lower than that of the most sophisticated fixed laboratory devices.

Non-invasive Comparisons of Compositions

Elemental analysis, that is to say the identification of chemical elements (in fact for only some of the chemical elements) uses the interaction of an X-ray beam with the electronic layers, close to the atomic nucleus (the so-called X-ray Fluorescence spectroscopy: XRF), and is therefore very sensitive to the nature of this atom. This interaction leads to the emission of new X-rays at different energies, characteristic of the chemical element, over a wide energy range. At low energy we mainly observe the peaks of low mass atoms (e.g. argon, magnesium, aluminium, silicon, sulphur). Hydrogen, oxygen, carbon, lithium, boron and even generally sodium cannot be measured with mobile instruments. These light elements remain difficult to measure non-invasively by alternative methods, whether fluorescence excited by an electron beam (electron microscope) or an ion beam accelerator. Only destructive methods such as mass spectrometry measure several of these elements completely and precisely to the point of differentiating their isotopes.

In X-ray fluorescence, light elements are measured in the first microns up to ten microns from the surface, the range of penetration of the excitation beam. However, the skin of a ceramic material is always a little different from the volume [47, 110]. The slightly heavier transition metals, that is to say the colouring agents, are measured over a thickness of several hundred microns, which corresponds quite well to the thickness of the enamels. Comparisons of the contents of these elements will be reliable. On the other hand, the measurement of heavy elements using the most intense characteristic peak—which is what most calculation software does—is carried out over a large depth (mm) which induces pollution of the measurement by the material located under enamel. The result will be distorted. It would be necessary to take the measurement on a (polished) section of the object, transversely, in the configuration where the enamel is homogeneous over several mm, that is to say, make an excision of the object! (or use SEM or TEM for a very local analysis).

Even if all the elements present are measurable, the non-invasive analysis of an object rarely allows compositions to be quantitatively determined; the obstacles are of different natures:

- The positioning must not endanger the object and the shape of the objects means that only the convex parts are generally accessible to a measurement.
- A composition calculation assumes that the medium is homogeneous in the surface analysed and in the thickness up to a depth greater than the penetration of the electromagnetic excitation beam. This hypothesis is not valid for thin enamels ($<$200–300 μm) and even more so for enamels coloured by pigments (composite effect) or presenting variations in tone and enamels (required to make a sophisticated decoration) and/or with complex stratigraphy. No surface corrosion should exist. In that case only comparison of the signal (peak area after subtraction of the background) is possible.
- Remember that important elements in enamels, the most effective fluxes, that is to say sodium, lithium and boron, cannot be measured. On the other hand, X-ray fluorescence is effective for measuring certain impurities in small quantities (0.01–0.5% by mass of oxide or metal) while others in similar quantities will escape the measurement.

It will therefore be possible to only make comparisons for the elements whose signal is obtained in a manner relevant to the question asked.

18.2 Conceptual and Experimental Tools

We will specify certain definitions, operations and practices necessary for the comparison of glazing technologies.

18.2.1 Definitions

The body of the object, glass, ceramic or metal, constitutes the substrate on which is produced a layer or several layers based on vitreous silicate called by the generic term enamel whatever the type of substrate. Placed raw on a porcelain body which is in its also raw ('green' state), the enamel will be called glaze - '*couverte*' in French—after the whole has been fired. Placed on an already fired paste (but remaining porous) the enamel is also called glaze—but '*glaçure*' in French—(the case of earthenware/majolica and terracotta). The English language uses 'glaze' in all these cases, without distinction. Note that a 'porcelain glaze' also corresponds to the enamel on the steel sheets of gas stoves, bathtubs and washing machines! The layers of coloured glass made on a non-porous, i.e. waterproof, support (porcelain covered with a glaze, metal, glass object and already made enamel) are called enamels or 'overglaze'. If the decor is sophisticated, similar to a painting, we speak of 'painted enamels'.

Concerning the numerous categories of Chinese porcelain, we distinguish white porcelain (*qinbai*), celadon (green to yellow in colour like jades), *doucai* 斗彩 or 'contrasting colours' (combination of a blue design placed under cover glaze ('underglaze') or in the glaze ('inglaze') and a decoration made on the glaze in red and/or green (overglaze)), and the *wucai* 五彩, strictly 'five colours' (the overglaze decoration is made up of large coloured areas which can also be yellow, even purple and blue). *Falangcai* 珐琅彩 (*falang* corresponds phonetically to 'Franks' and more broadly to European foreigners, 'coming from beyond the seas') are porcelains with sophisticated decoration of enamels placed on already fired porcelain (with or without cover) and produced in the imperial workshops of Beijing. Produced in the imperial workshops of Jingdezhen, they are then called *yangcai* 洋彩 ('foreign' colour). The name *fencai* 粉彩 corresponds to 'soft, pale' colours. The term *ruancai* 軟彩 or 软彩 which translates also as soft colours is given mainly to productions made under the Emperor Yongzheng in opposition to *yingcai* 硬彩, hard colours. These classifications are partly subjective and partly overlap, the last two for example with *wucai*.

The French names given by Albert Jacquemart in 1862 [106] of *Famille rose* and *Famille verte* also cover these Chinese names, the *Famille rose* being a *wucai, fencai or ruancai* using colloidal gold nanoparticles for the pink colour and the *Famille verte* largely using green enamels but not pink colour. The *Famille verte* (circa 1720–1730) therefore chronologically precedes the *Famille rose*.

18.2.2 Raw Materials, Their Sources and the Production of Enamelled Objects

The production of ceramics and glass is carried out from natural raw materials, rocks extracted in the environment close to the workshop and water. Geological context being very different between Europe and China, some differences will be present in the raw materials regarding minor phases/elements and traces. Grinding, drying, firing (ceramics) and melting (glass, enamels) require a lot of energy and when a heat treatment makes it possible to directly obtain fine powders, this way is used (plant ashes, astonishment of flints, frits)[36].

The enamels are the most sophisticated part and their production requires great know-how [42, 85]. Their colouring is obtained by the addition of a very small quantity of rare material: colouring agents and opacifiers. Typically the volume of enamel corresponds to a few % of that of the object and the colouring agents themselves to a few % of the enamel, so in total less than 1 part per thousand. Consequently, colouring raw materials can be rare, expensive and have always been the subject of long, even very long distance trade: thus during the Yuan (Mongol) Dynasty the raw material providing cobalt was imported from Iran, a country that was part of the Mongol/Yuan Empire [48].

The distribution and concentrations of chemical (inorganic) elements on earth other than silicon, aluminium and iron are very variable and the geological sites allow even more exploitation. This is particularly true for cobalt and tin but also lead. Each site can be characterized by associated elements in major quantities or in traces (and specific distribution of isotopes). The exploitation involves a selection of the 'good' minerals, a certain purification by crushing or grinding, washing and sequences of chemical (dissolution by acids) and thermal (calcination leading to departure by volatilization, fusion by heating with addition of flux) treatments to obtain the 'ceramic raw material', for example smalt (a potassium glass containing up to 20% by mass of CoO made from the slag obtained in the exploitation of silver, bismuth and cobalt ores) from cobalt ores. Depending on the application, for a given time different commercial grades are produced, for different uses—and costs! It was only from the second half of the nineteenth century that sophisticated chemical treatments led to the commercialization of purified nitrates, carbonates, sulphates, and oxides of metallic colouring elements that could be described as 'pure' [48]. Note that in a ceramic (or metallurgical) process certain impurities or minor elements have a beneficial role or even are essential. Using 'impure' ingredients is therefore necessary. This is why many raw materials are named after their place of origin or marketing (Armenian bole, Bristol blue, Umbrian earth, etc.), each having specific associated elements. A recipe indicates the quantity of each raw material and not a chemical composition, each raw material always providing several chemical elements.

18.2.3 Firing Temperatures and Enamel Compositions

For a comparison of enamelling technologies it is necessary to know the temperature ranges of the different thermal operations.

Typical firing temperatures for porcelains and enamels are given in Tables 18.2a and 18.2b. These temperatures are indicative. It is recalled that the transformation of an inorganic material is a dynamic process which depends on the combination of

Table 18.2a Typical firing temperatures

Porcelain paste	Firing temperature (°C)	Atmosphere	Remarks
French soft-paste	1200	Oxidizing	'Fritware'
English soft paste	1250	Oxidizing	Phosphatic
French hard-paste (18^{e} s.)	<1400	Reducing	Lime rich
Meissen hard paste	>1400	Reducing	Alumina/mullite rich
French hard-paste (>end of nineteenth c.)	1280	Reducing	Potash rich
Chinese hard-paste	1300 to 1400	Reducing	Potash rich

Table 18.2b Enamelling temperature

Enamel/glaze	Firing temperature (°C)	Atmosphere	In French
Hard-paste glaze	1280–1450	Reducing	'*Couverte*'
English soft-paste glaze	1120	Variable	'*Feu de moufle*'
French soft-paste glaze	1000	Oxidizing	'*Feu de moufle*'
Enamel on glaze Painted enamel	600–850	Oxidizing	'*Feu de réverbère*' '*Feu de peinture*'
Enamel on metal	500 Toa 800 °C	Oxidizing	'*Four de verrier*'

temperature—time (speed of rise in temperature, level)—geometry of the particles (contact-reaction surface, therefore dimension)—compaction (surfaces in contact) and the '*Phase rule*' (the temperature of a material remains stable as long as the reaction is not complete: for example the temperature at the bottom of a pan remains at 100 °C as long as there is has water!). Liquefaction, viscosity and wettability, and hence the gloss and thickness of enamel, therefore depend as much on grinding as on the firing temperature [47, 101, 108, 137].

18.2.4 Colouring and Opacifying Agents

Colouring agents are classified into 3 groups [21, 42, 85]:

– Ions exhibiting electronic transitions in the visible, namely ions of transition metals (Fe, Co, Cu, Mn, Cr, V, etc.) and rare earths (Pr, Nd, etc.) and complexes combining several ions (e.g. Fe-S amber chromophore). Depending on the element, the concentration necessary for colouring will vary according to the colouring power, typically between 0.1% by mass of oxide for cobalt to several % for iron oxide. These ions are dissolved in the silicate network forming the glassy matrix.

- Nanoparticles of metals (Au°, Cu°) or semiconductor (CdS-CdSe) with dimensions varying between 20 and 100 nm; the colouring power is very strong and the quantities required are less than 0.1%. These particles are dispersed in the vitreous matrix.
- Grains of a crystalline phase coloured by the techniques mentioned below dispersed or formed in the vitreous matrix. The grain size varies between a few microns and several tens of microns. The phase (pigment) can be prepared in advance or formed by controlled nucleation-precipitation during the thermal cycle.

The colour depends on the chromophore ion and the site it occupies in its host structure. The variability of possible sites is reduced as the firing temperature increases and therefore the range of possible colours is also greatly reduced.

The opalescence and opacification of a glass are obtained by dispersing a phase whose dimension is of the order of the resolution of the human eye (~10 μm) having an optical index different from that of the glass. For white opacification, these are generally bubbles, calcium phosphates, certain silicates, cassiterite (SnO_2) and lead alkali/earth-alkali arsenates [35].

Additionally, a blue colour can be obtained from scattered light (interferences) due to formation of a complex sequence of alternated metal nanoparticles-rich and –free layers [23, 32].

18.2.5 Fluxing Agents and Control of Thermal Processes

Glasses and ceramics are silicates, more precisely aluminosilicates, and made of the main elements of the earth's crust. The corresponding oxides, silica and alumina, melt at 1810 and 2072 °C respectively [128]. The whole art of glassmakers and potters is to replace silicon (Si^{4+}) and aluminium (Al^{3+}) ions with (fluxing) ions replacing very strong Si-O and Al-O bonds with weaker Na-O bonds, Pb-O, B-O, etc. to lower the melting point of an enamel composition. It can be lowered to less than 600 °C. At the structural level we speak of depolymerization of the silicate network: in pure silica SiO_2 all the SiO_4 tetrahedron—the structural and vibrational unit—share their oxygen ion and form a full polymerized three-dimensional network [33]. The substitution of a Si^{4+} ion by 4 Na^+ or Li^+ ions, or by 2 Pb^{2+} ions or 4/3 of a B^{3+} ion will cut strong metal-oxygen bonds and thus lower the 'solidity' of the network (depolymerisation) causing a drop in the melting temperature, a decrease in viscosity but an increase in the expansion coefficient and reactivity. The art of enamellers is therefore to adjust these effects. Note that the reactions take place at the grain scale, at grain contact [47]. The local composition of each grain is therefore decisive, more important than the average composition for the processes. Thus, two rocks of the same average composition will not require the same heating conditions. We can see the importance of combinations of raw materials and grinding.

18.2.6 Firing Atmosphere

The heat treatment is carried out in a furnace, that is to say an enclosure where there is a flow of heat generated before the appearance of electricity (if we exclude the solar ovens with mirrors or lenses used from the eighteenth century) by a fuel, generally wood or coal: in fact in the furnace the effective fuel is always gas (issued of primary fuel): combustion is a reaction producing methane (CH_4) and hydrogen (H_2) then giving by oxidation CO, CO_2 and water (H_2O). We don't consider nitrogen (N_2) and its derivatives (NO, NO_2) in this process. Depending on the ratio of the gas volumes we speak of an oxidizing atmosphere (oxygen O_2 is the majority) or reducing atmosphere (H_2 and/or CO the majority). Dry wood will mainly produce methane; green/wet wood will give a high quantity of H_2. Hydrogen is the most efficient fuel. Most ions, particularly colouring ones, have several oxidation states (we speak of valence): thus, iron can be in the Fe^{2+} state (it colours pale blue) or Fe^{3+} (colours yellow). The atmosphere only acts on the surface because the diffusion of reducing gases in molten glass is very weak—with the exception of hydrogen which diffuses easily due to its small size. The reduction can only occur at the core thanks to the presence of go-betweens, impurities formed from multivalent ions of small diameter and diffusing rapidly in a molten glass such as iron (Fe), tin (Sn), antimony (Sb) or arsenic (As) which are reduced or oxidized on the surface by the atmosphere will go through a redox balance to modify the oxidation state and therefore the colour [174].

The control parameters are therefore multiple and the production of enamelled objects is a high-tech process. Oxydizing atmosphere is generally obtained in muffle furnace (the combustion gas flux goes around from the chamber in which the artefacts are placed).

18.3 Comparison of European and Asian Chinese Glazing Techniques in the Seventeenth Century

Table 18.3 compares the enamel colouring know-how practiced in Europe and China in the seventeenth century. It was commented on in a preliminary article [60]. The Japanese processes are quite similar to those used in China, except that opacification with fluorite is not used—to our knowledge.

The main differences concern both the glass opacification technique (use of fluorite in China, use of cassiterite, calcium phosphates and complex lead arsenates in Europe) and certain colouring techniques such as for reds (colloidal gold in Europe), yellows and greens (solid pyrochlore solutions - Naples Yellow in Europe). Added to this are the different types of cobalt sources in Europe and Asia resulting from the great differences in geological context and the qualities required due to the types of

Table 18.3 Comparison of glass/enamel colouring agents in Europe and China at the turn of the 17th–18th centuries. Phases detected by mobile Raman micro-spectroscopy are marked in bold and those whose elements are detected by mobile X-ray fluorescence (XRF) in italics. European recipes imported into China appear underlined, according to [60]

Colour	*Europe (characteristics/ formula)*	Innovation date (century)	China (characteristics/ formula)	Innovation date (Dynasty)
White	***Cassiterite (SnO_2)***	<5th	<u>Imported</u>	
			Fluorite (CaF_2)	Tang (618–907)
	K/Ca/Pb Arsenates	>16th	<u>Imported</u>	
	$Ca_3(PO_4)_2$	~9th	<u>Rare occurrences, imported</u>?	
Blue	***As-rich cobalt***	Antiquity	*Mn-rich cobalt*	Yuan (1271–1368)
Yellow	***Solution solide pyrochlore $PbSnO_4$-$PbSb_{2-x}M_xO_{7-d}$***	Antiquity (pure Sn) Renaissance (solid solution)	*Fe^{2+}/Fe^{3+} ions* *Tin-lead pigment*[a]	Han (206 BCE-220)
Green	*Cu^{2+}*	Antiquity	*Cu^{2+}*	Han
	Blue + ***yellow***	Renaissance	<u>Imported</u>	
Red	***Hematite ($Fe_{2-x}M_xO_3$)***	Roman	***Hematite ($Fe_{2-x}M_xO_3$)***	<Ming (Wanli, 17e s.)
	Hercynite ($Fe_{2-x}M_xO_3$)	Roman		
	Cu[a]	Celts	***Cu***[a]	Yuan
	Au[a]	Roman (?) Renaissance	<u>Imported</u>	

[a]since the Wanli reign (1563–1620, Ming Dynasty) that corresponds to their introduction in Japan at Arita [147]

firing practiced in different atmospheres. Firing in a reducing atmosphere is essential to obtain a white hard-paste to convert the iron into Fe^{2+} ions. Cobalt ore extraction sites in Asia are of the primary type, geologically little transformed [48, 49] which combine the different transition metals, in particular iron, manganese and cobalt in the ores. Only in a reducing atmosphere are cobalt ions effective blue chromophores. In oxidative firing the contribution of Fe^{3+} and Mn^{3+} or Mn^{4+} ions becomes dominant and the colour becomes greenish to black. On the contrary, European mining sites are of hydrothermal types, characteristic of old massifs where the primary sites have been dissolved, with a new distribution of metallic elements (the series of 5 elements Ni-Co-As-Ag-Bi in the form of sulphides, selenides or arsenides [118]). None of the metallic elements really compete with the chromophore effect of cobalt in an oxidizing atmosphere (there is little nickel and the ore is also tolerant of a reducing atmosphere). The arsenic reacts with lead to form an opacifying phase that contributes to the quality of the blue colour.

We will review the main colouring agents in Table 18.3 in conjunction with enamel painting techniques.

18.3.1 *Traditional Transparent Colours*

Colouring by a dispersion of chromophore ions in a glass is called 'transparent colour' ('*couleurs transparentes*') because of its low colouring power, suitable for solid or thick glass (stained glass, bodies of glass objects, thick *cloisonnés* enamels). As shown in the *Famille verte* sherd in Fig. 18.2a or the detail of the *doucai* bowl in Fig. 18.2g, the colouring power of the green enamel whose chromophore is the Cu^{2+} ion in an enamel rich in lead to obtain the green colour jade (an alkaline enamel would give a turquoise colour [85]) is weak. Note the numerous cracks due to the too high coefficient of expansion of green overglaze compared to that of the glaze substrate (Fig. 18.2a'''). Although blue colouring is also obtained with Co^{2+} ions without pigment formation [40, 41], due to the exceptional colouring power of these ions, the blue colour is dense and vivid.

Traditionally enamel is prepared by combining the *corpo* (major ingredients) also called 'flux' which will form the vitreous matrix and therefore determines the melting point, the expansion coefficient, etc. and the *anima* (minor) which contains the colouring agent and therefore determines the colour. *Corpo* is a mixture of crushed and powdered raw materials, or even with the addition of ashes. *Anima* is a molten material prepared before, rich in colouring agent, cooled suddenly in water to facilitate fine grinding (also called frit).

18.3.2 *Opaque Colours*

For these enamels the colouring agent is a pigment: the enamel is then a composite material combining a vitreous matrix and crystalline phases, one or a mixture of different pigments.

The use of certain pigments dates back to Antiquity [12, 35] but three periods correspond to the development of new technical solutions:

(i) a first period which goes from the Renaissance to the eighteenth century linked to the dissemination of recipes by the first printed technical works,
(ii) a second period in the middle of the nineteenth century with the development of industrial chemistry and patent filings, etc. with a new productive, industrial organization,
(iii) and finally the years 1960–1980 with the disappearance of internal production at each factory and the recourse to the supply of pigments 'ready to use', traditional and new, by a few companies active at the global level [51].

Examples of opaque paint are visible in Figs. 18.2b and 18.3, for instance with the red background obtained with hematite (α Fe_2O_3) as pigment, the yellow area coloured with tin yellow ($PbSnO_{4-d}$), white area obtained with lead

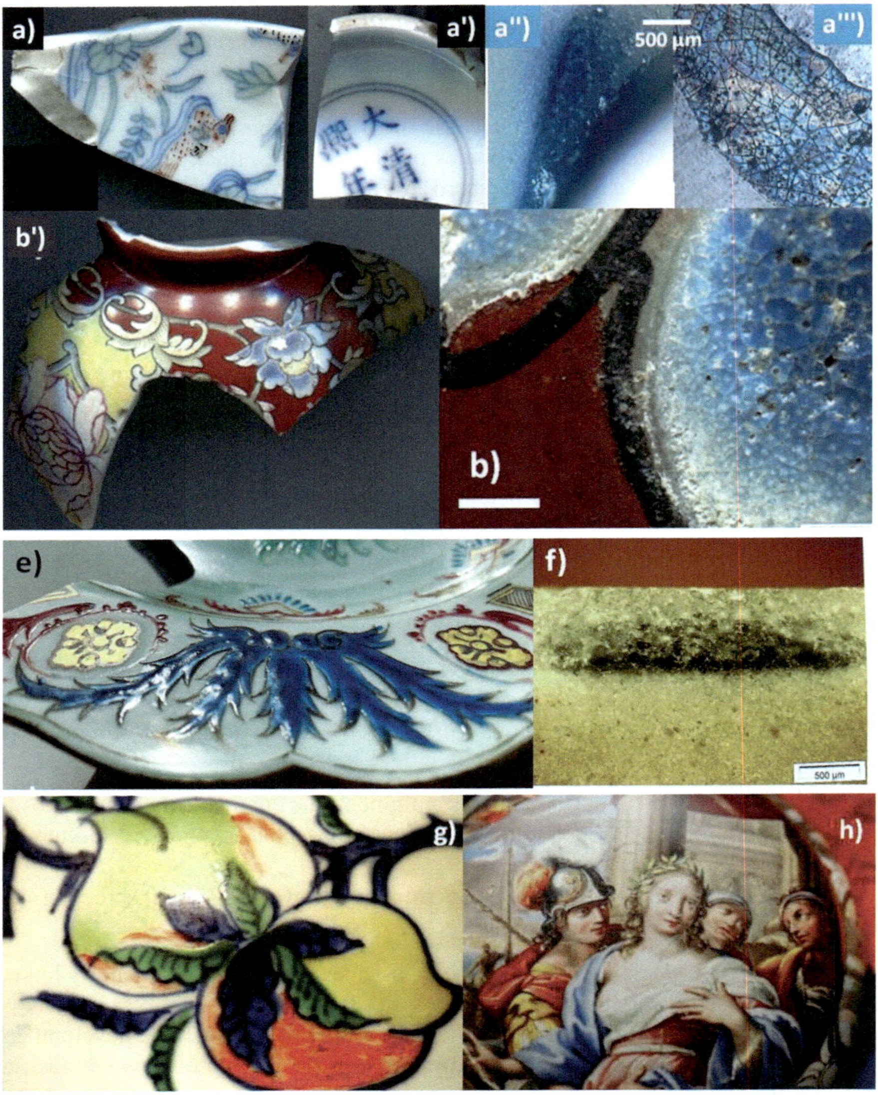

Fig. 18.2 Comparison between shards from the reign of Emperor Qianlong of *Famille verte/ wucai* (**a–a‴**) decorated with mainly 'transparent colours' and *falangcai* (bottom,b, b′) decorated with opaque enamels. Section **a″** of the *wucai* shard clearly shows the cobalt concentration at the paste-cover interface (underglaze blue) and the slight diffusion of cobalt in the cover glaze (bar = 500 µm); (**a‴**) top view of the green, cracked enamel (overglaze) on the colourless glaze; (**e**) see the thickness of some overglaze; (**f**) typical blue underglaze section; (**g**) use of 'transparent colours' in a *doucai* decoration on porcelain from the Kangxi reign (inv. G5696, Coll mnaa-Guimet) using 'transparent colours' and (**h**) a decoration of painted, opaque enamels on gold (watch, '*The Embarkation of Chariclée*', character from the Greek novel by Heliodorus of Emesa (third or fourth century), mid-seventeenth century, mechanism by G. Gamot, Paris, inv. OA8318, Coll Musée du Louvre)

Fig. 18.3 Examples of the gradual use of opaque enamels and their mixture: (1) inv. AR10218 (reign of Qianlong, Coll. Ariana Museum, Geneva), (2) inv. AR4608 (Yongzheng reign, idem), (3) inv. 2007-196 (Yongzheng reign, idem), (4) Qianlong reign bowl, (5) inv. 1937–615 (Yongzheng reign, Coll. Baur Foundation, Geneva), (6) inv. G3668 (Qianlong reign, Yixing kiln, Coll. mnaa-Guimet), (7) inv. 1930.630 (Daogang, idem)

arsenate- potassium-calcium: ($Pb_2Sb_{2-x-y}Sn_xM_yO_7$) and blue area coloured with cobalt ions and opacified with arsenates from the *falangcai* bowl fragment (Fig. 3, one of the other fragment of this bowl bears the imperial mark [40], coming from civil engineering works carried out in the Palace grounds, Beijing) [75, 130, 132]. Other examples are given in Fig. 18.2h (decor of a French watch, eighteenth century) [16, 43] and Fig. 18.4 (a, decor of a seventeenth century French watch; b, decor of a eighteenth century French watch; c, eighteenth *century Chinese cloisonné* enamels; d, painted decoration on ferruginous paste from Yixing) [58, 59].

The vitreous material is made opaque either by an opacifier (phase which does not give colouring), or by a pigment, or by combinations of these phases. The difference with transparent enamels is obvious when comparing with the *Famille verte* shard (Fig. 18.2a) where the blue was placed on the dry but unfired shard, covered by the glaze fired at high temperature before the affixing of the red and green decoration ('overglaze').

Examination of the sections of the *falangcai* bowl enamels shows the concentration of pigments on the upper surface due to the already fired glaze painting. The red background does not appear everywhere, the painting was done in reserve and we see that the black lines were done last to give sharper to the delimitation of the coloured areas. We also see that the blue was painted on a white layer but that the red was placed over the yellow, white being placed on the red. The decoration is therefore made both by reserve and by superposition of layers. Figure 18.3 gives illustration of the decors obtained using painted opaque enamels, technique introduced by the missionaries. It is reported that with their help the number of colours increased from 6 (blue, green, red, yellow, black, gilding) to more than 30 [178, 179].

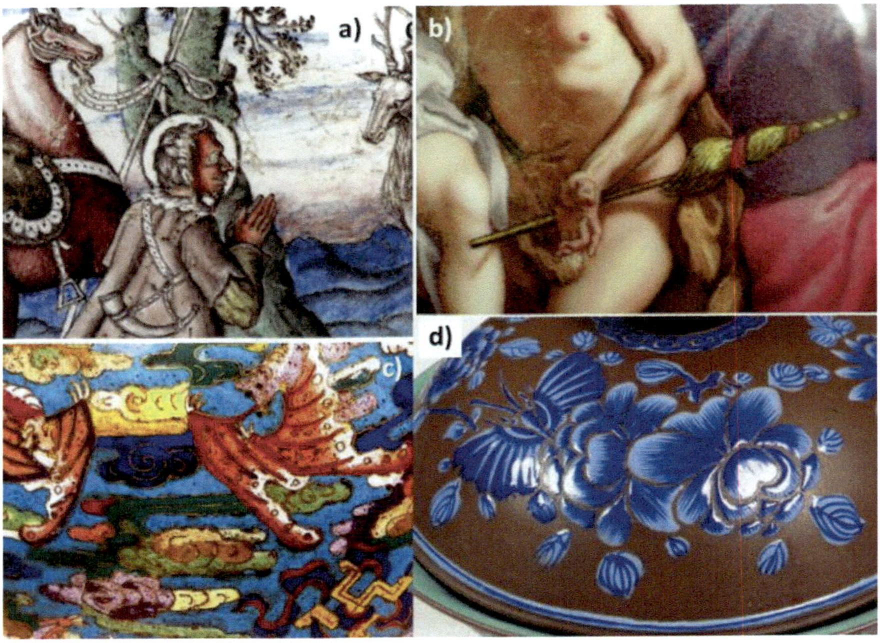

Fig. 18.4 Comparison of the variety of enamelled decoration techniques: (**a**) enamel painted with superimposed touches (watch, '*Vision of St-Hubert*', circa 1640, mechanism by Jehan Augier, Paris, inv. OA7074, Coll. musée du Louvre); (**b**) coloured dotted enamel (watch, '*Hercule et Omphale*', mechanism by Julien Le Roy, Paris, 1832–1838, inv. OA8337, Coll. musée du Louvre); (**c**) *cloisonné* enamel on metal alloy, Qianlong reign (Coll. national museum of Asian arts-Guimet); (**d**) enamels painted directly on ferruginous sandstone paste, Yixing kiln, Qianlong reign (inv. G8062, Coll. mnaa-Guimet)

18.3.3 Painting Touch Technique

Before considering the spectrometric analyzes for identification and comparison of the chemical elements present (XRF) and the phases formed (Raman), we will compare more precisely the visual appearance of the enamelled decorations.

Figure 18.2a and g show two examples of transparent colours; their mixture cannot give rise to a tone scale. Also *doucai* and *wucai/Famille verte* porcelain are characterized by flat areas of colour, in a way comparable to watercolour [52, 61].

Figure 18.3 (3–7) shows enamelled decorations still mainly using the solid colour technique (black or gilded) delimiting the coloured areas. On the contrary, the decor of Fig. 18.3 (1 and 2) does not use this technique, with for example the modelling of the mountains in blue. The solid line technique allows the creation of sophisticated floral decorations. The enamellers use the technique developed 2 centuries previously by the potters of Iznik [26] of separating the coloured areas by a black or red line or sometimes a gold line (Fig. 18.3(4))

Fig. 18.5 Examples of dotted painting on metal (**a**, reign of Qianlong, inv. F1467C, Musée de l'impératrice Eugénie, Château de Fontainebleau; **b**, on porcelain, reign of Qianlong, inv. 1935.608, Coll. Fondation Baur, Genève)

The most sophisticated decorations use painting in very small dots, the density of dots as well as the colour of each dot offers an infinite range of colour and pattern. This technique was used in France from the eighteenth century (Figs. 18.2h, 18.4b and 18.5) by miniaturist painters creating illustrations with a magnifying glass lens, often including several characters on a surface area of a few cm^2 [127]. Figure 18.5 shows two examples of Chinese production from the Qianlong reign, one on porcelain, the other on metal (gold ewer), similar to what watch enamellers in France produced a century earlier (Figs. 18.4b and 18.2h). We can see the finesse of the design. This European technique of painting associating mixable opaque pigment and dot painting was revolutionary for Chinese craftsmen.

18.4 Early Evidence of the Use of European Techniques During Kangxi's Reign

The analytical results will make it possible to document the compositions and the nature of the phases constituting the enamels in order to identify the uses of recipes and/or ingredients coming from Europe, trying if possible to find out if the recipes come from France, Italy or Germany and how they have been modified by Chinese craftsmen. The final objective would be to identify artefacts made with European ingredients and establish a comparative chronological scale of their use in the imperial workshops and in the private kilns of Jingdezhen or Canton compared to uses in Europe.

18.4.1 Enamel on Glass

It is reported that the first productions of objects decorated with opaque enamels in China were made on a glass body. Unfortunately, artefacts are very rare. Only one study is available in the literature [134]. It concerns 13 objects analysed in a micro-destructive manner (sampling) using an electron microscope, including only 3 fragments attributed to the Qianlong period, the others earlier (Twelfth to fourteenth centuries), later or without attribution. Recall that the glass imperial workshop was established in 1696 by Killian Stumpf [165] and directed until his death in 1720. Then the activity was led by two Frenchmen, Brossard and d'Incarville, until 1757 and 1758, the dates of their deaths [134]. The main novelty compared to previous Chinese glasses regarding ingredients is the large addition of borax and arsenic. The production also concerns lenses for astronomical telescopes and the addition of boron is well known to modify the optical index, which is necessary for the different types of lenses required to build an astronomical telescope. Ma et al. report that borax was already used during the Tang and Liao Dynasties (Tibet produced borax) and that the Beijing Palace Museum only has one glass object with the Kangxi reign mark and a few dozen with the Yongzheng mark. This rarity is attributed to poor conservation due to the use of a lead-free composition, without supporting this assertion experimentally. It is reasonable to assume that production volume was also low. The objects analysed correspond to productions subsequent to the direction by the French craftsmen. These potassium-calcium type glasses (*boshan*) contain a little fluorine (F) and the recipes indicate the use of fluorite (called purple stone, purple is a classic colour of this mineral), saltpetre, quartz crystals (called *maya* in the form of horse teeth, as used for Venetian Cristallo) [166, 167] and undoubtedly feldspars. Knowledge on Qing glass production therefore remains very limited and all the analytical work remains to be done for enamels on glass.

Studies concerning European enamelled glass are also rare, for the same reason of the low number of 'accessible' objects. Analytical studies of glass objects have focused almost exclusively on the body of the objects and not on the enamels. In 2009 the first study by Raman microspectroscopy of enamelled glass was made by [166, 167] (Coll. Musée national de Céramique, Sèvres) and more recently a campaign of analysis of the productions of the glassmaker Bernard Perrot (Bernardo Perrotti, originally from Altare near Genova, Italy) from the period ~1666–1709, supplier to Louis XIV, was made in Orléans at the Hotel Cabu museum [45, 94]. Certain compositional data have been previously obtained for an exhibition dedicated to this glassmaker [94]. The main results are the confirmation that Perrot's gold red ('ruby' red) is obtained with gold nanoparticles prepared by precipitation of the solution in *aqua regia* through the addition of arsenic and that traces of arsenic are detectable by Raman spectroscopy through the precipitation of lead arsenates [45]. Large arsenic amounts (~% wt) are associated with cobalt in blue enamel. Opacification with arsenic is also widely used by Perrot who simultaneously uses opacification with cassiterite [45].

18.4.2 Enamels on Metals

Enamels on metals produced during the Ming and Qing have been the subject of in-depth analytical work for several decades [7, 102, 115, 191] as well as those produced in Limoges in Europe [8, 37, 116, 159, 168]. They have also been the subject of numerous stylistic studies, both concerning Chinese *cloisonné* productions [8, 164] than those of Limoges, Blois and Paris [16]. In 2008–2009, a study of Limoges enamels from their period considered to produce the best quality of painted enamels, that is to say the Renaissance, was made [116] as well as that of Chinese *cloisonné* enamels from the Ming and Qing Dynasties (Fig. 18.6b), without knowing the historical data concerning the role of the Jesuits at the Manchu Court. Kırmızı et al. noted [115] that certain enamels, mainly green in colour, of objects attributed to the seventeenth century (Qing) used enamels with an alkaline base, such as Limoges enamels while all other Chinese enamels had an alkaline base that was rich in lead (Fig. 18.6a). We will see later that it is also the different varieties of greens which are used for opaque green painted enamels on porcelain which use European recipes, namely a pyrochlore type pigment of complex composition (Napes Yellow) in a coloured glass matrix with cobalt.

The work of Limoges enamels from the 17th and 18th centuries was the first comprehensive analysis relative to pigment identification [37], the artefacts of this period that have been practically never been analysed. The chart (Fig. 18.7) is

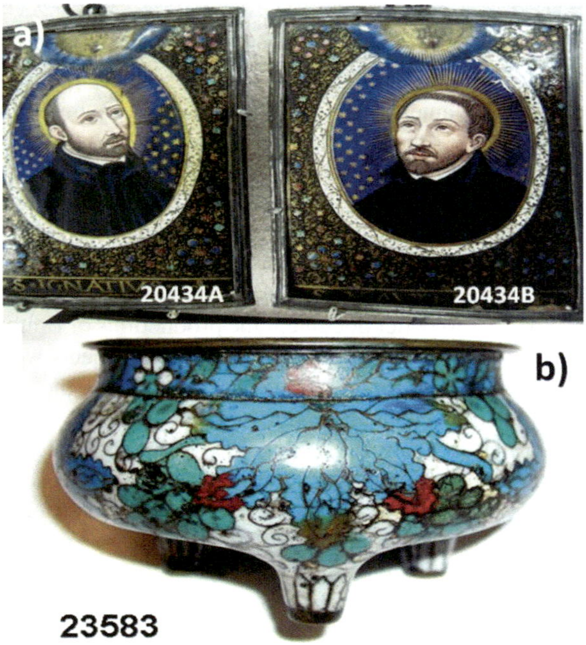

Fig. 18.6 Comparison of know-how in enamelling on metal in the seventeenth century in France (**a**) and in China (**b**): (**a**) Limoges enamels from the first century representing the founders of the Company of the Jesuits, St-Ignacius (Ignacius de Loyola, inv. 20434A) and St-Xavirus (François-Xavirus, inv. 20434B) attributed to Leonard III Limosin (circa 1635–1650); (**b**) vase with cloisonné enamel decoration attributed to the seventeenth century (reign of Kangxi, inv. 10,976)

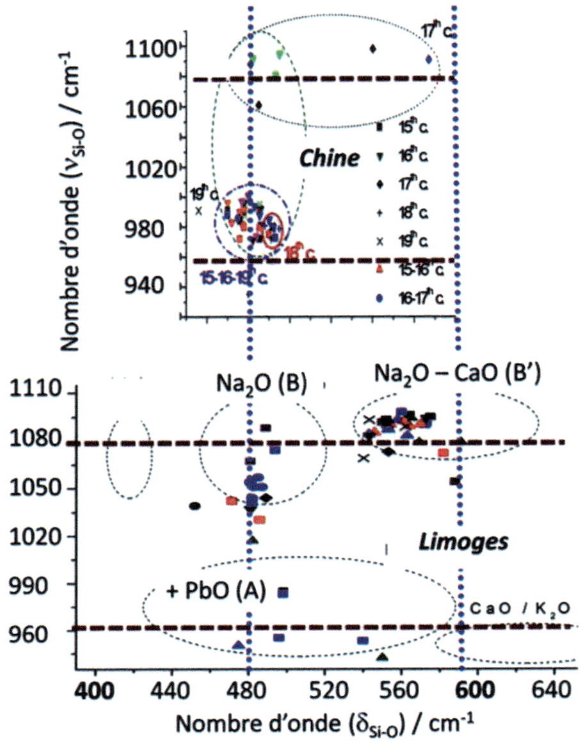

Fig. 18.7 Chart for classification of types of vitreous matrix of enamels on metal constructed from the two main parameters of the Raman spectrum, the wave numbers of the maxima of the bands relating to the modes of deformation (δSi-O) and elongation (νSi -O) of the SiO_4 vibrational unit (from [37]); flux are indicated: (top) Chinese *cloisonné* from the Ming and Qing Dynasties; (bottom) Limoges enamels (fifteenth to nineteenth century)

constructed from the main parameters of the Raman signature of the matrix of Limoges enamels and Chinese *cloisonné* enamels, that is to say the position in wavenumber (quantity equivalent to an energy) of the centre of gravity of the modes of deformation (δSi-O) and elongation (νSi-O) of the vibrational and structural unit SiO_4 of the glass classifies the different types of enamels: three types for Limoges enamels, at namely soda (the main flux is Na_2O), lime-soda (Na_2O and Cao fluxes) and lead (PbO flux); the latter type of glass was not used for this purpose until the eighteenth century.

In order to better document the productions of Limoges from the seventeenth and eighteenth centuries, a series of enamelled objects were analysed [37–39] in detail, including the portraits of the two founders of the Company of the Jesuits, Saint Ignacius de Loyola and Saint François Xavirus (Fig. 18.6a). These measurements confirmed the almost exclusive use of alkaline and alkaline earth enamels for the seventeenth century but highlighted the use of lead enamels from the eighteenth century in Limoges for the blue, yellow and red colours. Cassiterite remains the opacifier, hematite the red pigment and Naples Yellow with antimony (and zinc) are used. Arsenic in blue enamel arises from the cobalt ore characteristics: European cobalt ores come from hydrothermal veins associating the 'Five' elements according Kissin [118]. Figures 18.6, 18.7, and 18.8 show the variety of enamels on metal,

Fig. 18.8 Examples of enamelled metal objects: **a,c** and right ewer, (from a pair expected coming from Qing Summer Palace (Beijing), inv. F1467C, Empress Eugénie museum, Château de Fontainebleau), in gold decorated with *cloisonné* and painted enamel medallions, with Qianlong mark engraved on the feet back side; **b,d**: inv. F1448C, cloisonné incense burner vase, Empress Eugénie Museum

painted or *cloisonné*, the two techniques can be used simultaneously for the same piece (Fig. 18.8). These pieces combining several techniques appear to be very rare.

The finesse of the painted decorations is visible in Figs. 18.5 and 18.6. On the contrary, *cloisonné* enamels almost always have pores and fractures (Fig. 18.8) due to the removing of the glassy original surface by polishing. This is due to their manufacturing technique: the production requires several cycles of filling the spaces between partitions with enamel powder followed by firing until fusion; after the last cycle, the voids being approximately filled, a polishing step is carried out which leads to abrading the natural, smooth and bubble-free surface obtained by cooling the molten glass. In fact, the surface of a glass or ceramic, for various physical phenomena arising from the laws of diffusion and wettability, is always denser and smoother than that obtained by removal of this natural surface layer. In the case of the F1467C ewer, the examination of the enamels (observation of burrs [49] and the production of radiographs [153]) highlights the manufacturing technique from a hammered gold leaf. We can think that the *cloisonné* enamels were first made, fired and polished then the enamels of the medallions were made after the prior application (and firing?) of a white arsenic-based opacified base coat. The painted polychrome decoration was then made and fired, at a lower temperature than the *cloisonné* enamels, probably in several firing cycles. This low temperature (700 °C or less) would explain why these enamels maintain a matt appearance (Figs. 18.5a and 18.8a).

Comparisons between decorations painted on metal and on porcelain show great similarity and certain authors [191] consider that enamelling painted on porcelain is a transposition of the techniques used on metal. We disagree (the various coefficient

Table 18.5 Chemical elements and phases identified by pXRF and Raman in enamels on metal (minor and (traces)). All phases are rich in lead. The details of the attributions of the Raman and pXRF spectra have been discussed in [44, 49] for these and other objects (NY: Naples yellow)

Object	Reign	Analysis		
		Blue	Yellow	Green
F1448C *cloisonné*	Kangxi	Mn,Fe,Co, (Ni,Cu,Sn)	Sn,Sb,Fe,Zn[a]	Cu, Sn,Sb,Fe,Zn[a]
		CaF_2 (chromate)	$PbSnO_4$ Cassiterite	Pyrochlore NY Cassiterite
R957 painted	Qianlong	–	–	–
		CaF_2?	$PbSnO_4$ Lead arsenate	$PbSnO_4$ (lead arsenates traces)
R958 painted	Qianlong	–	–	–
		Lead arsenate	$PbSnO_4$ Lead arsenate	
R975 painted	Qianlong	Fe,Co,As, (Ni,Cu,Sn)	Sn,Zn(Ni)	Cu,Sn
		Lead arsenate	$PbSnO_4$ Lead arsenate	Lead arsenate
F1467C *cloisonné*	Qianlong	Co,Fe,(Cu,Sn)	Sn,Fe,Ni,(Zn)	Sn,Cu,Fe
			Pyrochlore NY Cassiterite	Pyrochlore NY Cassiterite
F1467C painted		Co,Fe,(As)	Sn,Fe,As,(Cu,Ni)	Sn,Fe,Cu,As(Mn)
		Lead arsenate Wollastonite	Pyrochlore NY Lead arsenate	Pyrochlore NY

[a]Possible contribution of Cu-Zn alloy walls; + possible contribution from a lower layer

of thermal expansion requires adaptation of the enamel composition) and will discuss this further after comparing the compositions and pigments.

The nature of the metal support was analyzed by XRF for the F1448C incense burner and one of the F1467C ewers [49]. In the first case it is a golden brass (by weight Cu: 65%, Zn: 28%, Sn: 6%, Fe: 1%) and in the second an alloy of gold and silver (Au: 84%, Ag: 14%, Cu: 1%, Fe: 1%).

Table 18.5 summarizes the elements identified by XRF other than the major ones of the vitreous matrix (Si, K, Ca and Pb) as well as the phases identified by Raman microspectroscopy on this first group of enamelled metalwares, the first analysed both by the two mobile methods. These analyses were used to refine the working procedure that will be implemented for the porcelains. The common characteristics of the set of elements associated to European cobalt (As, Ni, more rarely Bi and Ag) can be distinguished from the 'domestic' Chinese cobalt characterized by a high level of Mn and sometimes of Fe. The association of Cu with Co is characteristic of seventeenth century European cobalt [48, 49].

The white colour of the *cloisonné* enamels is obtained by CaF_2 fluorite [44, 49]. Detection is difficult because the main Raman peak at ~320 cm^{-1} of this phase is always weak and the fluorine element (F) is too light to be detected by XRF. This confirms the measurements made on the *cloisonné* from the collection of the Museum of Decorative Arts [115].

The blue *cloisonné* enamels of F1448C incense burner (Kangxi reign) do not show a crystalline phase, except fluorite in white areas. Cobalt is associated with a large quantity of manganese, which is typical of the 'domestic' cobalt used under the Ming [48]. On the other hand, we detect Naples yellow of complex composition (Sn, Sb, Fe, Zn) for the yellow and green areas. The use of a complex Naples yellow is new (end of Kangxi reign correspond to ca. 1700–1722) and should be linked to the use of an imported ingredient. Green associates European cobalt (not manganese but arsenic, and often nickel) with Naples yellow, a typical European technical solution. Tin yellow is detected in the yellow and green enamels attributed to Canton, e.g. Fig. 18.3 [54, 56] while green colours preferentially use complex Naples Yellows, that is to say with substitution of tin by antimony, zinc and undoubtedly also iron and silicon. Only imperial ewers use cassiterite opacification for certain greens (F1467C ewer). This is an indication that the use of ingredients and recipes imported from Europe still continues under Qianlong for exceptional objects.

The Raman examination of enamels painted by small points shows in part of these points (the diameter of the analysis spot is a few microns) the coexistence of numerous colouring agents and phases: for instance the Raman spectrum obtained on a green-yellow dot of the Fig. 18.5a medallion [49] identifies cassiterite (SnO_2), Naples yellow of complex composition (Sn, Sb, Zn, etc.), lead arsenate with apatite structure ($(Na_{1-x-y/2}K_xCa_yPb_4(AsO_4)_3)$), hematite ($\alpha$ Fe_2O_3), wollastonite ($CaSiO_3$) and the vitreous silicate matrix that contain Co^{2+} ions. It is the mixture at the micron scale of the different pigments which offers, with the association of points, an almost infinite palette. This simultaneous observation of multiple pigments in a volume of a few μm^3 is the characteristic of painted enamels that can compete with oil paint [52, 61] for realistic representation.

18.4.3 Painted Enamels on Glazed Porcelain

18.4.3.1 Imperial Artefacts

Imperial objects are identified by the conjunction of different criteria:

(i) the presence of marks indicating the year of the reign corresponding to the manufacture of the object and/or the name of the building of the Imperial Palace to which the object belongs,
(ii) the quality of the creation and
(iii) sometimes the descriptions in the palace archives.

The mark being one of the important criteria, its authenticity is questionable.

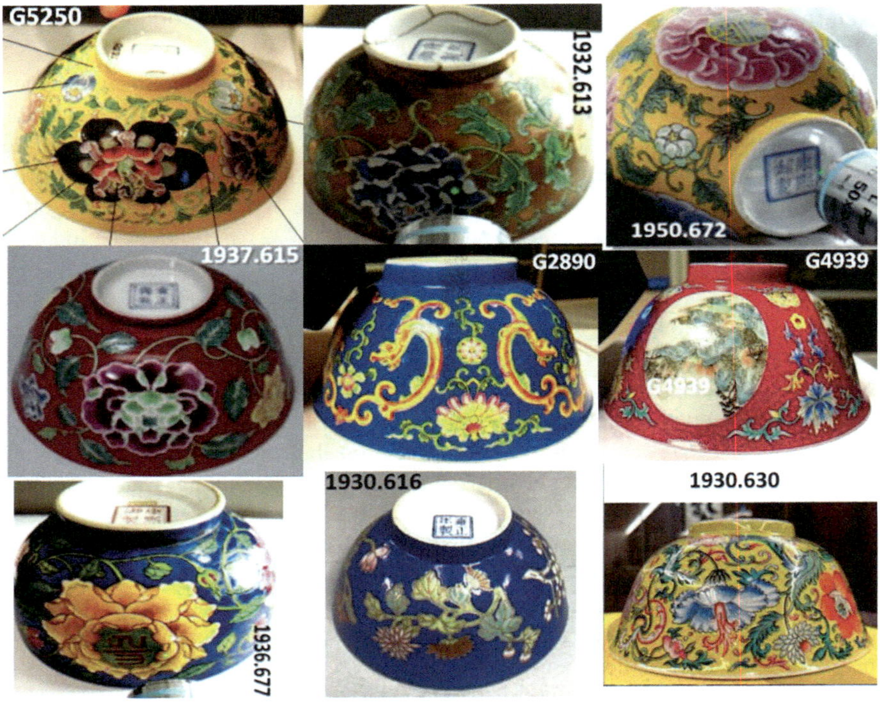

Fig. 18.9 Glazed porcelain with opaque painted enamels bearing imperial marks shown in Fig. 18.10. See Table 18.6 (Coll. Baur Fondation, Geneva)

Figure 18.9 presents some bowls that could be assigned as imperial productions. These analyzes have been the subject of several publications [38, 54, 55, 58, 59, 129, 130, 177]. Their marks are shown in Fig. 18.10.

These marks are generally painted blue directly on the paste before covering with glaze powder and firing. Sometimes they are made by overglazing, in blue (e.g. Fig. 18.10 bottom), red or rarely in black. More specific inscriptions indicate manufacture for an important event such as an emperor's birthday and lead to a more precise dating date. Ordered porcelain, often emblazoned and enamelled in Canton can also be dated by European documents attesting to the order and/or the delivery [83].

The decorations of the bowls in Fig. 18.9 are characterized by the use of opaque enamels. The constituent elements of the coloured areas identified by X-ray fluorescence are listed in Table 18.6 and the phases identified by Raman in Table 18.7.

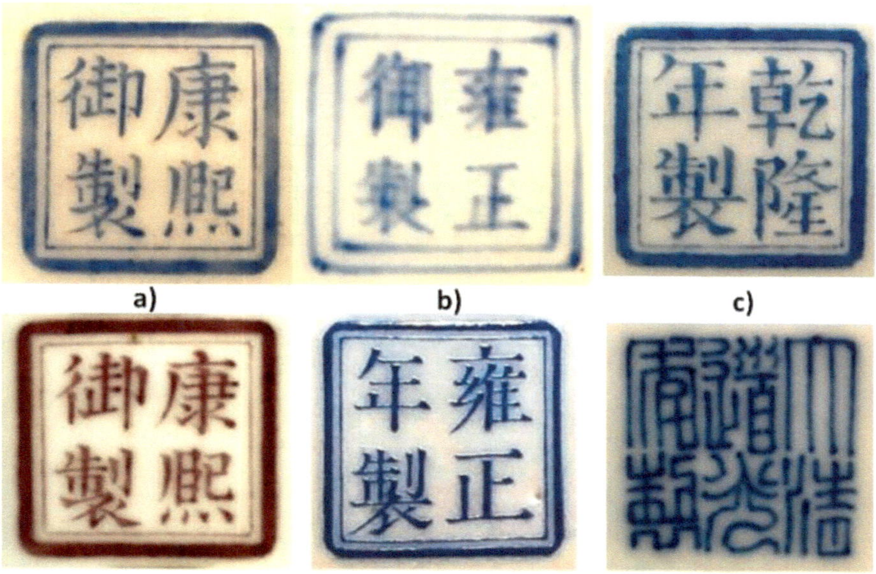

Fig. 18.10 Imperial marks, from left to right and bottom to top respectively in standard *kaishu* style relating to Kangxi ((**a**), the right column of the seal is the name of the emperor, the left column indicating 'made under'), Yongzheng (**b**), Qianlong (**c**, top) and in archaic *zhuanshu* Daogang style (**c**, bottom) (Coll. Baur Foundation, Geneva). (Note the different colours and the underglaze (painted on the paste) or overglaze (a slight relief is then visible))

18.4.3.2 Indications of the Use of Imported Ingredients

There are several colours for which decisive arguments make it possible to attest the use of ingredients imported from Europe and not only the use of European recipes or adapted from these recipes.

Concerning the presence of characteristic elements detected by pXRF (Table 18.6), blue, yellow and green are the colours that provide information. The numerous studies on Ming blue-and-white (see e.g. in references [48, 57, 73]) have well established the use of a cobalt ore very rich in manganese and iron, several times the amount of cobalt. It is reasonable to think that as it is indicated in the imperial archives that the objects enamelled in the imperial workshop at Beijing were made on 'blanks' produced in the kilns of Jingdezhen, therefore already marked in blue underglaze with the 'cobalt' used classically. Highlighting the use of 'cobalt' without manganese but associated with or with some of the five elements characteristic of hydrothermal veins in the decor will therefore be a criterion of importation. These objects will be noted in bold in the tables. Use of this 'European' cobalt for the mark will indicate a special order; the object will be in bold underlined in the tables.

Table 18.6 Chemical elements identified by FRX in porcelain enamels; (low intensity peaks); the objects using European ingredients are in *italics*, in **bold underlined** the objects maximizing the use of European ingredients, reign mark included

Object/Mark	Coll.	Date	Mark	Blue	Yellow	Green
G5696/Kangxi	Mnaa-Guimet		Mn,Fe,(Co)	Mn,Fe, (Co,Ni)	Fe	Cu
G823/Kangxi		<1713	Mn,Fe,(Co)			Cu
G5615/Kangxi		1715–1722?			Fe,Sn,Cu	Cu
G3361/Kangxi		<1723–1735	Mn,Fe,Co	Fe,Co,Cu,As	Cu,Sn	Cu
G5250/Kangxi		1715–1722	Fe,Co,Ni, As	Fe,Co,Ni,As	Fe,Sn,Sb,(Zn)	Sn,Cu
613/Kangxi	Baur	<1722	Fe,Co,Ni, As	As,Co,Ni,Sn	As,Fe,Ni,Cu,Sn,Sb	Cu,As,Ni,Sn
672/Kangxi		<1722	Fe,Co,Ni, As	As,Fe,Co,Ni	Sn,Sb,Fe,Ni	Cu,As,Ni,Sn
G823/Kangxi	Mnaa-Guimet	<1713	Mn,Fe,(Co)			Cu
677/Kangxi	Baur	<1722		Mn,Fe,Co,Ni,Cu,Zn	Sn,As	
G913/Yongzheng	Mnaa-Guimet	1723–1735	Fe,Mn,(Co)	Fe,Co,Mn,Ni,Cu,As	Fe	Cu,(Sn,Sb)
G4806/Yongzheng		1723–1735	Mn,Fe, (Co)	Fe,(Co,Cu,As)	Fe	Cu
615/Yongzheng	Baur	1723–1735	Mn,Fe,Co,Sn,Ni	As,Fe,Co,Ni,Cu,Mn,Sn	Sn,Ni,Zn	Sn,Cu,Ni,Zn,Mn
G2789/Yongzheng	Mnaa-Guimet	1723–1735	Fe			Cu,Sn,As,Fe
616/Yongzheng	Baur	1723–1735	Fe,Ni,Cu	Sn,Co,Mn,Fe,Ni,As		
TH487/Yongzheng	Louvre	1730–1735		Fe,Co,As,Ni, (Mn)	Fe,Sn,Sb,Ni	Cu,Mn,Fe,Ni,As,Sn, (Sb)
R1175/Yongzheng		ca. 1735		Fe,Co,Ni,As,Cu,Zn		
R1177/Yongzheng		ca. 1735			Zn,Sn	
R1041/Qianlong		1720–1740		Co,Fe,Ni,As	Fe,Zn,Sn	
R1045/Qianlong		1730–1745		Co,Fe,Ni,Cu,Zn,As		
R1048/Qianlong		1740–1760		Mn,Fe,Co,Ni,Zn,As	Mn,Sn	
G2890/Qianlong	Mnaa-Guimet	1735–1796	Fe	Fe,Co,Cu,Ni,Zn,As	Sn,Fe,Zn	Cu,Fe,Sn
G4939/Qianlong?		1735–1796?	Mn,Fe,Co			

608/Qianlong	Baur	1735–1796		Co,Ni,As		
R1175/Qianlong	Louvre	1735		Fe,Co,Ni,(As)		Mn,Fe,Cu
G907/Qianlong?	Mnaa-Guimet	1735–1796	Fe,Mn,(Co)	Co,Fe,Cu,As,Ni	Sn,Fe,Cu	Cu,Co,Fe,Cu
G8062/Qianlong				Fe,Co,As,(Cu)		
G3668/Qianlong?				Fe,Ni,Co,As	Fe,Zn	Cu,S,As,Fe
G907/Qianlong?			Fe,Mn, (Co)	Co,Fe,Cu,As,Ni	Sn,Fe,Cu	Cu,Co,Fe,As,Sn
630/Daogang	Baur		Fe,Mn,Ni,(Co)	Sn,Ni,Fe,Cu	Sn,Ni,Fe,Cu	Sn,Ni,Fe,Cu,Sb
G5615	Mnaa-Guimet				Fe,Sn,Cu	Cu
G4939			Mn,Fe,Co			
G2789			Fe			Cu,Sn,As,Fe

The use of gold nanoparticles, characteristic of *Famille rose*, is also a criterion for using a European recipe. There are two techniques for preparing gold NPs, the Perrot ('French') or Kunckel ('German') method [55, 58, 59], both techniques seeming to come from Italy [95].

The last determining criterion is the use of Naples yellow of complex composition (pyrochlore solid solution with Sb, Zn, Fe, etc.).

Table 18.7 Phases detected by Raman microspectroscopy (Fluo: fluorescence; NY: Naples yellow pyrochlore; na: not analyzed; Coll.: Collection). See previous caption

Object/Mark	Coll.	Date	Blue	Yellow	Green
G5696/Kangxi	Mnaa-Guimet		Na	Na	Na
G823/Kangxi		<1713			
G5615/Kangxi		1715–1722?		Na	Na
G3361/Kangxi		<1723–1735	Lead arsenate	NY(Sn)	
G5250/Kangxi		1715–1722		**NY(Sb)**	NY(Sn)
613/Kangxi	Baur	<1722	Lead arsenate	NY(Sb)	NY, **Cassiterite**
672/Kangxi		<1722	Lead arsenate	NY(Sb)	NY, **Cassiterite?**
676/Kangxi				NY, **Cassiterite**	NY
G823/Kangxi	Mnaa-Guimet	<1722	Na	Na	Na
677/Kangxi	Baur	<1722		Lead arsenate	Lead arsenate Quartz
G913/Yongzheng	Mnaa-Guimet	1723–1735	Lead arsenate	NY(Sn)	
G4806/Yongzheng		1723–1735		NY?	
615/Yongzheng	Baur	1723–1735	Lead arsenate	NY(Sb?), Lead arsenate	
G2789/Yongzheng	Mnaa-Guimet	1723–1735	Na	Na	Na
616/Yongzheng	Baur	1723–1735	Na	Na	Na
596/Yongzheng	Baur			NY, Cassiterite	
TH487/Yongzheng	Louvre	1730–1740	Lead arsenate	NY(Sb?) traces	
R1175/Yongzheng		Ca. 1735	Lead arsenate		
R1177/Yongzheng		Ca. 1735		NY	
R1045/Qianlong		1730–1745	Lead arsenates	NY(Sn)	
R1048/Qianlong		1740–1760	Lead arsenates	NY(Sn)	

(continued)

Table 18.7 (continued)

Object/Mark	Coll.	Date	Blue	Yellow	Green
G2890/ Qianlong	Mnaa-Guimet	1735–1796	Na	Na	Na
G4939/ Qianlong?		1735–1796?	Na	Na	Na
608/Qianlong	Baur	1735–1796	Lead arsenate	NY, Lead arsenate	Lead arsenate
G907/ Qianlong?	Mnaa-Guimet	1735–1796			
G8062/ Qianlong			Lead arsenate		
G3668/ Qianlong		>1750	Lead arsenates	NY-(Sn)	Lead arsenate NY(Sn)
G907		Na	Na	Na	Na
630/Daogang	Baur			NY	NY,

Concerning the first period, reign of Kangxi and perhaps also for that of Yongzheng, opacification by adding arsenic is without questions an European technique. But it is becoming widespread very quickly. More arsenic in dark blue areas than in light blue areas will be a proof of the use of imported As-rich European cobalt. High potassium level will prove the use of smalt. Furthermore, the practice of using a background As-opacified in white leads to the detection by pXRF of arsenic in colours covering this white background if the cover layer thickness is smaller than the beam penetration at the energy of the measurement (~mm). We note that the objects identified by one of the criteria are also identified by at least one of the others.

Table 18.7 lists for the main colours the phases corresponding to pigments or opacifiers identified by Raman spectroscopy. The significant phases are cassiterite and the various lead arsenates for the opacifiers and the blue areas, Naples yellow of complex composition (in particular with Sb and Zn) and the presence of a strong fluorescence characteristic of metallic NPs (but identification of the metal NPs requires elemental composition).

18.4.3.3 Semi-quantitative Analysis

Due to the energy dependence of penetration and absorption of X-rays, variations in thickness of enamels and the density distribution of colouring agents depending on the decoration, it is not possible to calculate a composition of thin glaze/enamel from measurement made through the surface. On the other hand, the comparisons of the intensities of the characteristic signals (after corrections of the continuous background and taking into account the superimpositions of neighbouring energy peaks) visualized in different ternary diagrams highlight the enamels whose relative proportions of the chemical elements considered, it is i.e. the compositions are similar.

This similarity in the location within ternary diagram indicates the use of the same raw materials when the elements considered are elements characteristics of geological context, for example transition metals, arsenic, bismuth or uranium are characteristic of certain cobalt mines. The minor and trace elements also depend from the ore or pigment processing. For instance the precipitation of Au° NPs with arsenic or tin or the preparation of different yellow pigments, 'simple' lead-tin yellow, complex Naples yellow, etc.

Examination of the diagrams concerning cobalt shows the differences in composition of the blue colour precursor, containing arsenic for the decorations and manganese for the marks, due to the use of two different sources of cobalt, that of the decorations having all the characteristics of the cobalts used in France [54], while that of the marks is identical to that used in Jingdezhen under the Ming Dynasty [181, 182]. Figure 18.11 shows the results of the imperial objects from the mnaa-Guimet collection (left side) [59] but the same results were obtained with the imperial objects from the Baur Foundation (right side) [54]. The presence of nickel in association with cobalt is one of the criteria for identifying European origin. A comparison is made with the rare data concerning Chinese mines producing 'domestic' ore [197]. This confirms its high concentration of manganese but the data are expressed in composition which distorts the comparison with the intensity diagrams and does not allow more precise conclusions. However, we see that the cobalt used under Yongzheng was the richest in Mn. It can be thought that the selection technique has been improved under Qianlong. These data are roughly in agreement with previous studies of domestic cobalt and confirm the use of imported cobalt (in the form of smalt) from Europe for objects located on (or near) the Co-Fe line, (almost) free of Mn.

Examination of all the data highlights a few cases where the cobalt is abnormally pure to come from smalt or from a 'Chinese' source; this indicates the use of 'purified' cobalt, therefore a production made after the second half of nineteenth century [50, 55].

Raman analyses allow more subtle classification to be made due to its sensitivity to structural changes. Four types of Raman signature (i.e. of As-based phases) are identified using multivariate statistical analysis in Principal Components and hierarchical Euclidian classification of the numerous Raman spectra recorded on the objects or available in the literature (Fig. 18.12) [15]. Subgroups are present.

The Euclidean hierarchical classification visualized in Fig. 18.12 shows 4 levels corresponding to at least four structural types of different compositions. If some of these types of pigments are used both in France and in China or as well for enamels on metal as on porcelain: each type of pigment corresponds preferentially to a family of objects, probably due to the different temperatures and (small) modifications to the composition of the vitreous matrices specific to the production of these different objects (processing temperature, expansion agreement). Refer to the detailed study [13].

Certain trace impurities (a few tens to hundreds of ppm, i.e. typically 0.01–0.1%) have a significant XRF signal which allows reliable measurements. It mainly concerns on the one hand rubidium, an impurity of sodium and potassium, on the other

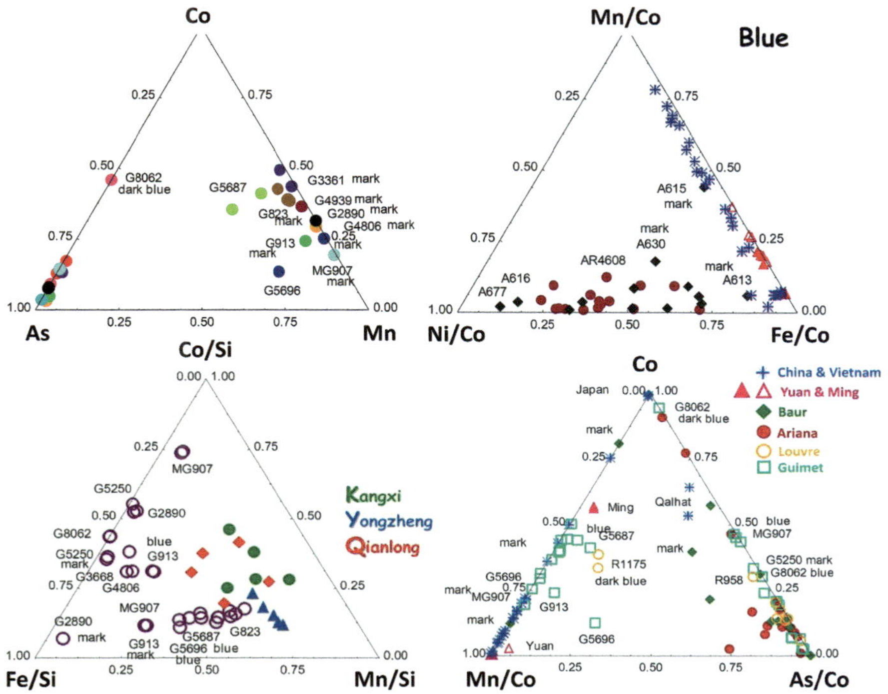

Fig. 18.11 Comparison of the areas of the XRF peaks characteristic of the elements Co, Fe and Mn (left, open circle: mnaa-Guimet objects, measurements for the blue of the marks and decorations; solid symbols: data for Chinese cobalt ores from mining places exploited during Kangxi, Yongzheng and Qianlong reigns [197]), and Co, Mn, As (objects from the Baur Foundation (diamonds, Axxx), from the Ariana Museum (rounds, ARxxxx), the stars and triangles correspond to blue-and-white colours respectively from the Ming and Yuan Dynasties) measured on the marks and areas blue of objects whose inventory number is given (see Tables 18.6 and 18.7, according to [55]). On the right the values have been normalized vs Co signal because they were obtained with different instruments; the blue star marked Japan located at the nCo pole corresponds to a blue and white porcelain from the twentieth century coloured with a high purity cobalt salt

hand strontium, an impurity of calcium, that is to say impurities of fluxes and finally yttrium and zirconium; zirconium is an impurity of magmatic minerals such as quartz (in the form of zircon ($ZrSiO_4$) and yttrium an impurity of zirconium). These are therefore impurities of sand. Added to this is uranium which is measured well even at low levels. For instance uranium is associated to some cobalt mining places of Ore Mountains (Erzgebirge) [58]. These impurities will therefore be a signature of the raw materials used for the *corpo*, the silicate matrix of the enamels, whatever their colouring and of some ores. Figure 18.13 compares Y-Rb-Sr diagrams constructed with all the data collected by XRF on both European and Chinese objects.

The Yttrium-Rubidium-Strontium diagram (normalized relative to the peak of the rhodium source to avoid small variations in measurement conditions) shows a

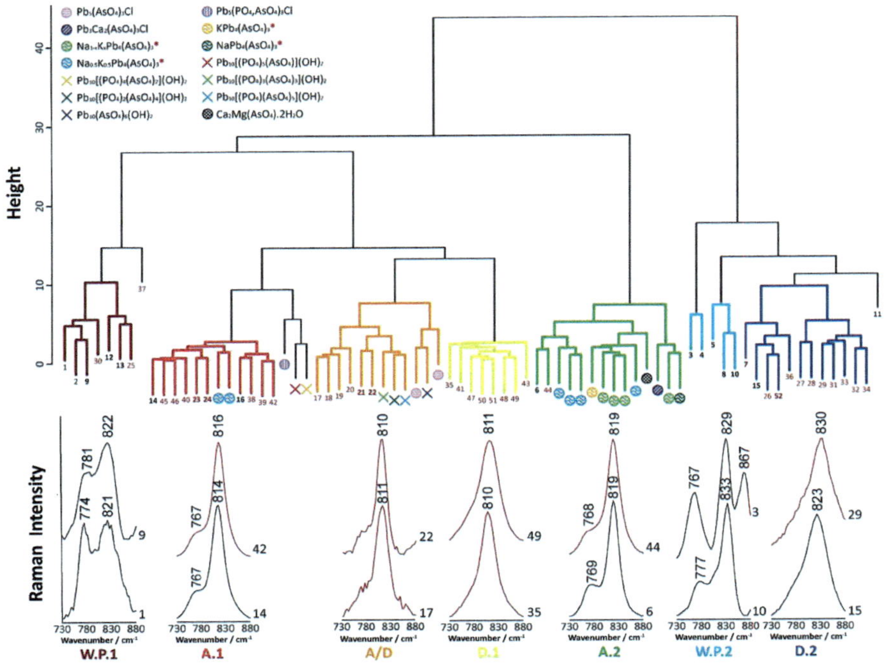

Fig. 18.12 Hierarchical Euclidean classification, according to [13]. Representative spectra in the As-O stretching mode range are shown

distribution on a line parallel to the Y-Rb side due to a certain constancy of the relative proportions silica/lime (SiO_2)/CaO and CaO/(Na_2O + K_2O) in the glazes/glazes and mixtures made between the two poles, the one rich in yttrium where the data for French soft porcelains are found, the sand used being rich in traces of yttrium, and the rubidium-rich pole where data from Chinese Ming porcelains [182] and contemporary ones from Vietnam [181] are located. We observe that the data acquired for the enamels of productions attributed to Canton are more in the group of Ming references. Those which correspond to European ingredients are located on or near pole Y. The data between are attributed to a use of mixing Asian ingredients and those imported from Europe.

Examination of the diagram concerning the colours blue, pink (violet, pink purple and similar) and gold shows a slightly greater dispersion. Five groups are identifiable:

- group 1 of the blue zones using European cobalt like that used for the very first European porcelain (French workshops and factories in the Ile de France).
- group 3 of Ming productions but also blue marks using Asian cobalt of the Ming Dynasty.

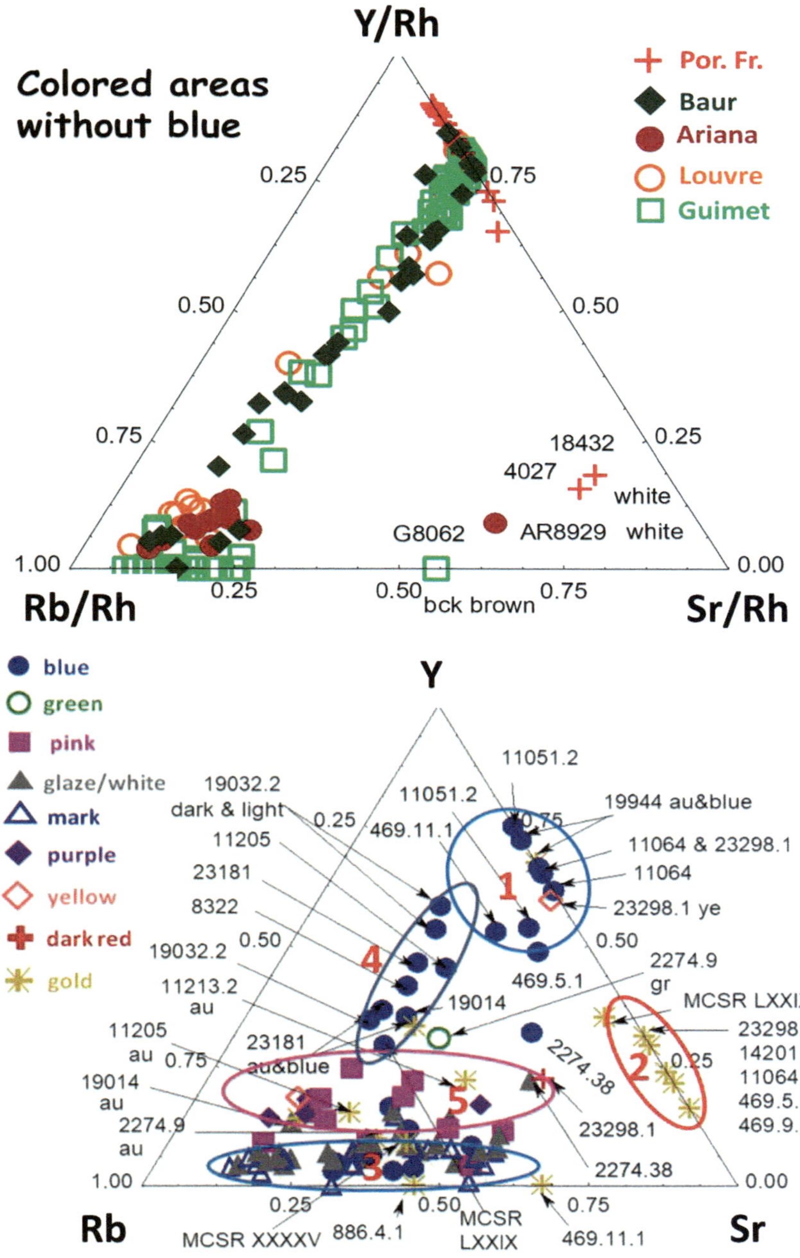

Fig. 18.13 Ternary diagrams of the intensities of the characteristic peaks of the impurities of fluxes (Rb, Sr) and sands (Y) on the left for the zones coloured with colours other than blue [55] and on the right for the zones coloured blue (after [58]).

- group 4 which corresponds either to the use of a mixture of 'French' and 'Chinese' cobalts, or to the use of 'Meissen' cobalt [58] mixed or not with 'Chinese' cobalt.

Remember that several cobalt mines were exploited at that time in Europe [48, 118, 142, 158], mainly in the Ore Mountains (Erzgebirge) forming the border between Saxony and Bohemia but also in other German regions (Harz, Schwartz Forest), France (Vosges Mountains: Sainte-Marie-aux-Mines), and the Spanish Pyrenees near the French border (Gistain or Gisthain valley, Catalogne) [40, 41, 48, 104, 135, 136, 158]. Many authors only consider the mines of the Ore Mountains without taking into account

(i) the large variations in the levels of associated elements for neighbouring mines such as Freiberg (uranium rich) and Schneeberg (no uranium),
(ii) the variations due to the preparation processing [142]
(iii) the preparation of different grades.

Indeed, Valmont de Bomare in his book of 1774 [186] indicates the existence of around ten qualities of smalt commercially available (therefore with different compositions, grain size, etc.). The Sèvres archives indicate the use of cobalt from the Spanish Giftain valley [48, 58, 104, 135, 136, 158]. Refer to the references for details of the discussions. If the distinction between Asian cobalt(s) very rich in Mn and Fe and European cobalts rich in As, Ni and/or Bi and/or Cu and/or U, is obvious, the identification of the right mining place in each continent is delicate given the current state of knowledge [48, 141, 142]. Furthermore, if the differentiation is clear between 'European' and 'Chinese' (or Indonesian, etc.) cobalts, it is difficult distinguishing between 'European' and 'Iranian' sources.

More work is needed to refine the understanding of the relationships between levels of impurities and the provenance of raw materials.

Figure 18.14 shows how the maps of the different chemical elements refine the understanding of the creation of enamelled decorations.

The high level of arsenic associated with the absence of manganese in the blue mantle attests to the use of imported 'European' cobalt. Black is obtained by a spinel based on Mn, Ni and Cu. The quality of the work is seen by the contrast in the distribution of lead only present in the overglaze.

18.4.3.4 Glassy Silicate Matrix

Examination of the ternary diagram of the intensities for the elements Pb, K and Ca (normalized with the silicon signal to avoid variations due to the use of different XRF instruments) for the blue zones clearly shows the separation between underglaze decorations (blue-and-white Yuan and Ming) [181, 182] and lead 'overglaze' enamels whether for European (example of Meissen [58]) or Qing productions [59]. Chinese enamels are more calcium-rich than Meissen enamels. We also observe lead pollution during firing when we analyze the blue of the marks but we note that

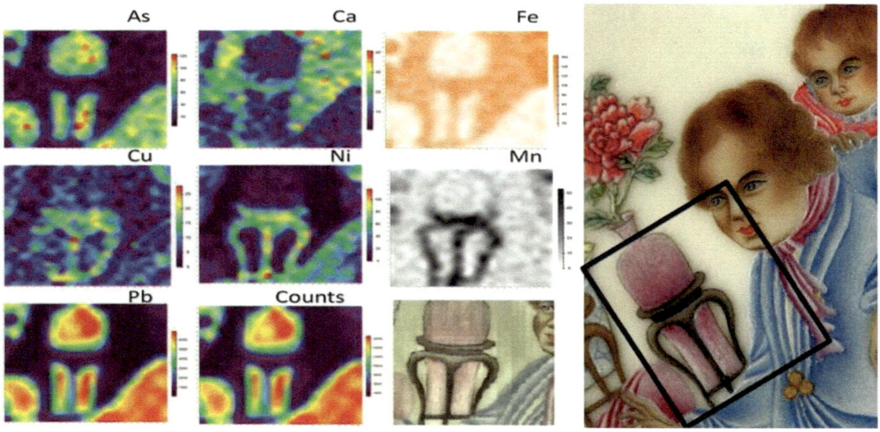

Fig. 18.14 Example of distribution maps of characteristic elements As, Ca, Fe, Cu, Ni, Mn, and Pb; on the right the black square delimits the mapped area. (Coll. Fondation Baur, Geneva, inv 1935.608)

for some objects (G5250/Kangxi, G8062/Qianlong, R1175/Yongzheng, G3361/Kangxi, G5696/Kangxi *wucai*, G5687) the measurement gives a lower lead composition. This is understandable for the *wucai* decoration of G5696 (Fig. 18.6), the small thickness of transparent enamel inducing a disturbance of the measurement by the contribution of the paste under the enamel. This does not seem to be the case for the other objects and this would indicate changes in the composition of the enamel matrices (Fig. 18.15).

Indeed, as shown both by the ternary diagrams comparing the intensities of the FRX peaks characteristic of fluxes and by the selection of Raman spectra (Fig. 18.16), the compositions of the enamels are variable. Glaze fired with the paste at high temperature, free or almost free of lead (lead surface pollution occurs by evaporation-condensation-reaction of very volatile PbO during the firing of enamelled decorations), corresponding at the values distribution along the Ca-K side of the ternary diagram show a Raman spectrum with a band of deformation modes of the SiO4 vibrational entity intense around 490 cm^{-1} and bands of elongation modes between 700 and 1300 cm^{-1}. The highest intensity of the deformation modes area (the A_{500}/A_{1000} ratio, called the polymerization index, versus the area of the deformation and stretching modes is >3), is in agreement with firing at more than 1200 °C [22, 29, 30, 33]. This ratio reaches ~0.5 for green enamel very rich in lead fired at much lower temperature (<800 °C). Simultaneously, the centre of gravity of the stretching modes shifts from approximately 1050–1100 cm^{-1} for items fired with the paste at high temperature to ~940 cm^{-1} for enamels fired at low temperature.

The peak around 815–820 cm^{-1} is characteristic of the As-O chemical bond (lead arsenates). Several types of arsenates are identified based on the wavenumber position and the shape and width of the peak. The exhaustive study by multivariate

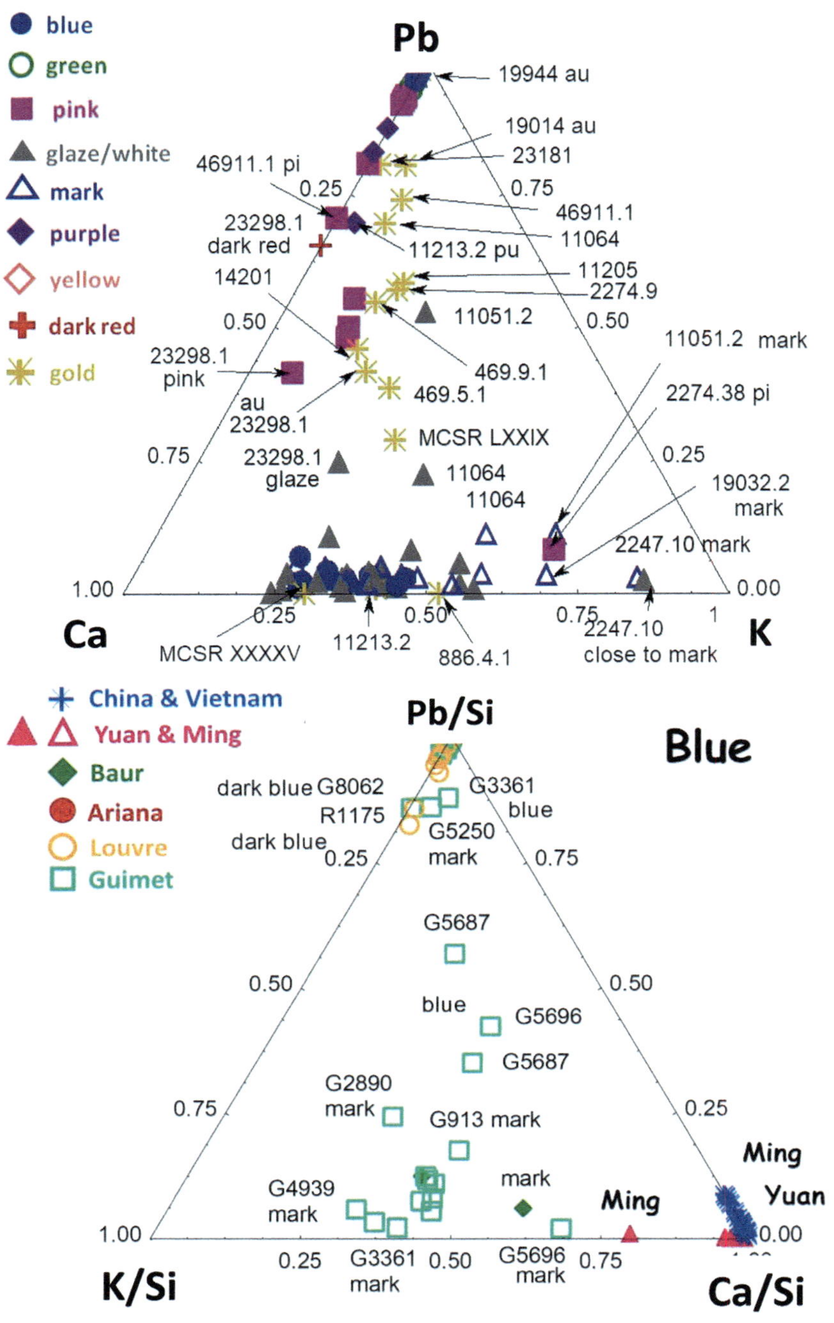

Fig. 18.15 Diagram comparing the intensities of the peaks characteristic of fluxes (Pb, Ca and K) for decorations of Meissen objects (on the left, for different coloured areas and the blue marks) [59] and Chinese objects (right, for blue zones and marks)

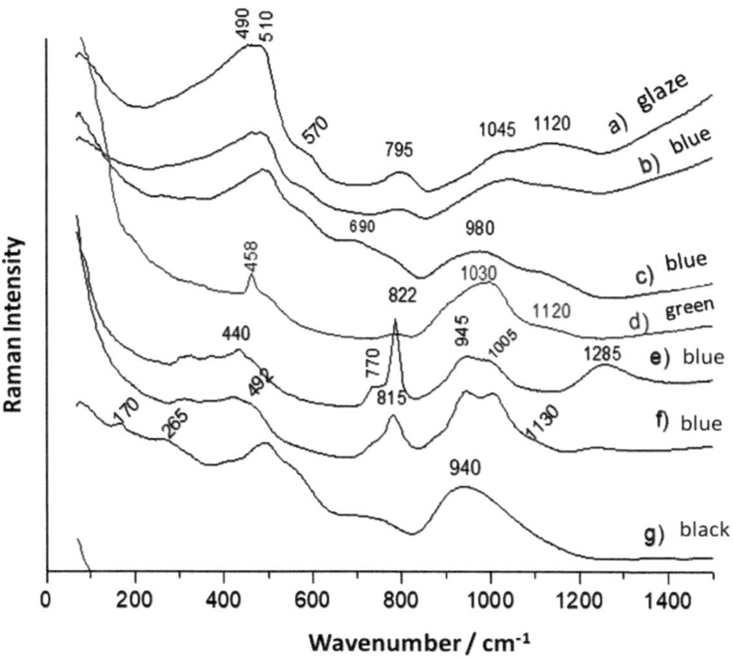

Fig. 18.16 Raman spectra representative of colourless glaze (**a**), blue underglaze areas (**b**, **c**), green overglaze enamel (**d**), blue overglaze enamel (**e**, **f**) and a uncoloured area of a black area (**g**) chosen from those collected on commissioned porcelain. (After [56])

analysis [14] demonstrates the existence of 4 types of arsenates with different compositions and/or structures.

The band at 1285 cm^{-1}, observed under blue laser excitation, is characteristic of the addition of borax to the enamel, in agreement with the recipe descriptions [135]. Borax provides boron but also a little lithium and sodium, elements that are difficult to measure whether the analysis is non-invasive or uses heavy and expensive instruments like AGLAE [13]. It very effectively lowers the melting point, therefore reduces viscosity and improves the 'tightness' and shine of enamels. The precise work of classifying the different types of silicate matrices whose stratigraphy is known remains to be done using the tools of multivariate analysis.

Direct Enamels on Paste or on Glaze

Non-invasive analyses, in particular carried out on site, do not allow us to study the stratigraphy and in particular to know whether the enamels are all over-covered or may have been carried out directly on biscuit as was observed for certain shards at the Palace museum [177].

18.5 *'Chine de Commande'*/Armorial Porcelain and Canton Productions

The collections of the Louvre Museum (Paris) and those of the Ariana Museum (Geneva) include objects attributed to the beginning of the eighteenth century but enamelled in the Canton Customs workshops, open district to foreigners and where enamelling workshops were set up [81]. Added to this are series of objects belonging to private collections made available for analysis (Fig. 18.17).

The first comparison between the productions attributed to the imperial workshops and those of Canton which were accessible to measurements is that almost all of these objects use both transparent enamels and opaque enamels, which links them to the definitions of the *Famille rose*.

Fig. 18.17 Examples of '*porcelaines de commande*' (armorial porcelain) produced for export to Europe during Yongzheng (Y) and Qianlong (Q) reigns and attributed by visual examination to Canton workshops. Comparison with European copies (Coll. Musée de l'Ariana, Geneva: 1) inv. AR08929 (Arita, Japan, 18th c.); 2) inv. AR2001-225 (Meissen copy); 3) inv. AR9168 (19th century copy); 4) inv. AR4530 (Y); 5) inv. AR4608 (Y); 6) inv. AR 10091 (Y); 7) inv. AR2007-196 (Y); 8) inv. AR 2007-213-2 (Y); 9) inv. AR2007-202 (Y); 10) inv. AR3646 (Q); 11) inv. AR10818 (Meissen copy, 18th c.); 12) inv. AR3680; 13) a dish from the so-called Service of Louis XV (circa 1720–1740), Private Coll. Refer to [53–57] for detailed information

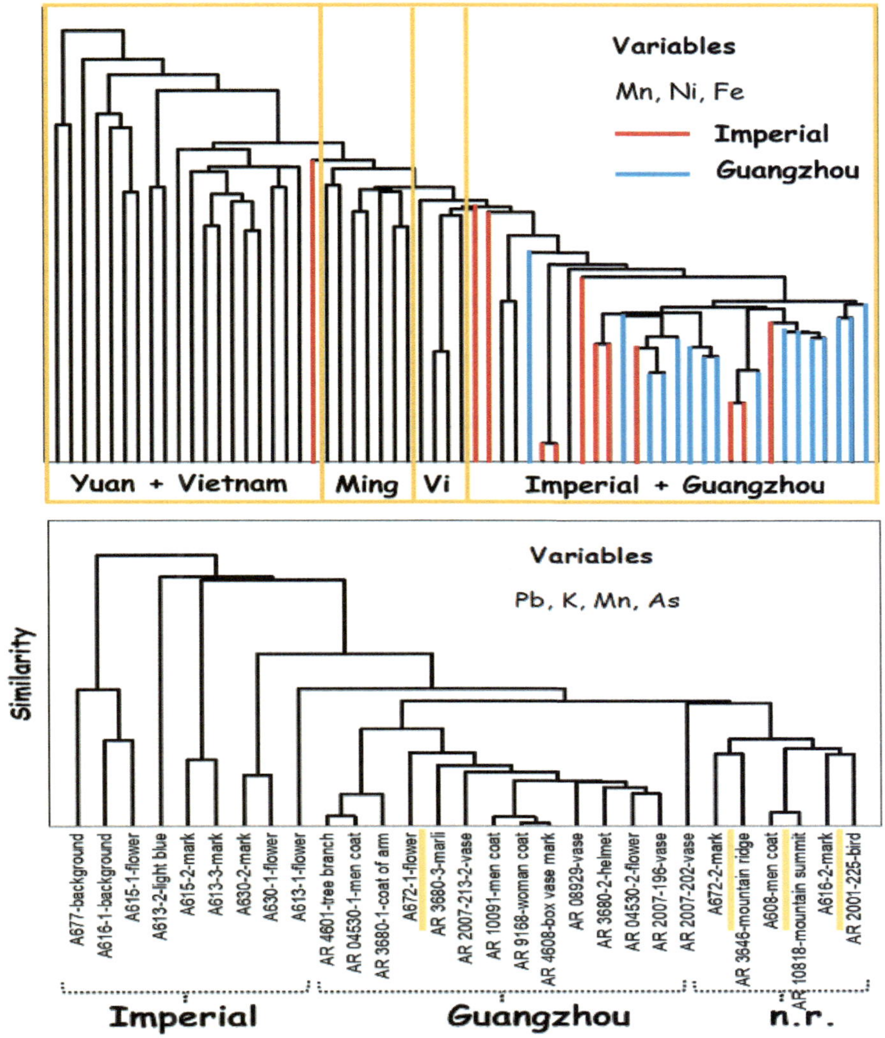

Fig. 18.18 Hierarchical classification based on the intensities of the characteristic peaks at the top of the elements Mn, Ni and Fe (associated with cobalt) measured on the blue areas of objects from the collections of the Baur Foundation and the Ariana Museum including productions of imperial and Canton (Guangzhou) kilns in comparison with blue-and-white porcelain from the Yuan and Ming Dynasties (including Vietnamese productions from the Ming period) and at the bottom elements Pb, K, Mn and As

Figure 18.18 visualizes the classification obtained with the main elements differentiating cobalt ores, imported or domestic: Mn, Ni and Fe. The classification is very effective, the marks of imperial objects made with Ming cobalt are in the same group as Ming productions. We also observe a fairly good separation between objects attributed 'visually' to Canton (private kilns) or to the imperial kilns at Jingdezhen or Beijing.

It is the same if we use the characteristic elements not only of the cobalt source but also of the silicate matrix Pb and K in addition to Mn and As which characterize the cobalt sources. In this classification only bowl A672 (from the nineteenth century, reign of Daogang) is poorly classified appearing among those from Canton. On the far right we find enamels whose thickness is very thin (and not reliable). For these objects (underlined) it is probable that their grouping into a group is solely due to the measurement bias due to the dominant contribution of the substrate, the paste. Differences may arise from attributions to be reviewed. Comparisons made on a larger number of objects should make it possible to identify different enamelling technologies and perhaps to differentiate the enamels made in Canton from those made in Jingdezhen.

Raman analyses of the blue 'overglaze' enamels of commissioned porcelain [56] suggest a constant use of borax (in accordance with Hellot's recipe [135]). Different types of hematite red and spinel-based black are highlighted which may correspond to different workshops or quality of raw materials.

18.6 Fakes and Embellishments

The knowledge acquired both concerning the elements associated with cobalt and the impurities present in Qing productions provides tools for identifying enamelled objects presenting the use of 'too' pure raw materials either for decoration or for brands, or for both. Thus, clarifications could be made concerning objects whose visual examination did not allow clear conclusions [50, 55], the use of raw materials too pure for the period which may concern the decoration or only the apocryphal mark.

18.7 Conclusions and New Questions

The findings concerning the analysis of a significant number of porcelains showed the effectiveness of the Raman—pFRX combination for comparing glazing techniques. In particular, the data processing procedures, whether Raman or pFRX, have been greatly improved. A fairly good statistic exists concerning the productions attributed to the reign of Kangxi. The number of pieces studied in relation to the number produced is certainly lower for the following reigns. It is the same for European productions: if the first Parisian soft porcelains and the productions of Meissen are quite well documented, those of other French manufactures such as Vincennes-Sèvres are less so. This will make it possible to compare the evolution of techniques in Europe and China. To this day it seems that the Canton workshops were very quickly informed of developments in Europe. Regarding enamels on metal, statistics are very limited. The same goes for knowledge of cobalt sources. We have shown that the bijection made by previous authors between the Ore Mountain mines is a gross simplification because each mine had specific compositional signatures and the treatment processes evolved significantly in the eighteenth century. Texts indicate the use of cobalt from the Giftain valley and not from the Ore Mountains. Knowledge is also very imprecise for Asian sources, none of the previous studies using our semi-quantitative approach which alone allows to date the identification of decorations made with the same raw materials. Our measurements provide arguments for a French connexion of blue enamelling technology, at least under Kangxi, the contribution of a German sector being possible under Qianlong, which is in agreement with the chronological data (the Meissen Manufacture is later than the Parisian manufactures). The French connexion is also proved by the use of Perrot' route to prepare Au° NPs and not the Kunckel' German one.

As the texts indicate boron is used by the addition of borax which adds not only boron but lithium and sodium. Boron is known to lower the melting temperature but also to improve the shine and the texture (surface quality). However, a significant part of the boron evaporates or diffuses into the substrate. In addition, its non-invasive detection is not very effective.

The great weakness of the work remains the extreme rarity of sherds for in-depth studies.

References

1. Agresti, G., Baraldi, P., Pelosi, C., Santamaria, U.: Yellow pigments based on lead, tin and antimony: ancient recipes, synthesis, characterization and hue choice in artworks. Colour Res. Appl. **41**(3), 226–231 (2016). https://doi.org/10.1002/col.22026
2. Alonge, G.: A History of Jesuit Missions in Japan: Evangelization, Miracles and Martyrdom, 1549–1614. Taylor & Francis (2023)
3. Bellemare, J., Carò, F., Stamm, K.: Persimmon and peonies: orange-coloured glass and enamels from the qing imperial workshops. Metrop. Mus. J. **58**(1), 96–106 (2023). https://doi.org/10.1086/728880
4. Bellemare, J.: A new palette: reassessing the development of enamel colours in early-eighteenth-century China. J. Glass Stud. **64**, 147–167 (2022). https://www.jstor.org/stable/48703407
5. Berg, M., Gottmann, F., Hodacs, H., Nierstrasz, C.: Goods from the East, 1600–1800. Palgrave MacMillan Springer, New York (2015)
6. Bertran, H., Reboulleau Magnier, A., Romain, A.: Nouveau Manuel Complet de la Peinture sur verre, sur porcelaine et sur émail. Encyclopédie-Roret, L. Mulo, Paris (1913). https://books.google.fr/books?hl=en&lr=&id=18Pd9it6eFEC&oi=fnd&pg=PA280&dq=H.+Bertran,+Reboulleau,+Magnier,+A.+Romain,+Nouveau+Manuel+Complet+de+la+Peinture+sur+verre,+sur+porcelaine+et+sur+%C3%A9mail,+Encyclop%C3%A9die-Roret,+L.+Mulo:+Paris,+1913.&ots=IaEM4Yt5We&sig=lFH-KEZN5yO0jfNYM746IAOXFc4&redir_esc=y#v=onepage&q&f=false
7. Biron, I., Quette, B.: Les premiers émaux chinois. Techné. **6**, 35–40 (1997)
8. Blanc, M., Biron, I., Colomban, P., Notin, V.: Emaux peints de Limoges, XVe-XVIIIe siècles – La collection du Musée des arts décoratifs. Les Arts Décoratifs, Paris (2011)

9. Bonneau, A., Moreau, J.-F., Hancock, R.G.V., Karklins, K.: Archaeometrical analysis of glass beads: potential, limitations, and results. Beads J. Soc. Bead Res. **26**, 35–46 (2014a). http://surface.syr.edu/beads/vol26/iss1/7
10. Bonneau, A., Moreau, J.-F., Auger, R., Hancock, R.G.V., Émard, B.: Analyses physicochimiques des perles de traite en verre de facture européenne: quelles instrumentations pour quels résultats? Archéologiques. **26**, 109–132 (2014b). https://www.academia.edu/download/32838510/Bonneau_et_al._archeologiques_2013.pdf
11. Bourson, P., Simon, G., Chapron, D., Kokanyan, N., Colomban, P.: Spectroscopie et spectrométrie Raman. In: Simon, G. (dir.) Spectroscopies vibrationnelles. Théorie, aspects pratiques et applications, pp. 109–154. eac: Paris, ISBN 97828 13002556 (2020)
12. Burgio, L., Clark, R.J.H.: Library of FT-Raman spectra of pigments, minerals, pigment media and varnishes, and supplement to existing library of Raman spectra of pigments with visible excitation. Spectrochim. Acta A Mol. Biomol. Spectrosc. **57**(7), 1491–1521 (2001). https://doi.org/10.1016/S1386-1425(00)00495-9
13. Burlot, J., Vangu, D., Bellot-Gurlet, L., Colomban, P.: Raman identification of pigments and opacifiers. Interest and limitation of multivariate analysis by comparison with solid state spectroscopical approach. I. Lead-tin and Naples yellow. J. Raman Spectrosc. **55**(2), 161–183 (2024a). https://doi.org/10.1002/jrs.6600
14. Burlot, J., Vangu, D., Bellot-Gurlet, L., Colomban, P.: Raman identification of pigments and opacifiers: interest and limitation of multivariate analysis by comparison with solid state spectroscopical approach. II. Arsenic-based opacifiers and relation with cobalt ores. J. Raman Spectrosc. **55**(2), 184–199 (2024b). https://doi.org/10.1002/jrs.6612
15. Burlot, J., Colomban, P., Bellot-Gurlet, L., Lemasson, Q., Pichon, L.: Non-invasive analyze of boron and lithium in 18th century Chinese porcelain enamel and glaze: a PIXE/PIGE study. J. Eur. Ceram. Soc. **44**(15), 116746 (2024c). https://doi.org/10.1016/j.jeurceramsoc.2024.116746
16. Cardinal, C.: Les montres et horloges de table du musée du Louvre. RMN, Paris (2000)
17. Castellucio, S.: Le prince et le marchand. Le commerce de luxe chez les marchands merciers parisiens pendant le règne de Louis XIV Paris: Kronos n°73. Editions S.P.M., Paris (2014) ISBN: 978-2-917232-16-3
18. Chan, T.-Y.: Mémoire d'un empire: La longue marche des collections du Musée du Palais dans un contexte de guerre. Etudes chinoises. **31**(2), 45–76 (2012). https://www.persee.fr/doc/etchi_0755-5857_2012_num_31_2_1464
19. Charles-Gaffiot, J. (Dir.): Des Samouraïs aux mangas. L'épopée chrétienne au Japon, Missions étrangères de Paris, Editions Invenit, Lille (2024). ISBN 9782376801252
20. Colomban, P., Treppoz, F.: Identification and differentiation of ancient and modern European porcelains by Raman macro-and micro-spectroscopy. J. Raman Spectrosc. **32**(2), 93–102 (2001). https://doi.org/10.1002/jrs.678
21. Colomban, P., Sagon, G., Faurel, X.: Differentiation of antique ceramics from the Raman spectra of their coloured glazes and paintings. J. Raman Spectrosc. **32**(5), 351–360 (2001). https://doi.org/10.1002/jrs.704
22. Colomban, P.: Polymerisation degree and Raman identification of ancient glasses used for jewellery, ceramics enamels and mosaics. J. Non-Crystalline Solids. **323**, 180–187 (2003). https://doi.org/10.1016/S0022-3093(03)00303-X
23. Colomban, P.: Secrets retrouvés du lustre Abbasside. La Revue de la Céramique et du Verre. **139**, 13–19 (2004)
24. Colomban, P., Milande, V., Lucas, H.: On-site Raman analysis of Medici porcelain. J. Raman Spectrosc. **35**(1), 68–72 (2004a). https://doi.org/10.1002/jrs.1085
25. Colomban, P., Robert, I., Roche, C., Sagon, G., Milande, V.: Identification des porcelaines «tendres» du 18e siècle par spectroscopie Raman: Saint-Cloud, Chantilly, Mennecy et Vincennes/Sèvres. Archeosciences, Revue d'Archéométrie. **28**, 153–167 (2004b)
26. Colomban, P., Milande, V., Le Bihan, L.: On-site Raman Analysis of Iznik pottery glazes and pigments. J. Raman Spectrosc. **35**(7), 527–535 (2004c). https://doi.org/10.1002/jrs.1163

27. Colomban, P.: Avant-Propos – Chine- Méditerranée. Routes et échanges de la céramique avant le XVIe siècle. Taoci. **4**, 5–7 (2005). https://www.researchgate.net/publication/270272013_Chine-Mediterranee_Routes_et_echanges_de_la_ceramique_avant_le_XVIe_siecle
28. Colomban, P., Milande Véronique, V.: On site analysis of the earliest known Meissen Porcelain and Stoneware. J. Raman Spectrosc. **37**(5), 606–613 (2006). https://doi.org/10.1002/jrs.1494
29. Colomban, P., Calligaro, T., Vibert-Guigue, C., Liem, N.Q., Edwards, H.G.M.: Accrochage des dorures sur les céramiques et tesselles anciennes. Revue d'Archéométrie-Archeosciences. **29**, 7–20 (2006a). https://doi.org/10.4000/archeosciences.424
30. Colomban, P., Tournié, A., Bellot-Gurlet, L.: Raman identification of glassy silicates used in ceramic, glass and jewellery: a tentative differentiation guide. J. Raman Spectrosc. **37**, 841–852 (2006b). https://doi.org/10.1002/jrs.1515
31. Colomban, P.: On-site Raman identification and dating of ancient glasses: procedures and tools. J. Cult. Herit. **9**, e55–e60 (2008). https://doi.org/10.1016/j.culher.2008.06.005
32. Colomban, P.: The use of metal nanoparticles to produce yellow, red and iridescent colour, from Bronze Age to present times in lustre pottery and glass: solid state chemistry, spectroscopy and nanostructure. J. Nano Res. **8**, 109–132 (2009). https://doi.org/10.4028/www.scientific.net/NYanoR.8.109
33. Colomban, P.: Non-destructive Raman analysis of ancient glasses and glazes. In: Janssens, K. (dir.) Modern Methods for Analysing Archaeological and Historical Glass, 1st edn, pp. 275–300. Wiley, London (2012a)
34. Colomban, P.: The on-site/remote Raman analysis with mobile instruments: a review of drawbacks and success in cultural heritage studies and other associated fields. J. Raman Spectrosc. **43**(11), 1529–1535 (2012b)
35. Colomban, P.: The destructive/non-destructive identification of enamelled pottery and glass artifacts and associated pigments—A brief overview. Arts. **2**, 77–110 (2013). https://doi.org/10.3390/arts2030077
36. Colomban, P.: Natural nanosized raw materials and sol-gel technology: The base of pottery since millenniums. In: Dillmann, P., Bellot-Gurlet, L., Nenner, I. (eds.) Nanoscience and Cultural Heritage. Atlantis Press, Paris (2016). https://doi.org/10.2991/978-94-6239-198-7_3
37. Colomban, P., Arberet, L., Kırmızı, B.: On-site Raman analysis of 17th and 18th century Limoges enamels: Implications on the European cobalt sources and the technological relationship between Limoges and Chinese enamels. Ceram. Int. **43**(13), 10158–10165 (2017a). https://doi.org/10.1016/j.ceramint.2017.05.040
38. Colomban, P., Zhang, Y., Zhao, B.: Non-invasive Raman analyses of *huafalang* and related porcelain wares. Searching for evidence for innovative pigment technologies. Ceram. Int. **43**(15), 12079–12088 (2017b). https://doi.org/10.1016/j.ceramint.2017.06.063
39. Colomban, P., Ambrosi, F., Ngo, A.-T., Lu, T.-A., Feng, X.-L., Chen, S., Choi, C.-C.: Comparative analysis of *wucai* Chinese porcelains using mobile and fixed Raman microspectrometers. Ceram. Int. **43**(16), 14244–14256 (2017c). https://doi.org/10.1016/j.ceramint.2017.07.172
40. Colomban, P., Lu, T.-A., Milande, V.: Non-invasive on-site Raman study of blue-decorated early soft-paste porcelain: The use of arsenic-rich (European) cobalt ores–comparison with huafalang Chinese porcelains. Ceram. Int. **44**(8), 9018–9026 (2018c). https://doi.org/10.1016/j.ceramint.2018.02.105
41. Colomban, P., Maggetti, M., d'Albis, A.: Non-invasive Raman identification of crystalline and glassy phases in a 1781 Sèvres Royal Factory soft paste porcelain plate. J. Eur. Ceram. Soc. **38**(15), 5228–5233 (2018b). https://doi.org/10.1016/j.jeurceramsoc.2018.07.001
42. Colomban, P.: Glaze and enamels. In: Richet, P. (dir.) Encyclopedia of glass science, technology, history and culture, 1st edn, pp. 1309–1326. Wiley, Hoboken (2020). https://doi.org/10.1002/9781118801017.ch10.6
43. Colomban, P., Kırmızı, B., Gougeon, C., Gironda, M., Cardinal, C.: Pigments and glassy matrix of the 17th–18th century enamelled French watches: A non-invasive on-site Raman and pXRF study. J. Cult. Herit. **44**, 1–14 (2020a). https://doi.org/10.1016/j.culher.2020.02.001

44. Colomban, P., Kırmızı, B., Zhao, B., Clais, J.-B., Yang, Y., Droguet, V.: Non-invasive on-site Raman study of pigments and glassy matrix of 17th–18th Century painted enamelled Chinese metal wares: Comparaison with French enamelling technology. Coatings. **10**, 471 (2020b). https://doi.org/10.3390/coatings10050471
45. Colomban, P., Kırmızı, B.: Non-invasive on-site Raman study of polychrome and white enamelled glass artefacts in imitation of porcelain assigned to Bernard Perrot and his followers. J. Raman Spectrosc. **51**(1), 133–146 (2020). https://doi.org/10.1002/jrs.5745
46. Colomban, P., Kırmızı, B., Zhao, B., Clais, J.B., Yang, Y., Droguet, V.: Investigation of the pigments and glassy matrix of painted enamelled Qing Dynasty Chinese porcelains by non-invasive on-site Raman microspectroscopy. Heritage. **3**(3), 915–940 (2020). https://doi.org/10.3390/heritage3030050
47. Colomban, P. Glass, ceramics and enamelled objects. In: Garside, P., Richardson, E. (dir.) Conservation science: heritage materials, 2nd edn, pp. 200–247. Royal Society of Chemistry, Londres (2021). https://doi.org/10.1039/BK9781788010931-00200
48. Colomban, P., Kırmızı, B., Simsek Franci, G.: Cobalt and associated impurities in Blue (and green) glass, glaze and enamel: relationships between raw materials, processing, composition, phases and international trade. Minerals. **11**, 633 (2021a). https://doi.org/10.3390/min11060633
49. Colomban, P., Gironda, M., Vangu, D., Kırmızı, B., Zhao, B., Cochet, V.: The technology transfer from Europe to China in the 17th–18th centuries: Non-invasive on-site XRF and Raman analyses of Chinese Qing Dynasty enameled masterpieces made using European ingredients/recipes. Materials. **14**, 7434 (2021b). https://doi.org/10.3390/ma14237434
50. Colomban, P.: Décors émaillés: Comprendre les secrets des productions du XVIIIe siècle en Chine. Le Secrets des Couleurs, Bulletin Fondation Baur. **80**, 11–23 (2022a). https://hal.science/hal-03940619/
51. Colomban, P.: Full spectral range Raman signatures related to changes in enameling technologies from the 18th to the 20th Century: guidelines, effectiveness and limitations of the Raman analysis. Materials. **15**(9), 3158 (2022b). https://doi.org/10.3390/ma15093158
52. Colomban, P.: The discovery and comparison of the manufacturing secrets of enamelled masterpieces. Orientations. **53**(1), 92–96 (2022c). https://ocean.exacteditions.com/issues/98880/spread/97?rc=1b40d5a3-8887-43e7-87cc-c861b62aa336
53. Colomban, P., Gironda, M., Edwards, H.G.M., Mesqui, V.: The Enamels of the First (Soft-paste) European Blue-and-white Porcelains: Rouen, Saint-Cloud and Paris Factories: complementarity of Raman and X-ray Fluorescence analyses with Mobile Instruments to identify the cobalt ore. J. Raman Spectrosc. **52**(12), 2246–2261 (2022a). https://doi.org/10.1002/jrs.6111
54. Colomban, P., Simsek Franci, G., Gironda, M., d'Abrigeon, P., Schumacher, A.-C.: pXRF data evaluation methodology for on-site analysis of precious artifacts: Cobalt used in the blue decoration of Qing dynasty overglazed porcelain enameled at customs district (Guangzhou), Jingdezhen and Zaobanchu (Beijing) workshops. Heritage. **5**, 1752–1778 (2022b). https://doi.org/10.3390/heritage5030091
55. Colomban, P., Simsek Franci, G., Gironda, M., d'Abrigeon, P.: Distinguishing genuine imperial Qing Dynasty Porcelain from Ancient Replicas y on-site non-invasive XRF and Raman Spectroscopy. Materials. **15**, 5747 (2022c). https://doi.org/10.3390/ma15165747
56. Colomban, P., Ngo, A.T., Fournery, N.: Non-invasive Raman analysis of 18th Century Chinese export/armorial overglazed porcelain: identification of the different enameling techniques. Heritage. **5**(1), 233–259 (2022d). https://doi.org/10.3390/heritage5010013
57. Colomban, P., Ngo, A.-T., Edwards, H.G.M., Prinsloo, L.C., Esterhuizen, L.V.: Raman identification of the different glazing technologies of Blue-and-White Ming porcelains. Ceram. Int. **48**, 1673–1681 (2022e). https://doi.org/10.1016/j.ceramint.2021.09.246
58. Colomban, P., Simsek Franci, G., Gerken, M., Gironda, M., Mesqui, V.: Non-invasive on-site XRF and Raman classification and dating of ancient ceramics: application to 18th and 19th Century Meissen Porcelain (Saxony) and comparison with Chinese Porcelain. Ceramics. ***6***(4), 2178–2212 (2023a). https://doi.org/10.3390/ceramics6040134

59. Colomban, P., Simsek Franci, G., Burlot, J., Gallet, X., Zhao, B., Clais, J.-B.: Non-invasive on-Site pXRF analysis of colouring agents, marks and enamels of Qing imperial and non-imperial Porcelain. *Ceramics*. **6**(1), 447–474 (2023b). https://doi.org/10.3390/ceramics6010026
60. Colomban, P.: Tracer l'utilisation de recettes et/ou d'ingrédients européens dans les Objects émaillés: Stratégie et premiers résultats. Artefact. Techniques, histoire et sciences humaines. **18**, 161–193 (2023). https://doi.org/10.4000/artefact.13945
61. Colomban, P.: The Quest for the Western Colours in China under the Qing Emperors, J1817 (2024). https://www.journal18.org/issue17/the-quest-for-the-western-colours-in-china-under-the-qing-emperors/
62. Corcuf, S.: Taipei, Refuge de la cite Interdite, https://www.historia.fr/histoire-du-monde/asie-oceanie/taipei-refuge-de-la-cite-interdite-2063478. Accessed 15 July 2024
63. Curtis, E.B.: A Plan of the Emperor's Glassworks. Arts Asiatiques. **56**, 81–90 (2001). https://www.persee.fr/doc/arasi_0004-3958_2001_num_56_1_1465
64. Curtis, E.B.: Glass Exchange between Europe and China, 1550–1800. Diplomatic, Mercantile and Technological Interactions. Ashgate, Aldershot (2009)
65. Curtis, E.B.: Kilian Stumpf, SJ at Kangxi's court, 1695-1720: A new perspective on his tenure as a missionary-artisan. In: Acta Pekinensia. Western Historical Sources for the Kangxi Reign, pp. 241–255. Instituto Ricci de Macau, Macau (2013)
66. Curtis, E.B.: Aspects of a multi-faceted process: the circulation of enamel wares between the Vatican and Kangxi's court. In Zhao, B., Landry-Deron Isabelle, I., Simon, F. (dir.) Des arts diplomatiques. Les cadeaux diplomatiques entre la Chine et l'Europe aux XVIIe-XVIIIe siècles. Pratiques et enjeux, Extrême-Orient Extrême-Occident 43, pp. 29–39 (2020). https://journals.openedition.org/extremeorient/1596
67. d'Albis, A.: Steps in the manufacture of the soft-paste porcelain of Vincennes, according to the books of Hellot. In: Kingery, W.D. (ed.) Ancient technology to modern science, Vol. I, ceramics and civilization, pp. 257–271. The American Ceramic Society, Westerville (1985)
68. d'Albis, A.: Traité de la porcelaine de Sèvres. Editions Faton, Dijon (2003)
69. d'Albis, A.: Différences entre pâte dure et pâte tendre, grès et faïence. Sèvres. Revue de la Société des Amis du musée national de Céramique. **20**, 21–34 (2011). https://doi.org/10.3406/sevre.2011.956
70. Deck, T.: *La Faïence*. Maison Quantin, Paris (1887). https://books.google.fr/books?hl=en&lr=&id=qoguSHlUhS0C&oi=fnd&pg=PA300&dq=T.+Deck,+La+Fa%C3%AFence,+Maison+Quantin:+Paris,+France,+1887.&ots=6b8HnZ4rug&sig=XuDhJXoBlNfJfIeL_Os0z8sIi94&redir_esc=y#v=onepage&q&f=false. Accès le 9 Mai 2024
71. Demirsar Arli, B., Simsek Franci, G., Kaya Sennur, D., Colomban, P.: Portable X-ray Fluorescence (p-XRF) uncertainty estimation for glazed ceramic analysis: case of Iznik Tiles. Heritage. **3**, 1302–1329 (2020). https://doi.org/10.3390/heritage3040072
72. Dentrecolles, F.-X.: Lettre du Pere d'Entrecolles, Missionnaire de la Compagnie de Jésus: au P... de la même Compagnie, etc. In: Vernarel, J., Cabin, E. (eds.) Lettres édifiantes et curieuses, écrites des missions étrangères, Tome 10, pp. 131–175, Lyon (1724). https://gallica.bnf.fr/ark:/12148/bpt6k9763708t/f583.image.r=Dentrecolles%20lettres%20porcelaine%20Chine?rk=128756;0. Accesed 3rd 2024
73. Dias, M.I., Prudêncio, M.I., De Matos, M.A.P., Rodrigues, A.L.: Tracing the origin of blue and white Chinese Porcelain ordered for the Portuguese market during the Ming dynasty using INAA. J. Archaeol. Sci. **40**(7), 3046–3057 (2013). https://doi.org/10.1016/j.jas.2013.03.007
74. Dik, J., Hermens, E., Peschar, R., Schenk, H.: Early production recipes for lead antimonate yellow in Italian art. Archaeometry. **47**(3), 593–607 (2005). https://doi.org/10.1111/j.1475-4754.2005.00221.x
75. Duan, H., Zhang, X., Kang, B., Wang, G., Qu, L., Lei, Y.: Non-destructive analysis and deterioration study of a decorated Famille Rose porcelain bowl of Qianlong Reign from the Forbidden City. Stud. Conserv. **64**(6), 311–322 (**2019**). https://doi.org/10.1080/00393630.2019.1617589

76. du Halgouet, H.: Pages Coloniales. Relations Maritimes de la Bretagne et de la Chine au XVIIIe Siècle, Lettres de Canton. Mem. Soc. Histoire et Archéologie de Bretagne, 331–442 (1934a). https://www.shabretagne.com/scripts/files/63cfb59ff1bef3.93206183/1934_06.pdf
77. du Halgouët, H.: Lettres de Canton (1934b). https://www.shabretagne.com/document/article/2410/Pages-coloniales-Relations-maritimes-de-la-Bretagne-et-de-la-Chine-au-XVIIe-siecle-Lettres-de-Canton
78. Dupaigne, B.: Le commerce du thé. De la Chine à l'Europe, XVIIe-XXIe siècle – M. Raibaud, F. Souty (dir.). Outre-Mers. Revue d'histoire. **95**(360–361), 356–358 (2008a). https://www.persee.fr/doc/outre_1631-0438_2008_num_95_360_4369_t1_0356_0000_2
79. Dupaigne, B.: Collectors, Collections and Museums. The Field of Chinese Ceramics in Britain, 1560–1960 S.- Pierson Stancey. Outre-Mers. Revue d'histoire. **95**(360–361), 358–360 (2008b). https://www.persee.fr/doc/outre_1631-0438_2008_num_95_360_4369_t1_0358_0000_2
80. Edwards, H.G.M.: Porcelain Analysis and its Role in the Forensic Attribution of Ceramics Specimens. Cham, Springer Nature (2021a). https://doi.org/10.1007/978-3-030-42192-2
81. Edwards, H.G.M.: Chinese porcelains and their Early European Competitors. In: Porcelain Analysis and its Role in the Forensic Attribution of Ceramics Specimens, pp. 65–98. Springer Nature, Cham (2021b). https://doi.org/10.1007/978-3-030-80952-2_2
82. Edwards, H.G.M.: The Birth of the Porcelain Trade with China. In: Welsh Armorial Porcelain: Nantgarw and Swansea Crested China, pp. 1–47. Springer, Cham (2022a)
83. Edwards, H.G.M.: Welsh armorial porcelain: Nantgarw and Swansea crested China. Springer, Cham (2022b)
84. Edwards, H.G.M., Vandenabeele, P., Colomban, P.: Raman spectroscopy in cultural heritage preservation. Cham, Springer Nature (2022)
85. Eppler, R.A., Eppler, D.R.: Glazes and Glass Coating. Westerville, The American Ceramic Society (2000). https://archive.org/details/glazesglasscoati0000eppl
86. Ferchault de Réaumur, A.R.: Idée générale des différentes manières dont on peut faire la porcelaine et quelles sont les véritables matières de celle de la Chine. Mémoires Académie des Sciences, Paris (1727). https://www.google.com/url?sa=t&source=web&rct=j&opi=89978449&url=https://www.academie-sciences.fr/pdf/dossiers/Reaumur/Reaumur_pdf/p185_203_vol3589m.pdf&ved=2ahUKEwiT1qrxkoCGAxVERqQEHQ67CnoQFnoECBMQAQ&usg=AOvVaw2_33YgOseTCyfmDzSB5APN
87. Ferchault de Réaumur, A.R.: Second mémoire sur la porcelaine ou suite des principes qui doivent conduire dans la composition des porcelaines de différents genres et qui établissent les caractères des matières fondantes qu'on ne peut choisir pour tenir lieu de celle qu'on emploie à la Chine, Mémoires Académie des Sciences, Paris (1729). https://www.google.com/url?sa=t&source=web&rct=j&opi=89978449&url=https://www.academie-sciences.fr/pdf/dossiers/Reaumur/Reaumur_pdf/p325_344_vol3527m.pdf&ved=2ahUKEwiFpa2dk4CGAxXqfKQEHVIGCLQQFnoECA8QAQ&usg=AOvVaw2mikucmiULn5hXd99Azeae
88. Ferchault de Réaumur, A.R.: Mémoire sur l'art de faire une nouvelle espèce de porcelaine par des moyens extrêmement simples et faciles ou de transformer le verre en porcelaine. Mémoires Académie des Sciences, Paris (1739). https://www.google.com/url?sa=t&source=web&rct=j&opi=89978449&url=https://www.academie-sciences.fr/pdf/dossiers/Reaumur/Reaumur_pdf/p370_388_vol3536m.pdf&ved=2ahUKEwiBkYeqk4CGAxUKUaQEHQPaCfUQFnoECAoQAQ&usg=AOvVaw0drF0SP0INYM8D4BKsXUUV
89. Freestone, I.C., Barber, D.J.: The development of the colour of sacrificial red glaze with special reference to a Qing Dynasty saucer dish. In: Scott, R.E. (dir.) Chinese Copper Red Wares Percival David Foundation of Chinese Art, Monograph Series n°3, pp. 53–62. University of London, School of Oriental and African Art, London (1992)
90. Figueiredo, M.O., Silva, T.P., Veiga, J.P.: A XANES study of cobalt speciation state in blue-and-white glazes from 16th to 17th century Chinese porcelains. J. Electron Spectrosc. Relat. Phenom. ***185***, 97–102 (2012). https://doi.org/10.1016/j.elspec.2012.02.007
91. Finn, M., Smith, K. (eds.): The East India Company at Home, 1757–1857. UCL Press (2018) ISBN: 9781787350274

92. Froment, F., Tournié, A., Colomban, P.: Raman identification of natural red to yellow pigments: ochre and iron-containing ores. J. Raman Spectrosc. **39**(*5*), 560–568 (2008). https://doi.org/10.1002/jrs.1858
93. Gauthier, M.-M.: Emaux. In: Encyclopédie Universalis, vol. 6, pp. 939–960, Paris (1985)
94. Geyssant, J. (ed.): *Bernard Perrot (1640–1709), Secrets et chefs d'œuvre des verreries royales d'Orléans* Catalogue, Paris. Musée des Beaux-Arts d'Orléans—SOMOGY Editions d'Arts, Paris (2013a)
95. Geyssant, J.: Secret du verre rouge transparent de Bernard Perrot et comparaison avec celui de Johann Kunckel. In: *Bernard Perrot (1640–1709): Secrets et chefs-d'œuvre des verreries royales d'Orléans*, Catalogue, Musée des Beaux-Arts d'Orléans, pp. 51–54. SOMOGY Editions d'Arts, Paris (2013b)
96. Giannini, R., Freestone, I.C., Shortland, A.J.: European cobalt sources identification in the production of Chinese Famille Rose porcelain. J. Archaeol. Sci. **80**, 27–36 (2016). https://doi.org/10.1016/j.jas.2017.01.011
97. Gioffredi Superbi, E.F., Pagliarulo, G.: Medieval mosaics: light, colour, materials. Borsook, E. (dir.), pp. 47–64. Silvana Editoriale, Florence (2000)
98. Gratuze, B., Soulier, I., Blet, M., Vallauri, L.: De l'origine du cobalt: du verre a la céramique. Revue d'Archéometrie. **20**, 77–94 (1996). https://hal.science/hal-01829843
99. Guo, Y.: Raw materials for making porcelain and the characteristics of porcelain wares in north and south China in ancient times. Archaeometry. ***29***(*1*), 3–19 (1987). https://doi.org/10.1111/j.1475-4754.1987.tb00393.x
100. Hancock, R.G.V., McKechnie, J., Aufreiter, S., Karklins, K., Kapches, M., Sempowski, M., Moreau, J.F., Kenyon, I.: Non-destructive analysis of European cobalt blue glass trade beads. *J. Radioanal. Nucl. Chem.* ***244***, 567–573 (2000). https://doi.org/10.1023/A:1006705014455
101. Haussonne, M.: Technologie Céramique Générale: Faïences, grès, porcelaine – vol. II, J.-B. Baillière & Fils Editeurs, Paris (1969)
102. Henderson, J., Tregear, M., Wood, N.: The technology of sixteenth- and seventeenth century Chinese Cloisonné enamels. Archaeometry. **31**(*2*), 133–146 (1989). https://doi.org/10.1111/j.1475-4754.1989.tb01009.x
103. Honey, W.B.: The origin and affiliations of the European porcelain-factories. Burlingt. Mag. Connoiss. ***87***(*511*), 245–247 (1945). https://www.jstor.org/stable/869196
104. Hoppensack, M.: Account of the mines of Spain. Mon. Mag. **6**(38), 320–322 (1798)
105. Huard, P., Wong, M.: Un album chinois de l'époque Ts'ing consacré à la fabrication de la porcelaine. In: Arts asiatiques, tome 9 fascicule 1–2. pp. 3–60 (1962). https://doi.org/10.3406/arasi.1962.892
106. Jacquemart, A.: *Histoire Artistique, Industrielle Et Commerciale de la Porcelaine: Sujets Et Emblèmes Qui La Décorent, Marks Et Inscriptions* qui font reconnaître les fabriques d'où elle sort, les variations de prix qu'ont obtenus les principaux Objects connus & les collections où ils sont conservés aujourd'hui, J. Techener, Paris, 1862.
107. Jami, C.: An 8 de l'ère Kangxi: le jésuite Ferdinand Verbiest au Bureau de l'astronomie. In: L'exploration du monde – Une autre histoire des Grandes DécouGreenes, pp. 278–282. Editions du Seuil, Paris (2019)
108. Jouenne, C.A.: Céramique Générale 2 Vol. Gauthier-Villars, Paris (1960)
109. Joutard, P.: Réseaux huguenots et espace Européen (XVIe–XXIe siècle). Rev. Synth. ***123***, 111 (2002). https://doi.org/10.1007/BF02963323
110. Kingery, W.D., Bowen, H.K., Uhlmann, D.R.: Introduction to Ceramics, 2nd edn. Wiley, New York (1976)
111. Kingery, W.D., Vandiver, P.B.: Song Dynasty Jun(Chung) ware glazes. Am. Ceram. Bull. **62**, 1269–1274 (1983)
112. Kingery, W.D.: The development of European porcelain. In: Kingery, W.D. (ed.) High-Technology Ceramics – Past, Present, and Future, Vol. III, Ceramics and Civilization, pp. 153–180. The American Ceramic Society, Westerville (1985)

113. Kingery, W.D., Smith, D.: The development of European soft-paste (Frit) porcelain. In: Kingery, W.D. (ed.) *Ancient Technology to Modern Science*, vol. I, pp. 273–292. Ceramics and Civilization, The American Ceramic Society, Westerville (1985)
114. Kingery, W.D., Vandiver, P.B.: The Eighteenth-Century Change in Technology and Style from the *Famille-Greene* Palette to the *Famille-Rose* Palette. In: Kingery, W.D. (ed.) Technology and Style Ceramics and Civilization Series, vol. 2, pp. 363–381. The American Ceramic Society, Colombus (1986)
115. Kırmızı, B., Colomban, P., Quette, B.: On-site analysis of Chinese *Cloisonné* enamels from fifteenth to nineteenth centuries. J. Raman Spectrosc. **41**(7), 780–790 (2010a). https://doi.org/10.1002/jrs.2516
116. Kırmızı, B., Colomban, P., Blanc, M.: On-site analysis of Limoges enamels from sixteenth to nineteenth centuries: an attempt to differentiate between genuine artefacts and copies. *J. Raman Spectrosc.* ***41****(10)*, 1240–1247 (2010b). https://doi.org/10.1002/jrs.2566
117. Kırmızı, B., Göktürk, H., Colomban, P.: Colouring agents in the pottery glazes of Western Anatolia: a new evidence for the use of Naples yellow pigment variations during the late Byzantine period. Archaeometry. ***57***, 476–496 (2015). https://doi.org/10.1111/arcm.12101
118. Kissin, S.A.: Five-element (Ni-Co-As-Ag-Bi) veins. Geosci. Can. ***19****(3)*, 113–124 (1992). https://journals.lib.unb.ca/index.php/gc/article/view/3768/4282
119. Kruta, V.: The first Celtic expansion: prehistory to history. In: Catalogue, E., Grassi, P., Mario Andreose, V., Tanzi, C. (eds.) The Celts. Milano, Bompiani (1991)
120. Landry-Deron, I.: Les Mathématiciens envoyés en Chine par Louis XIV en 1685. Arch. Hist. Exact Sci. ***55****(5)*, 423–463 (2001). https://www.jstor.org/stable/41134120
121. Landry-Deron, I.: La Preuve par la Chine. La «Description» de J.-B. Du Halde, jésuite, 1735. Paris, Éditions de l'EHESS, Paris (2002)
122. Landry-Deron, I. (ed.): *L'Europe missionnaire en Chine sous l'empereur Kangxi. Lettres du père jésuite Jean-François Foucquet écrites de Chine à sa famille (1698–1721)*, Coll. Bibliothèque de l'Institut des hautes études chinoises. Collège de France/Institut des hautes études chinoises, Paris (2022)
123. Landry-Deron, I.: La contribution des missionnaires aux arts du feu à la cour de Pékin (XVIIe-XVIIIe siècles). Artefact. Techniques, histoire et sciences humaines. **18**, 117–137 (2023)
124. Landry-Deron, I. 2024 unpublished.
125. Landry-Deron, I. (藍莉): Faguo zhuanjiaoshi wenxian suojian Kangxi nianjian dai wang Zhongguo de qiju法国传教士文献所见康熙年间带往中国的器具» (Les Objects apportés en Chine sous Kangxi d'après les archives missionnaires françaises), Gugong bowuguan yuankan, 4(264), 18–26. Traduction Qiu Dandan (2024b).
126. Le Corbeiller, C.: Eighteenth-century Italian porcelain. Metropolitan Museum of Art, New York (1985). https://libmma.contentdm.oclc.org/digital/collection/p15324coll10/id/58497/
127. Lécosse, C.: Quelle légitimité pour les peintres en miniature? Le petit format à l'épreuve des discours académiques. Études Épistémè. Revue de littérature et de civilisation (XVIe–XVIIIe siècles). **36** (2019). https://doi.org/10.4000/episteme.5384
128. Levin, E.M., Robbins, C.R., McMurdie, H.F.: Phase Diagrams for Ceramists, vol. 1–5. The American Ceramic Society, Westerville (1975)
129. Li, Y., Zhu, J., Ji, L., Shan, Y., Jiang, S., Chen, G., Sciau, P., Wang, W., Wang, C.: Study of arsenic in Famille rose porcelain from the Imperial Palace of Qing Dynasty, Beijing, China. *Ceram. Int.* ***44***(2), 1627–1632 (2018). https://doi.org/10.1016/j.ceramint.2017.10.084
130. Li, Y., Sciau, P., Zhu, J., Ji, L., Shan, Y., Song, G.: Microscopic analysis of overglaze green pigment on Chinese Famille rose porcelain from the Imperial Palace. Microsc. Res. Tech. **84**(6), 1106–1114 (2021). https://doi.org/10.1002/jemt.23669
131. Lili, F.: La Céramique chinoise. China Intercontinental Press, Beijing (2011)
132. Liu, Z., Jia, C., Li, L., Li, X., Ji, L., Wang, L., Lei, Y., Wei, X.: The morphology and structure of crystals in Qing Dynasty purple-gold glaze excavated from the Forbidden City. J. Am. Ceram. Soc. **101**(11), 5229–5240 (2018). https://doi.org/10.1111/jace.15759

133. Long, D.A.: Raman spectroscopy. McGraw-Hill, New York (1977)
134. Ma, H.J., Henderson, J., Cui, J.F., Chen, K.L.: Glassmaking in the Qing Dynasty: a review, new data, and new insights. Adv. Archaeomater. **1**(*1*), 27–35 (2020). https://doi.org/10.1016/j.aia.2020.11.001
135. Magetti, M., d'Albis, A.: Phase and compositional analysis of a Sèvres soft paste porcelain plate from 1781, with a review of early porcelain techniques. Eur. J. Mineral. ***29***(3), 347–367 (2017). https://doi.org/10.1127/ejm/2017/0029-2627
136. Manuel, J., Subías, I., Fanlo, I., Arranz, E., Gervilla, F.: Multi-isotope survey on the metallogenesis of the hydrothermal Co-Ni deposits in the Alpine Central Pyrenees of Spain. Ore Geol. Rev. **94**, 225–238 (2018). https://doi.org/10.1016/j.oregeorev.2018.01.030
137. Menneret, J.: Cours sur les émaux. Sèvres, ENSCI (1971)
138. Ménard, P.: Marco Polo, Le Devisement du monde: études littéraires et philologiques. Orléans, Editions Paradigmes (2023)
139. Meynard, T.: Ricci et les intellectuels chinois aujourd'hui. In: La Chine des Ming et de Matteo Ricci (1552–1610): Le premier dialogue des savoirs avec l'Europe, pp. 107–131. Bruxelles, Fidélité – Editions Jésuites (2013)
140. Miao, J., Jung, B., Mu, D.: Identification and differentiation of opaque Chinese overglaze yellow enamels by Raman spectroscopy and supporting techniques. Archaeometry. ***52***(*1*), 146–155 (2010). https://doi.org/10.1111/j.1475-4754.2009.00466.x
141. Mimoso, J.: Origin, early history and technology of the blue pigment in azulejos (2015). https://repositorio.lnec.pt/handle/123456789/1007962
142. Molera, J., Climent-Font, A., Garcia, G., Pradell, T., Vallcorba, O., Zucchiatti, A.: Experimental study of historical processing of cobalt arsenide ore for colouring glazes (15–16th century Europe). J. Archaeol. Sci. Rep. **36**, 102797 (2021)
143. Montanari, R., Alberghina, M.F., Casanova Municchia, A., Massa, E., Pelagotti, A., Pelosi, C., Schiavone, S., Sodo, A.: A polychrome Mukozuke (1624–1644) porcelain offers a new hypothesis on the introduction of European enameling technology in Japan. J. Cult. Herit. **32**, 232–237 (2018). https://doi.org/10.1016/j.culher.2017.12.010
144. Montanari, R., Murakami, N., Alberghina, M.F., Pelosi, C., Schiavone, S.: The Origin of overglaze-blue enameling in Japan: New discoveries and a reassessment. J. Cult. Herit. **37**, 94–102 (2019). https://doi.org/10.1016/j.culher.2018.11.008
145. Montanari, R., Murakami, N., Colomban, P., Alberghina, M.F., Pelosi, C., Schiavone, S.: European ceramic technology in the Far East: enamels and pigments in Japanese art from the 16th to the 20th century and their reverse influence on China. Herit. Sci. **8**, 48 (2020). https://doi.org/10.1186/s40494-020-00391-2
146. Montanari, R., Alberghina, M.F., Schiavone, S., Pelosi, C.: The Jesuit painting Seminario in Japan: European Renaissance technology and its influence on Far Eastern art. X-Ray Spectrom. **51**(*1*), 64–85 (2022). https://doi.org/10.1002/xrs.3256
147. Montanari, R., Colomban, P., Alberghina, M.F., Schiavone, S., Pelosi, C.: European *Smalt* in 17th-century. Porcelain decoration and Sacred Art Heritage, Japan (2024)
148. Morishita, S.: L'Art des missions catholiques au Japon (xvie-xviie siècle), collection Patrimoines – *Archives de sciences sociales des religions*. Éditions du Cerf, Paris (2020)
149. Morishita, S.: Le catholicisme au Japon: art et histoire du XVIe au XIXe siècle, Conférence des Missions Etrangères de Paris, 23 mars 2024, Paris, France (2024)
150. Neri, E., Morvan, C., Colomban, P., Guerra, M.F., Prigent, V.: Late Roman and Byzantine mosaic opaque "glass-ceramics" tesserae (5th–9th century). Ceram. Int. **42**(*16*), 18859–18869 (2016). https://doi.org/10.1016/j.ceramint.2016.09.033
151. Nijman, J.: The VOC and the expansion of the world-system 1602–1799. Polit. Geogr. **13**(3), 211–227 (1994). https://doi.org/10.1016/0962-6298(94)90027-2
152. Norris, D., Delbey, T.: The influence of Qing glass technology on Qianlong and Jiaqing painted enamel copperwares. J. Cult. Herit. **61**, 160–167 (2023). https://doi.org/10.1016/j.culher.2023.03.008
153. Ouairy, C.: Examen Radiographique, Rapport d'analyse Archipel, (unpublished) 2021.

154. Panighello, S., Kavcic, A., Vogel-Mikus, K., Tennent, N.H., Wallert, A., Hocevar, S.B., van Elteren, J.T.: Investigation of smalt in cross-section of 17th century paintings using elemental mapping by laser ablation ICP-MS. Microchem. J. **125**, 105–115 (2016). https://doi.org/10.1016/j.microc.2015.11.015
155. Papot, E.: Le Musée chinois: trésor du château de Fontainebleau, le rêve d'une impératrice. Editions Le Charmoiset, Blois (2023)
156. Perazzoli-t'Serstevens, M.: La céramique chinoise en Italie, XIIIe-début XVIe siècle. *Taoci*. ***4***, 75–86 (2005)
157. Pereira, M., de Lacerda-Aroso, T., Gomes, M.J.M., Mata, A., Alves, L.C., Colomban, P.: Ancient Portuguese ceramic wall tiles («Azulejos»): characterization of the glaze and ceramic pigments. J. Nano Res. **8**, 79–88 (2009). https://www.scientific.net/NYanoR.8.79
158. Pérez-Arantegui, J., Montull, B., Resano, M., Ortega, J.M.: Materials and technological evolution of ancient cobalt-blue-decorated ceramics: Pigments and work patterns in tin-glazed objects from Aragon (Spain) from the 15th to the 18th century AD. J. Eur. Ceram. Soc. **29**(*12*), 2499–2509 (2009). https://doi.org/10.1016/j.jeurceramsoc.2009.03.004
159. Perez y Jorba, M., Rommeluere, M., Bahezre, C.: Microstructure d'une plaque d'émail peint de Limoges du XVIe siecle. Stud. Conserv. **36**(2), 76–84 (1991). https://doi.org/10.1179/sic.1991.36.2.76
160. Pierson, S.: Collectors, collections and museums: the field of Chinese ceramics in Britain, 1560–1960. Bern, Peter Lang AG (2007)
161. Pierson, S.: True or False? Defining the Fake in Chinese Porcelain. Cah. Fram. **31** (2019). https://doi.org/10.4000/framespa.6168
162. Pollard, A.M.: Letters from China: a history of the origins of the chemical analysis of ceramics. Ambix. **62**(1), 50–71 (2015). https://doi.org/10.1179/1745823414Y.0000000008
163. Querzola, J.: Le Quadraturiste. Giovanni Gherardini en Chine sous le règne de Kangxi. GH Research, Paris/Londres (2024)
164. Quette, B., Bartholomew, T.T.: Cloisonné: Chinese Enamels from the Yuan, Ming and Qing Dynasties. Bard Graduate Center, New York (2011)
165. Reil, S.: Kilian Stumpf 1655–1720: Ein Würzburger Jesuit am Kaiserhof zu Peking. Aschendorff, Münster (1978)
166. Ricciardi, P., Colomban, P., Tournié, A., Macchiarola, M., Ayed, N.: A non-invasive study of Roman age mosaic tesserae by mean of Raman spectroscopy. J. Archaeol. Sci. **36**, 2551–2559 (2009a). https://doi.org/10.1016/j.jas.2009.07.008
167. Ricciardi, P., Colomban, P., Tournié, A., Milande, V.: Nondestructive on-site identification of ancient glasses: genuine artefacts, embellished pieces or forgeries? J. Raman Spectrosc. **40**(6), 604–617 (2009b). https://doi.org/10.1002/jrs.2165
168. Röhrs, S., Stege, H.: Analysis of Limoges painted enamels from the 16th to 19th centuries by using a portable micro x-ray fluorescence spectrometer. X-Ray Spectrom. **33**(6), 396–401 (2004). https://doi.org/10.1002/xrs.713
169. Rosi, F., Manualli, V., Milliani, C., Brunetti, B.G., Sgamellotti, A., Grygar, T., Hradil, D.: Raman scattering features of lead pyroantimonate compounds. Part I: XRD and Raman characterization of $Pb_2Sb_2O_7$ doped with tin and zinc. J. Raman Spectrosc. **40**, 107–111 (2009). https://doi.org/10.1002/jrs.2092
170. Roulière-Lambert, M.J. (ed.): Les émaux celtiques. Centre Jurassien du Patrimoine, Lons-le-Saunier (2008)
171. Sakellariou, K., Miliani, C., Morresi, A., Ombelli, M.: Spectroscopic investigation of yellow majolica glazes. J. Raman Spectrosc. **35**(1), 61–67 (2004). https://doi.org/10.1002/jrs.1084
172. Samoyault-Verlet, C., Bayou, H.: Le musée chinois et les salons de l'impératrice Eugénie – Collection petit guide n°132. RMN, Paris (1991)
173. Sandalinas, C., Ruiz-Moreno, S., Lopez-Gil, A., Miralles, J.: Experimental confirmation by Raman spectroscopy of a Pb-Sn-Sb triple oxide yellow pigment in sixteenth-century Italian pottery. J. Raman Spectrosc. **37**(10), 1146–1153 (2006). https://doi.org/10.1002/jrs.1580

174. Schreiber, H.: Properties of redox ions in glasses: An interdisciplinary perspective. J. Non-Crystalline Solids. **42**(1–3), 175–183 (1980). https://doi.org/10.1016/0022-3093(80)90019-8
175. Sciau, P., Leon, Y., Goudeau, P., Fakra, C.S., Webb, S., Mehta, A.: Reverse engineering the ancient ceramic technology based on X-ray fluorescence spectromicroscopy. J. Anal. Atomic Spectrom. **26**(5), 969–976 (2011). https://escholarship.org/uc/item/6rt7n961
176. Sciau, P., Noé, L., Colomban, P.: Metal nanoparticles in contemporary potters' masterpieces: lustre and red "pigeon blood" potteries as models to understand the ancient pottery. Ceram. Int. **42**(14), 15349–15357 (2016). https://doi.org/10.1016/j.ceramint.2016.06.179
177. Shih, N.: Analyse scientifique des couleurs falang d'Objects conservés au Palais ou retrouvés en contexte archéologique. Septembre Collège de France, Paris (2019)
178. Shih, C.-F.: Evidence of East-West Exchange in the Eighteenth Century: The Establishment of Painted Enamel Art at the Qing Court in the Reign of Emperor Kangxi. Natl. Palace Mus. Res. Q. **24**(3), 45–94 (2007)
179. Shih, C.-F.: L'Avènement d'une nouvelle palette de couleurs à Jingdezhen au XVIIIe siècle. In d'Abrigeon, P. (dir.) Le Secret des couleurs. Céramiques de Chine et d'Europe du XVIIIe siècle à nos jours, pp. 21–59. Fondation Baur-5 Continents, Genève (2022)
180. Simon, F.: Les cadeaux diplomatiques entre la Chine et l'Europe aux xviie-xviiie siècles. Pratiques et enjeux. Extrême-Orient Extrême-Occident. **43**, 5–24 (2019). https://doi.org/10.4000/extremeorient.1596
181. Simsek, G., Colomban, P., Wong, S., Zhao, B., Rougeulle, A., Liem, N.Q.: Toward a fast non-destructive identification of pottery: the sourcing of 14th-16th century Vietnamese and Chinese ceramic shards. J. Cult. Herit. **16**(2), 159–172 (2015). https://doi.org/10.1016/j.culher.2014.03.003
182. Simsek Franci, G.: Handheld X-ray fluorescence (XRF) versus wavelength dispersive XRF: Characterization of Chinese blue-and-white porcelain sherds using handheld and laboratory-type XRF instruments. Appl. Spectrosc. **74**(3), 314–322 (2020). https://doi.org/10.1177/0003702819890645
183. Sougy, N.: Liberté, légalité, qualité: le luxe des produits d'or et d'argent à Genève au XIXe siècle. Entrep. Hist. **46**(1), 71–84 (2007). https://doi.org/10.3917/eh.046.0071
184. Su, Y., Qu, L., Duan, H., Tarcea, N., Shen, A., Poppe, J., Hu, J.: Elemental analysis-aided Raman spectroscopic studies on Chinese Cloisonné Wares and painted enamels from the Imperial palace. Spectrochim. Acta A. **153**, 165–170 (2016). https://doi.org/10.1016/j.saa.2015.08.005
185. Tang, H.: "The finest of Earth": The English East India Company's enamelled porcelain trade at Canton during the eighteenth century. Artefact. **8**, 69–88 (2018). https://doi.org/10.4000/artefact.1964
186. Valmont de Bomare, M.: Minéralogie, ou Nouvelle exposition du règne minéral, vol. 2, 2nde edn, pp. 76–96. Vincent Imprimeur-Libraire, Paris (1774)
187. Van Pevenage, J., Lauwers, D., Herremans, D., Verhaeven, E., Vekemans, B., De Clercq, W., Vincze, L., Moens, L., Vandenabeele, P.: A combined spectroscopic study on Chinese porcelain containing Ruan-Cai colours. Anal. Methods. **6**, 387–394 (2014). https://doi.org/10.1039/C3AY41072B
188. Verlet, P.: Le commerce des Objects d'art et les marchands merciers à Paris au XVIIIe siècle. Annales. **13**(1), 10–29 (1958). https://www.persee.fr/doc/ahess_0395-2649_1958_num_13_1_2705
189. Wood, N.: Chinese Glazes: Their Origins, Chemistry and Recreation. University of Pennsylvania Press (1999) SBN 10: 0812234766
190. Wood, N.: An AAS study of Chinese imperial yellow porcelain bodies and their place in the history of Jingdezhen's porcelain development. Adv. Archaeomater. **2**, 1–17 (2021). https://doi.org/10.1016/h.aia.2021.09.002
191. Wood, N.: Chinese cloisonné revisited. Adv. Archaeomaterials. **3**(2), 92–108 (2022). https://doi.org/10.1016/j.aia.2023.03.001

192. Wypyski, M.T., Richter, R.W.: Preliminary compositional study of 14th and 15th century European Enamels. Techné. **6**, 48–57 (1997)
193. Xu, X.: Europe-China-Europe: The Transmission of the Craft of Painted Enamel in the Seventeenth and Eighteenth Centuries. In: Maxime, B. (dir.) Goods from the East, 1600–1800 Trading Eurasia, pp. 92–106. Palgrave Macmillan, London (2015)
194. Yang, Y., Feng, M., Ling, X., Mao, Z., Wang, C., Sun, X., Guo, M.: Micro-structural analysis of the colour-generating mechanism in Ru ware, modern copies and its differentiation with Jun ware. J. Archaeol. Sci. **32**, 301–310 (2005)
195. Yanyi, G.: Raw materials for making porcelain and the characteristics of porcelain wares in north and south China in ancient times. Archaeometry. **29**(1), 3–19 (1987). https://doi.org/10.1111/j.1475-4754.1987.tb00393.x
196. Zhang, F.: The origin and development of traditional Chinese glazes and decorative ceramic colours. In: Kingery, W.D. (dir.) Ancient technology to modern science, ceramics and civilization, pp. 163–180. American Ceramic Society, Westerville (1984)
197. Zhang, R., Gethin, P.: Provenance of the cobalt pigment used for Jingdezhen minyao blue-and-white porcelain in the early Qing dynasty. Ceram. Int. **47**(18), 25763–25768 (2021)
198. Zhao, B., Jami, C.: Objets émaillés, techniques et pouvoir. Artefact. **18**, 9–21 (2023). https://doi.org/10.4000/artefact.13614
199. Zhao, B., Wang, G., Biron, I., Colomban, P.: La circulation des techniques de l'émail entre la France et la Chine du XVIIème au XIXème siècle. Le CNRS en Chine Bulletin. **21**, 20–25 (2016)
200. Zhou, Y., Jin, Y., Wang, K., Sun, J., Cui, Y., Hu, D.: Opaque ancient K_2O-PbO-SiO_2 glass of the Southern Song Dynasty with fluorite dendrites and its fabrication. Herit. Sci. **7**, 56 (2019). https://doi.org/10.1186/s40494-019-0300-z
201. Zum Kolk, C.: La sédentarisation de la cour à Paris d'après les itinéraires des derniers Valois (1515–1589). In: Bove, B., Gaude-Ferragu, M., Michon, C. (eds.) Paris, ville de cour (XIIIe-XVIIIe siècle). PUR, Rennes (2017). https://hal.archives-ouGreenes.fr/hal-01829675/

Chapter 19
Conclusions and Summary

Abstract A conclusions chapter which gives a resume of the text of this book, considers the historical evolution of blue pigments, their descriptors in Greek and Roman times and changes in the terminology made in the Renaissance. Green pigments manufactured by synthetic changes incurred in the production of blue pigments. The influence of missionary delegations such as the Jesuits on the release of compositional information and recipes of Chinese and Japanese porcelains in the West. Analytical chemcial data and their usage in determining the origin of cobalt ores in the manufacture of cobalt blues and the adoption of Western sources in Chinese blue decorations summarises the findings and also gives some proposals for future work in this area: some future ideas for analytical studies of decorated ceramics, enamels and glasses.

Keywords Blue · Advanced technology · Fire arts · Chemicals · Culture · Religion · Europe · Asia · America · Missionary · Technology transfer

The objective of the book was to draw together the state of the art knowledge on the various paths used since the beginnings of the arts of Fire to obtain blue decorations on artefacts. This colour, which is common among plants (hydrangea, veronica, cornflowers, lilac, iris, violet, morning glories, hyacinth, clematis, etc.) but rare among minerals, plays a particular role due to the Tyndall effect (the diffusion of light is in proportion to ν^4): therefore the sky and the sea, almost infinite worlds for the human eye, are blue and its colour is appreciated in most cultural contexts, colour being both a scientific and cultural object.

The content of the nineteen chapters in this book is determined by the work carried out on the different colouring agents. Egyptian blue is the first inorganic synthetic pigment (excluding from the notion of synthesis the 'simple' heat treatments that were practiced since almost prehistoric times at least for iron and manganese oxide/hydroxides) and several chapters deal with this pigment and associated phases. The second widely treated colouring agent that has also been used since Antiquity is the cobalt ion 'simply' dissolved in the vitreous silicate network.

H. G. M. Edwards, P. Colomban, *Blue by Fire: A Marker of the Technical History of Glass and Ceramics*, Cultural Heritage Science,
https://doi.org/10.1007/978-3-031-77122-4_19

It is clear that the weight of archaeological work distorts the representativeness of studies, both through the interest in certain ancient civilizations, mainly Mediterranean—'the Ancient World'—and the availability of samples that can be studied destructively. Indeed, for a long time the destructive nature, macroscopically but even microscopically, of analyses prohibited or limited our knowledge of quality objects, those that are most representative of the excellence of artistic and technical creations. Since colour is related to the electronic transitions determined by minor elements and structural defects, comprehensive analytical studies require a deep physical and chemical understanding of the coloured materials being used.

For centuries, objects of art, in particular those of the Fire Arts, were first and foremost objects of high technology, at the cutting edge of knowledge, the artists being the best professionals or *vice versa*. It is perhaps trivial to report that Leonardo da Vinci was a scientist, an engineer, a musician and an artist, a true polymath. Later the availability of industrial chemicals meant that artists no longer had the obligation to be scientists or to associate with scholars. But the technicality of the production of glass and ceramics meant that this quality lasted longer for these artefacts than in painting. This means that technical innovations concerning pigments are good temporal markers. This technical research appears in numerous chapters here and has been revealed by the use of new analytical instruments. For instance, if analysts had not been able to demonstrate the presence of cobalt in certain Islamic blue glasses and had attributed this to the exceptional colouring power of the cobalt ions, 0.01% wt CoO being sufficient for this purpose, a value corresponding to the detection limit of the methods used such as SEM-EDS or XRF, they did not consider for a long time after the first evidence published that lapis lazuli could also be a pigment for glasses. It is Raman microspectroscopy combined with the resonant nature of this Raman Effect for the green laser excitations used which made it possible to establish the use of lapis lazuli in this context. Unfortunately the laboratories dedicated to cultural heritage are often quite conservative and prefer traditional analysis techniques, assimilating only after a certain delay the conceptual and instrumental developments of techniques considered marginal. It is clear that only a combination of techniques makes it possible to understand the reality of material at its different scales is essential. Some ancient recipes in fact used a word that could be understood as identifying the use of lapis lazuli but experts judged that this was a metaphor and thereby missed its significance!

Consideration of ancient recipes must be carried out with caution because the meaning of the words is often obscure historically and our understanding therefore remains very partial. More surprisingly, multitechnical analyses showed that several colouring agents producing the same colour, namely lapis lazuli and cobalt, could be used together, without it being possible to understand a technical reason for their combination. Creating a decoration, an art object is not just a technical action; it is also a cultural, even a religious, action.

The attention paid in this area appears directly proportional to the time spent in the history of mankind. Studies of colouring agents in the Modern Era (in the classical historical sense, after the discovery of America, the fall of Constantinople and of al-Andalus and the establishment of Maritime Trade with Asia) are more limited, although the arrival of blue-and-white porcelain, both Japanese and Chinese, can be considered as the establishment of blue-and-white decoration as a reference with majolica, azulejos, earthenware and porcelain, these objects having a determining influence on European arts (paintings, bacci, porcelain pavilions, ...). Those concerning nineteenth and twentieth century productions are minimal, with a few exceptions concerning the art deco period. Contemporary works in glass, ceramic or enamels do not pose the acute conservation problems of those using polymers, and studies on these artefacts are marginal.

This hiatus between the scientific studies of colour and the variety of objects in Fire Arts has its counterpart between Western productions and other cultures. Aesthetic attention to the productions of the Oriental world in Fire Arts is however old, triggered with the taste in Orientalism and the Universal Exhibitions that were held in the middle of the nineteenth century. An explosion of analytical work has concerned the productions of the world of Chinese culture in the broad sense for several decades (due to the significant increase in the contribution of China in academic research), but it is only since the 2000s that the work, which was previously centred on the study of pastes—imposed by the volume of material necessary for analyses (destructive) –has moved on to the most sophisticated part, the enamelled decoration.

Very little work concerns American productions. Africa, without specific large production, only appears through the circulation of glass beads, traded for exchanges for tens of centuries.

A big absence is that of works concerning the economic flywheel of colouring agents; quantities produced or exchanged, marketing networks, counterfeits, etc. Some works concern the processing of cobalt ores, the role of Dutch merchants in financing the exploitation of cobalt and its trade in Europe. The determining role of Portuguese missionaries mainly at the end of the sixteenth century, French at the end of the 17th and beginning of the eighteenth century, respectively in Japan and China in the innovation of polychrome enamelling, in particular blue and green colours (green being obtained by mixing yellow + blue) is now well established, showing that already in the eighteenth century China—the Canton workshops—were producing artefacts for global trade. Studies on glass beads show that the globalization of large-scale trade took place around the Indian Ocean thanks to the Monsoon Winds from the tenth century.

Ceramics and glass are well-established markers for dating ancient civilizations due to their remarkable preservation whatever the weather or climatic conditions. Advances in measuring instruments allowing non-invasive analyses on site, and therefore the study of the most relevant objects, has opened up a field of research that is still underused due to the possibility of working on all the multi-scale components of the coloured material. Thus the role of ceramics and glass as witnesses to history and innovation remains decisive but requires small-scale studies. The very poor knowledge of ancient mining places is a big handicap for tracing provenances and circulation. The determination of isotopic fingerprints could be the breadcrumb trail for progress in the history of techniques and commercial exchanges.

We have dealt here with the colour blue and partly with certain greens—in many Asian cultures blue and green are variants of the same colour, the same work must now be done for the other colouring agents, producing yellows and greens and red to purple colours.

Glossary

Armorial china High-fired ceramics decorated with the heraldic achievements of an arms-bearing family which could constitute the full coat-of-arms comprising the escutcheon, motto, supporters, helm, lambrequins, crest and torse or any number of these or more usually perhaps just consisting of a crest and torse. In some cases, the crest and torse, with or without a motto, is encompassed in other decoration and may be accompanied by additional gilding and other geometric motifs.

Biscuit porcelain or earthenwares The unglazed product of the high-temperature firing process in the manufacture of ceramics using a "biscuit kiln", which operated typically at a temperature in the region of 600–1400 °C. The resultant ceramic is of a varied colour, which is dependent upon the amount of iron oxide present in the raw material clays, and possessed a texture which, if blemish-free and perfectly shaped, was very highly prized by ceramic artists and modellers particularly for the construction of ornamental figurines in simulation of statuary as first perfected by the Meissen porcelain manufactory in the mid-eighteenth century. In the mid- to late -eighteenth century, the finest biscuit porcelain artefacts were commonly placed on dining tables for admiration by the guests and to stimulate conversation during dinner. Generally, the biscuit porcelain, after cooling from the kiln firing process, was painted, glazed and gilded—during which process any small defects or blemishes in the body or the glaze could often be masked by strategically placed enamelling: hence, painted and enamelled figurines are found much more commonly than their biscuit analogues, which if found to be imperfect after their firing, were usually destroyed or sent to be decorated and glazed. The figures were designed by a "modeller" or sculptor and assembled in the workshop by a "repairer".

Blemish A blemish is a fault in the fired ceramic, whether this is in biscuit or glazed form, arising from several problems which could arise during the firing process such as inadequate kiln temperature control leading to firing cracks, the

H. G. M. Edwards, P. Colomban, *Blue by Fire: A Marker of the Technical History of Glass and Ceramics*, Cultural Heritage Science,
https://doi.org/10.1007/978-3-031-77122-4

introduction of contaminants such as carbonaceous matter which then burned to leave soot deposits and the emission of gases during the firing process which created bubbles in the vitrified paste or the glaze. Serious blemishes resulted in ceramic items being unsaleable, whereas it was a common practice to mask minor blemishes, such as small spots of soot or little bubbles in an otherwise perfectly shaped piece, by the shrewd and strategic placement of motifs prior to a final glazing process taking place.

Body paste The wet mixture of raw material components which comprise the formulaic recipes for ceramic bodies. Texturally moist to the touch, the paste could be moulded or formed into shapes preparatory to the firing process taking place in the biscuit kiln. This term is also applied to the fired body prior to its chemical analysis, when of course it is then not a wet "paste" at all but rather a hard ceramic material.

Burnishing The gentle polishing of gilt decoration on a glazed ceramic surface to a highly polished reflective coating. Early gilding was accomplished using "honey gilding" whereby 24-carat gold leaf was applied to the surface with an adhesive medium of honey, or with the use of applied resins such as gum arabic, which were both replaced in the late eighteenth century by mercury gilding using an amalgam of mercury which had been triturated with the gold leaf. During the final firing of the ceramic piece in the "glost" kiln at low temperatures the organic adhesive carrier component or the mercury in the amalgam was volatilised off into the atmosphere, leaving a dull golden finish which was then hand-polished by skilled gilders using a finely powdered jeweller's rouge, iron (III) oxide Fe_2O_3, to give a brilliant finish. The skill of the gilder was much admired in porcelain workshops as it offset and complemented the applied enamelling decoration but it was always an expensive part of the finishing process. Some service commissions received only a gilded decoration without accompanying enamelling decoration whereas others were not gilded at all. A classic example of fine gilding is the stoneware bust of Prince Rupert of the Rhineland Palatinate, by John Dwight of eth Fulham Pottery, whose Order of the Garter chain and pendant is gilded.

Chemical descriptors Care should be taken to differentiate between the descriptors of chemical compounds and their component elements. For example, silica is SiO_2 and contains silicon and oxygen: likewise, calcia is calcium oxide CaO (lime), alumina is aluminium oxide Al_2O_3, magnesia is magnesium oxide MgO, potash is potassium oxide K_2O and soda is sodium oxide Na_2O. Other chemicals are described precisely as they occur, such as phosphorus pentoxide P_2O_5—also known as phosphorus (V) oxide or phosphorus pentoxide, relative to phosphorus (III) oxide P_2O_3—and sulfur dioxide SO_2. Iron oxide is troublesome as this can mean both ferrous oxide FeO and ferric oxide Fe_2O_3, where the iron is in the oxidation state (II) and (III), respectively, and the former is a green colour and the latter orange-red, and a conversion between these can be undertaken in the oxidising or reducing atmospheres of a firing kiln.

China A generic term for high-fired ceramics, which may be glazed or unglazed, and used for decorative or utilitarian purposes around the home. Believed to originate from descriptors of ceramics that were made in simulation of imported Chinese porcelain, this term covers a wide range of ceramics including porcelain, earthenwares, creamwares, stonewares, celadons, majolica and delftwares. The term "china" has been used freely to describe ceramic articles of a range of compositions since the importation of porcelain from China in the sixteenth century and some of the artefacts do not actually contain china clay (kaolinite).

Chinese porcelain A generic term which historically describes the characteristic hard paste porcelain comprising *kaolinite* and *petuntse* as major raw materials fired at a high kiln temperature; termed "true porcelain" by some historians to differentiate it from "artificial porcelain" which is retained for the soft paste body variants which included synthetic components as raw materials, for example calcined bone ash, potash, lime and glass frit. During the early eighteenth century, the Chinese also perfected a soft paste body variant which has been recognised as a comparative rarity and also made a glazed earthenware to simulate tin-glazed delftwares. Chinese porcelains have a range of compositions which varied both with their production sites geographically and also with the specified quality: the highest quality porcelains from the Imperial kilns at Jingdezhen were composed of an equal ratio of china clay (kaolin) and petuntse.

Cipher See Monogram

Coade stone Historians tend to group Coade stone with stonewares and creamwares, which analytically it resembles closely. An artificial stone, developed by Eleanor Coade from an earlier recipe of Daniel Pincot in 1769, it is an extremely robust and hard-wearing material which was moulded into statuary, garden ornaments and artefacts for the embellishment of buildings, such as pediments, pilasters, plinths, cornices, columns, paterae, porticos, heraldic crests and family coats-of-arms. It was marketed as a better alternative to natural sandstone and marbles for survival and defeating erosion in the environment. Demand was eventually diminished with the advent of Portland cement in 1828, a cheaper cold-cured and unfired alternative that could be cast and moulded on site without the use of a furnace or kiln. Coade stone was exported globally from the manufactory in Lambeth, London, an dis now to be found in many parts of the world, including Russia, the USA, South America and Australasia as well as in Europe. In several lands it is referred to as terracotta, although Eleanor Coade deliberately used raw materials that were deficient in iron oxide content (less than 1%) so her wares were greyish-cream in colour, as were many marbles.

Crested china A term reserved for a partial coat-of-arms depicted upon porcelain or earthenwares which may exhibit the crest solely, with or without its torse, and perhaps accompanied by a motto, usually placed in either the verge or reserve or centrally located on the artefact. It may be enamelled in polychrome pigments, or it is more usually found in gilt or occasionally and more rarely in a monochrome pigment.

Diplomatic Gifts The use of porcelain as a diplomatic gift in Europe can be traced back to the early fourteenth century with Nestorian emissaries from the Chinese Emperors: the first manufacture of armorial porcelain for this particular purpose can be credited to Francisco Medici, Duke of Tuscany, who in the mid to late sixteenth century used his own porcelain as diplomatic gifts to other European rulers. Porcelain, and later delftwares, was very highly regarded and was ideally suited for this purpose, giving it a status similar to that of religious relics, jewels and bullion. The rendering of armorial bearings on such porcelain gifts afforded them a personality that was specific to the receiver.

Electromagnetic spectrum This describes the range of wavelengths which define the characteristics of electromagnetic radiation from the high-energy, low wavelength, high frequency cosmic and X rays, through the ultraviolet (200–400 nm) and then the visible spectrum (400–700 nm) from the violet (through indigo, blue, green, yellow and orange) to the red into the infrared, and thence through to the low-energy, high wavelength, low frequency microwave radiation. Analytical techniques probe different ranges of the electromagnetic spectrum to interrogate the molecular and elemental compositions of ceramic specimens and derive quantitative and qualitative data which can be correlated with the materials used in the ceramic bodies, glazes and pigments. The unit of the wavelength used analytically to define the electromagnetic spectrum is the nanometre, 10^{-9} metre or nm. As the wavelength of the radiation increases, the associated energy becomes smaller, through from the Xray to the microwave region, according to the Planck-Einstein radiation law, $E = h\nu$, where E is the energy of the radiation, h is the Planck constant and ν is the frequency of the electromagnetic radiation, which is inversely proportional to the wavelength of the radiation, λ. Generally, the elemental composition is interrogated in the Xray region and the molecular composition is found from molecular energy transitions occurring in the visible and near infrared regions of the spectrum. Analysis of the elemental and molecular spectral data gives a complementary picture of the porcelain body, glaze and pigments used in the artefact.

Faience The application of a glaze to an biscuit earthenware fired ceramic body results in a generic class termed faience—typical glazes are formed from tin oxides, lead oxides, salt and an alumino-calcic glaze with silicaceous materials. Faience specifically when tin-glazed is associated with delftwares and majolicas, but even then some bodies in this category were lead glazed. A variety of substrates is covered, from terracottas, stonewares, artificial stone and creamwares.

Gilding The application of a gilt decoration to a fired ceramic involving 24-carat gold leaf in a carrier such as mercury, honey or an organic resin followed by burnishing, see under ***burnishing.*** Gilding was a highly specialist task in decorator's workshops, the best gilders being permitted to sign their work with a numeral on the underside of the piece or being mentioned specifically in the service commission documentation and gilding was an expensive component of the decorative process.

Glaze The application of a coating to a fired ceramic biscuit body to seal the ceramic pores and to enhance the appearance with a glossy, smooth texture. Sometimes applied over the decorative pigments, but not exclusively, it acted as a protectant for the decoration. Overglaze decoration, known technically as painting rather than enamelling in the ceramic workshops, was subject to paint loss and wear during usage. The glaze was usually applied as a slip, a thick suspension of clays, alkaline earths and other additives with lead oxide component (or a tin oxide additive, such as cassiterite, for a lead -free glaze). Fired at a lower temperature than the biscuit body, this final stage of preparation enhanced the beauty of the porcelain or earthenwares: it was essential that the glaze composition mechanically and chemically was compatible with the fired porcelain otherwise firing faults were created and an unsightly glaze "craquelure" could form, whereby the glaze physically separates from the porcelain substrate in contraction from the kiln upon cooling. Certain decorative pigments were favoured for their enhancement properties upon interaction with the glaze application, cobalt blue being one of these, and it was equally important that the pigment maintained its integrity and was able to resist "running" on application of the glaze.

Glost kiln Used in the final stage of preparation before and/or after applying the enamelled decoration and involving the application of an alkaline "slip" containing china clay, soda, potash and a lead oxide component usually in the form of a powdered flint glass, which at lower temperatures will form a hard, transparent glaze coating. Occasionally, a glost kiln was used for the calcination of raw material components at lower temperatures and for the drying of raw materials in which variable amounts of water were found to occur. Initially, glazes were lead-based but in the first decade of the nineteenth century, when the toxicity of lead compounds was fully appreciated, reversion to a less toxic, tin-based glaze was effected (as patented by John Rose at the Coalport China Works in 1820, using cassiterite, SnO_2, instead of lead oxide, litharge or massicot, PbO). Tin glazes had been used for many years hitherto in the manufacture of faience and delftwares. China sold from manufactories "in the white" could be unglazed or glazed for decoration in ateliers, with subsequent glazing being undertaken by the enamellers.

Initials see Monogram.

Instrumental Chemical Analysis the use of instrumentation to detect qualitatively and to estimate quantitatively the presence of chemical elements and their compounds in ceramic bodies and glazes. The basis of instrumental analysis is the use of special signatures or properties of chemical species or chemical bonds such as silica, SiO_2, or the Si=O bonds in complex macromolecular structures, such as silicaceous ceramic bodies, from their behaviour towards the scattering or absorption of electromagnetic radiation in the electromagnetic spectrum from the low wavelength (high wavenumber) Xray region to the high wavelength (low wavenumber) microwave region. When coupled with a microscope, an analytical instrument is capable of interrogating very small specimens or domain regions in a larger specimen of the order of cubic microns or cubic nanometres. Instrumentation that has been commonly used in the assessment of the constitution of porcelain and ceramic bodies, glazes and pigments includes scanning electron microscopy/energy dispersive and X Ray spectrometry (SEM/EDAXS), X Ray fluorescence spectroscopy (XRF) for the elemental oxide composition and X Ray diffraction spectrometry (XRD), laser Raman Spectroscopy (RS) and Fourier-transform infrared spectroscopy (FTIR) for mineralogical and pigment identification.

Jasperwares in the 1790s, Josiah Wedgwood of the Etruria Pottery, Staffordshire, produced a biscuit creamware predominantly coloured with a small amount of cobalt blue pigment, upon which was applied cameos of classical and mythological figures. Various colours from pale blue to dark blue, green, and yellow were formed by the addition of heavy metal oxides. Jasperwares were unglazed.

Monogram Initials which are arranged in a design so that the removal of one initial compromises the design and renders one or more of the other initials indecipherable. A cipher is a design in which initials which are merely arranged together whereby the removal of one does not affect the others. Monograms, ciphers and initials all appear on applied ceramic decoration and when accompanied by an heraldic device such as a coronet or a crest then constitute an armorial representation. Otherwise, they are not strictly considered as armorial although historically they have often been included as such by many ceramics historians.

Moulding The shape and size of ceramic items can be a reliable stylistic indicator of a particular factory's output and a clue to the attribution of unmarked pieces. Mouldings were generally unique to each manufactory and comprised individual and idiosyncratic cross-sections of flatwares such as saucers and plates, the shapes of cups, the presence and pattern of footrims, the shape of spill vases (e.g. cylindrical, fluted, tulip-shape, trumpet-shaped, flared), the impressed moulded and embossed verges of plates and the number and type of indentations at the rim. Many factories characteristically used the application of moulded flowers which were integral to the item and also applied reliefs such as feet, the heads of animals, floral decoration and beads, as well as knops or handles on tureens and comports.

Named service A porcelain service which has been commissioned by a particular client or individual from the china factory or one of its agents, especially where factory records support the attribution and where occasionally a specific reason for its purchase may be stated, such as the *Duke of Cambridge* service in Nantgarw porcelain which was ordered from the Nantgarw agents, John Mortlock, Oxford Street, London, as a wedding present from George, the Prince Regent, for the marriage of his younger brother, Adolphus, Duke of Cambridge, to Princess Augusta, Landgravine of Hesse-Cassel, at Buckingham Palace in June, 1818. The term also includes services which have been purchased from the later sale of effects or estates and have thereby acquired the name of their new owners by association. Occasionally, the named services do not bear the name of the person who commissioned them but rather the place where they were to be used or perhaps referred to even the decoration adopted: an example of the former is the *Marino Ballroom* service in Swansea porcelain commissioned by John Vivian for his residence in Sketty Hall, Swansea, and of the latter, the *Garden Scenery* service of Swansea porcelain originally commissioned by Lewis Weston Dillwyn, proprietor of the Swansea China Works.

Pigments In ceramics decoration, these are coloured minerals which are thermally stable at the operating temperatures adopted in the glost kiln after application of the glaze slip and firing up to 600 °C or greater. Mineral pigments are usually metal oxides, sulfates and sulfides such as haematite, gypsum and orpiment.

Care needs to be taken in the interpretation of old recipes for the decoration of ceramics as minerals were often confusingly assigned an incorrect nomenclature, such as *minium*, which has been used historically to describe both red lead, trilead tetroxide, Pb_3O_4, and cinnabar, mercury sulfide, HgS, and red oxide of lead can mean both red lead ($2PbO.PbO_2$) and litharge (PbO). Analytically, the interrogation of pigments on enamelled ceramics is a fertile area of study as manufacturers often changed their pigment sources and the composition of their pigments with time and the presence of trace quantities of unusual elements such as zinc, niobium, rubidium and copper can indicate valuable information about the timelines of production and sources of the pigment by the manufactory concerned. For example, the presence of a minor elemental component of bismuth in a cobalt blue pigment on an Iznik glazed faience tile from the sixteenth/seventeenth century indicated that the source of the cobalt ore was European (from the Erzengebirge mine in Saxony) rather than the more usually encountered Asian mines. Arsenic traces in the cobalt blue pigment on Chinese eighteenth century porcelains are indicative of the use of European cobalt ores in the preparation of the pigment rather than the Asian sources which are arsenic – free.

Porcelain types commencing with hard paste and soft paste porcelains historically (also termed rather confusingly as “true” and “artificial” porcelains), this inadequate description generated a further differentiation into the porcelain as hard paste (highly siliceous), phosphatic, bone china, magnesian, glassy and hybrid types. The categorisation of a porcelain type is dependent not only upon the body composition but also upon the process methodology and firing sequences adopted. However made, all types of porcelain must possess a degree of translucency to qualify for its appellation and inclusion in the category.

Raw Materials The raw materials are components comprising the body and glaze recipes used in porcelain syntheses. Mostly these were natural minerals and rocks which were sourced from precise mines and locations which gave rise to several alternative names. Natural raw materials include feldspars, china clays, ball clays and soapstone. Synthetic raw materials include bone ash, pearl ash, soda ash, smalt, lime and magnesia.

Service A ceramic service could be commissioned directly from the factory, from an appointed agent, bought at auction from “chinamen” or acquired through an import agency such as the Honourable East India Company who traded with Chinese porcelain agencies and their brokers or factors (Hongs) through the port of Canton (Guangzhou). A porcelain service was normally defined as a dinner, dessert or tea service, but combinations of dinner-dessert, tea-coffee and special breakfast services were quite commonly ordered. The usual dessert or dinner service comprised 42 pieces approximately, but special orders expanded this to over 100 pieces, with the inclusion of soup dishes, tureens, centre fruit comports and stands, ice pails, spill vases, muffin dishes, guglets (water bottles), vegetable dishes, egg-cups and carrier, sauce boats and trios (coffee can/cup, tea cup and saucer). There are instances of very large services being commissioned, such as those from Meissen in the eighteenth century of over 1000 pieces with some 50 or 60 place settings. The service was usually decorated according to the specific requirements of the commission and the inclusion of a coat of arms or heraldic crest centrally or at the verge would confer upon it the description of an "armorial" service. An armorial service is especially valuable for provenancing and historical purposes as it facilitates the assignment of the commission to a particular person in an arms-bearing family and often enables one to attribute a definitive date for a commission, such as the granting of arms or perhaps a family succession to a title or elevation in the peerage which may have been celebrated accordingly with the ordering of a porcelain service.

Shipwrecks Recently, cargoes from shipwrecks have provided a rich source of ceramic artefacts: sometimes the cargoes are very extensive and number many thousands of objects being exported for trade and in passenger personal belongings. A range of ceramic wares encompassing porcelains, celadons and earthenwares is usually encountered, some of which extend over several dates of manufacture and from different regions and kilns. A major advantage of studying the ceramics recovered from a shipwreck is that associated objects can afford a reliable date for the manufacture --- especially wooden artefacts which can be used to estimate the age of the vessel from radiocarbon dating methods. Care must he exercised as it is possible to find significantly older ceramic artefacts in relatively more modern cargoes: for example the *Tek Sing* cargo, from the ship which foundered in 1822, although comprising comtemporary Chinese porcelain, celadons and earthenwares dating from the early nineteenth century, also yielded examples of Ming porcelains that were already 200 or more years older.

Translucency This is perhaps the greatest achievable asset of porcelain manufacture to which every manufacturer subscribed and hoped to attain in emulation of the Chinese "eggshell" wares which were imported to Europe in the mid-eighteenth century. It describes the transmission of visible radiation, or light, through a solid object: at one end of the spectral scale is glass, which is usually *transparent* (100% transmission) and at the other is earthenware which is *opaque* (0% transmission). However, see entry above for "Glass", where the occurrence of a milk glass is noted in which the transparency is less than 100% and the glass is then termed translucent. Porcelain is measured by its translucency, which is the clarity for transmission when viewed with background lighting. An intermediate descriptor is "semi-opaque" or even "semi-porcelain", which was applied to some china during the mid-nineteenth century, which is rather indefinite and conveys little information about their true category. Other descriptors such as "vitreous china" and "ironstone china" likewise do little to assist in their classification. By definition, porcelain must be able to transmit light to some extent unlike stonewares, faience, terracotta and earthenwares.

Index

H. G. M. Edwards, P. Colomban, *Blue by Fire: A Marker of the Technical History of Glass and Ceramics*, Cultural Heritage Science, https://doi.org/10.1007/978-3-031-77122-4

T

U

GPSR Compliance

The European Union's (EU) General Product Safety Regulation (GPSR) is a set of rules that requires consumer products to be safe and our obligations to ensure this.

If you have any concerns about our products, you can contact us on ProductSafety@springernature.com

In case Publisher is established outside the EU, the EU authorized representative is:

Springer Nature Customer Service Center GmbH
Europaplatz 3
69115 Heidelberg, Germany

Batch number: 10726133

Printed by Printforce, the Netherlands